A Matilde

Antonio Machì

Groups

An Introduction to Ideas and Methods of the Theory of Groups

Antonio Machì
Department of Mathematics
University La Sapienza, Rome

The translation is based on the original Italian edition: A. Mach : Gruppi. Una introduzione a idee e metodi della Teoria dei Gruppi, Springer-Verlag Italia 2007

UNITEXT La Matematica per il 3+2
ISSN print edition: 2038-5722 ISSN electronic edition: 2038-5757

ISBN 978-88-470-2420-5 ISBN 978-88-470-2421-2(eBook)
DOI 10.1007/978-88-470-2421-2

Library of Congress Control Number: 2011941284

Springer Milan Heidelberg New York Dordrecht London

9 8 7 6 5 4 3 2 1

Cover-Design: Simona Colombo, Milano

Typesetting with LaTeX: PTP-Berlin, Protago TeX-Production GmbH, Germany
(www.ptp-berlin.eu)

Springer-Verlag Italia S.r.l., Via Decembrio 28, I-20137 Milano
Springer is a part of Springer Science+Business Media (www.springer.com)

Preface

This book contains an exposition of the basic ideas and techniques of the theory of groups, starting from scratch and gradually delving into more profound topics. It is not intended as a treatise in group theory, but rather as a text book for advanced undergraduate and graduate students in mathematics, physics and chemistry, that will give them a solid background in the subject and provide a conceptual starting point for further developments. Both finite and infinite groups are dealt with; in the latter case, the relationship with logical and decision problems is part of the exposition.

The general philosophy of the book is to stress what according to the author is one of the main features of the group structure: groups are a means of classification, via the concept of the action on a set, thereby inducing a partition of the set into classes. But at the same time they are the object of a classification. The latter feature, at least in the case of finite groups, aims at answering the following general problem: how many groups of a given type are there, and how can they be described? We have tried to follow, as far as possible, the road set for this purpose by Hölder in his program. It is possible to give a flavor of this research already at an elementary level: we formulate Brauer's theorem on the bound of the order of a group as a function of the order of the centralizer of an involution immediately after the introduction of the basic concepts of conjugacy and that of centralizer.

The relationship with Galois theory is often taken into account. For instance, after proving the Jordan-Hölder theorem, which is the starting point for the classification problem, we show that this theorem arises in Galois theory when one considers the extensions of a field by means of the roots of two polynomials. When dealing with the theory of permutation groups we will have an opportunity to stress this relationship.

Groups of small order can be classified using the concepts of direct and semidirect product; moreover, these concepts provide a first introduction to the theory of extensions that will be considered in greater detail in the chapter on the cohomology of groups. The problem of finding all the extensions of a group by another is one half of Hölder's program, the other half being the classification of all finite simple groups.

A large section is devoted to the symmetric and the alternating groups, and to the relevance of the cycle structure for the determination of the automorphism group of the symmetric group. The notion of an action of a group on a set is thoroughly dealt with. An action provides a first example of a representation of a group, in the particular case of permutation matrices, and of notions like that of induced representation (Mackey's theorem is considered in this framework). These notions are usually introduced in the case of linear representations over the complex numbers (a whole chapter is devoted to this topic); we think that important concepts can already be introduced in the case of permutation representations, allowing a better appreciation of what is implied by the introduction of linearity. For instance, two similar actions have the same character, but the converse is false in the case of permutation representations and true in the case of the linear ones.

Among other things, the notion of an action is used to prove Sylow's theorem. Always in the vein of the classification problem, this theorem can be used to prove that many groups of small order, or whose order has a special arithmetic structure, cannot be simple. A section is devoted to prove that some projective groups are simple, and to the proof of the simplicity of the famous Klein group of order 168.

The classification of finite abelian groups has been one of the first achievements of abstract group theory. We show that the classification of finitely generated abelian groups is in fact a result in linear algebra (the reduction of an integer matrix to the Smith normal form). This allows us to introduce the concepts of a group given by generators and relations, first for abelian groups and then in the general case.

Infinite groups play an important role both in group theory and in other fields of mathematics, like geometry and topology. The concept of a free group is especially important: a large section is devoted to it, including the Nielsen-Schreier theorem on the freeness of a subgroup of a free group. As already mentioned, infinite groups allow the introduction of logical and decision problems, like the word problem. Among other things, we prove that the word problem is solvable for a finitely presented simple group; Malcev's theorem on hopfian groups is also proved. The analogy between complete theories in logic, simple groups and Hausdorff spaces in topology is explained.

This kind of problem is also considered in the case of nilpotent groups; the word problem is solvable for these groups. A detailed analysis of nilpotent groups is given, also in the finite case. The notion of nilpotence is shown to be the counterpart of the same notion in the theory of rings. In the finite case, nilpotent groups are generalized to p-nilpotent groups: the transfer technique is employed to prove Burnside's criterion for p-nilpotence. The relationship between the existence of fixed-point-free automorphisms and p-nilpotence is also illustrated.

A whole area of problems concerns the so-called local structure of groups, a local subgroup being the normalizer of a p-subgroup. A major breakthrough in this area is Alperin's theorem, which we prove; it shows that conjugacy has a local character.

Solvable groups are a natural generalization of nilpotent groups, although their interest goes far beyond this property, especially for the role they play in Galois theory. Among other things, we explain what is the meaning in Galois theory of the result on the solvability of the transitive subgroups of the symmetric group on a prime number of elements.

The technique of using a minimal normal subgroup for proving results on solvable groups is fully described and applied, as well as the meaning of the Fitting subgroup as a counterpart of the center in the case of nilpotent groups. Carter subgroups are also introduced, and the analogy with the Cartan subalgebras of Lie algebras is emphasized.

The last two chapters are more or less independent of the rest of the book. Representation theory is a subject in itself, but its applications in group theory are often unavoidable. For instance, we give the standard proof of Burnside's theorem on the solvability of groups divisible by at most two primes, a proof that only needs the first rudiments of the theory, but that is very difficult and involved without representation theory.

Finally, the cohomology theory of groups is dealt with. We give a down to earth treatment mainly considering the aspects related to the extensions of groups. We prove the Schur-Zassenhaus theorem, both for the interest it has in itself, and to show how cohomological methods may be used to prove structural results. Schur's multiplier and its relationship with projective representations close the section and the book.

Every section of the book contains a large number of exercises, and some interesting results have been presented in this form; the text proper does not make use of them, except where specifically indicated. They amount to more than 400. Hints to the solution are given, but for almost all of them a complete solution is provided.

Rome, September 2011 *Antonio Machì*

Notation

Contents

1

Introductory Notions

1.1 Definitions and First Theorems

Definition 1.1. A *group* G is a non empty set in which it is defined a binary operation, i.e. a function:

$$G \times G \to G,$$

such that, if ab denotes the image of the pair (a, b),

i) the operation is associative: $(ab)c = a(bc)$, for all triples of elements $a, b, c \in G$;
ii) there exists an element $e \in G$ such that $ea = a = ae$, for all $a \in G$. This element is unique: if e' is also such that $e'a = a = ae'$, for all $a \in G$, $ea = a$ implies, with $a = e'$, that $ee' = e'$, and $a = ae'$ implies, with $a = e$, that $ee' = e$. Thus $e' = e$;
iii) for all $a \in G$, there exists $b \in G$ such that $ab = e = ba$.

We then say that the above operation endows the set G with a *group structure.*

The set G *underlies* the group defined in it. The element ab is the *product* of the elements a and b (in this order); one also writes $a \cdot b$, $a * b$, $a \circ b$, $a + b$ and the like, and $G(\cdot)$, $G(+)$ if the operation is to be emphasized.

The element e is the *identity element* of the operation (it leaves an element unchanged when combined with it). Other names are *neutral element, unity, zero* (the latter when the operation is denoted additively); we shall mainly use the notation 1, I or 0 (zero). If the group G reduces to a single element, necessarily the identity, then G is the *trivial group* or the *identity group*, denoted $G = \{1\}$ or $G = \{0\}$.

From *i*), *ii*) and *iii*) follows the *cancellation law*:

$$ab = ac \Rightarrow b = c \text{ and } ba = ca \Rightarrow b = c.$$

Machì A.: Groups. An Introduction to Ideas and Methods of the Theory of Groups.
DOI 10.1007/978-88-470-2421-2_1, Springer-Verlag Italia 2012

Indeed, if $ab = ac$, let $x \in G$ be such $xa = e$; from $x(ab) = x(ac)$ it follows, by *i*), $(xa)b = (xa)c$, so that $eb = ec$ and $b = c$. The other implication is similar. In particular, if $ax = e = ay$, then $x = y$, that is, given a, the element x of *iii*) is unique. This element is called the *inverse* of a, and is denoted a^{-1} (in additive notation it is the *opposite* of a, denoted $-a$). From $aa^{-1} = e$ it follows that a is an inverse for a^{-1}, and therefore the unique one; hence $(a^{-1})^{-1} = a$. The cancellation law implies that *multiplication by an element of the group is a bijection*, meaning that if S is a subset of G, and x an element of G, the correspondence $S \rightarrow Sx$ between S and the set of products $Sx = \{sx,\ s \in S\}$ given by $s \rightarrow sx$ is a bijection. It is onto (sx comes from s), and it is injective because if $sx = s'x$, then by the cancellation law $s = s'$. In particular, with $S = G$ we obtain, for each $x \in G$, a bijection of G with itself, that is, a permutation of the set G (see Theorem 1.8).

When a binary operation is defined in a set, and this operation is associative, then one can speak of the product of any number of elements in a fixed order, meaning that all the possible ways of performing the product of pairs of elements yield the same result. For instance, a product of five elements a, b, c, d, e, in this order, can be performed as follows: first $ab = u$ and $cd = v$, then uv, and finally the product of this element with e: $(uv)e$. On the other hand, one can first perform the products $bc = x$ and $de = y$, then the product xy, and finally the product of a with the latter element, $a(xy)$. By the associative law, $a(xy) = (ax)y$, i.e. $(a(bc))y = ((ab)c)y = (uc)y$, and $(uc)y = u(cy) = u(c(de)) = u((cd)e) = u(ve) = (uv)e$, as above.

We now prove this in general.

Theorem 1.1 (Generalized associative law). *All the possible products of the elements $a_1, a_2, \ldots, a_n$ of a group G taken in this order are equal.*

Proof. Induction on n. For $n = 1, 2$ there is nothing to prove. Let $n \geq 3$; for $n = 3$ the result is true because the operation is associative. Let $n > 3$, and assume the result true for a product of less than n factors. Any product of the n elements is obtained by multiplying two elements at a time, so that it will eventually reduce to a product of two factors: $(a_1 \cdots a_r) \cdot (a_{r+1} \cdots a_n)$. If $(a_1 \cdots a_s) \cdot (a_{s+1} \cdots a_n)$ is another such reduction, we may assume without loss that $r \leq s$. By induction,

$$a_{r+1} \ldots a_n = (a_{r+1} \cdots a_s) \cdot (a_{s+1} \cdots a_n)$$

is a well defined product, as well as

$$a_1 \cdots a_s = (a_1 \cdots a_r)(a_{r+1} \cdots a_s).$$

It follows:

$$(a_1 \cdots a_r) \cdot ((a_{r+1} \cdots a_s) \cdot (a_{s+1} \cdots a_n)) = u(vt)$$
$$((a_1 \cdots a_r)(a_{r+1} \cdots a_s))(a_{s+1} \cdots a_n) = (uv)t,$$

and the associativity of the operation gives the result. ◇

In particular, if $a_i = a$ for all i, we denote the product by a^n, and for m, n we have:

$$a^m a^n = a^{m+n},\ (a^m)^n = a^{mn},$$

setting $a^0 = e$. Writing $a^{-m} = (a^{-1})^m$ (the product of m times a^{-1}), the first equality also holds for negative m and n; if only one is negative, the same equality is obtained by deleting the elements of type aa^{-1} (or $a^{-1}a$) that appear. Moreover,

$$(a^{-m})^n = ((a^{-1})^m)^n = (a^{-1})^{mn} = a^{(-m)n},$$

and

$$(a^{-m})^{-n} = (((a^{-1})^m)^{-1})^n = (a^m)^n = a^{(-m)(-n)}.$$

Thus, the above equalities hold in all cases, with positive, negative or zero exponents; in additive notation they become:

$$\begin{aligned} ma + na &= (m+n)a, \\ m(na) &= (mn)a. \end{aligned}$$

Moreover,

$$ab \cdot b^{-1}a^{-1} = a(bb^{-1})a^{-1} = a \cdot 1 \cdot a^{-1} = aa^{-1} = 1,$$

and therefore

$$(ab)^{-1} = b^{-1}a^{-1}.$$

More generally,

$$(a_1 a_2 \cdots a_n)^{-1} = a_n^{-1} a_{n-1}^{-1} \cdots a_1^{-1}.$$

Definition 1.2. The *order* of a group G is the cardinality of the set underlying it, denoted $|G|$. If this is finite, G is a *finite* group; otherwise, G is an *infinite* group.

Definition 1.3. A group G is *commutative* or *abelian* if $ab = ba$, for all pairs of elements $a, b \in G$.

If G is a finite group of order n, the *multiplication table* of G is a table like the following:

$\cdot$	a_1	a_2	$\dots$	a_n
a_1	$a_1 \cdot a_1$	$a_1 \cdot a_2$	$\dots$	$a_1 \cdot a_n$
a_2	$a_2 \cdot a_1$	$a_2 \cdot a_2$	$\dots$	$a_2 \cdot a_n$
$\dots$	$\dots$	$\dots$	$\dots$	$\dots$
a_n	$a_n \cdot a_1$	$a_n \cdot a_2$	$\dots$	$a_n \cdot a_n$

The table depends on the order chosen for the elements of the group; usually one begins with the identity ($a_1 = 1$). A group can be considered known when its multiplication table is known: the table shows the result of the application of the operation to any pair of elements a_i, a_j of the group (infinite tables can also be considered).

If $G(\cdot)$ is a group, and G_1 is a set whose elements are in a bijection φ with those of G, a group structure can be given to G_1 starting from that of G as follows. Let $x_1, y_1 \in G_1$, $x_1 = \varphi(x), y_1 = \varphi(y)$, and let $z = x \cdot y$. Define the product $x_1 * y_1$ as $\varphi(z)$, that is, $\varphi(x) * \varphi(y) = \varphi(x \cdot y)$. The two groups can only differ by the nature of the elements and by the type of operation: the way they combine is the same. Conversely, given two groups $G(\cdot)$ and $G_1(*)$, if there is a bijection φ between them, such that for all $x, y \in G$,

$$\varphi(x \cdot y) = \varphi(x) * \varphi(y), \tag{1.1}$$

then two group structures are determined by each other.

Definition 1.4. Two group $G(\cdot)$ and $G_1(*)$ are *isomorphic* if there is a bijection $\varphi : G \to G_1$ such that (1.2) holds; write $G \simeq G_1$, and φ is called an *isomorphism.*

It is clear the the isomorphism relation is an equivalence relation in the set of all groups; therefore, we can speak of *isomorphism classes.* We shall say that a property is a *group theoretic property* if together with a group G, all groups isomorphic to G enjoy it. An isomorphism class is also called an *abstract group.*

Theorem 1.2. *If $\varphi : G \to G_1$ is an isomorphism, then $\varphi(1) = 1'$, where 1 and $1'$ are the units of G and G_1, respectively, and $\varphi(g^{-1}) = \varphi(g)^{-1}$.*

Proof. We have:

$$\varphi(1)\varphi(g) = \varphi(1g) = \varphi(g),$$

so that, by the uniqueness of the neutral element, $\varphi(1) = 1'$. Moreover,

$$\varphi(g^{-1})\varphi(g) = \varphi(g^{-1}g) = \varphi(1),$$

and $\varphi(g^{-1})$ is the inverse of $\varphi(g)$, i.e. $\varphi(g^{-1}) = \varphi(g)^{-1}$. ◇

Two isomorphic groups have the same multiplication table, up to the symbols that denote the elements of the two groups and the order in which they appear in the two tables.

Examples 1.1. 1. The integers are a group under addition, with 0 as neutral element and $-n$ the opposite of n. It is denoted by **Z**[1]. The multiples (positive, negative or zero) of a fixed integer m also form a group, with respect to

[1] From the German *Zahl.*

the same operation of addition: $hm + km = (h+k)m$, and $-(hm) = h(-m)$ (by hm we mean the result of adding m to itself h times). The correspondence $\varphi : n \longrightarrow nm$ is a bijection, and

$$\varphi(h+k) = (h+k)m = hm + km = \varphi(h) + \varphi(k).$$

Therefore, φ is an isomorphism. The set consisting of the multiples of m is denoted by $\langle m \rangle$. Observe that $\langle m \rangle = \langle -m \rangle$. In particular, the whole group $\mathbf{Z}$ is the set of the multiples of 1, or of -1: $\mathbf{Z} = \langle 1 \rangle = \langle -1 \rangle$. With respect to multiplication, however, the integers do not form a group: although the product of two integers is again an integer, and a neutral element exists (the unity 1) an integer $n \neq 1, -1$ does not admit an inverse in the integers.

2. The rational numbers, the real numbers, and the complex numbers form a group under the usual addition. They are denoted by $\mathbf{Q}$, $\mathbf{R}$, and $\mathbf{C}$, respectively (if the operation is to be emphasized, then one writes $\mathbf{Q}(+)$, etc.) By removing the zero element, we have three groups under the usual ordinary multiplication $\mathbf{Q}^*$, $\mathbf{R}^*$ and $\mathbf{C}^*$, with neutral element 1 and $1/a$ the inverse of a.

3. The positive real numbers also form a group under ordinary multiplication. This group, denoted by $\mathbf{R}^+(\cdot)$, is isomorphic to the group $\mathbf{R}$: given a positive real number a, the mapping φ_a which associates with a real number r its logarithm, $\varphi_a : r \longrightarrow \log_a r$, is an isomorphism $\mathbf{R}^+(\cdot) \longrightarrow \mathbf{R}(+)$. Observe that a is the element that under φ_a goes to 1. The inverse of φ_a is the exponential function $r \longrightarrow a^r$.

4. Similarly, the positive rationals form a group with respect to multiplication, $\mathbf{Q}^+(\cdot)$. However, unlike the case of the reals, this group is not isomorphic to $\mathbf{Q}(+)$. Indeed, let $\varphi : \mathbf{Q}^+(\cdot) \longrightarrow \mathbf{Q}(+)$ be an isomorphism, and let $\varphi(2) = x$. If $x/2$ is the image of a under φ, then $\varphi(a^2) = x$, and therefore, φ being injective, $a^2 = 2$, and a is not rational.

5. The congruence modulo an integer n splits the integers $\mathbf{Z}$ into equivalence classes: two integers belong to the same class if they give the same remainder when divided by n, that is, if they differ by a multiple of n. The set of these classes becomes an abelian group, denoted $\mathbf{Z}_n$, when equipped with the operation

$$[i] + [j] = [i+j].$$

If $h \in [i]$ and $k \in [j]$ then by definition $[h] + [k] = [h+k]$. But $h = i + sn$ and $k = j + tn$, so that $h + k = i + j + (s+t)n$, and therefore $h+k$ and $i+j$ belong to the same class: $[h+k] = [i+j]$. In other words, the operation is well defined (see Remark 1.1).

The possible remainders of the division by n are $0, 1, \ldots, n-1$; accordingly, the classes are denoted by $[0], [1], \ldots, [n-1]$ (see Theorem 1.10 below). This is the group of *residue classes* mod n.

6. The plane rotations r_k with a given center and angle $2k\pi/n,\ k = 0, 1, \ldots, n-1$, form a group under composition of rotations: the "product" of two rotations is the rotation obtained by performing one rotation followed by the other one, and then taking the result mod 2π. The mapping

$$k \to r_k,\ k = 0, 1, 2, \ldots, n-1,$$

is an isomorphism between the group $\mathbf{Z}_n$ and this group. The n-th roots of unity in the complex field $\{e^{\frac{2k\pi}{n}i},\ k = 0, 1, 2, \ldots, n-1\}$ form a group isomorphic to both the group $\mathbf{Z}_n$ and the group of rotations.

Remark 1.1. When one says that a mapping f, that is a function, is *well defined*, what is meant is that one is actually defining a function, that is, an element cannot have two different images under f. In other words, if x and y denote the same element, then it must be $f(x) = f(y)$, i.e. $x = y \Rightarrow f(x) = f(y)$. In *Ex.* 5[2] above, the mapping $f : ([i], [j]) \to [i+j]$ defining the sum of two classes is defined after a choice of representatives of the two classes has been made. If different representatives h and k are chosen, the classes being the same, the image must be the same.

In the examples above all the groups are commutative. Next, let us see a non commutative group.

Definition 1.5. A *permutation* of a set Ω is a bijection of Ω to itself. The set of permutations of Ω is denoted by S^Ω.

Using a geometric language, the elements of Ω are often called *points*.

Let us denote by α^σ the image of $\alpha \in \Omega$ under $\sigma \in S^\Omega$. If $\sigma, \tau \in S^\Omega$, we define the *product* $\sigma\tau$ as the composition of σ and τ, that is, the mapping obtained by first applying σ and then τ[3]:

$$\alpha^{\sigma\tau} = (\alpha^\sigma)^\tau,$$

(see the product of rotations in *Ex.* 1.1 5). Let us show that the mapping $S^\Omega \times S^\Omega \to S^\Omega$ defined by $(\sigma, \tau) \to \sigma\tau$ is again a permutation (that is, it has values in S^Ω). The mapping $\sigma\tau$ is surjective: if $\alpha \in \Omega$, there exists $\beta \in \Omega$ such that $\alpha = \beta^\tau$, because τ is surjective. Since σ is also surjective, there exists $\gamma \in \Omega$ such that $\beta = \gamma^\sigma$. Thus,

$$\alpha = \beta^\tau = (\gamma^\sigma)^\tau = \gamma^{\sigma\tau},$$

so that $\sigma\tau$ is surjective. If $\alpha^{\sigma\tau} = \beta^{\sigma\tau}$, then $(\alpha^\sigma)^\tau = (\beta^\sigma)^\tau$, and therefore, τ being injective, $\alpha^\sigma = \beta^\sigma$, and σ being injective, $\alpha = \beta$. This proves that $\sigma\tau$ is injective.

---

[2] Here and in what follows, *Ex.* means "example" and *ex.* "exercise".

[3] Some authors compute the product by first applying τ and then σ (see Remark 1.2).

We now prove that with this product, S^{Ω} is a group. The product is associative:

$$\alpha^{(\sigma\tau)\eta} = (\alpha^{(\sigma\tau)})^{\eta} = ((\alpha^{\sigma})^{\tau})^{\eta} = (\alpha^{\sigma})^{\tau\eta} = \alpha^{\sigma(\tau\eta)},$$

for all $\alpha \in \Omega$, and therefore $(\sigma\tau)\eta = \sigma(\tau\eta)$.

The identity permutation $I : \alpha \to \alpha$ is a neutral element:

$$\alpha^{I\sigma} = (\alpha^{I})^{\sigma} = \alpha^{\sigma},$$

for all $\alpha \in \Omega$, so $I\sigma = \sigma$; similarly, $\sigma I = \sigma$.

As far as the inverse is concerned, given σ and α there exists a unique β such that $\alpha = \beta^{\sigma}$. The mapping τ defined for all α by $\alpha^{\tau} = \beta$ (that is, τ sends an element α to the element whose image under σ is α) is again a permutation of Ω. It is surjective: given α, τ sends α^{σ} to the unique element β such that $\beta^{\sigma} = \alpha^{\sigma}$; it follows that $\beta = \alpha$, so that α is the image under τ of α^{σ}. It is injective: if $\alpha^{\tau} = \gamma^{\tau}$, let $\alpha = \beta^{\sigma}$ and $\gamma = \eta^{\sigma}$. Then $\alpha^{\tau} = \beta$ and $\gamma^{\tau} = \eta$, so that $\beta = \eta$; it follows that $\alpha = \beta^{\sigma} = \eta^{\sigma} = \gamma$. Finally, $\alpha^{\sigma\tau} = (\alpha^{\sigma})^{\tau} = \alpha$, for all α, and similarly $\alpha^{\tau\sigma} = (\alpha^{\tau})^{\sigma} = \alpha$. Therefore, $\sigma\tau = I = \tau\sigma$, that is, $\tau = \sigma^{-1}$, and we have *iii*) of Definition 1.1.

Definition 1.6. The group S^{Ω} of all permutations of a set Ω is called the *symmetric group* on Ω. A *permutation* group of Ω is a subgroup of S^{Ω}.

If $\alpha^{\sigma} = \alpha$, α is a *fixed point* of σ.

The case in which Ω is finite is especially important. If $|\Omega| = n$, its element are denoted by 1,2,... n. The order of $|S^{\Omega}|$ is $n!$. Indeed, $\sigma \in S^{\Omega}$ is determined once the images of the elements of Ω are assigned. There are n choices for the image of 1; once such a choice has been made, there are $n-1$ choices left for the image of 2. Proceeding in this way, we have $n-2$ choices for the image of 3,..., 2 for the image of $n-1$ and only one for that of n, for a total of $n(n-1)(n-2)\cdots 2 \cdot 1 = n!$ choices.

The usual notation for a permutation σ in the finite case is

$$\sigma = \begin{pmatrix} 1 & 2 & \dots & n \\ i_1 & i_2 & \dots & i_n \end{pmatrix},$$

where i_k is the image of k under σ. It is clear that the order of the columns is of no importance.

Let $\Omega = \{1, 2, 3\}$, and consider the two permutations

$$\sigma = \begin{pmatrix} 1\,2\,3 \\ 2\,1\,3 \end{pmatrix}, \ \tau = \begin{pmatrix} 1\,2\,3 \\ 3\,2\,1 \end{pmatrix}.$$

Then the product is $\sigma\tau = \begin{pmatrix} 1\,2\,3 \\ 2\,3\,1 \end{pmatrix}$, whereas $\tau\sigma = \begin{pmatrix} 1\,2\,3 \\ 3\,1\,2 \end{pmatrix}$. This provides us a first example of a non commutative group. If now Ω is any set, $|\Omega| \geq 3$, even infinite, and we consider three element α, β and γ, let σ be a permutation that exchanges α and β and fixes all the other points, and τ one that exchanges α and γ and fixes all the other points. Then $\sigma\tau \neq \tau\sigma$, exactly as in the case $|\Omega| = 3$ we have just seen; it follows:

Theorem 1.3. *If $|\Omega| \geq 3$, the group S^Ω is a non commutative group.*

If Ω and Ω_1 are two sets with the same cardinality, and ϕ is a bijection $\Omega \to \Omega_1$, then ϕ induces an isomorphism $\overline{\phi}$ between the two groups S^Ω and S^{Ω_1} given by:

$$\overline{\phi} : \sigma \to \phi^{-1}\sigma\phi,\ \sigma \in S^\Omega.$$

In other words, the mapping $\overline{\phi}(\sigma) : \Omega_1 \to \Omega_1$ is defined by:

$$\Omega_1 \overset{\phi^{-1}}{\to} \Omega \overset{\sigma}{\to} \Omega \overset{\phi}{\to} \Omega_1.$$

This shows that the isomorphism class of S^Ω only depends on the cardinality of Ω. This allows us to write $S^{|\Omega|}$ instead of S^Ω, and S^n in the finite case $|\Omega| = n$.

Note that if $\Omega = \Omega_1 \cup \Omega_2$, and $\sigma, \tau \in S^\Omega$ are such that σ fixes all the points of Ω_2 and τ all those of Ω_1, then it is clear that it makes no difference to let σ act before τ or conversely. In other words, σ and τ commute: $\sigma\tau = \tau\sigma$. More generally, if $\Omega = \Omega_1 \cup \Omega_2 \ldots \cup \Omega_t$ and $\sigma_1, \sigma_2, \ldots, \sigma_t$ are elements of S^Ω, such that σ_i fixes all the points of $\Omega \setminus \Omega_i$, then $\sigma_i\sigma_j = \sigma_j\sigma_i$.

Let $\sigma \in S^n$, and consider, starting from any element i_1, the images $\sigma(i_1) = i_2$, $\sigma(i_2) = i_3$, and so on. Ω being finite, there will eventually be a repetition: for some i_k, the element $\sigma(i_k) = i_{k+1}$ has already been encountered. But the first repetition can only be i_1, i.e. $i_{k+1} = i_1$, because all the others already appear as images under σ. We write

$$\gamma_1 = (i_1, i_2, \ldots, i_k)$$

for the restriction of σ to the subset $\{i_1, i_2, \ldots, i_k\}$ of Ω. Since σ permutes cyclically the elements of this subset, in the written order, γ_1 is called a *cycle of σ of length k* or a *k-cycle of σ*. A cycle of length 2 is a *transposition.* Clearly, the same information on the action of σ is obtained if γ_1 is written as $(i_2, i_3, \ldots, i_k, i_1)$ or $(i_k, i_1, \ldots, i_{k-2}, i_{k-1})$, and this means that there are k ways of writing a k-cycle. Observe that

$$\gamma_1 = (i_1, \sigma(i_1), \sigma^2(i_1), \ldots, \sigma^{k-1}(i_1)).$$

When $k = n$, all the elements of Ω appear in γ_1. In this case, $\sigma = \gamma_1$ and σ is itself a cycle, an *n-cycle* or a *cyclic permutation.* Otherwise, let j_1 be an element not appearing in γ_1, and form the cycle

$$\gamma_2 = (j_1, j_2, \ldots, j_h).$$

No j_s can be equal to any i_t of the cycle γ_1, otherwise

$$\sigma^{s-1}(j_1) = j_s = i_t = \sigma^{t-1}(i_1)$$

from which $j_1 = \sigma^{t-s}(i_1)$, and j_1 would belong to γ_1. By proceeding in this way, all the elements of Ω are exhausted, and σ can be written as the union of its cycles:

$$\sigma = \gamma_1\gamma_2\ldots = (i_1, i_2, \ldots, i_k)(j_1, j_2, \ldots, j_h)\ldots$$

All the information needed to determine σ is contained in this writing: for each i we know its image under σ. A fixed point yields a cycle of length 1, a 1-*cycle*. Summing up:

Theorem 1.4. *Let $\sigma \in S^n$. Then the set $\Omega = \{1, 2, \ldots, n\}$ splits into the disjoint union of subsets:*

$$\Omega = \Omega_1 \cup \Omega_2 \cup \ldots \cup \Omega_t$$

on each of which σ is a cycle.

If γ_i is the cycle of σ on Ω_i, we can extend γ_i to a permutation of the whole set Ω, by setting $\gamma_i(j) = j$ if $j \notin \Omega_i$. If we keep denoting by γ_i this permutation of Ω, σ turns out to be the product of the γ_i's, and these commute in pairs. The previous theorem can then be rephrased by saying that *every permutation of S^n is the product of its own cycles.* Such an expression is unique up to the order in which the cycles appear and to circular permutations inside them.

Example 1.2. Let $n = 7$ and let

$$\sigma = \begin{pmatrix} 1\,2\,3\,4\,5\,6\,7 \\ 4\,2\,7\,5\,1\,6\,3 \end{pmatrix}.$$

Then $\Omega = \{1, 2, \ldots, 7\}$ splits into four subsets:

$$\Omega_1 = \{1, 4, 5\}, \Omega_2 = \{2\}, \Omega_3 = \{3, 7\}, \Omega_4 = \{6\},$$

on each of which σ is a cycle, and we write:

$$\sigma = (1, 4, 5)(2)(3, 7)(6).$$

Clearly, the same information is contained in any other expression like $\sigma = (6)(5, 1, 4)(7, 3)(2)$, or any other obtained as explained above.

Definition 1.7. A nonempty subset H of a group G is a *subgroup* of G if H is a group with respect to the restriction to H of the operation of G. In other words, H is a subgroup of G if it is a group with respect to the same operation of G. We write $H \leq G$.

Lemma 1.1. *Let $H \leq G$; then:*
i) *the unit of H is the same as that of G;*
ii) *if $h \in H$, the inverse of h in H equals the inverse of h in G.*

Proof. *i*) If $1'$ is the unit of H, and $h \in H$, then $1' \cdot h = h$. But $h \in G$, so $1 \cdot h = h$, where 1 is the unit of G. Hence $1' \cdot h = 1 \cdot h$, and by the cancellation law $1' = 1$.

ii) If k is the inverse of h in H, then $hk = 1$. But $hh^{-1} = 1$, where h^{-1} is the inverse of h in G; cancelling h yields $k = h^{-1}$. ◇

A group always admits two subgroups: the subgroup $\{1\}$ reduced to the identity, called the *identity* or the *trivial subgroup*, and the whole group G. If H is a subgroup, and $H \neq G$, then H is a *proper subgroup*, and we write $H < G$.

Lemma 1.2. *Let H be a nonempty subset of a group G. Then H is a subgroup of G if and only if:*
i) $h, h_1 \in H \Rightarrow hh_1 \in H$ *(closure of H);*
ii) *if 1 is the identity of G, then $1 \in H$;*
iii) $h \in H \Rightarrow h^{-1} \in H$.

Proof. If H is a subgroup, then, by definition, it satisfies the closure property, that is *i*). For *ii*) and *iii*) see the previous lemma. Therefore, these conditions are necessary for H to be a subgroup. Conversely, if the three conditions are satisfied, *i*) says that the operation of G is defined in H, and *ii*) and *iii*) say that the identity and the inverse in G of an element of H belong to H. Hence, H satisfies the requirements for the definition of a group, and since the operation is the same as that of G, H is a subgroup of G. ◇

Theorem 1.5. *A nonempty subset H of a group G is a subgroup of G if, and only if, for each pair of elements $h, h_1 \in H$, possibly equal, one has $hh_1^{-1} \in H$.*

Proof. If $H \leq G$ and $h_1 \in H$ then $h_1^{-1} \in H$ and, by the closure property, $hh_1^{-1} \in H$. Conversely, if the condition holds, with the pair h, h we have $1 \in H$, and with the pair $1, h$, we have $1 \cdot h^{-1} \in H$, so that the inverse of every element of H belongs to H. As to the closure, if $h, h' \in H$, then $h' = h_1^{-1}$, so $hh' = hh_1^{-1} \in H$. ◇

Theorem 1.6. *If H and K are subgroups of a group G, their intersection $H \cap K$ is again a subgroup of G. Their set-theoretic union $H \cup K$ is a subgroup if, and only if, one of the two subgroups is contained in the other.*

Proof. As far as the intersection is concerned, the requirements of the previous theorem are met. Now consider $H \cup K$; if $H \not\subseteq K$ and $K \not\subseteq H$ let $x \in H \setminus K$ and $y \in K \setminus H$. Then $xy \notin H$, otherwise, together with $x \in H$ and therefore also $x^{-1} \in H$, $x^{-1} \cdot xy \in H$ and $y \in H$, contrary to assumption. Similarly, $xy \notin K$, and $H \cup K$ cannot be a subgroup. If $H \subseteq K$, then $H \cup K = K$, which is a subgroup, and similarly in the other case. ◇

The same proof shows that the intersection of any family of subgroups is a subgroup. In order to define a subgroup starting from the union of a family of subgroups we need the notion of subgroup generated by a subset as given in the following definition.

Definition 1.8. Let S be a subset of a group G. The subgroup *generated* by S, denoted $\langle S\rangle$, is the intersection of all the subgroups of G containing S. The set S is a *set* or *system* of generators for the subgroup $\langle S\rangle$. If S is finite, the subgroup $\langle S\rangle$ is *finitely generated.*

Note that the intersection of Definition 1.8 is never empty because there is at least the group G that contains S. In every group G there exists a set S such that $\langle S\rangle = G$: just take $S = G$. The subgroup $\langle S\rangle$, being contained in every subgroup of G that contains the subset S, is often called the *smallest* subgroup of G containing S.

Given a family of subgroups we can now define their *group-theoretic union* as the subgroup generated by the set-theoretic union of the subgroups of the family.

The following theorem describes the nature of the elements of the subgroup generated by a set.

Theorem 1.7. *The subgroup generated by a nonempty set S in a group G is the set of all the elements of G obtained as finite products of elements of S and their inverses:*

$$\langle S\rangle = \{s_1 s_2 \cdots s_n,\ s_i \in S \cup S^{-1}\},$$

$n = 0, 1, 2, \ldots$, *and* $S^{-1} = \{s^{-1},\ s \in S\}$. *If S is empty,* $\langle S\rangle = \{1\}$.

Proof. Let $\overline{S} = \{s_1 s_2 \cdots s_n,\ s_i \in S \cup S^{-1}\}$. If H is a subgroup containing S, then it contains all products between the elements of S and their inverses, so $H \supseteq \overline{S}$, and $\langle S\rangle \supseteq \overline{S}$. Conversely, if $s = s_1 s_2 \cdots s_n \in \overline{S}$, then $s^{-1} = s_n^{-1} \cdots s_2^{-1} s_1^{-1}$ is also an element of $\overline{S}$, and if $s' = s'_1 s'_2 \cdots s'_n \in \overline{S}$, then $ss' \in \overline{S}$ as well. Moreover, $1 \in \overline{S}$ (the empty product). This shows that $\overline{S}$ is a subgroup, and therefore $\overline{S} \supseteq \langle S\rangle$. If $S = \emptyset$, all the subsets, and so all the subgroups, contain S; in particular, $\{1\}$ contains S. But all subgroups contain $\{1\}$, and so does their intersection. The result follows. ◇

If S is a singleton, $S = \{s\}$, then $\langle S\rangle$ is the set of all powers (multiples) positive, negative and zero of s. Such a group is a *cyclic group*, and since powers (multiples) of an element commute, a cyclic group is commutative. As we have already seen, this is the case of the group of integers $\mathbf{Z}$, with $S = \langle 1\rangle$ or $S = \langle -1\rangle$. However, $S = \{2, 3\}$ also generates $\mathbf{Z}$, ($1 = -1 \cdot 2 + 1 \cdot 3$, so that every n is the sum of multiples of 2 and 3, $n = -2n + 3n$) and similarly any finite set of integers having gcd equal to 1. The expression in the elements of S is by no means unique in general; in the previous example, we have $6 = 3 \cdot 2 + 0 \cdot 3 = 6 \cdot 2 + (-2) \cdot 3$, etc.

Examples 1.3. **1.** *Subgroups of the integers.* We have already observed (*Ex.* 1.1, 1) that the multiples of an integer m form a group $\langle m\rangle$ with respect to addition; therefore, this set a cyclic subgroup of $\mathbf{Z}$. Let us show that these are the only subgroups of $\mathbf{Z}$. Let $H \leq \mathbf{Z}$; if $H = \{0\}$ there is nothing to prove (H can be thought of as the set of multiples of 0). Let $H \neq \{0\}$; since H is a subgroup, together with an integer it contains its opposite, and therefore contains a positive integer. By the least integer principle there exists a least integer $m > 0$ in H. If $n \in H$, by dividing n by m we have $n = qm + r$, $0 \leq r < m$, (qm means q times m). H being a subgroup, qm belongs to H, and so does $-qm$; therefore, $n - qm \in H$, and $r \in H$. If $r > 0$, this contradicts the minimality of m. Thus $r = 0$, that is n is a multiple of m and so $H \subseteq \langle m \rangle$. Since H contains m, it contains all the multiples of m, that is, $\langle m \rangle \subseteq H$, and therefore $H = \langle m \rangle$. We have seen (*Ex.* 1.1 1) that $\langle m \rangle$ is isomorphic to $\mathbf{Z}$.

2. *Linear groups.* Let V be a finite or infinite dimensional vector space over a field K. The invertible linear transformations of V in itself form a group. It consists of the permutations φ of the set V that preserve the vector space structure

$$\varphi(v + w) = \varphi(v) + \varphi(w), \text{ and } \varphi(kv) = k\varphi(v),\ v, w \in V,\ k \in K.$$

This group is the *general linear group over* K, denoted $GL_K(V)$ or simply $GL(V)$. We now prove that the order of $GL(V)$ equals the cardinal number of the ordered bases of V. Fix a basis $B = \{v_\lambda\}$, $\lambda \in \Lambda$ of V and let $\{w_\lambda\}$, $\lambda \in \Lambda$, be another ordered basis. A mapping $\varphi : v_\lambda \to w_\lambda$ uniquely extends to an invertible linear transformation of V. Conversely, if $\varphi \in GL(V)$, then φ determines the ordered basis $B' = \{\varphi(v_\lambda)\}$. This proves that there is a bijection between the elements of $GL(V)$ and the the ordered bases of V.

If the field is finite and of order q (it is well known that q is a power of a prime) it is denoted by $\mathbf{F}_q$ or, if $q = p$, a prime, also by $\mathbf{Z}_p$. If V is of finite dimension n over $\mathbf{F}_q$ then it is finite and consists of q^n elements. Let us determine the number of ordered bases in this case. The first element of a basis can be chosen in $q^n - 1$ ways (any non zero vector), the second in $q^n - q$ ways (any vector not depending on the first),..., the n-th one in $q^n - q^{n-1}$ ways (any vector not in the subspace generated by the first $n - 1$ vectors). Thus,

$$|GL(V)| = (q^n - 1)(q^n - q) \cdots (q^n - q^{n-1}).$$

The $n \times n$ invertible matrices over a field K form a group $GL(n, K)$ under ordinary matrix multiplication. Once a basis has been chosen, with a linear transformation of a space V of dimension n there is associated a non singular matrix. This correspondence is an isomorphism $GL(V) \to GL(n, K)$. The matrices having determinant 1 form a subgroup, denoted $SL(n, K)$[4] ($SL(n, q)$), and the corresponding subgroup in $GL(V)$ is the *special linear group*, denoted $SL(V)$.

Matrices can also be defined over rings. The integer matrices having determinant 1 or –1 form a group $GL(n,\mathbf{Z})$.

---

[4] In practice, the notation $GL(n, K)$ is also used in place of $GL(V)$, and similarly for the other linear groups.

3. *Permutation matrices.* Among the linear transformations φ of a space of dimension n there are those which permute the elements of a fixed basis $\{v_i\}$. The corresponding matrices have only one 1 in each row and column and 0 elsewhere ((i,j)=1 if $\varphi(v_i) = v_j$). If $\varphi(v_i) = v_i$, then the diagonal entry (i,i) equals 1, so that the sum of the diagonal entries (the trace) gives the number of basis vectors fixed by φ. Clearly, this matrix group is isomorphic to S^n: the correspondence associating with φ the permutation taking i in j if φ takes v_i in v_j (see *ex.* 26). In this way, $GL(n,K)$ contains a copy of S^n.

4. *Klein group.* The reflections with respect to the axes x and y and to the origin 0 of a Cartesian coordinate system of the plane, denoted a, b and c, respectively, form, together with the identity transformation I, an abelian group of order 4 with product given by composition. The product of any two nonidentity elements equals the third:

$$ab = ba = c, ac = ca = b, bc = cb = a.$$

The four transformations take a point of coordinates (x, y) to the points:

$$I : (x,y) \to (x,y), a : (x,y) \to (x,-y), b : (x,y) \to (-x,y), c : (x,y) \to (-x,-y)$$

respectively. This group is the *Klein group*, or *Klein four-group*, and is denoted V[5].

5. More generally, in 3-dimensional Euclidean space the products of reflections with respect to three mutually orthogonal planes and to the origin make up, together with the identity, an abelian group of order 8, the eight elements being the transformations $(x, y, z) \to (\pm x, \pm y, \pm z)$. Similarly, in n-dimensional Euclidean space, one has the abelian group of order 2^n given by the transformations:

$$(x_1, x_2, \ldots, x_n) \to (\pm x_1, \pm x_2, \ldots, \pm x_n).$$

6. *Dihedral groups D_n.* It is known from geometry that a regular polygon with n vertices admits $2n$ symmetries (*isometries*), n of which are *rotatory* around the center of the polygon, and are obtained as powers of a rotation of $2\pi/n$. The other n are *axial* symmetries. For n odd, the latter are flips across the n axes passing through a vertex and the mid-point of the opposite side; if n is even, we have $n/2$ axes passing through pairs of opposite vertices (*diagonal symmetries*), and $n/2$ through the mid-points of opposite sides. In any case, we have a total of n symmetries. Under the product given by composition these $2n$ symmetries form a group, the *dihedral group*, denoted D_n. Observe that D_n, considered as a group of permutations of the n vertices of the polygon, is isomorphic to a subgroup of S^n. Let $1, 2, \ldots, n$ be the n vertices. A symmetry takes a fixed vertex, 1 say, to a vertex k, and since it preserves distances, it takes sides to sides, so that the images of two consecutive vertices

[5] From the German *Vierergruppe*.

are again consecutive. If the order of the vertices is preserved, the symmetry is obtained by a rotation of $2(k-1)\pi/n$ around the center of the polygon; if it is inverted, a flip is performed across the axis passing through the vertex 1 followed by a rotation of the same angle. The $2n$ symmetries can be described as follows:

$$I, r, r^2, \ldots, r^{n-1}, a, ar, ar^2, \ldots, ar^{n-1},$$

a being the flip across an axis passing through a fixed vertex, r the rotation of $2\pi/n$ and I the identity symmetry (also denoted by 1). Therefore, the elements all have the form

$$a^h r^k, \ h = 0, 1, \ k = 0, 1, \ldots, n-1, \tag{1.2}$$

where $a^h r^k$ is the symmetry obtained by first performing a^h and then r^k ($a^0 = r^0 = 1$ is the identity symmetry). In particular, D_n is generated by the two elements a and r. If, after ar, one performs once more flip a, the final result will be the same as that obtained by rotating the polygon in the opposite sense, that is, $ara = r^{-1}$. In general, the same argument gives $ar^k a = r^{-k}$, and since an axial symmetry equals its inverse, $a = a^{-1}$, we have $r^k a = ar^{-k}$.

All this should be clear for geometric reasons. In particular, D_n is a non-commutative group. The previous equality allows the reduction of the product of two elements of the form (1.2) to an element of the same form:

$$a^h r^k \cdot a^s r^t = \begin{cases} a^h r^{k+t} & \text{if } s = 0, \\ a^{h+1} r^{t-k} & \text{if } s = 1 \end{cases}$$

(the exponents of a are taken modulo 2, and those of r modulo n).

D_n is also generated by the two symmetries a and ar. Indeed, the smallest subgroup of D_n containing a and ar also contains the product $a \cdot ar = r$, and therefore also a and r that generate D_n. If n is odd, a and ar are symmetries with respect to axes joining two adiacent vertices; if n is even, ar is a symmetry with respect to axes joining mid-points of opposite sides.

In Euclidean plane, with the polygon centered at the origin and the vertices at the points $(\cos\frac{2k\pi}{n}, \sin\frac{2k\pi}{n})$, the n (counterclockwise) rotations are obtained as linear transformations with matrices:

$$r^k = \begin{pmatrix} \cos\dfrac{2k\pi}{n} & -\sin\dfrac{2k\pi}{n} \\ \sin\dfrac{2k\pi}{n} & \cos\dfrac{2k\pi}{n} \end{pmatrix}, \ k = 0, 1, \ldots, n-1$$

and the axial symmetries (across an axis making an angle of $k\pi/n$ with the x−axis) with matrices:

$$ar^k = \begin{pmatrix} \cos\dfrac{2k\pi}{n} & \sin\dfrac{2k\pi}{n} \\ \sin\dfrac{2k\pi}{n} & -\cos\dfrac{2k\pi}{n} \end{pmatrix}, \ k = 0, 1, \ldots, n-1.$$

7. *Rotational isometries of the cube.* If $1, 2, \ldots, 8$ are the vertices of a cube, an isometry takes 1 to i, say. There are eight choices for i, and for each of these there are three choices corresponding to the three vertices lying on the three sides meeting at i. Indeed, the cube can rotate of $2k\pi/3$, $k = 1, 2, 3$, around the diagonal through i. This determines the isometry completely, for a total of $8 \cdot 3 = 24$ isometries. (By composing a rotation with a reflection one obtains 24 additional isometries.)

Let now G be a group, x a fixed element of G. We have already seen that by multiplying all the elements of G by x, the set of products one obtains runs over all the elements of G, i.e. this multiplication is a permutation of the set underlying the group. Let σ_x be the permutation induced by multiplying on the right by x, and consider the mapping $G \longrightarrow S^G$ given by $x \longrightarrow \sigma_x$. This mapping is injective: if $\sigma_x = \sigma_y$ then $gx = gy$ for all $g \in G$, and therefore $x = y$ (observe that it is sufficient that the equality $gx = gy$ holds for one element g in order that it holds for all the elements). From the associative property $g(xy) = (gx)y$ it follows that $\sigma_{xy} = \sigma_x\sigma_y$. We have proved the next theorem.

Theorem 1.8 (Cayley). *Every group is isomorphic to a group of permutations of its underlying set. In particular, if the group is finite, of order n, then it is isomorphic to a subgroup of S^n.*

Definition 1.9. The isomorphism of Theorem 1.8 is called the *right regular representation of the group.*

Corollary 1.1. *S^n contains a copy of every group of order n.*

The number of groups of order n is at most equal to the number of binary operations that can be defined on a finite set G of order n, that is the number of functions $G \times G \longrightarrow G$, i.e n^{n^2}. The previous corollary yields a bound of the same order of magnitude. Actually, the sought number is at most equal to that of the subsets of cardinality n in S^n, i.e $\binom{n!}{n} \sim (n!)^{n-1} \sim n^{n(n-1)}$. We shall see (Theorem 1.23) that this bound can be lowered to $n^{n \log_2 n}$.

Although multiplication on the left by an element x of the group also gives a permutation τ of the set G, the mapping $x \longrightarrow \tau_x$ is not an isomorphism $G \longrightarrow G$ since from $(xy)g = x(yg)$ it follows $\tau_{xy} = \tau_y\tau_x$ (the mapping is an *anti-isomorphism*). In order to obtain an isomorphism it is necessary to associate with x the permutation τ' obtained by multiplying on the left by the inverse x^{-1}, because then $(xy)^{-1}g = (y^{-1}x^{-1})g = y^{-1}(x^{-1}g)$, and therefore $\tau'_{xy} = \tau'_x\tau'_y$. One obtains in this way the *left regular representation.*

Remark 1.2. We have defined the product of two permutations $\sigma\tau$ as $\alpha^{\sigma\tau} = (\alpha^\sigma)^\tau$, i.e first σ, then τ: this is the *right action* or *action on the right* . If the *left action* is chosen, $\alpha^{\sigma\tau} = (\alpha^\tau)^\sigma$, one has two perfectly analogous theories: the right action of $\sigma\tau$ is the left action of $\tau\sigma$. If σ and τ commute, $\tau\sigma = \sigma\tau$ and the two actions coincide. A similar situation is that of the right and left modules over a ring: if the ring is commutative, the distinction disappears.

If G is a finite group, the powers a^k of an element $a \in G$ cannot be all distinct. Let $a^h = a^k$ with $h > k$; then $a^{h-k} = 1$. By the least integer principle there exists a least n such that $a^n = 1$, and a has n distinct powers $a, a^2, \ldots, a^{n-1}, a^n = 1$.

Definition 1.10. Let $a \in G$; the smallest integer n, if any, such that $a^n = 1$ is called the *order* or *period* of a, denoted $o(a)$. In this case, a is of *finite order*, and has n distinct powers $a, a^2, \ldots, a^{n-1}, a^n = 1$ which form the cyclic subgroup generated by a. If for no $k > 0$ $a^k = 1$, then a is of *infinite order.* Note that $o(1) = 1$, and that 1 is the only element of order 1.

In the group of integers, if $km = 0$, $k > 0$, then $m = 0$: all non-zero elements are of infinite order. In the multiplicative group of rationals, -1 has period 2, and all the other elements (except 1) have infinite period. If $o(a)$ is finite, a is also called a *torsion* element. A group is a *torsion group* (or a *periodic group*) if its elements are torsion elements, and a *torsion free* group (or *aperiodic*) if no nontrivial element has finite order. The group is *mixed* if there are nontrivial elements of both types. If the order of the elements are uniformly bounded, their lcm is the *exponent* of the group.

Theorem 1.9. *Let $a \in G$; then:*

i) *if $o(a) = n$ and $a^k = 1$, then n divides k;*

ii) *an element and its inverse have the same order;*

iii) *if a and b are of finite order and commute, then $o(ab)$ divides the* lcm *of $o(a)$ and $o(b)$;*

iv) *if a and b are of finite order and commute, and have no nontrivial powers in common (this happens in particular if $(o(a), o(b)) = 1$), then $o(ab)$ equals the* lcm *of $o(a)$ and $o(b)$ (which equals the product $o(a)o(b)$ if $(o(a), o(b)) = 1$)*[6]*;*

v) *a nonidentity element is equal to its inverse if, and only if, it has order 2 (such an element is called an* involution*);*

vi) *ab and ba have the same order;*

vii) *a and $b^{-1}ab$ have the same order;*

viii) *if $o(a) = n$, then $o(a^k) = n/(n, k)$, and in particular the order of a power of an element divides the order of the element;*

ix) *if $o(a) = n = rs$ with $(r, s) = 1$, then a can be written in a unique way as the product of two commuting elements of order r and s.*

Proof. *i)* Divide k by n: $k = nq + r$, with $0 \leq r < n$. Then:

$$1 = a^k = a^{nq+r} = a^{nq}a^r = (a^n)^q a^r = a^r.$$

$r > 0$ contradicts the minimality of n. Thus $r = 0$ and k is a multiple of n.

ii) Let $o(a) = n < \infty$; then $(a^{-1})^n = (a^n)^{-1} = 1^{-1} = 1$, so $o(a^{-1})|n$ and $o(a^{-1})$ is finite. With $(a^{-1})^{-1} = a$ we have, by symmetry, $n = o(a)|o(a^{-1})$, and the result.

[6] Here and in the sequel (a, b) stands for $\gcd(a, b)$.

iii) For all t, $(ab)^t = abab\cdots ab = aa\cdots abb\cdots b = a^tb^t$. If $t = m = \mathrm{lcm}(o(a), o(b))$, then $a^m = b^m = 1$ and therefore $(ab)^m = 1$, and $o(ab)|m$. It may happen that $o(ab) < m$ (take $b = a^{-1}$).

iv) From $(ab)^t = 1$ it follows that $a^tb^t = 1$, $a^t = b^{-t}$ and, if a and b have no nontrivial powers in common, $a^t = 1$ e $b^{-t} = b^t = 1$. Therefore t is a multiple of $o(a)$ and $o(b)$, and therefore $t \geq m = \mathrm{lcm}(o(a), o(b))$. If $t = o(ab)$, then obviously $(ab)^t = 1$, and $o(ab) \geq m$. However, $o(ab)|m$, so that $o(ab) = m$. If $(o(a), o(b)) = 1$, a and b have no nontrivial powers in common. Indeed, if $a^h = b^k$, then $(a^h)^{o(b)} = b^{ko(b)} = (b^{o(b)})^k = 1$, so $o(a)|ho(b)$, and $o(a)$ must divide h. Thus $a^h = 1$.

v) If $a = a^{-1}$, then by multiplying both sides by a we have $a^2 = 1$. Conversely, if $a^2 = 1$, multiplying both sides by a^{-1} gives $a = a^{-1}$.

vi) Let $o(ab) = n < \infty$; then:

$$1 = (ab)^n = abab\cdots ab = a(ba\cdots ba)b = a(ba)^{n-1}b.$$

Multiply on the right by b^{-1} and on the left by a^{-1}; then $(ba)^{n-1} = a^{-1}b^{-1}$, and therefore, after multiplication of both sides by ba, $(ba)^n = 1$ and $o(ba)|n = o(ab)$. The same argument applied to ba yields $o(ab)|o(ba)$, and we have equality. Let $o(ab) = \infty$; if $o(ba) = n < \infty$, we would have $o(ab)|n$.

vii) If $o(a) = n$, then

$$\begin{aligned}(b^{-1}ab)^n &= b^{-1}abb^{-1}ab\cdots b^{-1}abb^{-1}ab\\ &= b^{-1}a(bb^{-1})a\cdots a(bb^{-1})ab\\ &= b^{-1}a^nb = b^{-1}b = 1,\end{aligned}$$

and therefore $o(b^{-1}ab)$ is finite and $n|o(b^{-1}ab)$. If $(b^{-1}ab)^k = 1$, $k < n$ then, as above, $1 = (b^{-1}ab)^k = b^{-1}a^kb$, from which $a^k = bb^{-1} = 1$, against $o(a) = n$.

viii) We have $(a^k)^{\frac{n}{(n,k)}} = (a^n)^{\frac{k}{(n,k)}} = 1$, $o(a^k)|\frac{n}{(n,k)}$. But $(a^k)^{o(a^k)} = a^{ko(a^k)} = 1$, and therefore $n|ko(a^k)$. It follows $ko(a^k) = nt$; division of both members by (n,k) gives $\frac{k}{(n,k)}o(a^k) = \frac{n}{(n,k)}t$. Then $\frac{n}{(n,k)}$ divides $\frac{k}{(n,k)}o(a^k)$, and being relatively prime to $\frac{k}{(n,k)}$, it must divide $o(a^k)$.

ix) With two integers u and v such that $ru + sv = 1$, we have $a = a^1 = a^{ru+sv} = a^{ru}a^{sv}$. Now, $o(a^{sv}) = \frac{n}{(n,sv)} = \frac{rs}{(rs,sv)} = \frac{rs}{s} = r$, because of *viii)* and the fact that r and v are relatively prime. Similarly, $o(a^{ru}) = s$, and being $(r,s) = 1$ we have the result. As for uniqueness see *ex.* 15. ◇

If $o(a) = n$, then the n powers of a: $a, a^2, \ldots, a^{n-1}, a^n = 1$ are all distinct. Indeed, if $a^h = a^k$, with $h, k < n$ and $h > k$, we have $a^{h-k} = 1$, with $h-k < n$, against the minimality of n. Observe that the n powers include the negative ones since, if $0 \leq t < n$, then $a^{-t} = a^{n-t}$. Moreover, if $k > n$, let $k = nq + r$, with $0 \leq r < n$. Then $a^k = a^{nq+r} = a^{nq}a^r = a^r$, with $r < n$. Therefore, if

$o(a) = n$ the above mentioned powers are all the powers of a. Conversely, if a has n distinct powers, the same argument shows that $o(a) = n$.

Examples 1.4. 1. The elements a, b and c of the Klein group are involutions, as well as those of the group of order 2^n of *Ex.* 1.3, 5.

2. In the dihedral group D_n, the n rotations are the n-th powers r^k of r, the rotation of $2\pi/n$, $k = 1, 2, \ldots, n$. Therefore, r^k is the rotation of $2k\pi/n$, of period $n/(n,k)$. Axial symmetries have period 2.

3. If two elements do not commute, *iii*) of the preceding theorem does not necessarily hold. Indeed, the two permutations on three elements mentioned before Theorem 1.3 have order 2, whereas their product has order 3. The order of the product of two non commuting elements of finite order can even be infinite. For instance, the two matrices of $SL(n, \mathbf{Z})$:

$$\begin{pmatrix} 0 & -1 \\ 1 & 0 \end{pmatrix}, \begin{pmatrix} 0 & 1 \\ -1 & -1 \end{pmatrix},$$

have orders 4 and 3, respectively; their product is $\begin{pmatrix} 1 & 1 \\ 0 & 1 \end{pmatrix}$, whose order is infinite:

$$\begin{pmatrix} 1 & 1 \\ 0 & 1 \end{pmatrix}^k = \begin{pmatrix} 1 & k \\ 0 & 1 \end{pmatrix}, \quad k = 1, 2, \ldots.$$

4. Let $\sigma = (1, 2, \ldots, n) \in S^n$ be an n-cycle; σ^k takes i to $i + k$, $i + k$ to $i + 2k, \ldots$, and if r is the least integer such that $rk \equiv 0 \bmod n$, then $i + (r-1)k$ goes to i, and the cycle to which i belongs closes. We know that $r = n/(n,k)$ (Theorem 1.9, *viii*)), so that the cycles of σ^k all have the same length $n/(n,k)$; therefore, their number is (n,k). Conversely, let τ be a permutation whose cycles all have the same length m:

$$\tau = (i_1, i_2, \ldots, i_m)(j_1, j_2, \ldots, j_m) \ldots (k_1, k_2, \ldots, k_m);$$

then τ is a power of an n-cycle; more precisely,

$$\tau = (i_1, j_1, \ldots, k_1, i_2, j_2, \ldots, k_2, \ldots, i_m, j_m, \ldots, k_m)^{\frac{n}{m}}.$$

5. Let $(a_1, a_2, \ldots, a_n)$ be an n-tuple of elements, not necessarily distinct. If by circularly permuting k times the a_i's one obtains k distinct n-tuples, with the $(k+1)$-st equal to the initial n-tuple, $(a_1, a_2, \ldots, a_n) = (a_{k+1}, a_{k+2}, \ldots, a_{k+n})$, then $a_1 = a_{k+1} = a_{2k+1} = \ldots = a_{sk+1}$, and similarly for the other a_i's. This means that the a_i's are equal n/k by n/k ($s + 1 = \frac{n}{k}$, that is, $sk + 1 = n - (k-1)$), i.e. the n n-tuples obtained by the n circular permutations are equal k by k; in particular, k divides n. Therefore, if $n = p$, a prime, $k = 1$ or $k = p$. If $k = 1$, then $a_1 = a_2 = \ldots = a_p$, and there is only one p-tuple (the converse is obvious: if the a_i's are all equal there is only one p-tuple); if

$k = p$, there are p distinct p-tuples. We shall see later an application of this result (Theorem 1.20).

6. If $m < n$, S^n contains subgroups isomorphic to S^m (fix $n - m$ digits and consider all the $m!$ permutations of the remaining ones).

7. The non zero elements of a field K, considered as elements of the additive group $K(+)$, all have the same order. If the characteristic of K is $p > 0$ this common order is p, otherwise is infinite. Conversely, if in an abelian group all the elements $a \neq 1$ have order the same prime p, then the group can be given the structure of a vector space over the field of residue classes mod p by defining $[h] \cdot a = a + a + \cdots + a$, h times.

Definition 1.11. A group G is *cyclic* if all its elements are powers a^k of some one element a of G, $k \in \mathbf{Z}$. In this case a is a *generator* of G, and G is said to be *generated* by a.

If G is cyclic, $G = \langle a \rangle$, and $x, y \in G$, then $x = a^h, y = a^k$, some h and k, so $xy = a^{h+k} = a^{k+h} = yx$, that is, *a cyclic group is commutative* (commutativity reduces to that of the sum of the integers).

If a group is cyclic, of order n, an element that generates the group has n distinct powers, and therefore is of order n. We have already seen examples of cyclic groups: the additive group $\mathbf{Z}_n$, generated by the residue class 1, and the additive group of integers $\mathbf{Z}$, generated by 1 or -1. There are no other examples, as the following theorem shows.

Theorem 1.10. *A finite cyclic group of order n is isomorphic to $\mathbf{Z}_n$. An infinite cyclic group is isomorphic to $\mathbf{Z}$.*

Proof. If $G = \langle a \rangle$ is a finite cyclic group of order n, then a has n distinct powers: $a, a^2, \ldots, a^{n-1}, a^n = 1$. The mapping $G \to \mathbf{Z}_n$ given by $a^i \to i$ is the required isomorphism. If G is infinite, since the powers of a run over G, a is of infinite order, and therefore $G = \{a^k,\ k \in \mathbf{Z}\}$. The mapping $G \to \mathbf{Z}$ given by $a^k \to k$ is the required isomorphism. ◊

The groups of n–th roots of unity and that of finite rotations are cyclic groups. A finite cyclic group of order n is also denoted C_n.

Theorem 1.11. *The subgroups of a cyclic group G are cyclic.*

Proof. Let $H \leq G$ and let $G = \langle a \rangle$. The proof is similar to that for the integers (1.21, *ex.* 1; but here we use multiplicative notation, so that multiples become powers). Like all the elements of G, the elements of H are powers of a (in the integers, every integer is a multiple of 1). Let a^m be the smallest positive power of a that belongs to H (in the integers, this would be the smallest positive multiple m of 1 that belongs to H, i.e. the smallest positive integer belonging to H), and let a^s be an element of H. Divide s by m: $s = mq + r$, with $0 \leq r < m$, and therefore $a^s = a^{mq+r} = a^{mq}a^r$, i.e. $a^r = a^s(a^{-m})^q$. This is a product of two elements of H, and therefore $a^r \in H$. If $r > 0$, this contradicts the minimality of m; then $r = 0$ and $a^s = a^{mq} = (a^m)^q$. Therefore, every element of H is a power of a^m, i.e. H is cyclic, generated by a^m. ◇

For every $m \neq 0$, $\mathbf{Z}$ is isomorphic to $\langle m \rangle$, so that $\mathbf{Z}$ is isomorphic to every proper subgroup.

However, if all proper subgroups of a group are cyclic, the group is not necessarily cyclic (and not even abelian). Here is an example.

Example 1.5. *The quaternion group.* The eight elements:

$$1, -1, i, -i, j, -j, k, -k$$

form a nonabelian group w.r.t. the product defined as follows:

$$\begin{gathered} ij = k, jk = i, ki = j \\ ji = -k, kj = -i, ik = -j \\ (\pm i)^2 = (\pm j)^2 = (\pm k)^2 = -1. \end{gathered}$$

This group is the *quaternion group*[7], denoted $\mathcal{Q}$. Note that the inverse of i is $-i = i^3$, and similarly for j and k. Moreover, $\{1, i, -1, -i\}$, $\{1, j, -1, -j\}$ and $\{1, k, -1, -k\}$ are subgroups (cyclic, of order 4, generated by i or $-i$, etc.) having the subgroup $\{1, -1\}$ in common. If $\{1\} \neq H < \mathcal{Q}$ and $H \neq \{1, -1\}$, let $i \in H$, say; then all the powers of i belong to H, and therefore $|H| \geq 4$. If $|H| > 4$, H must contain either j or k; however, if it contains one of them, j say, then it contains the other, because it must contain the product ij. It follows $H = \mathcal{Q}$. Thus, the proper non trivial subgroups of $\mathcal{Q}$ are those seen above, and are all cyclic.

Theorem 1.12. *A cyclic group of order n contains one and only one subgroup of order m, for each divisor m of n.*

Proof. Let $G = \langle a \rangle$ and $|G| = n$. If $m|n$, the element $a^{\frac{n}{m}}$ has order $n/(n, \frac{n}{m}) = n/(\frac{n}{m}) = m$, and therefore generates a subgroup of order m. Conversely, if $H \leq G$ and $|H| = m$, let $H = \langle a^k \rangle, k \leq n$. Since $o(a^k) = m$, and $o(a^k) = n/(n,k)$, we have $m = n/(n,k)$, $(n,k) = n/m$, so that n/m divides k. Let $k = (n/m)t$; then

$$a^k = a^{\frac{n}{m}t} = (a^{\frac{n}{m}})^t \in \langle a^{\frac{n}{m}} \rangle,$$

and therefore $\langle a^k \rangle \subseteq \langle a^{\frac{n}{m}} \rangle$. But both these subgroups have order m, and one being contained in the other they coincide. ◇

This result admits the following converse: if a finite group admits *at most* one subgroup for every divisor of the order, then it is cyclic (and therefore, by the theorem, it has has only one such subgroup). See Theorem 2.4.

The group of integers is generated by 1 and also by its opposite -1, and only by these. A finite cyclic group is generated by an element a and also by

[7] More precisely, the *group of the quaternion units of the skew field of quaternions.*

its inverse a^{-1} (a and a^{-1} have the same order). However, there may be other elements that generate the group. For instance, $\mathbf{Z}_8$ is generated by 1 and 7 (7 is congruent to -1 mod 8) and also by 3 and 5. We shall see in a moment that in the group $\mathbf{Z}_p$, p a prime, all non zero elements are generators.

Definition 1.12. The *Euler totient function* $\varphi(n)$ is defined for all positive integers n as:

i) $\varphi(1) = 1$;
ii) $\varphi(n)$=number of integers less than n and coprime to n, if $n > 1$.

Theorem 1.13. *A finite cyclic group of order n has $\varphi(n)$ generators.*

Proof. Let $G = \langle a \rangle$. An element a^k generates G if, and only if, it is of order n, and $o(a^k) = n/(n,k)$. This equals n if, and only if, $(n,k) = 1$, i.e. if and only if k is coprime to n. ◊

Examples 1.6. **1.** *A group of order $\varphi(n)$.* A group structure can be given to the set of integers less than n and coprime to n under the usual product followed by reduction mod n. Indeed, if $a, b < n$ and $(a,n) = (b,n) = 1$, then $(ab,n) = 1$. Dividing ab by n we have $ab = nq + r$, with $0 \leq r < n$ and $(r,n) = 1$; then define $a \cdot b$ to be r. This is the group $U(n)$ of invertible elements (*units*) of the ring $\mathbf{Z}_n$. For example, for $n = 8$, the group $U(8)$ consists of the elements 1, 3, 5 and 7; we have $3^2 = 9 \equiv 1 \bmod 8$, $5^2 = 25 \equiv 1 \bmod 8$, and $7^2 = 49 \equiv 1 \bmod 8$: the non identity elements have order 2. Moreover, $3 \cdot 5 = 15 \equiv 7 \bmod 8$, $3 \cdot 7 = 21 \equiv 5 \bmod 8$ and $5 \cdot 7 = 35 \equiv 3 \bmod 8$. The product of two non identity elements equals the third: it is clear that this group is isomorphic to the Klein group.

Thus, the generators of the additive group $\mathbf{Z}_n$ form a multiplicative group, which is abelian and of order $\varphi(n)$. If $n = p$, a prime, $\mathbf{Z}_p$, is a field, so that its multiplicative group is cyclic (see Corollary 2.2). In other words $U(p)$ is cyclic.

2. If p is a fixed prime, the p^n-th roots of unity form a cyclic multiplicative group C_{p^n}, of order p^n, subgroup of $\mathbf{C}^*$ and generated by a primitive p^n-th root. If $H \leq C_{p^n}$, H is generated by a p^h-th root z, for some h, and if z_1 is a p^k-th root, $h \leq k \leq n$, then $H = C_{p^h} \subseteq C_{p^k}$. This shows that the subgroups of C_{p^n} form a chain:

$$\{1\} \subset C_p \subset C_{p^2} \subset \ldots \subset C_{p^n}.$$

3. *The group C_{p^∞}.* Consider the set-theoretic union of the groups C_{p^n}, $n = 1, 2, \ldots$. This is again a multiplicative group, subgroup of $\mathbf{C}^*$. It consists of all p^n-th roots of unity, and its subgroups form an infinite chain:

$$\{1\} \subset C_p \subset C_{p^2} \subset \ldots \subset C_{p^n} \subset \ldots.$$

This is the *Prüfer group* or the p^∞ *group*, and is denoted C_{p^∞}. It is an infinite ascending union of subgroups, and as such it cannot be finitely generated (*ex.* 27). If $C_{p^i} = \langle w_i \rangle$, then

$$w_1 = w_2^p, w_2 = w_3^p, \ldots, w_{i-1} = w_i^p, \ldots$$

4. The set of rationals m/p^n, $n \geq 0$, of the interval $[0,1]$ having denominator a power of the prime p is a group under the sum modulo 1 (for instance, if $p = 2$, then $\frac{1}{2} + \frac{1}{2} = 1 \equiv 0$, $\frac{1}{2} + \frac{3}{4} = \frac{5}{4} = 1 + \frac{1}{4} \equiv \frac{1}{4}$, etc.) In this group, denoted $\mathbf{Q}^p$, all the elements have order a power of p. Indeed, $p^n(m/p^n) = m \equiv 0$. For a given n, the elements

$$0, \frac{1}{p^n}, \frac{2}{p^n}, \ldots, \frac{p^{n-1}-1}{p^n}$$

form a cyclic subgroup C_{p^n} of order p^n. As in the case of C_{p^∞}, as n grows these subgroups form a chain: $C_{p^m} \subset C_{p^n}$, if $m < n$. The two groups are isomorphic: if w is a primitive p^n–th root of unity, then the mapping $w^k \to k/p^n$ is an isomorphism.

Exercises

1. Prove that *i*), *ii*) and *iii*) of Definition 1.1 are equivalent to *i*) and *iv*) given $a, b \in G$ there exist $x, y \in G$ such that $ax = b$ and $ya = b$ (*quotient axioms*).

2. Under an isomorphism, an element and its image have the same order.

3. *i*) If $(ab)^2 = a^2b^2$, then $ab = ba$;
ii) if $(ab)^n = a^nb^n$ for three consecutive integers, then $ab = ba$.

4. Let a and b be as in Theorem 1.9, *iv*), and let $d = (o(a), o(b))$ and $m = \mathrm{lcm}(o(a), o(b))$. Prove that $o((ab)^d) = m/d$.

5. If a product $a_1a_2 \cdots a_n$ equals 1, the same holds for any cyclic permutation of the factors: $a_ia_{i+1} \cdots a_na_1a_2 \cdots a_{i-1} = 1$.

6. If an element $a \in G$ is unique of its order, then either $a = 1$ or $o(a) = 2$.

7. A group of even order contains an involution; more generally, it contains an odd number of involutions. [*Hint*: if $a \neq a^{-1}$ for all a, then the group has odd order.]

8. If $o(a) = o(b) = o(ab) = 2$ then $ab = ba$. In particular, if all non identity elements are involutions, the group is abelian.

9. A group of order 2, 3 or 5 is cyclic. [*Hint*: if $|G| = 3$, let $G = \{1, a, b\}$; the group must contain the product ab, and the only possibility is $ab = 1$ (if $ab = a$ then $b = 1$, and if $ab = b$ then $a = 1$); hence $b = a^{-1}$. Moreover, $a^2 \in G$, etc.]

10. The integers 1 and -1 form a group with respect to the ordinary product, which is not a subgroup of the group of integers.

11. Give an example of a group containing two commuting elements a and b, $b \neq a^{-1}$, such that $o(ab) < \mathrm{lcm}(o(a), o(b))$.

12. An infinite group has infinite subgroups. [*Hint*: a group is the set-theoretic union of the subgroups generated by its elements.]

13. Let $H \neq \emptyset$ be a subset of a group G, and assume that every element of H is of finite period. Then H is a subgroup of G if, and only if, $a, b \in H \Rightarrow ab \in H$.

14. Let $\varphi(n)$ be Euler's function. Using theorems 1.33 and 1.35 show that

$$\sum_{d|n} \varphi(d) = n,$$

where the sum is over all the divisors d of n (1 and n included). [*Hint*: divide up the elements of C_n according to their orders.]

15. Let $n = p_1^{h_1} p_2^{h_2} \cdots p_t^{h_t}$. Show that:

i) if a is an element of a group G and $o(a) = n$, then a can be written in a unique way as a product of pairwise commuting elements of orders $p_1^{h_1}, p_2^{h_2}, \ldots, p_t^{h_t}$;

ii) if $\varphi(n)$ is Euler's function, then:

$$\varphi(n) = \varphi(p_1^{h_1})\varphi(p_2^{h_2}) \cdots \varphi(p_t^{h_t}) \text{ and } \varphi(p^k) = p^k - p^{k-1} = p^{k-1}(p-1).$$

16. The dihedral group D_n is isomorphic to S^n only for $n = 3$.

17. D_4 has three subgroups of order 4, one cyclic and two Klein groups, and five subgroups of order 2.

18. D_6 contains two subgroups isomorphic to S^3. [*Hint*: consider the two triangles obtained by taking every second vertex of a hexagon.]

19. Determine a group of 2×2 complex matrices isomorphic to the quaternion group.

20. The matrices $\begin{pmatrix} \pm 1 & k \\ 0 & 1 \end{pmatrix}$, where $k = 0, 1, \ldots, n-1$, form a group isomorphic to D_n (integers mod n).

21. The matrices $\begin{pmatrix} a & b \\ -b & a \end{pmatrix}$, with a and b real numbers not both zero, form a group isomorphic to the multiplicative group of non zero complex numbers.

22. Let $\mathbf{Z}[x]$ be the additive group of polynomials with integer coefficients. Prove that this group is isomorphic to the multiplicative group of positive rationals. [*Hint*: order the prime numbers, and consider the n-tuple of the exponents of the prime numbers occurring in a rational number r/s.]

23. Determine the orders of the elements of the group $U(n)$ for $n = 2, 3, \ldots, 20$.

24. Let $H \leq G$, $a \in G$, $o(a) = n$, and let $a^k \in H$ with $(n, k) = 1$. Prove that $a \in H$. [*Hint*: $mn + hk = 1$, $a = a^1 = a^{mn+hk}$, etc.]

25. Prove that, for $k = 1, 2, \ldots, n$, the number of k-cycles of S^n is

$$\frac{n(n-1)\cdots(n-k+1)}{k}.$$

In particular, the number of n-cycles is $(n-1)!$.

26. Prove that the determinant of a permutation matrix is either 1 or –1 and that the matrices with determinant 1 form a subgroup containing half of the permutation matrices[8].

27. Prove that a group which is an infinite ascending union of subgroups cannot be finitely generated.

28. Prove that the only subgroups of C_{p^∞} are the C_{p^n} of the chain.

29. If $H < G$, then G is generated by the elements outside H.

1.2 Cosets and Lagrange's Theorem

Let H be a subgroup of a group G, finite or infinite. Let us introduce in the set underlying G the following relation:

$$a\rho b \ \text{ if } \ ab^{-1} \in H.$$

Theorem 1.14. *The above relation is an equivalence relation. The three properties of such a relation correspond to the three properties of a subgroup (Lemma 1.16).*

Proof. ρ is:

i) reflexive: $a\rho a$. Indeed, $aa^{-1} = 1$ and $1 \in H$ (because of *ii*) of 1.16);

ii) symmetric: if $a\rho b$ then $b\rho a$; indeed, if $a\rho b$ then $ab^{-1} \in H$, and therefore (by property *iii*) of 1.16) the inverse $(ab^{-1})^{-1} \in H$, i.e. $ba^{-1} \in H$, that is $b\rho a$;

iii) transitive: if $a\rho b$ and $b\rho c$, then $a\rho c$; indeed, if ab^{-1} and $bc^{-1} \in H$ their product also belongs to H (by property *i*) of 1.16): $ab^{-1}bc^{-1} = ac^{-1} \in H$, and this means $a\rho c$. ◇

The classes of this equivalence are the *right cosets* of H in G. Two elements $a, b \in G$ belong to the same coset if and only if there exists $h \in H$ such that $a = hb$. Thus, the coset of an element $a \in G$ is $\{ha, \ h \in H\}$; we shall also simply write Ha. The elements of the right coset to which a belongs are obtained by multiplying on the right the elements of H by a[9].

If we write Ha, we choose a as *representative* of the coset to which a belongs. Every element of the coset can be chosen as a representative of the coset. Indeed, if $b \in Ha$, i.e. $b = ha$, then $Hb = H \cdot ha = \{h'ha, h' \in H\} = Ha$ ($h'h$ runs over all elements of H as h' does). A set of representatives of all the right (left) cosets of a subgroup is called a right (left) *transversal.*

The set of cosets is the *quotient set* G/ρ. Similarly, one defines the *left cosets* through the relation:

$$a\rho' b \ \text{ if } \ a^{-1}b \in H.$$

[8] This subgroup is the *alternating group* (see 2.79).

[9] According to some authors, for example M. Hall jr., these are left cosets.

Thus, the coset of an element $a \in G$ will be $[a]' = \{ah,\ h \in H\} = aH$. Note that both in case of ρ and of ρ', the coset containing the identity of G is the subgroup H.

Theorem 1.15. *i) The right and left cosets of the subgroup H all have the same cardinality, namely $|H|$;*
ii) the quotient sets G/ρ and G/ρ' have the same cardinality.

Proof. *i)* The mapping $H \to Ha$, given by $h \to ha$, is injective (if $ha = h'a$, then $h = h'$) and surjective (an element $ha \in Ha$ comes from $h \in H$); similarly for the correspondence $H \to aH$. Hence,

$$|Ha| = |H| = |aH|.$$

ii) The mapping:

$$G/\rho \to G/\rho',$$

given by $Ha \to a^{-1}H$, is well defined. If, instead of a, a different representative for Ha is chosen, b say, then $b = ha$ and $b^{-1} = a^{-1}h^{-1}$, so that

$$Ha = Hb \Rightarrow b^{-1}H = a^{-1}h^{-1}H = a^{-1}H.$$

Moreover, since $h^{-1}H = H$, we have:

$$a^{-1}H = b^{-1}H \Rightarrow b^{-1} \in a^{-1}H \Rightarrow ab^{-1} \in H \Rightarrow Ha = Hb,$$

i.e. the mapping is injective. It is obvious that it is surjective. ◇

Remark 1.3. In Remark 1.1 we have seen that a function is well defined if

$$x = y \Rightarrow f(x) = f(y).$$

The other implication:

$$f(x) = f(y) \Rightarrow x = y$$

means that f is injective. An example of the fact that the two implications are inverse of each other was given in the proof of part *ii)* of the preceding theorem.

Definition 1.13. If H is a subgroup of a group G, the cardinality of the quotient set G/ρ (or G/ρ') is the *index of H in G*, and is denoted $[G : H]$. The index of the identity subgroup is the cardinality of G.

Theorem 1.15 is especially important in the case of finite groups.

Theorem 1.16 (Lagrange). *If H is a subgroup of a finite group G the order of H divides the order of G.*

Proof. Under the relation ρ (or ρ') G has $[G : H]$ equivalence classes, each having $|H|$ elements; therefore,

$$|G| = |H|[G : H].$$

Then $|H|$ divides $|G|$ and the quotient is the index of H in G. ♢

Remark 1.4. The above theorem has already been proved in the case of cyclic groups (Theorem 1.12). The converse of this theorem: if m is a divisor of the order of a finite group then the group admits a subgroup of order m, is true for cyclic groups (again by Theorem 1.12, with in addition the uniqueness of such a subgroup), and for other classes of groups (see *ex.* 35), but is false in general (Theorem 2.36).

Corollary 1.2. *The order of an element a of a finite group G divides the order of the group. In particular, $a^{|G|} = 1$.*

Proof. If $a \in G$ and $o(a) = m$, then a generates a cyclic subgroup of order m. By the theorem, m divides $|G|$. ♢

Corollary 1.3. *A group of prime order is cyclic.*

Proof. A non identity element generates a subgroup whose order must divide the order of the group. ♢

In particular, a group of prime order admits no nontrivial proper subgroups. The converse holds too.

Theorem 1.17. *A group G admitting no nontrivial proper subgroups is finite and of prime order.*

Proof. Let $1 \neq a \in G$. Then $\langle a \rangle = G$, otherwise $\langle a \rangle$ would be a nontrivial proper subgroup. Thus G is cyclic. If G is infinite, G is isomorphic to $\mathbf{Z}$, and $\mathbf{Z}$ admits nontrivial proper subgroups. Hence, G is finite. If its order is not a prime, then for every proper divisor m of its order the group has a subgroup of order m (Theorem 1.12). ♢

Examples 1.7. 1. The congruence relation modulo an integer n partitions $\mathbf{Z}$ into n equivalence classes, according to the remainders in the division by n: two integers belong to the same class if they give the same remainder when divided by n, i.e. if their difference is divisible by n. Thus,

$$a \equiv b \bmod n \text{ if } a - b = kn,\ k \in Z$$

i.e.

$$a\rho b \text{ if } a - b \in \langle n \rangle.$$

We have n classes, in each of which we may take as representative the remainder in the division by n. The subgroup $\langle n \rangle$ has index n (see *Ex.* 1.1, 5).

2. Using Lagrange's theorem, let us show that, up to isomorphisms, there are only two groups of order 4, the cyclic one and the Klein group. Let $1, a, b, c$ be the four elements of the group G. The order of the three nonidentity elements is 2 or 4 (Corollary 1.2). If there is an element of order 4, its powers run over the whole group, and the group is cyclic. Otherwise, a, b and c have order 2; the product ab must belong to the group, and there are four possibilities: $ab = 1$, $ab = a$, $ab = b$ and $ab = c$. The first three lead to a contradiction ($a = b^{-1} = b$, $b = 1$, $a = 1$), so that $ab = c$, and similarly $ba = c$. The product of two nonidentity elements equals the third, and it is clear that this group is (isomorphic to) the Klein group.

3. Let $S^3 = \{I, (1,2)(3), (1,3)(2), (2,3)(1), (1,2,3), (1,3,2)\}$, and consider the subgroup $H = \{I, (2,3)(1)\}$. H consists of the permutations of S^3 fixing 1. The right cosets of H are:

$$\begin{aligned} H &= \{I, (2,3)(1)\}, \\ H(1,2)(3) &= \{(1,2)(3), (1,2,3)\}, \\ H(1,3)(2) &= \{(1,3)(2), (1,3,2)\}. \end{aligned}$$

Hence, the three cosets consists of the elements taking 1 to 1, 1 to 2 and 1 to 3, respectively. The left cosets are:

$$\begin{aligned} H &= \{I, (2,3)(1)\}, \\ (1,2)(3)H &= \{(1,2)(3), (1,3,2)\}, \\ (1,3)(2)H &= \{(1,3)(2), (1,2,3)\}. \end{aligned}$$

They take 1 to 1, 2 to 1 and 3 to 1. Note that the right and left cosets containing $(1,2)(3)$ (and (1,3)(2)) are different.

In the group S^n, there are $(n-1)!$ elements fixing a digit, 1 say, (all the permutations on the remaining digits) and form a subgroup isomorphic to S^{n-1} and of index n. Here too the n cosets consist of the elements taking 1 to 1, 1 to 2, $\ldots$, 1 to n. The proof is the same as that for S^3. If σ is a permutation leaving 1 fixed, and $\tau_i = (1,i)$ is a transposition, then $\sigma\tau_i$ takes 1 to i. On the other hand, if η is a permutation taking 1 to i, then $\eta\tau_i$ leaves 1 fixed. But $\eta = (\eta\tau_i)\tau_i$, and therefore all the elements taking 1 to i are obtained by multiplying by τ_i a permutation leaving 1 fixed . Therefore, the coset $S^{n-1}\tau_i$ consists of the elements taking 1 to i; this leads to the decomposition of S^n into the n right cosets of S^{n-1}:

$$S^n = S^{n-1} \cup S^{n-1}\tau_2 \cup \ldots \cup S^{n-1}\tau_n.$$

Similarly for the left cosets.

4. In a dihedral group D_n there is only one nonidentity symmetry leaving a given vertex v of the regular $n-$gon fixed, the one w.r.t. the axis through v. The subgroup H fixing v has two elements, and is of index n. The elements of H take v to v, and those of the other cosets of H take v to the other vertices.

5. The pairs of real numbers (x, y) (points of the cartesian plane $\mathbf{R}^2$) form an additive group w.r.t. the parallelogram rule: $(x_1, y_1) + (x_2, y_2) = (x_1 + x_2, y_1 + y_2)$. Given a point (x, y), with x and y not both zero, the set of points $(\lambda x, \lambda y)$, $\lambda \in \mathbf{R}$, is a line r passing through the origin, and also a subgroup of $\mathbf{R}^2$. If $(x_1, y_1) \notin r$, the points $\{(x_1, y_1) + (\lambda x, \lambda y)\ , \lambda \in \mathbf{R}\} = (x_1, y_1) + r$ make up a line r through (x_1, y_1) and parallel to r, and also the coset of the subgroup to which (x_1, y_1) belongs. Hence, the lines through the origin are subgroups, and the lines parallel to them are their cosets.

Remark 1.5. The points of a line r through the origin have coordinates that are multiples by the elements of $\mathbf{R}$ of any of its points (x, y), and the vectors lying on r are multiples of the vector v with starting point (0,0) and endpoint (x, y). However, the subgroup thus obtained is not cyclic: in λv, λ does not vary only among the integers. It contains a cyclic subgroup, the one given by the λv with λ an integer.

6. Given the rational function $x_1x_2 + x_3x_4$ of four commuting variables, how many values are obtained if the x_i are permuted in all possible ways? It is easily seen that three values are obtained:

$$\varphi_1 = x_1x_2 + x_3x_4,\ \varphi_2 = x_1x_3 + x_2x_4,\ \varphi_3 = x_1x_4 + x_2x_3.$$

Each of these values remains unchanged when the x_i are permuted under the three subgroups of S^4:

$$\begin{aligned}
D^{(1)} &= \{I, (1,2), (3,4), (1,2)(3,4), (1,3)(2,4), (1,4)(2,3),\\
&\quad (1,3,2,4), (1,4,2,3)\},\\
D^{(2)} &= \{I, (1,3), (2,4), (1,3)(2,4), (1,4)(2,3), (1,2)(3,4),\\
&\quad (1,2,3,4), (1,4,3,2)\},\\
D^{(3)} &= \{I, (1,4), (2,3), (1,4)(2,3), (1,3)(2,4), (1,2)(3,4),\\
&\quad (1,2,4,3), (1,3,4,2)\},
\end{aligned}$$

respectively (fixed digits are omitted). One then says that the function $\varphi_i(x)$ *belongs* to the subgroup $D^{(i)}$, and this subgroup interchanges the other two values. These $D^{(i)}$ are isomorphic to the dihedral group D_4, as it can be seen by numbering 1 a vertex of a square and 3,2,4; 2,3,4 and 2,4,3, in a circular order, the other vertices, and have in common the subgroup

$$V = \{I, (1,2)(3,4), (1,3)(2,4), (1,4)(2,3)\},$$

a Klein group. Each of the groups D_4 has three right cosets in S^4: the elements of one of the D_4 fix one of the three values, and those of its cosets cyclically permute the three values. The cosets of V either fix all three values (V itself), or fix one value and interchange the other two, or else they cyclically permute the three values. Note that the right and left cosets of V coincide (V is a *normal* subgroup; see next chapter).

7. We have seen that in the group of integers all nontrivial subgroups have finite index (*Ex.* 1.7, 1). An example of the opposite case is that of the additive group of rationals, in which all proper subgroups have infinite index. Observe first that, in any group, if two distinct powers of an element a belong to the same coset of a subgroup H, then a non trivial power of a belongs to H. Indeed, if $Ha^s = Ha^t$, with $s \neq t$, then $ha^s = h'a^t$ and $a^{s-t} \in H$. Now let $H \leq \mathbf{Q}$ be of finite index m:

$$\mathbf{Q} = H \cup (H + a_2) \cup \ldots \cup (H + a_m).$$

If no integer multiple of one of the a_i belongs to H, then by what we have just observed the cosets $H + ta_i, t \in \mathbf{Z}$, are all distinct, and therefore H would be of infinite index. Let $n = \text{lcm}(n_i)$, where $n_i a_i \in H, i = 2, \ldots, m$. If $q \in \mathbf{Q}$, consider q/n; this element belongs to a coset $H + a_i$, so that $q/n = h + a_i$, with $h \in H$. But then $q = n(q/n) = nh + na_i \in H$, and therefore $H = \mathbf{Q}$.

Theorem 1.18. *Let $K \subseteq H$ be two subgroups of a group G, and let their indices $[G : H]$ and $[H : K]$ be finite. Then the index of K in G is finite and*

$$[G : K] = [G : H][H : K].$$

Proof. Let $G = Hx_1 \cup Hx_2 \cup \ldots \cup Hx_n$, $H = Kh_1 \cup Kh_2 \cup \ldots \cup Kh_m$. Then for $i = 1, 2, \ldots, n$, $Hx_i = Kh_1x_i \cup Kh_2x_i \cup \ldots \cup Kh_mx_i$. All the cosets of K thus obtained as Hx_i varies are disjoint: if $kh_jx_i = k'h_sx_t$, then, since $kh_j, k'h_s \in H$, it would follow $Hx_i = Hx_t$. Hence G is the union of the Kh_jx_i, and the number of these is $nm = [G : H][H : K]$. ♢

Theorem 1.19. *Let H and K be two subgroups of a group G. Then:*

$$[H : H \cap K] \leq [\langle H \cup K \rangle : K],$$

and if $[\langle H \cup K \rangle : H]$ and $[\langle H \cup K \rangle : K]$ are coprime, then equality holds.

Proof. Let $H = \bigcup_{i=1}^{r} (H \cap K)h_i$, be the partition of H into the cosets of $H \cap K$. If $Kh_i = Kh_j$ then $h_i = kh_j$, $k \in H \cap K$, and therefore $(H \cap K)h_i = (H \cap K)h_j$. It follows that K has at least r cosets in $H \cup K$. By the previous theorem,

$$[H \cup K : H \cap K] = [H \cup K : K][K : H \cap K] = [H \cup K : H][H : H \cap K],$$

so that $[H \cup K : K]$ divides $[H \cup K : H][H : H \cap K]$, and being coprime to $[H \cup K : H]$ it divides $[H : H \cap K]$. Together with the above inequality we have the result. ♢

Theorem 1.20. *The intersection of two subgroups of finite index is of finite index. More generally, the intersection of any finite number of subgroups of finite index is of finite index.*

Proof. Let H and K be the two subgroups, and let $y \in (H \cap K)x$; we have $y = zx$, with $z \in H \cap K$. Then $z \in H$, $y \in Hx$, and $z \in K$, so that $y \in Kx$; it follows $y \in Hx \cap Kx$ and therefore $(H \cap K)x \subseteq Hx \cap Kx$. Conversely, if $y \in Hx \cap Kx$, then $y = hx = kx$, so that $h = k$ and $h \in H \cap K$. Thus $y = hx \in (H \cap K)x$ and $Hx \cap Kx \subseteq (H \cap K)x$. Hence,

$$(H \cap K)x = Hx \cap Kx,$$

i.e. a coset of $H \cap K$ is obtained as the intersection of a coset of H with one of K, so that if these cosets are finite in number, their intersections are also finite. The extension to a a finite number of subgroups is immediate. ◇

Given a finite group G and an integer m dividing its order, there are two ways to approach the problem of the existence in G of a subgroup of order m: to make hypotheses on the structure of G (e. g. G cyclic, abelian, and so on), or else to make hypotheses on the arithmetic nature of m (m prime, power of a prime, product of distinct primes, and so on.) An example of the first way has already been seen: the converse of Lagrange's theorem holds for cyclic groups. The following theorem approaches the problem in the second way.

Theorem 1.21 (Cauchy). *If p is a prime dividing the order of a group G, then there exists in G a subgroup of order p.*

Proof. First observe that if p is a prime, the existence of a subgroup of order p is equivalent to that of an element of order p; let us then prove the existence of such an element. Consider the following set of p-tuples of elements of G:

$$S = \{(a_1, a_2, \ldots, a_p) | a_1 a_2 \cdots a_p = 1\}.$$

The set S contains n^{p-1} elements, where n is the order of G (the first $p-1$ elements can be chosen in n ways, and there is only one choice for the last one: it must be the inverse of the product of the first $p-1$). Let us call equivalent two p-tuples when one is obtained from the other by cyclically permuting the elements; under such a permutation the product is again 1, (*ex.* 5). If the a_i are all equal, then the p-tuple is the unique element of its equivalence class; if at least two a_i are distinct, the class contains p p-tuples (*Ex.* 1.4, 5). If r is the number of elements $x \in G$ such that $x^p = 1$, r equals the number of one-element classes; if there are s p-element classes, then $r + sp = n^{p-1}$. Now, $r > 0$ because there is at least the p-tuple $(1, 1, \ldots, 1)$, p divides n and therefore also n^{p-1}, and divides sp. It follows that p divides r, so $r > 1$, and this means that there exists at least one p-tuple of the form $(a, a, \ldots, a)$, with $a \neq 1$. This element a is of order p. ◇

Remark 1.6. The special case $p = 2$ has already been seen in *ex.* 7. More generally, we shall see in Chapter 3 that if p^k is a power of the prime p dividing the order a group then there exists in the group a subgroup of order p^k.

Definition 1.14. Let p be a prime number. A *p-element* in a group is an element of order a power of p. A p' element is an element whose order is not divisible by p. A group is a p-group if all its elements have order a power of p. A *p-subgroup* is a subgroup which is a p-group.

Examples of p-groups are the cyclic groups of order a power of p. The dihedral groups D_n, where n is a power of 2, and the quaternion group are examples of 2-groups. The group C_{p^∞} is an example of an infinite p-group.

Theorem 1.22. *A finite group G is a p-group if, and only if, its order is a power of p.*

Proof. If $|G| = p^n$, since $a^{p^n} = 1$ for all $a \in G$, we have $o(a)|p^n$ and therefore $o(a) = p^k$, $k \leq n$; therefore G is a p-group. Conversely, if G is a p-group, and $q||G|$, q a prime, $q \neq p$, by Cauchy's theorem there exists in G an element of order q, a contradiction. ◇

Examples 1.8. 1. Let us show that there are only two groups of order 6: the cyclic one, and the symmetric group S^3 (or D_3). If G has an element of order 6 then it is cyclic. If this is not the case, let $a, b \in G$ with $o(a) = 2$ and $o(b) = 3$; these two elements exist by Cauchy's theorem. If $ab = ba$, then $o(ab) = 6$ and G is cyclic, which we have excluded. The six elements $1, b, b^2, a, ab, ab^2$ are distinct (any equality leads to a contradiction: for instance, if $b^2 = a$, then $b^4 = 1$, while $o(b) = 3$). The element ba must be one of the above, and it is easily seen that it can only be ab^2. As in the case of the dihedral groups (*Ex.* 1.3, 6), the equality $ba = ab^2$ determines the product of the group, and in fact the correspondence $a^h b^k \to a^h r^k$, with $h = 0, 1$ and $k = 0, 1, 2$, establishes an isomorphism between G and the group D_3.

2. *Groups of order* 8. We know the groups $C_8, D_4, \mathcal{Q}$, the group of reflections w.r.t the three coordinate planes, that we denote $\mathbf{Z}_2^{(3)}$, and the group $U(15)$.

In the following table, the numbers appearing in the first column denote the orders of the elements; at the intersection of the row labeled i with column j the elements of order i of the group j is given:

	C_8	$\mathbf{Z}_2^{(3)}$	$U(15)$	D_4	$\mathcal{Q}$
1	1	1	1	1	1
2	1	7	3	5	1
4	2	0	4	2	6
8	4	0	0	0	0

Since two isomorphic groups have the same number of elements of a given order[10] the five groups above are not isomorphic. We will see (*Ex.* 2.10, 5) that there are no other groups of order 8.

[10] But the converse is false (*Ex.* 2.10, 6).

Definition 1.15. A subgroup H of a group G is said to be *maximal* w.r.t. a property $\mathcal{P}$ if it has property $\mathcal{P}$ and there exist in G no subgroup having $\mathcal{P}$ and containing H. In other words, if H has $\mathcal{P}$, and $H \leq K < G$, then either $H = K$ or K does not have $\mathcal{P}$.

For example, if $\mathcal{P}$ is the property of being abelian, then "maximal abelian" means "not properly contained in an abelian proper subgroup of the group". By saying "maximal subgroup", without any further specification, we mean that $\mathcal{P}$ is the property of being a proper subgroup. In this case, "H maximal" means that if $H \leq K$, then either $H = K$ or $K = G$.

A finite group certainly contains maximal subgroups, and indeed every subgroup is contained in a maximal one. In the group of integers, the subgroups generated by prime numbers are maximal (and these only are maximal). An infinite group may not admit maximal subgroups. For example, in the group C_{p^∞} the subgroups form an infinite chain and therefore for every subgroup there is a larger one containing it.

Theorem 1.23. *In a group G of order n there always exists a system of generators whose cardinality does not exceed* $\log_2 n$.

Proof. Let M be a maximal subgroup of G and let $x \notin M$. The subgroup $\langle M, x\rangle$ properly contains M, and therefore equals G. Since $M < G$, by induction on n we can assume the theorem true for the group M, i.e. that M can be generated by at most $\log_2 m$ elements, where $m = |M|$. Then G can be generated by at most $\log_2 m + 1$ elements. On the other hand, by Lagrange's theorem m divides n, so that $m \leq n/2$. It follows $\log_2 m + 1 \leq \log_2 \frac{n}{2} + 1 = \log_2 n - \log_2 2 + 1 = \log_2 n$. ◊

The logarithm of previous theorem arises as follows. Let

$$G = H_0 \supset H_1 \supset H_2 \supset \cdots \supset H_{s-1} \supset H_s = \{1\}$$

be a chain of subgroups each maximal in the preceding one. Since $[H_i : H_{i+1}]$ is at least 2, $i = 0, 1, \ldots, s-1$, and since the product of these indices is equal to the order n of the group, we have $n \geq 2 \cdot 2 \cdots 2 = 2^s$, and taking logarithms $s \leq \log_2 n$. Let us show that G can be generated by s elements. Picking an element x_1 in $G \backslash H_1$, by the maximality of H_1 in G we have $G = \langle H_1, x_1\rangle$. Similarly, if $x_2 \in H_1 \setminus H_2$ we have $H_1 = \langle H_2, x_2\rangle$, and therefore $G = \langle x_1, x_2, H_2\rangle$. Proceeding in this way, we have $G = \langle x_1, x_2, \ldots, x_s\rangle$, as required.

Corollary 1.4. *The number of non isomorphic groups of order n is at most* $n^{n \log_2 n}$.

Proof. The product of two elements in a group G is determined once the product of each element of G by the elements of a set X of generators of G is known. This follows from the associative law. Indeed, given $g, h \in G$, if $h = x_1 x_2 \ldots x_t$, $x_i \in X$, we have:

$$gh = gx_1x_2 \ldots x_t = (gx_1)x_2 \ldots x_t = (g_1x_2)x_3 \cdots x_t = \ldots = g_{t-1}x_t,$$

where we have set $g_i = g_{i-1}x_i$ ($g_0 = g$). Therefore, the number of group operations on a set of $n = |G|$ elements is at most equal to the number of functions $G \times X \to G$, i.e. $|G|^{|G|\cdot|X|}$. By the previous theorem, we can choose a set X with at most $\log_2 n$ elements. The result follows. $\diamondsuit$

Exercises

30. An equivalence relation defined on a group G is said to be *compatible* on the right (left) with the operation of G if $a\rho b \Rightarrow ag\rho bg$ ($ga\rho gb$), for all $g \in G$. Show that a compatible equivalence relation is one whose classes are the cosets of a subgroup.

31. If $H \leq G$, H is the only coset of H which is also a subgroup of G.

32. If $H, K \leq G$ and $Ha = Kb$, for some $a, b \in G$, then $H = K$.

33. (Euler's theorem) If $n > 1$ and a are two integers such that $(a, n) = 1$, then $a^{\varphi(n)} \equiv 1 \bmod n$.

34. (Fermat's little theorem) If p is a prime number not dividing a, then $a^{p-1} \equiv 1 \bmod p$.

35. Prove that the converse of Lagrange's theorem holds for the dihedral groups D_p, p a prime, and that if $p > 2$ the subgroups of D_p are given by $\langle r \rangle$ and $\langle ar^i \rangle$, $i = 0, 1, 2, \ldots, p-1$. (The converse of Lagrange's theorem holds for all dihedral groups D_n; cf. *ex.* 36, *i*) of Chapter 2.)

36. Determine the subgroups of D_4 and D_p, p a prime.

37. In a group of odd order every element is a square.

38. In the case of a finite group, prove directly that two cosets of a subgroup H either coincide or are disjoint, and use this fact to prove Lagrange's theorem.

39. Prove that there exist only two groups of order $2p$, p a prime: the cyclic one and the dihedral one.

40. A subgroup of finite index in an infinite group has a nontrivial intersection with every infinite subgroup of the group.

41. Prove that the cosets of $\mathbf{Z}$ in $\mathbf{R}$ are in one-to-one correspondence with the points of the interval [0,1).

42. Prove that the group of order 8 of *Ex.* 1.3, 5 has 7 subgroups of order 2 and 7 of order 4.

The seven nonzero elements $a, b, c, a+b, a+c, b+c, a+b+c$ of the group of the previous exercise are the seven points of the *Fano plane*, the projective plane of order 2, and the seven subgroups of order 4 are the seven lines of this plane. There are three points on each line (three elements in each subgroup, plus zero), three lines pass through each point (three subgroups meet in a subgroup of order 2). This plane can be represented as follows (zero is not marked):

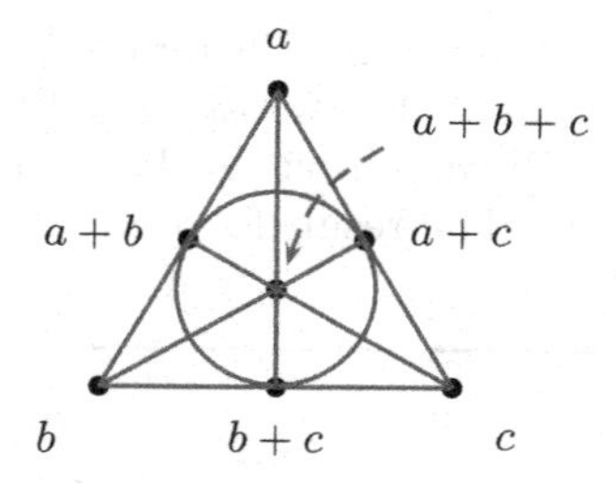

1.3 Automorphisms

Definition 1.16. An isomorphism of a group with itself is an *automorphism.*

The set of automorphisms of a group G is a group under composition, as is easily seen; it is denoted $\mathbf{Aut}(G)$. Moreover, an automorphism being a bijection and therefore a permutation of the set G, $\mathbf{Aut}(G)$ is a subgroup of the symmetric group S^G.

Examples 1.9. 1. Let G be a cyclic group, $G = \langle a \rangle$. If $g \in G$, then $g = a^k$, some k, and if $\alpha \in \mathbf{Aut}(G)$ we have $\alpha(g) = \alpha(a^k) = \alpha(a)^k$. Thus, α is determined by the value it takes on the generator a of G.

Let us distinguish two cases.

i) G finite, of order n. If $\alpha(a) = a^k$, we have $o(a^k) = o(a) = n$, since α preserves the orders of the elements, and therefore $(n, k) = 1$ (Theorem 1.9, *viii*)). Conversely, if $(n, k) = 1$, the mapping $\alpha : a^s \to a^{sk}$, $s = 0, 1, \ldots, n-1$, is injective because if $a^{sk} = a^{tk}$ with $s > t$ then $a^{(s-t)k} = 1$ and n divides $(s-t)k$, and being coprime to k it divides $s - t < n$, which absurd. Since G is finite, it is also surjective. Moreover, $a^s \cdot a^t = a^{s+t} \to a^{(s+t)k} = a^{sk} \cdot a^{tk}$, so that α preserves the operation, and therefore is an automorphism. The order of α is the smallest integer m such that $k^m \equiv 1 \bmod n$.

The mapping $\mathbf{Aut}(G) \to U(n)$, that associates with α the integer k less than n and coprime to n it determines, is an isomorphism. It follows that $\mathbf{Aut}(G)$ is abelian and has order $\varphi(n)$. In particular, if $n = p$, a prime, $\mathbf{Aut}(G) \simeq U(p)$, cyclic of order $p - 1$.

Consider $G = \langle a \rangle$, cyclic of order 7. Then $\mathbf{Aut}(G)$ is cyclic of order 6 and generated, for instance, by $\alpha : a \to a^k$, where $1 < k < 7$. Let $k = 3$; the automorphism $\alpha : a \to a^3$ induces the permutation $(1)(a, a^3, a^2, a^6, a^4, a^5)$ of the elements of G, its square $\alpha^2 : a \to a^2$ the $(1)(a, a^2, a^4)(a^3, a^6, a^5)$, and its third power $\alpha^3 : a \to a^6$ the $(1)(a, a^6)(a^3, a^4)(a^2, a^5)$. Similarly for the other powers of α.

ii) G infinite. If $\alpha \in \mathbf{Aut}(G)$ and $\alpha(a) = a^k$, since $a = \alpha(g)$ for some g (α is a bijection) and $g = a^h$ for a certain h, we have $a = \alpha(g) = \alpha(a^h) = \alpha(a)^h = (a^k)^h = a^{kh}$, and $a^{kh-1} = 1$. The order of a being infinite, this implies $kh - 1 = 0$, $kh = 1$, and therefore $k = \pm 1$. It follows that there are only two automorphisms, the identity and the one interchanging an element a with its inverse a^{-1}. In $\mathbf{Z}$, the nonidentity automorphism interchanges n and $-n$.

Remark 1.7. The following result of Gauss holds: if G is a cyclic group of order n, $\mathbf{Aut}(G)$ is abelian, and is cyclic if, and only if, $n = 2, 4, p^e, 2p^e$, p an odd prime, and $e \geq 1$. The symmetries of a polygon with vertices at the points $e^{2\pi i/n}$ of a unit circle given by the plane rotations form a cyclic group of order n. Following Gauss, the automorphisms of the latter group may be considered as "hidden symmetries" of the polygon.

2. It is clear that isomorphic groups have isomorphic automorphism groups. The converse is false: the groups C_3, C_4, C_6 and $\mathbf{Z}$ all have C_2 as automorphism group.

3. The invertible linear transformations of a vector space V over a field K are the automorphisms of the additive group of V that commute with the multiplication by scalars. If K is a prime field ($\mathbf{Z}_p$ or the field of rational numbers $\mathbf{Q}$), linearity and additivity are equivalent: all the automorphisms commute with the scalars, and therefore all are linear transformations. Indeed, if $K = \mathbf{Z}_p$, the multiplication $[h] \cdot v$ of $v \in V$ by a scalar $[h] \in \mathbf{Z}_p$ means the sum v with itself a number h of times (see *Ex.* 1.4, 7), and therefore for $\alpha \in \mathbf{Aut}(V)$ we have

$$\alpha([h] \cdot v) = \alpha(v + v \cdots + v) = \alpha(v) + \alpha(v) + \cdots + \alpha(v) = [h] \cdot \alpha(v).$$

Hence, $\mathbf{Aut}(V) \simeq GL(V)$, and if V is of dimension n over $\mathbf{Z}_p$,

$$\mathbf{Aut}(V) \simeq GL(n, \mathbf{Z}_p).$$

In particular, if $n = 1$, $V \simeq \mathbf{Z}_p$, cyclic of order p, and we have, as we know (see *Ex.* 1),

$$\mathbf{Aut}(\mathbf{Z}_p) \simeq U(p)$$

(the 1×1 invertibile matrices over $\mathbf{Z}_p$ are identified with its nonzero elements).

If $K = \mathbf{Q}$, α additive means $\alpha(s \cdot v) = s\alpha(v)$, where s is an integer. We now show that $\alpha(r/sv) = r/s\alpha(v)$ for all rationals r/s. Indeed,

$$s\alpha(\frac{r}{s}v) = \alpha(s \cdot \frac{r}{s}v) = \alpha(rv) = r(\alpha v),$$

and by dividing the first and the last member by s we have the result.

If V is of dimension n over $\mathbf{Q}$ we have

$$\mathbf{Aut}(V) \simeq GL(n, \mathbf{Q}),$$

and if V is of dimension 1, $V \simeq \mathbf{Q}$,

$$\mathbf{Aut}(\mathbf{Q}) \simeq \mathbf{Q}^*.$$

In other words, the automorphisms of the additive group of rationals are multiplications by nonzero rationals. Indeed, these automorphisms are all linear, so that $\alpha(x) = \alpha(x \cdot 1) = x\alpha(1)$, and given the image of 1 α is determined as the multiplication by $\alpha(1)$.

The rational field has no nonidentity field automorphisms (*ex.* 46).

Remark 1.8. Multiplication by a fixed element gives an automorphism also in the case of the additive group of the reals. It is clear that the function α defined by $\alpha(x) = cx$, $c \in \mathbf{R}$, is a continuous function. Conversely, if α is additive, for r rational we have as above $\alpha(r) = r\alpha(1)$. Now let $x = \lim r_i$, r_i rational; if α is continuous, then $\alpha(x) = \alpha(\lim r_i) = \lim \alpha(r_i) = \lim r_i\alpha(1) = \alpha(1)\lim r_i = \alpha(1)x$. In other words, the only additive and continuous functions are multiplications by a fixed real number. There are other automorphisms, none of which are Lebesgue measurable (indeed, an additive and measurable function is continuous). The real field has no nonidentity field automorphisms (*ex.* 47).

4. If $V = \{1, a, b, c\}$ is the Klein group, an automorphism fixes 1 and permutes the three nonidentity elements. Thus, $\mathbf{Aut}(V)$ is isomorphic to a subgroup of S^3. But every permutation of the three nonidentity elements is an automorphism. For example, let α be given by $\alpha(a) = b$, $\alpha(b) = a$, $\alpha(c) = c$. Then $\alpha(a \cdot b) = \alpha(c) = c = b \cdot a = \alpha(a) \cdot \alpha(b)$, $\alpha(b \cdot c) = \alpha(a) = b = a \cdot c = \alpha(b) \cdot \alpha(c)$, and similarly for the other products. If $\alpha(a) = b$, $\alpha(b) = c$, $\alpha(c) = a$, then $\alpha(a \cdot b) = \alpha(c) = a = b \cdot c = \alpha(a) \cdot \alpha(b)$, etc. Hence $\mathbf{Aut}(V) \simeq S^3 \simeq GL(2, \mathbf{Z}_2)$.

5. In any group, the inversion $\alpha : a \to a^{-1}$, is a bijection. In an abelian group it preserves the group operation, and therefore is an automorphism. Indeed, $\alpha(ab) = (ab)^{-1} = b^{-1}a^{-1} = a^{-1}b^{-1} = \alpha(a)\alpha(b)$. The converse is also true: if the inversion is an automorphism, the group is abelian. We have $\alpha(ab) = \alpha(a)\alpha(b) = a^{-1}b^{-1}$, so that $a^{-1}b^{-1} = b^{-1}a^{-1}$, i.e. $(ba)^{-1} = (ab)^{-1}$, and $ab = ba$. Hence *the correspondence* $\alpha : a \to a^{-1}$ *is an automorphism if, and only if, the group is abelian.*

If $\alpha \in \mathbf{Aut}(G)$ and $H \leq G$, then $\alpha(H)$ is again a subgroup of G and is isomorphic to H.

Definition 1.17. A subgroup H of a group G is a *characteristic* subgroups if every automorphism of G maps H to itself.

Examples 1.10. **1.** The identity subgroup and the group itself are characteristic subgroups.

2. The group of integers has a unique nonidentity automorphism, and this takes n to $-n$. Thus it maps every subgroup to itself. The same happens for a finite cyclic group, because then a subgroup is unique of its order, and therefore is mapped to itself. In conclusion, all subgroups of cyclic groups are characteristic.

3. In the Klein group, no nontrivial subgroup is characteristic, as *Ex.* 1.9, 4 shows.

4. The group $\mathbf{Q}$ of rationals has no nontrivial characteristic subgroups. Indeed, if $0 \neq a \in H < \mathbf{Q}$, let $b \notin H$; then the automorphism given by multiplication by b/a takes $a \in H$ to $b \notin H$.

5. The two previous examples are special cases of the fact that given a subspace W of a vector space V, two vectors $0 \neq v \in W$ and $v' \notin W$, are part of two bases B and B', respectively, and therefore there exists an invertible linear transformation of V, which in particular is an automorphism of the additive group of V, taking B to B' and v in v'.

6. In D_n the subgroup of rotations is characteristic: it is the unique cyclic subgroup of order n of D_n.

Exercises

43. Prove that in an abelian group of odd order the mapping sending every element to its square is an automorphism (see *ex.* 37).

44. Prove that the set of elements of a group left fixed by an automorphism is a subgroup.

45. Prove that in the group $GL(n, K)$, the mapping $\alpha : M \to (M^{-1})^t$ (inverse transpose) is an automorphism. The subgroup of the fixed elements consists of the orthogonal matrices ($M^t = M^{-1}$).

46. Prove the only automorphism of the rational field is the identity.

47. Prove that the only automorphism of the real field is the identity. [*Hint*: φ preserves squares, $\varphi(t^2) = \varphi(t)^2$, and therefore preserves order: if $r < s$, then $s - r > 0$ and therefore is a square, $s - r = t^2$, so that $\varphi(s-r) = \varphi(s) - \varphi(r) = \varphi(t^2) = \varphi(t)^2 > 0$. Hence $\varphi(s) - \varphi(r)$ is positive, that is, $\varphi(r) < \varphi(s)$. The set of elements fixed by φ is a subfield, hence it contains $\mathbf{Q}$; therefore $\varphi(q) = q$ for all rational numbers q. If $\varphi(r) \neq r$, and q is a rational number between r and $\varphi(r)$, then since $\varphi(q) = q$ we have $\varphi(r) < \varphi(q) = q$, so φ inverts the order between r and q.]

Remark 1.9. Unlike the reals, the complex field $\mathbf{C}$ has infinite automorphism. One of them is well known, and is conjugation; besides the identity, it is the only one fixing the reals elementwise. Moreover, every automorphism of a subfield of $\mathbf{C}$ extends to one of $\mathbf{C}$[11].

---

[11] Cf. the paper by Yale P.B.: Automorphisms of the complex numbers. Math. Magazine **39** (1966), 135–141.

2

Normal Subgroups, Conjugation and Isomorphism Theorems

2.1 Product of Subgroups

Definition 2.1. Let H and K be two subsets of a group G. The product of H by K is the set $HK = \{hk,\ h \in H, k \in K\}$.

This product is associative because such is the product of the group, but not necessarily commutative (commutativity means that given an element $hk \in HK, h \in H, k \in K$ there exist elements $k' \in K, h' \in H$ such that $hk = k'h'$). For example, in the group S^3, the product of the subgroups $H = \{I, (1,2)(3)\}$ and $K = \{I, (2,3)(1)\}$ is $HK = \{I, (1,2)(3), (2,3)(1), (1,3,2)\}$, while $KH = \{I, (1,2)(3), (2,3)(1), (1,2,3)\}$. This also shows that the product of two subgroups is not necessarily a subgroup (HK does not contain the inverse (1,2,3) of (1,3,2)). The next theorem gives a necessary and sufficient condition for the product of two subgroups to be a subgroup.

Theorem 2.1. *The product of two subgroups is a subgroup if, and only if, the two subgroups commute.*

Proof. Let H and K be two subgroups of a group. If HK is a subgroup, since $K, H \subseteq HK$ the products kh belong to HK, for all $k \in K$ and $h \in H$. Hence $KH \subseteq HK$. As to the other inclusion, if $x \in HK$ then $x^{-1} \in HK$, so that $x^{-1} = hk$, some h and k. Thus $x = k^{-1}h^{-1} \in KH$, and therefore $HK \subseteq KH$.

Conversely, let $HK = KH$. Then $1 \in HK$, and if $x \in HK$ then $x = hk$, and $x^{-1} = k^{-1}h^{-1} \in KH = HK$. Finally, if $x = hk$ and $y = h_1k_1$ then $xy = hkh_1k_1 = h(kh_1)k_1 = h(h_1'k')k_1 = (hh_1')(k'k_1) \in HK$. ♢

Theorem 2.2. *Let H and K be two subgroups of a finite group G. Then the number of cosets of H contained in HK equals the number of cosets of $H \cap K$ contained in K:*

$$[HK : H] = [K : H \cap K].$$

Machì A.: Groups. An Introduction to Ideas and Methods of the Theory of Groups.
DOI 10.1007/978-88-470-2421-2_2, Springer-Verlag Italia 2012

In particular, if HK is finite,

$$|HK| = \frac{|H||K|}{|H \cap K|}. \tag{2.1}$$

Proof. Let $g \in HK$, and let Hg be a coset of H. Then $g = hk, h \in H, k \in K$, so that $Hg = Hhk = Hk$, and we associate Hg with the coset $(H \cap K)k$ in K. If $g = h_1k_1$ is another expression of g, then $(H \cap K)k_1 = (H \cap K)k$ because $h_1k_1 = hk$ implies $k_1 = h_1^{-1}hk$, and $h_1^{-1}h = k_1^{-1}k \in H \cap K$. ♢

From this theorem we see that the cardinality of the intersection $H \cap K$ tells us how many times an element is repeated in the product HK, that is, the number of ways in which an element of HK may be expressed as a product of an element of H and one of K. In other words, all expressions of $g = hk$ are obtained as $(hx)(x^{-1}k)$ where x is any element of $H \cap K$.

Equality (2.1) reminds of $\mathrm{lcm}(a, b) = a \cdot b/\gcd(a, b)$.

Corollary 2.1. *If H and K are subgroups of a finite group G of coprime indices, then*

$$G = HK \quad \text{and} \quad [G : H \cap K] = [G : H][G : K].$$

More generally, if H_i, $i = 1, 2, \ldots, n$, are subgroups of coprime indices, then

$$[G : \cap_{i=1}^n H_i] = \prod_{i=1}^n [G : H_i].$$

Proof. We have $[G : H][H : H \cap K] = [G : K][K : H \cap K]$ because they are both equal to $[G : H \cap K]$ (Theorem 1.18). $[G : K]$ is coprime to $[G : H]$, so it divides $[H : H \cap K]$, and in particular $[G : K] \leq [H : H \cap K]$. It follows

$$|HK| = \frac{|H||K|}{|H \cap K|} = \frac{|H|}{|H \cap K|}|K| \geq [G : K]|K| = |G|,$$

so that $G = HK$. From the proof of Theorem 1.19 we have $[G : H \cap K] \leq [G : H][G : K]$; by what we have seen above, both factors divide the left hand side, and since they are coprime, their product also divides it. The general case follows by induction. ♢

2.2 Normal Subgroups and Quotient Groups

In this section we introduce one of the most important notions of the theory, that of a normal subgroup.

We begin by asking the following question: when is the product of two (right) cosets of a subgroup again a coset? (Note that a coset Ha is the product of H and the singleton $\{a\}$.) A product

$$Ha \cdot Hb = \{h_1a \cdot h_2b;\ h_1, h_2 \in H\}$$

contains the coset Hab (obtained for $h_2 = 1$); therefore, if it has to be a coset it will necessarily be Hab. Thus, the question becomes: find conditions under which

$$Ha \cdot Hb = Hab, \tag{2.2}$$

for all $a, b \in G$. It remains to prove the inclusion

$$Ha \cdot Hb \subseteq Hab. \tag{2.3}$$

If (2.3) holds for all $h_1, h_2 \in H$, there exists $h_3 \in H$ such that $h_1 a \cdot h_2 b = h_3 ab$, i.e. $ah_2 = h_1^{-1} h_3 a = h'a$. Therefore, a necessary condition for (2.3) to hold is that, given $h \in H$, there exists $h' \in H$ such that

$$ah = h'a. \tag{2.4}$$

But condition (2.4) is also sufficient. Indeed, it implies

$$h_1 a \cdot h_2 b = h_1(ah_2)b = h_1(h_2' a)b = h_1 h_2' \cdot ab = hab$$

i.e. (2.3).

Condition (2.4) may also be expressed as follows: for all $a \in G$,

$$aH = Ha \tag{2.5}$$

i.e., left and right cosets of H coincide, or as

$$a^{-1}ha \in H, \tag{2.6}$$

for $a \in G$ and $h \in H$, that is $a^{-1}Ha = H$.

Condition (2.4) is equivalent to the following: for $x, y \in G$,

$$xy \in H \Rightarrow yx \in H; \tag{2.7}$$

that is, if H contains the product of two elements of G in a given order then it also contains the product of the two elements in the reverse order. Indeed, by (2.4), with $a = x^{-1}$ and $h = xy$ we have $x^{-1}(xy) = h'x^{-1}$, that is $y = h'x^{-1}$, i.e. $yx = h' \in H$. Conversely, if (2.7) holds, consider $a \in G$, $h \in H$ and the element aha^{-1}. With $ha^{-1} = x$ and $a = y$ we have $xy = ha^{-1} \cdot a = h \in H$. By assumption, $yx = h' \in H$; but $yx = aha^{-1}$, so $aha^{-1} = h'$, i.e. (2.4).

The fact that the equivalent conditions (2.4) and (2.7) hold has the following noteworthy consequence: *the set of cosets of H is a group under operation* (2.2). Indeed,

i) associative law:

$Ha \cdot (Hb \cdot Hc) = Ha \cdot (Hbc) = Ha(bc)$,

$(Ha \cdot Hb) \cdot Hc = H(ab) \cdot Hc = H(ab)c$,

and by the associative law of G the two products are equal;

ii) identity element. It is the coset H:

$$Ha \cdot H = Ha \cdot H1 = H(a \cdot 1) = Ha$$

and similarly for $H \cdot Ha$;

iii) inverse. The inverse $(Ha)^{-1}$ of coset Ha is the coset to which the inverse of a belongs:

$$Ha \cdot Ha^{-1} = Haa^{-1} = H1 = H,$$

and similarly for $Ha^{-1} \cdot Ha$.

Definition 2.2. A subgroup H of a group G satisfying one of the equivalent conditions (2.2)–(2.7) is said to be a *normal* subgroup. We write $H \trianglelefteq G$ ($H \triangleleft G$ if H is proper). The group whose elements are the cosets of H under operation (2.2) is the *quotient group* or *factor group* of G w.r.t. H and is denoted G/H[1].

If H is a not necessarily normal subgroup of G, and K is a subgroup of G containing H, it may happen that (2.5) only holds for the elements a of K. In this case, H is a normal subgroup of the group K, $H \trianglelefteq K$.

Examples 2.1. 1. In an abelian group, every subgroup is normal: (2.4) holds with $h' = h$.

2. In any group, a subgroup of index 2 is normal. Indeed, for $a \notin H$, $G = H \cup Ha$ and $G = H \cup aH$, disjoint unions. Thus $aH = Ha$ because they are both equal to $G \setminus H$.

3. In the quaternion group $\mathcal{Q}$ every subgroup is normal. Indeed, the three cyclic subgroups of order 4 are normal (index 2), and the subgroup $N = \{1, -1\}$ is normal because it consists of elements commuting with all the others, so that (2.4) holds. Here we have an example of a nonabelian group in which all subgroups are normal; such a group is called *hamiltonian*

The quotient $\mathcal{Q}/N$ is a group of order 4, and therefore is either C_4 or the Klein group V. But the square of the six elements of order 4 is –1, and the square of –1 is 1: the square of every element belongs to N, and this means that every element of the quotient has order 2 ($(Ng)^2 = Ng^2 = N$). Thus $\mathcal{Q}/C_2 \simeq V$.

The subgroups of $\mathcal{Q}$ are all cyclic, and G/N is a Klein group. This shows that if $N \trianglelefteq G$, the group G need not contain a subgroup isomorphic to G/N. (This is true for abelian groups; see Theorem 4.20.)

4. In the dihedral group D_4, the three subgroups of order 4 are normal (index 2), they contain the eight elements of the group, and intersect in the

[1] Obviously, a group structure may be given to the set of cosets of H even if H is not normal (a group structure can be given to any set). However, in this case, the group operation cannot be that of (2.2). In other words, (2.2) is a group operation if, and only if, H is a normal subgroup, and only in this case one speaks of a quotient group.

subgroup $N = \{1, r^2\}$. The three subgroups being abelian, r^2 commutes with all the elements of the group. Thus (2.4) holds, and $N \lhd G$.

The square of the two elements of order 4 is r^2, the other elements have order 2. Thus, the square of every element belongs to N, so that $G/N \simeq V$.

No other subgroup of order 2 is normal. For example, if $H = \{1, d\}$, where d is a diagonal symmetry, then $r^{-1}Hr = \{1, d'\} \neq H$, where d' is the other diagonal symmetry.

5. In the dihedral group D_n the subgroup of rotations is normal. Indeed (see *Ex.* 1.3, 6) $ar^ka = r^{-k}$, $r^{-h}r^kr^h = r^k$ and therefore (recall that $a^{-1} = a$):

$$(ar^h)^{-1}r^k(ar^h) = r^{-h}(ar^ka)r^h = r^{-h}r^{-k}r^h = r^{-k},$$

and (2.6) holds. This can also be seen by the fact that subgroup of rotations is of index 2.

6. In the group S^3, the subgroup of order 3 is normal (index 2). No subgroup of order 2 is normal. For example, for $H = \{I, (1,2)(3)\}$ and $a = (1,2,3)$ we have

$$Ha = \{(1,2,3), (1,3)(2)\} \neq \{(1,2,3), (2,3)(1)\} = aH.$$

7. The group $\mathbf{Z}_n$ of residue classes modulo n with the operation of addition modulo n is the quotient group $\mathbf{Z}/n\mathbf{Z}$ (see *Ex.* 1.1, 5).

8. The group $SL(n, K)$ of matrices over a field K with determinant 1 is a normal subgroup of $GL(n, K)$ (see *Ex.* 1.3, 2). Indeed, if $A \in SL(n, K)$ and if B is any matrix of $GL(n, K)$, then $\det(B^{-1}AB)= \det(B)^{-1}\det(A)\det(B)=\det(A)=1$, so that $B^{-1}AB \in SL(n, K)$. Note that two matrices belong to the same coset of $SL(n, K)$ if, and only if, they have the same determinant, and since every non zero element of K is the determinant of a matrix of $GL(n, K)$ we have $GL(n, K)/SL(n, K) \simeq K^*$.

The above argument also shows that the matrices whose determinant belongs to a subfield of K form a normal subgroup of $GL(n, K)$.

9. A normal subgroup H commutes with every subgroup K. Indeed, if $H \trianglelefteq G$ then $ah = h'a$, for all $a \in G$. In particular, for $k \in K$, we have $kh = h'k$ and therefore $KH = HK$. Due to the normality of H, the condition $kh = h'k'$ is obtained here with $k' = k$. Therefore, by Theorem 2.1, the product of a normal subgroup by any other subgroup is again a subgroup.

10. The product of two normal subgroups is normal. If $hk \in HK$, with $H, K \trianglelefteq G$, then $x^{-1}hkx = x^{-1}hxx^{-1}kx = (x^{-1}hx)(x^{-1}kx) \in HK$. The product of any finite number of normal subgroups is normal (same proof).

11. The intersection of two normal subgroups is normal. If $x \in H \cap K$, with $H, K \trianglelefteq G$, then $a^{-1}xa \in H$, since $H \trianglelefteq G$, and $a^{-1}xa \in K$, because $K \trianglelefteq G$. Hence $a^{-1}xa \in H \cap K$. The intersection of any number of normal subgroups is normal (same proof).

12. From $H \trianglelefteq K$ and $K \trianglelefteq G$ it does not necessarily follow that $H \trianglelefteq G$ (*normality is not transitive*)[2]. For instance, in the group D_4, the subgroup $H = \{I, a\}$ is not normal (cf. *Ex.* 6); it is normal in one of the two Klein groups V because these are abelian, and $V \lhd G$.

13. The image of a normal subgroup under an automorphism of the group is again normal. Indeed, $a^{-1}H^{\alpha}a = (a^{-\alpha^{-1}}Ha^{\alpha^{-1}})^{\alpha} = H^{\alpha}$.

In spite of its simplicity, the next theorem is very important.

Theorem 2.3. *Two normal subgroups with trivial intersection commute elementwise.*

Proof. H being normal, $hk = kh'$; K being normal, $kh' = h'k'$. Therefore, $hk = h'k'$. On the other hand, by Theorem 2.2 $|H \cap K| = 1$ implies that the expression of a product of an element of H and one of K is unique. Thus $h = h'$ (and $k = k'$), i.e. $hk = kh$. This result may also be proved by considering the element $h^{-1}k^{-1}hk$. If we read it as $h^{-1}(k^{-1}hk)$, then it is a product of two elements of H because, by the normality of H, $k^{-1}hk \in H$; if we read it as $(h^{-1}k^{-1}h)k)$ it is the product of two elements of K, by normality of K. Hence $h^{-1}k^{-1}hk = 1$, and $hk = kh$[3]. ◇

In the quaternion group, the subgroups $H = \langle i \rangle$ and $K = \langle j \rangle$ are normal, and therefore commute, but not elementwise. Indeed, $ij = k$ and $ji = -k$. We have $|H \cap K| = 2$, and therefore the assumption $|H \cap K| = 1$ cannot be removed.

Theorem 2.4. *A finite group G having at most one subgroup for each divisor of its order is cyclic* (and therefore it has one and only one such subgroup).

Proof. Let us show that an element x of maximal order generates the whole group. Let $y \in G$; then $o(y)|o(x)$. Indeed, if $o(y)$ does not divide $o(x)$, for some prime p we have $o(x) = p^k s$ and $o(y) = p^h r$, with $h > k$, and r and s coprime to p. Let $x_1 = x^{p^k}$, $y_1 = y^r$; then $o(x_1) = s$, $o(y_1) = p^h$, and $(o(x_1), o(y_1)) = 1$. But $\langle x_1 \rangle$ and $\langle y_1 \rangle$ are normal subgroups: for all a, $a^{-1}x_1a$ has the same order as x_1 (Theorem 1.9, *vii*)), and therefore they generate subgroups of the same order; by uniqueness, the two subgroups are the same, so that $a^{-1}x_1a = x^k$, hence $x_1a = ax^k$, i.e. (2.4); similarly for $\langle y_1 \rangle$. By Theorem 2.3 the two subgroups $\langle x_1 \rangle$ and $\langle y_1 \rangle$ commute elementwise. It follows $o(x_1y_1) = o(x_1)o(y_1) = p^h s > p^k s = o(x)$, contrary to the maximality of $o(x)$. Hence $o(y)|o(x)$; $\langle x \rangle$ being cyclic, it contains a subgroup of order $o(y)$. The latter being unique, it is necessarily the subgroup $\langle y \rangle$; thus $y \in \langle x \rangle$[4]. ◇

---

[2] It does follow $H \trianglelefteq G$ if H is characteristic in K (cf. the observation before Theorem 2.25), or if K is a direct factor of G (Remark 2.3, 1).

[3] The element $h^{-1}k^{-1}hk$ is the *commutator* of h and k (see Section 2.9).

[4] One may say that, in this case, uniqueness implies existence.

Corollary 2.2. *A finite subgroup of the multiplicative group of a field is cyclic.*

Proof. In a field K, the multiplicative group K^* has at most one subgroup of order m, for all m. Indeed, if H is a subgroup of order m, then $h^m = 1$ for all $h \in H$, and therefore h is a root of the polynomial $x^m - 1$ which has at most m roots in K. Hence there cannot exist in K^* more than one subgroup of order m. In particular, the multiplicative group of a finite field is cyclic. ♢

The next theorem shows the relationship between the subgroups of a quotient group and those of the group.

Theorem 2.5. *There exists a one-to-one correspondence between the subgroups of a quotient group G/N and the subgroups of G that contain N. In this correspondence, normality and indices are preserved, i.e. $H/N \trianglelefteq G/N$ if, and only if, $H \trianglelefteq G$, and if $N \leq K \leq H$ then $[H : K] = [H/N : K/N]$.*

Proof. Let $S = \{N, Na, Nb, \ldots\}$ be a subgroup of G/N, and consider

$$K = N \cup Na \cup Nb \cup \ldots.$$

K is a subgroup of G: if $xa, yb \in K$, then $xa \cdot yb = xy'ab = zab$ with $z \in N$, and therefore K is closed with respect to the product. Moreover, $1 \in N$ so that $1 \in K$. Finally, if $xa \in K$, then $xa \in Na \in S$ and $(xa)^{-1} = a^{-1}x^{-1} = x_1a^{-1} \in Na^{-1}$; but $Na^{-1} \in S$ because $Na \in S$ and S is a subgroup.

Thus, S consists of the cosets of N to which the elements of K belong, and since $N \subseteq K$, we have $S = K/N$. In this way, the subgroup S of G/N determines a subgroup of G containing N, i.e. the subgroup K.

Conversely, if K is a subgroup of G containing N, then the cosets of N containing the elements of K form a group: $Nk_1Nk_2 = Nk_1k_2$, and $k_1k_2 \in K$ implies $Nk_1k_2 \in K/N$. Moreover $N \in K/N$, and if $Nk \in K/N$ then $(Nk)^{-1} = Nk^{-1} \in K/N$. It follows $K/N \leq G/N$. Note that for $K = N$, we have $N/N = \{N\}$, the identity of G/N.

As for normality, let $S = K/N \trianglelefteq G/N$; we show that $K \trianglelefteq G$. Indeed, let $k \in K$, $x \in G$, and consider the element $x^{-1}kx$. We have:

$$Nx^{-1}kx = Nx^{-1}NkNx = (Nx)^{-1}NkNx = Nk'$$

for a certain $k' \in K$ by normality of K/N. Then $x^{-1}kxk'^{-1} \in N \subseteq K$, and therefore $x^{-1}kx \in K$. In other words, if K/N is normal in G/N then K is normal in G. Conversely, if $K \supseteq N$ and $K \trianglelefteq G$, then

$$(Nx)^{-1}NkNx = Nx^{-1}kx = Nk' \in K/N$$

for all $k \in K$ and $x \in G$, i.e. $K/N \trianglelefteq G/N$.

Finally, if $N \leq K \leq H$ the correspondence $Kh \to (K/N)Nh$ is well defined:

$$Kh = Kkh \to (K/N)Nkh = (K/N)Nh$$

because Nkh and Nh differ by an element of K/N:

$$Nkh(Nh)^{-1} = Nkhh^{-1} = Nk.$$

It is injective:

$$(K/N)Nh = (K/N)Nh' \Rightarrow Nhh'^{-1} = NhNh'^{-1} \in K/N$$

and therefore $hh'^{-1} \in K$ and $Kh' = Kh$. It is obvious that it is surjective. ♢

If H is a subgroup of G not containing N, and S is the set of cosets of N to which the elements of H belong:

$$S = \{N, Nh_1, Nh_2, \ldots\},$$

then, by what we have seen in the proof of the preceding theorem, S is a subgroup of G/N, and the subgroup of G containing N that corresponds to it is $K = NH$ (from *Ex.* 2.1, 9 we know that NH is a subgroup). Hence $S = NH/N$.

If $N \trianglelefteq G$, the mapping:

$$\varphi : G \longrightarrow G/N, \tag{2.8}$$

obtained by associating with an element of G the coset modulo N to which it belongs is such that $\varphi(ab) = Nab = NaNb = \varphi(a)\varphi(b)$.

More generally:

Definition 2.3. Let G and G_1 be two groups. A *homomorphism* between G and G_1 is a mapping $\varphi : G \to G_1$ such that, for $a, b \in G$,

$$\varphi(ab) = \varphi(a)\varphi(b). \tag{2.9}$$

If $G_1 = G$, φ is an *endomorphism*. The set of endomorphisms of G is denoted $\mathbf{End}(G)$. If φ is surjective we say that G_1 is *homomorphic* to G. If φ is both surjective and injective, then (Definition 1.4) G_1 is isomorphic to G. We recover the notion of an isomorphism as a special case of that of a homomorphism.

The mapping (2.8) is a homomorphism, called the *canonical* homomorphism.

If φ is a homomorphism, $\varphi(1) = 1$, as follows by taking $a = b = 1$ in (2.9); moreover, $\varphi(a^{-1}) = \varphi(a)^{-1}$ (take $b = a^{-1}$).

Let K be the set of elements of G whose image is the identity element of G_1, $K = \{a \in G \mid \varphi(a) = 1\}$. Then K is a subgroup of G. Indeed,

i) $\varphi(1) = 1$, so that $1 \in K$;

ii) if $a \in K$, then $\varphi(a^{-1}) = \varphi(a)^{-1} = 1$, and therefore $a^{-1} \in K$;

iii) if $a, b \in K$ then $\varphi(ab) = \varphi(a)\varphi(b) = 1 \cdot 1 = 1$, and thus $ab \in K$.

Moreover, K is a normal subgroup of G: if $k \in K$, then $\varphi(x^{-1}kx) = \varphi(x^{-1})\varphi(k)\varphi(x) = \varphi(x)^{-1}\varphi(x) = 1$.

Definition 2.4. If $\varphi : G \to G_1$ is a homomorphism, the set of elements of G whose image is the identity element of G_1 is the *kernel* of φ, and is denoted by $ker(\varphi)$. As we have seen, $ker(\varphi)$ is a normal subgroup of G.

If $\varphi : G \to G_1$ is a homomorphism, and two elements of G have the same image under φ, $\varphi(a) = \varphi(b)$, then $\varphi(a)\varphi(b)^{-1} = 1 = \varphi(ab^{-1})$, i.e. $ab^{-1} \in ker(\varphi)$, so that a and b belong to the same coset of $ker(\varphi)$. Conversely, if a and b belong to the same coset of $ker(\varphi)$, then $ab^{-1} \in ker(\varphi)$. Hence, two elements of G have the same image under φ if, and only if, they belong to the same coset of the kernel of φ. In other words, the equivalence relation "$a\rho b$ if $\varphi(a) = \varphi(b)$" coincides with the relation "$a\rho b$ *if* $ab^{-1} \in ker(\varphi)$", and its classes are the cosets of $ker(\varphi)$. The mapping $\varphi(a) \to Ka$ between the elements of the image of φ and the elements of the quotient group G/K, $K = ker(\varphi)$, is therefore one-to-one. Moreover, this mapping is a homomorphism, because if $\varphi(a) \to Ka$, $\varphi(b) \to Kb$ then $\varphi(a)\varphi(b) = \varphi(ab) \to Kab = KaKb$. By denoting $Im(\varphi)$ the image of φ we have:

Theorem 2.6 (First isomorphism theorem[5]). *If G and G_1 are groups and $\varphi : G \to G_1$ is a homomorphism, then:*

$$Im(\varphi) \simeq G/ker(\varphi).$$

Hence, every homomorphism $G \to G_1$ factors through a surjective homomorphism, $G \to G/ker(\varphi)$, and an injective one $G/ker(\varphi) \to Im(\varphi) \subseteq G_1$.

Corollary 2.3. *Up to isomorphisms, the only groups homomorphic to a group G are the quotient groups G/N, where N is a normal subgroup of G.*

Proof. If G_1 is homomorphic to G under φ, then $G_1 = Im(\varphi)$; by the preceding theorem, $G_1 \simeq G/N$, where $N = ker(\varphi)$. Conversely, if $N \trianglelefteq G$, we already know that the group G/N is homomorphic to G (under the canonical homomorphism). ◇

A normal subgroup may be the kernel of many homomorphisms (the images will however be all isomorphic). For example, the identity subgroup is the kernel of all the automorphisms.

The image of a subgroup under a homomorphism φ is again a subgroup. Indeed, if H is a subgroup, then $\varphi(h_1)\varphi(h_2) = \varphi(h_1h_2) \in \varphi(H)$, and similarly for the identity and the inverse. In the canonical homomorphism $\varphi : G \to G/N$ we know (Theorem 2.5) that the image of a subgroup H of G is NH/N.

Theorem 2.7 (Second isomorphism theorem). *Let $H \leq G$ and $N \trianglelefteq G$. Then:*

i) $H \cap N \trianglelefteq H$;

ii) $NH/N \simeq H/H \cap N$.

[5] Also known as the "fundamental theorem of homomorphisms".

Proof. NH is a subgroup because N is normal. Consider the canonical homomorphism of the group NH, $\varphi : NH \to NH/N$. The image of the subgroup H of NH is NH/N, and therefore the restriction of φ to H is surjective, $\varphi|_H : H \to NH/N$. Obviously, $\varphi|_H$ is again a homomorphism. Its kernel consists of the elements of H that belong to the coset N of N in NH, and therefore the kernel of $\varphi|_H$ consists of the elements of $H \cap N$. $H \cap N$ being the kernel of a homomorphism, it is normal, and *i*) follows. Theorem 2.6 applied to the restriction $\varphi|_H$ gives *ii*). ◇

Theorem 2.8 (Third isomorphism theorem). *Let H and K be two normal subgroups of a group G with $K \subseteq H$. Then:*

i) $H/K \trianglelefteq G/K$;

ii) $(G/K)/(H/K) \simeq G/H$.

Proof. *i*) has already be seen (Theorem 2.5), but we recover it here by showing that H/K is the kernel of a homomorphism. The mapping $G/K \to G/H$ given by $Ka \to Ha$ is well defined: if $Ka = Kb$, then $a = kb$, some $k \in K$, and therefore $Ha = Hkb = Hb$ since $k \in K \subseteq H$. This mapping is a homomorphism because $Ka \cdot Kb = Kab \to Hab = Ha \cdot Hb$, and is obviously surjective. The kernel consists of the cosets Ka such that $a \in H$, and therefore is H/K, and we have *i*). *ii*) follows from Theorem 2.6. ◇

Theorem 2.9. *If φ is a homomorphism of a group G, and a is an element of G of finite order, then the order of $\varphi(a)$ divides the order of a.*

Proof. Let $o(a) = n$. Then $\varphi(a)^n = \varphi(a^n) = \varphi(1) = 1$, and therefore $o(\varphi(a))$ divides $o(a)$. ◇

Corollary 2.4. *If G is a finite group, and $N \trianglelefteq G$, the order of a coset of N, as an element of the quotient group G/N, divides the order of every element belonging to the coset.*

Proof. This is the preceding theorem with φ the canonical homomorphism $G \to G/N$. ◇

It may happen that the order of a coset properly divides that of every elements it contains. For example, in the quaternion group let $N = \langle i \rangle$; then $G/N = \{N, Nj\}$, and the coset $Nj = \{j, -j, k, -k\}$ is an element of order 2 while its elements have order 4.

Remark 2.1. Instead of the induction principle, it is sometimes useful to make use of the equivalent least integer principle. This applies as follows. Let $\mathcal{C}$ be a class of finite groups defined by some property (e.g. being abelian, cyclic, of order having a certain arithmetic nature, etc.) and let T be a theorem about the groups of $\mathcal{C}$. If T is false, there exists at least one group of $\mathcal{C}$ for which T is false; this group will have a certain order n. Hence, the set I of integers that are orders of groups of $\mathcal{C}$ for which the theorem T is false is nonempty. These orders being positive integers, by the least integer principle the set I contains a least integer m. This m is the order of a group G of $\mathcal{C}$; the group G is a *minimal counterexample* to the theorem. Hence, for all groups of $\mathcal{C}$ of order less than $|G|$ theorem T is true, and if we can prove that this implies that T is true for G we find the contradiction

$$\text{“}T \text{ false for } G\text{”} \Rightarrow \text{“}T \text{ true for } G\text{”}. \tag{2.10}$$

Thus, such a group G, and therefore the integer m, cannot exist, and the set I is empty. Hence T is true for all groups of the class $\mathcal{C}$.

Examples 2.2. 1. Let us prove Cauchy's theorem for abelian groups (Theorem 1.21) using the technique seen in the previous remark. The class $\mathcal{C}$ is now the class of finite abelian groups whose order is divisible by a given prime p, and theorem T now states that a group belonging to $\mathcal{C}$ contains an element (a subgroup) of order p. Let G be a minimal counterexample, and let $1 \neq a \in G$. If $o(a) = kp$, then $o(a^k) = p$, and (2.10) holds. If $p \nmid o(a)$, let $H = \langle a \rangle$. Then $H \trianglelefteq G$ because G is abelian, $p||G/H|$ and therefore $G/H \in \mathcal{C}$, and $|G/H| < |G|$. Hence, T is true for G/H, i.e. there exists in G/H an element Ha of order p. But the order of a coset divides the order of every element it contains (Corollary 2.4), so that $o(a) = kp$, and we are in the previous case. In any case, the existence of a minimal counterexample G leads to a contradiction. It follows that such a group G does not exist, hence T is true for all groups of the class $\mathcal{C}$.

2. As a further example of this technique let us prove that the converse of Lagrange's theorem holds for the class $\mathcal{C}$ of finite abelian groups. Let G be a minimal counterexample, and let $m||G|$. If $m = p$, a prime, just apply the previous example. If m is not a prime, and p is a prime dividing m, let H be a subgroup of order p. We have $G/H \in \mathcal{C}$, $|G/H| < |G|$ and $\frac{m}{p}||G/H|$, so that there exists in G/H a subgroup K/H of order $\frac{m}{p}$. But $\frac{m}{p} = |K/H| = |K|/|H|$ implies $|K| = m$, and we have the contradiction.

Exercises

1. A subset H of a group G is a subgroup of G if, and only if, the inclusion $H \to G$ is a homomorphism.

2. *i)* Operation (2.2) is well defined if, and only if, H is normal.
ii) If $H \trianglelefteq G$, (2.2) is the unique operation on the set of cosets of H for which the projection $G \to G/H$ is a homomorphism.

3. If G is finite, the only homomorphism of G to the integers is the trivial one sending the whole G to zero.

4. In a finite group, a normal subgroup of order coprime to the index is unique of its order.

5. If $p > 2$ is prime, the subgroup of rotations is the unique normal subgroup of D_p.

6. If G is a finite cyclic group, and p is a prime dividing $|G|$, then G contains $|G|/p$ elements that are p-th powers. [*Hint*: the mapping $x \to x^p$ is a homomorphism; consider the kernel.]

7. If G is finite and H is a subgroup of index 2, then H contains all the elements of odd order of G. [*Hint*: $x^{2k-1} = 1$ implies $x = (x^2)^k$.]

8. If a group has two normal subgroups of index p, p a prime, with trivial intersection, then it is a non cyclic group of order p^2.

9. If $A, B \leq G$, $H \trianglelefteq G$, $H \subseteq A \cap B$ and $(A/H)(B/H) \leq G/H$, then $AB \leq G$ and $(A/H)(B/H) = AB/H$.

2.3 Conjugation

If a and b are two elements of a group G we have seen in Theorem 1.9, *vi*) that the elements ab and ba, although distinct in general, have the same order. Note that ba can be obtained from ab as follows:

$$ba = a^{-1}(ab)a. \tag{2.11}$$

Definition 2.5. Two elements x and y of a group G are said to be *conjugate* if there exists $g \in G$ such that $y = g^{-1}xg$. We will then say that y is conjugate to x by g. The relation:

$$x \sim y \text{ if there exists } g \in G \text{ such that } y = g^{-1}xg$$

is an equivalence relation. Indeed, $x \sim x$ by 1 (or by any element commuting with x); if $x \sim y$ by g, then $y \sim x$ by g^{-1}, and finally, if $x \sim y$ by g, and $y \sim t$ by s, then $x \sim t$ by gs. We will also use the notation x^g to mean $g^{-1}xg$.

Definition 2.6. The equivalence relation just defined on a group is called *conjugation.* The classes of this equivalence are the *conjugacy classes* . Two elements are said to be *conjugate* if they belong to the same class. An element is *self-conjugate* if it is conjugate only itself. The conjugacy class to which x belongs is denoted $\mathbf{cl}(x)$.

Equality (2.11) then says that the elements ab and ba are conjugate. The next theorem shows that (2.11) expresses the only way in which two elements of a group can be conjugate.

Theorem 2.10. *Two elements x and y of a group G are conjugate if, and only if, there exist two elements a and b such that:*

$$x = ab \text{ and } y = ba.$$

Proof. The condition is sufficient by (2.11). As to necessity, let $y = g^{-1}xg$; setting $g = a$ and $g^{-1}x = b$ yields the result. ◇

Conditions (2.6) and (2.7) now say that a subgroup is normal if, and only if, together with an element it contains all its conjugates. Hence, *a subgroup is normal if, and only if, it is a union of conjugacy classes.*

From Theorem 1.9, *vi*) and *vii*) we have:

Corollary 2.5. *Two conjugate elements have the same order.*

Corollary 2.6. *A group is abelian if, and only if, the conjugacy classes all consist of a single element.*

We now ask ourselves the following question: how many elements are there in the conjugacy class of an element x? In order to know how many distinct elements belong $\mathbf{cl}(x)$ we must be able to tell when two elements g and g_1 give rise to the same conjugate of x, $g^{-1}xg = g_1^{-1}xg_1$. This happens if, and only if, the element gg_1^{-1} commutes with x.

Definition 2.7. The *centralizer* in G of an element $x \in G$ is the subgroup of the elements of G commuting with x:

$$\mathbf{C}_G(x) = \{g \in G \mid gx = xg\}.$$

If $y \in \mathbf{C}_G(x)$ we may say that y *centralizes* x. The centralizer allows us to answer the above question. This is the content of the following theorem.

Theorem 2.11. *If x is an element of a group G, then*

$$|\mathbf{cl}(x)| = [G : \mathbf{C}_G(x)], \tag{2.12}$$

In particular, if G if finite, $|\mathbf{cl}(x)|$ divides $|G|$.

Now fix an element g of G, and consider the mapping $\gamma_g : G \to G$ that sends an element $x \in G$ to its conjugate by g:

$$\gamma_g : x \to g^{-1}xg. \tag{2.13}$$

γ_g is injective (if $g^{-1}xg = g^{-1}yg$ then $x = y$) and surjective ($x \in G$ comes from gxg^{-1}); moreover,

$$\gamma_g(xy) = g^{-1}xyg = g^{-1}xg \cdot g^{-1}yg = \gamma_g(x)\gamma_g(y),$$

so that γ_g preserves the group operation. It follows that γ_g is an automorphism.

Definition 2.8. The automorphism γ_g of the group G given by (2.13) is called the *inner automorphism* induced by the element g of G.

The identity automorphism is inner (induced by $g = 1$). The inverse automorphism γ_g^{-1} is induced by g^{-1}: $\gamma_g^{-1} = \gamma_{g^{-1}}$, and the product $\gamma_g\gamma_{g_1}$ is induced by the product gg_1. Thus, the inner automorphisms of a group G

form a subgroup of $\mathbf{Aut}(G)$; it is denoted $\mathbf{I}(G)$. From (2.6) we have yet another definition of a normal subgroup: *a subgroup is normal if, and only if, it goes into itself under an inner automorphism* (it is *invariant* under inner automorphisms). Indeed, if $g^{-1}hg \in H$, all h and g, i.e. $g^{-1}Hg \subseteq H$ all g, we also have, with g^{-1} in place of g, that $gHg^{-1} \subseteq H$. But from $g^{-1}Hg \subseteq H$ it follows, by multiplying by g, that $H \subseteq gHg^{-1}$, from which the equality $gHg^{-1} = H$ follows, for all g, and $g^{-1}Hg = H$. (If G is infinite, it may happen that $g^{-1}Hg \subset H$ and $g^{-1}Hg \neq H$ (see *Ex.* 4.1, 4).)

Clearly, if a group G is abelian the only inner automorphism is the identity, $\mathbf{I}(G) = \{I\}$, and conversely. In a nonabelian group there always exist nonidentity inner automorphisms: if x and y are such that $xy \neq yx$, then the automorphism γ_x is not the identity because it sends y to $x^{-1}yx \neq y$.

If C is a conjugacy class of a group G, and $\alpha \in \mathbf{Aut}(G)$, then C^α, i.e. the set of images of the elements of C under α, is again a conjugacy class of G, since if x and y are conjugate by g, then x^α and y^α are conjugate by g^α. If α is inner, then by definition $C^\alpha = C$[6].

The subgroup $\mathbf{I}(G)$ is normal in $\mathbf{Aut}(G)$: if $\alpha \in \mathbf{Aut}(G)$ we have,

$$x^{\alpha^{-1}\gamma_g\alpha} = ((x^{\alpha^{-1}})^{\gamma_g})^\alpha = (g^{-1}x^{\alpha^{-1}}g)^\alpha = (g^{-1})^\alpha x g^\alpha = (g^\alpha)^{-1}xg^\alpha,$$

(we denote x^α the image of x under α), and therefore by conjugating the inner automorphism induced by g by $\alpha \in \mathbf{Aut}(G)$, one obtains the inner automorphism induced by the image of g under α.

Definition 2.9. The quotient group $\mathbf{Aut}(G)/\mathbf{I}(G)$ is called the group of *outer automorphisms* of G. It is denoted $\mathbf{Out}(G)$.

Two elements g and g_1 of G may induce the same inner automorphism:

$$g^{-1}xg = g_1^{-1}xg_1, \quad \forall x \in G; \tag{2.14}$$

this happens if, and only if, $gg_1^{-1}x = xgg_1^{-1}$, for all $x \in G$. In other words, $\gamma_g = \gamma_{g_1}$ if, and only if, gg_1^{-1} commutes with all the elements of G.

Next, consider the set $\mathbf{Z}(G)$ of elements $x \in G$ that commute with every element of G:

$$\mathbf{Z}(G) = \{x \in G \mid xy = yx, \;\; \forall y \in G\}.$$

It is clear that $1 \in \mathbf{Z}(G)$. If $xy = yx$, then $x^{-1}y = yx^{-1}$, and therefore if $x \in \mathbf{Z}(G)$ then $x^{-1} \in \mathbf{Z}(G)$, and finally if $x_1, x_2 \in \mathbf{Z}(G)$, then

$$(x_1x_2)y = x_1(x_2y) = x_1(yx_2) = (x_1y)x_2 = (yx_1)x_2 = y(x_1x_2),$$

and the product x_1x_2 also belongs to $\mathbf{Z}(G)$. Therefore, $\mathbf{Z}(G)$ is a subgroup of G. It equals G if, and only if, G is abelian.

[6] But there can be automorphisms fixing all conjugacy classes that are not inner (see Huppert, p. 22).

Definition 2.10. The subgroup $\mathbf{Z}(G)$ just defined is the *center* of G[7].

The following theorem is immediate.

Theorem 2.12. *i) The center of a group is the intersection of the centralizers of all the elements of the group;*

ii) an element belongs to the center if, and only if, its centralizer is the whole group.

Obviously, $\mathbf{Z}(G)$ is abelian and normal (and so are its subgroups). Its elements are self-conjugate. Moreover, if $x \in \mathbf{Z}(G)$ and $\alpha \in \mathbf{Aut}(G)$, then

$$x^\alpha y^\alpha = (xy)^\alpha = (yx)^\alpha = y^\alpha x^\alpha$$

for all $y \in G$, and therefore x^α commutes with all the images under α of the elements of G, and so with all the elements of G (α is a bijection), i.e. $x^\alpha \in \mathbf{Z}(G)$. Hence, *the center is a characteristic subgroup.*

The above discussion may be summarized as follows:

Theorem 2.13. *i) Two elements of a group G induce the same inner automorphism if, and only if, they belong to the same coset of $\mathbf{Z}(G)$.*
ii) $\mathbf{I}(G) \simeq G/\mathbf{Z}(G)$.

Proof. *i)* has been seen above. As for *ii)*, the mapping $G \to \mathbf{I}(G)$ given by $g \to \gamma_g$ is surjective, and the kernel is precisely the center. ◇

A group is a disjoint union of conjugacy classes; therefore if it is finite,

$$|G| = \sum_i |\mathbf{cl}(x_i)|, \tag{2.15}$$

where x_i is a representative of the i-th class. The union of the one-element classes is the center of the group; thus, (2.15) may be written as

$$|G| = |\mathbf{Z}(G)| + \sum_i |\mathbf{cl}(x_i)|, \tag{2.16}$$

where $|\mathbf{cl}(x_i)| \neq 1$. (2.16) is the *class equation.*

We now consider two applications of this equation. The first one is a property of finite p-groups.

Theorem 2.14. *The center of a finite p-group is nontrivial.*

Proof. If $|G| = p^n$, $n > 0$, then by Theorem 2.11 the order of a conjugacy class is a power of p. Every summand in the summation of (2.16) is greater than 1, and therefore is a nontrivial power of p. Hence, the sum is divisible by p, and $|G|$ being a power of p, $|\mathbf{Z}(G)|$ is divisible by p. In particular, $\mathbf{Z}(G) \neq \{1\}$. ◇

[7] Z is the initial of the German word Zentrum.

Theorem 2.15. *A group G of order p^2, p a prime, is abelian.*

Proof. By the preceding theorem, $Z = \mathbf{Z}(G)$ has order at least p, and therefore (Lagrange) either $|Z| = p$ or p^2. In the first case, let $x \notin Z$. Then $\mathbf{C}_G(x)$ contains Z and x and therefore its order is greater than p. It follows $|\mathbf{C}_G(x)| = p^2$, i.e. $\mathbf{C}_G(x) = G$, and this means that every element of G commutes with x, contrary to $x \notin Z$. Therefore, there are no elements outside the center, i.e. $Z = G$ and G is abelian. Another way of obtaining this result is the following. As above, let $x \notin Z$, and let $H = \langle x \rangle$. Then HZ is a subgroup of G properly containing Z, and therefore $HZ = G$. But this implies that G is abelian: $hz \cdot h'z' = hh' \cdot zz' = h'h \cdot zz' = h'z' \cdot hz$, contrary to $x \notin Z$. ◇

Examples 2.3. 1. Using *Ex.* 2.2, 1, let us prove Cauchy's theorem for any finite group G . We may assume G nonabelian, and let $x \notin \mathbf{Z}(G)$. If p divides $|\mathbf{C}_G(x)|$, since $|\mathbf{C}_G(x)| < |G|$ the theorem follows by induction. Therefore, we may suppose that p does not divide the order of the centralizer of any element $x \notin \mathbf{Z}(G)$, so that p divides the index of each of these subgroups. From $|\mathbf{cl}(x)| = [G : \mathbf{C}_G(x)]$, we have that p divides all the terms of the summation of (2.16), and therefore divides the sum; since it divides $|G|$, it also divides $|\mathbf{Z}(G)|$. $\mathbf{Z}(G)$ being abelian, the result follows from *Ex.* 2.2, 1.

2. The converse of Lagrange's theorem holds for p-groups; moreover, a p-group contains a *normal* subgroup for each divisor of the order. Indeed, if $|G| = p^n$, $n > 0$, the divisors of $|G|$ are the powers p^i, $i = 0, 1, \ldots, n$. We know that $Z = \mathbf{Z}(G) \neq \{1\}$, and therefore, by Cauchy, there exists in Z a subgroup H of order p. H is normal, and G/H has order p^{n-1}. By induction it has normal subgroups K_i/H of order p^i, $i = 0, 1, \ldots, n-1$. But then the K_i are normal in G and of order p^{i+1}.

Lemma 2.1. *Two involutions in finite group either they are conjugate, or they both centralize some third involution.*

Proof. Let x and y be two involutions; we distinguish two cases.

i) $o(xy) = m$, odd. Using $xy = yx^y$, the product $xyxy \cdots xy = 1$, m times, can be written as:
$$x^m y^{x^{m-1}} y^{x^{m-2}} \cdots y^x y = 1.$$
m being odd and $o(x) = 2$, the latter becomes:
$$xyy^x y \cdots y^x y = x(yy^x y \cdots y^x) y (y^x y \cdots y^x y) = 1;$$
Setting $a = yy^x y \cdots y^x$ and $b = y^x y \cdots y^x y$ we have $a = b^{-1}$, from which $b^{-1} y b = x^{-1} = x$, i.e. x and y are conjugate.

ii) $o(xy) = 2k$. Then $o((xy)^k) = 2$ and
$$x^{-1}(xy)^k x = (x^{-1}xyx)^k = (yx)^k = (y^{-1}x^{-1})^k = (xy)^{-k} = (xy)^k,$$
i.e. x centralizes the involution $(xy)^k$. Similarly for y. ◇

Theorem 2.16 (Brauer). *Let G be a finite group of even order in which not all the involutions are conjugate. If m is the maximal order of the centralizer of an involution, then $|G| < m^3$.*

Proof. Let y be an involution, $C_1 = \mathbf{C}_G(y)$, and let $y = y_1, y_2, \ldots, y_t$ be the involutions belonging to C_1. If $C_i = \mathbf{C}_G(y_i)$, $i = 1, 2, \ldots, t$, for the set-theoretic union of the C_i we have

$$|\bigcup_{i=1}^{t} C_i| \leq \sum_{i=1}^{t}(|C_i| - 1) + 1 \leq \sum_{i=1}^{t}(m-1) + 1 = (m-1)t + 1$$
$$= mt - (t-1) \leq mt,$$

and since $t < |C_1| \leq m$ (C_1 contains 1, which is not an involution),

$$|\bigcup_{i=1}^{t} C_i| \leq mt < m^2.$$

Let x be an involution such that $|\mathbf{C}_G(x)| = m$, and let $x = x_1, x_2, \ldots, x_h$ be the elements of G conjugate to x (all involutions!). Then $[G : \mathbf{C}_G(x)] = h$ (2.12) and therefore $|G| = mh$. Hence, it is sufficient to show that $h < m^2$. If y is an involution not conjugate to x, by the previous lemma there exists an involution z that centralizes x and y. Thus $z \in C_1$, and z is one of the y_k belonging to C_1. But $x \in \mathbf{C}_G(z) = \mathbf{C}_G(y_k)$, and therefore $x \in \bigcup_{i=1}^{t} C_i$, and this holds for all the h involutions conjugate to x. It follows $h < m^2$, and the theorem is proved. ◇

The hypothesis that the group has more than one conjugacy class of involutions is necessary in the above theorem. The dihedral group D_p, $p > 3$ a prime, has only one class of involutions and the centralizer of an involution has order 2 (see *ex.* 21). The theorem would give $2p < 2^3$.

Remark 2.2. It can be proved that there exist at most $m^2!$ finite simple groups in which the centralizer of an involution has order m (Brauer and Fowler, 1955). In particular, there exist only a finite number of finite simple group in which the centralizer of an involution is isomorphic to a given finite group. Results of this type fall within Brauer's program for the classification of finite simple groups.

2.3.1 Conjugation in the Symmetric Group

A permutation of S^n is a product of disjoint cycles, and therefore a conjugate of it by a permutation σ is the product of the conjugates of these cycles by σ. Conjugation of permutations thus reduces to conjugation of their cycles.

Theorem 2.17. *Let $c = (1, 2, \ldots, k)$ be a cycle of S^n, and let $\sigma \in S^n$. Then:*

$$(1, 2, \ldots, k)^\sigma = (1^\sigma, 2^\sigma, \ldots, k^\sigma).$$

In words: the conjugate of a cycle c by a permutation σ is the cycle in which there appear the images under σ of the digits of c in the same order .

Proof. Let $c_1 = (1^\sigma, 2^\sigma, \ldots, k^\sigma)$, and let $i \in c_1$, $i = j^\sigma$. We have:

$$i^{\sigma^{-1}c\sigma} = (j^\sigma)^{\sigma^{-1}c\sigma} = j^{c\sigma} = (j+1)^\sigma = (j^\sigma)^{c_1} = i^{c_1},$$

and therefore $\sigma^{-1}c\sigma$ and c_1 have the same value on the digits appearing in c_1. If $i \notin c_1$, i.e. $i \neq j^\sigma$ for all j, $j = 1, 2, \ldots, k$, then $i^{c_1} = i$,

$$i^{\sigma^{-1}c\sigma} = ((i^{\sigma^{-1}})^c)^\sigma = (i^{\sigma^{-1}})^\sigma = i,$$

i.e. $\sigma^{-1}c\sigma$ and c_1 have the same value also on the digits that do not appear in the cycle c_1. ♢

Definition 2.11. Two permutations $\sigma, \tau \in S^n$ are said to have the same *cycle structure* $[k_1, k_2, \ldots, k_n]$ if, whenever σ splits into k_i cycles of length i, $i = 1, 2, \ldots, n$, the same happens for τ.

We have:

$$1 \cdot k_1 + 2 \cdot k_2 + \cdots + nk_n = n$$

(obviously, some of the k_i may be zero). From Theorem 2.17 we have that two conjugate elements have the same cycle structure. Conversely:

Theorem 2.18. *If two elements S^n have the same cycle structure then they are conjugate.*

Proof. Let

$$\sigma = (i_1, i_2, \ldots, i_{r_1})(j_1, j_2, \ldots, j_{r_2}) \cdots (k_1, k_2, \ldots, k_{r_l}),$$
$$\tau = (p_1, p_2, \ldots, p_{r_1})(q_1, q_2, \ldots, q_{r_2}) \cdots (t_1, t_2, \ldots, t_{r_l}).$$

The permutation η defined as follows:

$$\eta = \begin{pmatrix} i_1\ i_2\ \ldots\ i_{r_1}\ j_1\ j_2\ \ldots\ j_{r_2}\ \cdots\ k_1\ k_2\ \ldots\ k_{r_l} \\ p_1\ p_2\ \ldots\ p_{r_1}\ q_1\ q_2\ \ldots\ q_{r_2}\ \cdots\ t_1\ t_2\ \ldots\ t_{r_l} \end{pmatrix},$$

which is obtained by writing one on top of the other the cycles of σ and τ of the same length, takes σ to τ: $\eta^{-1}\sigma\eta = \tau$. ♢

Example 2.4. In S^9,

$$\sigma = (1,3)(2,4,5,7)(8,9)(6) \text{ and } \tau = (1,3)(4,6,5,8)(7,9)(2)$$

have the same cycle structure $[1, 2, 0, 1, 0, 0, 0, 0, 0]$, and a permutation η that conjugates them is, for instance,

$$\eta = \begin{pmatrix} 1\,3 \vdots 2\,4\,5\,7 \vdots 8\,9 \vdots 6 \\ 7\,9 \vdots 4\,6\,5\,8 \vdots 1\,3 \vdots 2 \end{pmatrix} = (1,7,8)(2,4,6)(3,9)(5).$$

Since $(1,3) = (3,1)$, $(7,9) = (9,7)$ e $(4,6,5,8) = (5,8,4,6)$, then

$$\eta' = \begin{pmatrix} 1\,3 \vdots 2\,4\,5\,7 \vdots 6 \vdots 89 \\ 3\,1 \vdots 5\,8\,4\,6 \vdots 2 \vdots 97 \end{pmatrix} = (1,3)(2,5,4,8,9,7,6)$$

also conjugates σ and τ.

Theorem 2.19. *The number of elements of S^n having a given cycle structure $[k_1, k_2, \dots, k_n]$, and therefore the number of elements of a given conjugacy class, is*

$$\frac{n!}{1^{k_1} \cdot 2^{k_2} \cdots n^{k_n} k_1! k_2! \cdots k_n!}. \tag{2.17}$$

Proof. The digits $1, 2, \dots, n$ may appear in all the $n!$ possible ways to give a product of k_i disjoint i-cycles, $i = 1, 2, \dots, n$. But given one of these products, the resulting permutation is the same as that obtained by exchanging the k_i cycles in all possible ways, i.e. in $k_i!$ ways. Moreover, for each i-cycle there are i ways of writing it; since there are k_i cycles, there are i^{k_i} possible ways of writing the i-cycles. Thus, for every i, the same permutation is obtained i^{k_i} times. The result follows. (Set $0! = 1$). ◇

We now determine the conjugacy classes of S^n. Recall that a partition of a natural number n is a way of representing n as a sum of distinct natural numbers $n_1 \leq n_2 \leq \dots \leq n_k$. A conjugacy class of S^n is determined by a cycle structure $[k_1, k_2, \dots, k_n]$; conversely, given such a structure, the $k_i > 0$ yield the multiplicity of the positive integers i in a partition of n. Hence, given a partition of n in integers i each with multiplicity k_i, by considering all the products of k_i cycles of length i, we have, as i varies, all the permutations having cycle structure $[k_1, k_2, \dots, k_n]$ ($k_j = 0$ if j does not appear in the partition). Consequently:

Theorem 2.20. *The number of conjugacy classes of S^n equals the number of partitions of n.*

Example 2.5. Let us determine the conjugacy classes of S^4 and A^4. There being five partitions of 4, i.e.

$$4 = 1+1+1+1,\ 4 = 1+1+2,\ 4 = 1+3,\ 4 = 2+2,\ 4 = 4,$$

there are five classes. These partitions give five possible cycle structures: four cycles of length 1; two cycles of length 1 and one of length 2; etc., so that the five classes consist of the permutations:

$$\begin{aligned}
C_1 &= \{(1)(2)(3)(4)\} = \{I\}, \\
C_2 &= \{(1,2),(1,3),(1,4),(2,3),(2,4),(3,4)\}, \\
C_3 &= \{(1,2,3),(1,3,2),(1,2,4),(1,4,2),(1,3,4),(1,4,3),(2,3,4),(2,4,3)\}, \\
C_4 &= \{(1,2)(3,4),(1,3)(2,4),(1,4)(2,3)\}, \\
C_5 &= \{(1,2,3,4),(1,4,3,2),(1,2,4,3),(1,3,4,2),(1,4,2,3),(1,3,2,4)\}.
\end{aligned}$$

The union of C_1, C_3 and C_4 is a normal subgroup[8]. Class C_3 splits into two conjugacy classes of A^4, which are $\{(1,2,3),(1,3,4),(1,4,2),(2,4,3)\}$ and the squares (inverses) $\{(1,3,2),(1,4,3),(1,2,4),(2,3,4)\}$. The other two classes of A^4 are C_1 and C_3.

Exercises

10. If a finite group only has two conjugacy classes, the order of the group is 2. [*Hint*: use the class equation.]

11. If $x = a_1 a_2 \cdots a_n$, the cyclic permutations of the a_i are all conjugate to x. [*Hint*: ab and ba are conjugate.]

12. Prove that if $n \geq 3$ the center of S^n is the identity.

13. Show that the center of D_n has order 1 or 2 according to n being odd or even.

14. If $y^{-1}xy = x^{-1}$ then $y^2 \in \mathbf{C}_G(x)$. [*Hint*: conjugate twice by y.]

15. If $o(x) = p$, a prime, and $y^{-1}xy = x^k$, $(p,k) = 1$, then $xy^{p-1} = y^{p-1}x$. [*Hint*: conjugate $p-1$ times by y and apply Fermat's little theorem.]

16. If H is the unique subgroup of order 2 in a group G, then $H \subseteq \mathbf{Z}(G)$. If G is finite, then more generally if H is the unique subgroup of order p, where p is the smallest divisor of the order of the group, then $H \subseteq \mathbf{Z}(G)$.

17. Prove that the following are equivalent:
i) $(ab)^n = (ba)^n$, for all $a, b \in G$;
ii) $x^n \in \mathbf{Z}(G)$ for all $x \in G$.

18. If $(ab)^2 = (ba)^2$ for all $a, b \in G$, then every element of G commutes with all its conjugates.

19. *i)* If the product of two elements belongs to the center, the two elements commute;
ii) an element x belongs to the center if, and only, if $x = ab \Rightarrow x = ba$.

20. If $H \leq G$ and if $\mathbf{cl}(h) \cap H = \{h\}$ for all $h \in H$ then H is abelian. [*Hint*: $x^{-1}hx = h$, for all $x, h \in H$.]

21. A dihedral group D_p, $p > 2$ a prime, has only one conjugacy class of involutions, and the centralizer of an involution has order 2.

22. It C and C' are two conjugacy classes of a group then $CC' = C'C$. [*Hint*: $xy = y(y^{-1}xy)$.]

23. If the quotient group with respect to the center is cyclic, the group is abelian.

24. The group of integers cannot be the automorphism group of a group. The same is true for a non trivial finite cyclic group of odd order.

25. A group that contains a non identity element which is a power of every other non identity element (e.g. a cyclic 2-group or the quaternion group) cannot be the group of inner automorphisms of a group.

---

[8] It is the alternating group A^4 (see Section 2.8).

26. Prove that all the automorphisms of S^3 are inner[9].

27. In a finite p-group, a maximal subgroup is normal and has index p.

28. The center of a group is contained in all the maximal subgroup of composite index. [*Hint*: if $Z = \mathbf{Z}(G) \not\subseteq M$, then $MZ = G$; see the solution of the previous exercise.]

29. A conjugacy class of a subgroup of index 2 either is a conjugacy class of the group, or it contains half of the elements of it.

30. If H is a subgroup of index 2 of a finite group G such that $\mathbf{C}_G(h) \subseteq H$ for all $h \in H$, prove that the elements of $G \setminus H$ are involutions that are all conjugate.

31. (G.A. Miller) If a finite group admits an automorphism which inverts more than 3/4 of the elements then it is abelian. [*Hint*: let α be the automorphism. If $G \neq \mathbf{Z}(G)$, then since $|\mathbf{Z}(G)| \leq \frac{1}{4}|G|$, there exists $x \notin \mathbf{Z}(G)$ such that $x^\alpha = x^{-1}$, and since $|G \setminus \mathbf{C}_G(x)| \geq \frac{1}{2}|G|$ there exists $y \in G \setminus \mathbf{C}_G(x)$ such that $y^\alpha = y^{-1}$.] Give an example of a nonabelian group with an automorphism inverting exactly 3/4 of the elements.

32. If a group has more than two elements, its automorphism group is nontrivial.

33. If the center of a group is the identity, then the center of its automorphism group is also the identity.

34. In a group G, let $\rho(a)$ be the number of elements whose square is a. Show that:
i) $\sum_{a\in G} \rho(a)^2 = \sum_{a \in G} \rho(a^2)$;
ii) $\sum_{a\in G} \rho(a)^2 = \sum_{a \in G} I(a)$, where $I(a)$ is the set of elements $x \in G$ that invert a;
iii) a conjugacy class is *ambivalent* if it contains the inverses of all its elements. Show that $\frac{1}{|G|}\sum_{a\in G} \rho(a)^2 = c'(G)$, where $c'(G)$ is the number of ambivalent classes of G;
iv) verify *ii)* in the case of a group of odd order and the quaternion group.

2.4 Normalizers and Centralizers of Subgroups

If $H \leq G$, and γ_x is an inner automorphism of G, we denote H^x the image of H under γ_x, i.e. $H^x = x^{-1}Hx$.

Definition 2.12. The subgroup

$$H^x = x^{-1}Hx = \{x^{-1}hx,\ h \in H\}$$

is the *conjugate subgroup* to H by x. If $H, K \leq G$, and there exists x such that $H^x = K$, then H and K are said to be *conjugate*. We write $H \sim K$.

As in the case of elements, conjugacy is an equivalence relation, which is defined in the set of subgroups of the group. The conjugacy class to which H belongs is denoted $\mathbf{cl}(H)$.

---

[9] This is true for all the groups S^n, $n \neq 2, 6$ (Theorem 3.29).

Two isomorphic subgroups have the same order, but if the group is infinite not the same index in general. For example, in **Z** the subgroups $\langle 2\rangle$ and $\langle 3\rangle$ are isomorphic but are of index 2 and 3, respectively. However:

Theorem 2.21. *Two conjugate subgroups have the same index.*

Proof. The correspondence

$$H^x y \to Hxy$$

between the cosets of H^x and those of H is well defined and injective:

$$H^x y = H^x z \Leftrightarrow z = x^{-1}hxy \Leftrightarrow Hxz = Hxx^{-1}hxy = Hxy.$$

It is also surjective: Hy is the image of $H^x z$, $z = x^{-1}y$. ◊

If x and y give rise to the same conjugate of H, i.e. $H^x = H^y$, then $H^{xy^{-1}} = H$. We are led to the following definition.

Definition 2.13. The *normalizer* of a subgroup H in a group G is the subset of G:

$$\mathbf{N}_G(H) = \{x \in G \mid H^x = H\}.$$

It is immediate that the normalizer of a subgroup H is a subgroup. It consists of the elements $x \in G$ such that, given $h \in H$, there exists $h' \in H$ such that $xh = h'x$. Hence it is the largest subgroup of G in which H is contained as a normal subgroup. If $x \in \mathbf{N}_G(H)$ we say that x *normalizes* H, or that x commutes with H. Since $H \trianglelefteq \mathbf{N}_G(H)$, H is a normal subgroup of G if and only if $\mathbf{N}_G(H) = G$. If $\mathbf{N}_G(H) = H$ then H is *self-normalizing*.

Two elements $x, y \in G$ that give rise the same conjugate of H belong to the same coset of the normalizer of H. Hence:

Theorem 2.22. *If* $H \leq G$,

$$|\mathbf{cl}(H)| = [G : \mathbf{N}_G(H)].$$

In particular, if the group is finite, the number of subgroups conjugate to a given subgroup divides the order of the group.

The above discussion shows that the notion of the normalizer of a subgroup corresponds to that of the centralizer of an element (cf. 2.12). Among the elements of the normalizer of a subgroup H there are those that commute with H elementwise.

Definition 2.14. The *centralizer* in G of a subgroup H is the set of elements of G commuting with H elementwise:

$$\mathbf{C}_G(H) = \{x \in G \mid xh = hx,\ \forall h \in H\}.$$

If $H \trianglelefteq G$, and $x \in G$, the inner automorphism of G induced by x induces an automorphism of H (not inner, if $x \notin H$), given by:

$$\alpha_x : h \to x^{-1}hx.$$

Define $G \to \mathbf{Aut}(H)$ by $x \to \alpha_x$. This mapping is a homomorphism. The same argument shows that there is a homomorphism

$$\varphi : \mathbf{N}_G(H) \to \mathbf{Aut}(H).$$

The kernel of φ consists of the elements of $\mathbf{N}_G(H)$ that induce the identity automorphism of H, i.e. that commute with H elementwise. In other words, $ker(\varphi) = \mathbf{C}_G(H)$.

From the first isomorphism theorem it follows:

Theorem 2.23 (*N*/*C* theorem). *Let $H \leq G$. Then:*

i) $\mathbf{C}_G(H) \trianglelefteq \mathbf{N}_G(H)$. *In particular, if H is normal in G, its centralizer is also normal;*

ii) $\mathbf{N}_G(H)/\mathbf{C}_G(H)$ *is isomorphic to a subgroup of* $\mathbf{Aut}(H)$.

This theorem may be used to obtain information about the way in which a group H can be a subgroup of another group G. For example, let $H = C_4$; then H can be contained as a normal subgroup in a group G only if commutes elementwise with at least half of the elements of G. Indeed, since $\mathbf{Aut}(C_4) \simeq C_2$ we have $|\mathbf{N}_G(C_4)/\mathbf{C}_G(C_4)| = |G/\mathbf{C}_G(C_4)| = 1$ or 2. If it equals 1, then $\mathbf{C}_G(C_4) = G$ and therefore $\mathbf{C}_G(C_4) \subseteq \mathbf{Z}(G)$; if it equals 2, then $\mathbf{C}_G(C_4)$ contains half the elements of G. Other examples are C_3 and $\mathbf{Z}$.

We have seen in the *Ex.* 2.1, 12 that, in general, normality is not transitive. However, if H is not only normal, but also characteristic in K and K is normal in G, then H is normal in G. Indeed, for all $g \in G$ the mapping $k \to g^{-1}kg$ is an automorphism of K fixing H: $H = g^{-1}Hg$, so that $H \trianglelefteq G$.

Theorem 2.24. *Let $H \leq G$. The intersection $K = \bigcap_{x \in G} H^x$ of all the conjugates of H is a normal subgroup of G contained in H, and every normal subgroup of G contained in H is contained in K. (In this sense, K is the largest normal subgroup of G contained in H).*

Proof. Let $k \in K$, $x \in G$. Then $k \in H$ (H is a subgroup of the intersection) and therefore $k^x \in H^x$, for all $x \in G$, i.e. $k^x \in K$, from which the normality of K follows. If $L \trianglelefteq G$ and $L \subseteq H$, then $L = L^x \subseteq H^x$ for all x, and therefore $L \subseteq K$. ♢

Theorem 2.25 (Poincaré). *If a group has a subgroup of finite index, then it also has a normal subgroup of finite index.*

Proof. Let H be of finite index in G. The normalizer of H contains H, so its index is also finite and therefore, by Theorem 2.22, H has a finite number of conjugates. By Theorem 2.21 the latter also have finite index, and so does their intersection (Theorem 1.19). This intersection is normal (Theorem 2.24), and is the subgroup we seek. ◇

Let us now consider the above notions in the case of the symmetric group.

Theorem 2.26. *i) The only elements of* S^n *that commute with an* n*-cycle* c *are the powers of* c*, and therefore* $\mathbf{C}_{S^n}(c) = \langle c \rangle$*;*

ii) the normalizer S^n *of the subgroup generated by an* n*-cycle has order* $n \cdot \varphi(n)$*, where* $\varphi(n)$ *is Euler's function.*

Proof. i) The number of n-cycles is $(n-1)!$, and they are all conjugate. Thus, the centralizer of c has index $(n-1)!$ and therefore order n, and since the n powers of c commute with c we have the result.

ii) If σ normalizes $\langle c \rangle$ it sends c to a power of c having the same order. Hence, $\sigma^{-1}c\sigma = c^k$, with $1 \leq k < n$ and $(n,k) = 1$. The mapping $\mathbf{N}_{S^n}(\langle c \rangle) \to U(n)$ given by $\sigma \to k$ is a homomorphism, and is surjective: if $k \in U(n)$, then c^k is again an n-cycle, and therefore is conjugate to c. Then there exists σ such that $\sigma^{-1}c\sigma = c^k$, and since c^k generates $\langle c \rangle$, σ normalizes $\langle c \rangle$. The kernel of this homomorphism is the centralizer of c; by *i)*, this has order n. The result follows. ◇

Exercises

35. A group cannot be the product of two conjugate subgroups. [*Hint*: if $G = HH^x$, how do you write x?.]

36. *i)* Prove that the converse of Lagrange's theorem holds for the dihedral groups D_n, for all n.

ii) If H is a cyclic subgroup of a group having trivial intersection with all its conjugates, then H and all its non trivial subgroups have the same normalizer.

37. Let $G = HK$, $H, K \leq G$, and let $x, y \in G$. Prove that:
i) $G = H^x K^y$;
ii) there exists $g \in G$ such that $H^g = H^x$ and $K^g = K^y$.

38. Let $\alpha \in \mathbf{Aut}(G)$, $H, K \leq G$, $H^\alpha = K$. Prove that $\mathbf{C}_G(H)^\alpha = \mathbf{C}_G(K)$ and $\mathbf{N}_G(H)^\alpha = \mathbf{N}_G(K)$. In particular, if H is α-invariant ($H^\alpha = H$), its centralizer and normalizer are also α-invariant.

39. There is no finite group which is a union of self-normalizing subgroups having trivial intersections[10].

40. Prove that if $H \leq G$ and $K \trianglelefteq G$ then $\mathbf{N}_G(H)K/K \subseteq \mathbf{N}_{G/K}(HK/K)$. If $K \subseteq H$ then equality holds. Give an example to show that the above inclusion may be proper

[10] This result will be used in the proof of Theorem 5.54.

(in the canonical homomorphism $G \to G/K$ the image of a normalizer does not coincide with the normalizer of the image). [*Hint*: in D_4, let $H = C_2$ (not central) and K a Klein group not containing H.]

41. Let $|G| = n^2$, $H \leq G$ and $|H| = n$. Prove that H has a nontrivial intersection with all its conjugates.

42. Let $H \leq G$, and let H commute with every subgroup of G. Prove that:
i) every conjugate of H has the same property;
ii) if H is maximal w.r.t. this property, then H is normal.

43. A subgroup H of a group G is said to be *abnormal* if every element g of G belongs to the subgroup generated by H and $g^{-1}Hg$. Prove that H is abnormal if, and only if, it meets the following conditions:
i) if $H \leq K$ then $\mathbf{N}_G(K) = K$;
ii) H is not contained in two distinct conjugate subgroups.

44. Let M be the set of cosets of a subgroup H of a group G. The product of two cosets is not defined in M for all pairs, but only for the (ordered) pairs Ha, Hb such that $a \in \mathbf{N}_G(H)$. The cosets Ha with $a \in \mathbf{N}_G(H)$ form a group, the group $\mathbf{N}_G(H)/H$, and this is the largest subgroup contained in M. Moreover, the following associative law holds: $(Ha \cdot Ha') \cdot Hb = Ha \cdot (Ha' \cdot Hb)$, for $a, a' \in \mathbf{N}_G(H)$ and any $b \in G$.

45. Let $\mathcal{Q}$ be the quaternion group, $G = \mathbf{Aut}(\mathcal{Q})$, $H = \mathbf{I}(\mathcal{Q})$. Prove that:
i) $\mathbf{C}_G(H) = H$;
ii) $G/H \simeq S^3$;
iii) $G \simeq S^4$.

[*Hint to iii)*: the permutations $(i,j,k)(-i,-j,-k)(1)(-1)$ and $(i,j,-i,-j)(k)(-k)(1)(-1)$ give automorphisms of $\mathcal{Q}$ that generate S^4 as a subgroup of S^6.][11]

46. A subgroup is a maximal abelian subgroup if, and only if, it coincides with its own centralizer.

47. Let $A = \mathbf{Aut}(G)$, $\mathbf{I} = \mathbf{I}(G)$. Prove that if $\mathbf{Z}(G) = \{1\}$, then $\mathbf{C}_A(\mathbf{I}) = \{1\}$, and in particular $\mathbf{Z}(A) = \{1\}$.

48. If a group is a finite set-theoretic union of subgroups then one of these has finite index. [*Hint*: let $G = \bigcup_{i=1}^n H_i$, and let $H_1x^{-1} \neq H_1$ be a coset of H_1; then $H_1 \cap H_1x^{-1} = \emptyset$, so $H_1x^{-1} \subseteq \bigcup_{i=2}^n H_i$, from which $G = \bigcup_{i=2}^n (H_i \cup H_ix)$.]

49. $[H : H \cap K] = [H^x : H^x \cap K^x]$.

2.5 Hölder's Program

Definition 2.15. A group is said to be *simple* if it has no proper normal subgroups. Equivalently, a group is simple if the only groups homomorphic to it are either isomorphic to it or to the identity group.

[11] Label x and $-x$ two opposite faces of a cube, $x = i, j, k$. The 24 isometries of the cube give the 24 automorphisms of $\mathcal{Q}$.

Example 2.6. A group of prime order is simple: it has no subgroups different from the identity and the whole group. An abelian simple group is of prime order: all its subgroups are normal, so it has no subgroups and therefore is of prime order (Theorem 1.17). We shall see that the smallest nonabelian simple group has order 60 (see Section 2.8).

Every finite group uniquely determines a family of simple groups, in much the same way as a natural number determines its prime factors. This is the content of the Jordan-Hölder theorem which we prove in a moment.

A maximal normal subgroup of a group is a normal subgroup which is not properly contained in any proper normal subgroup (see Definition 1.14; here property $\mathcal{P}$ is that of being normal): if $H \leq K$, and $K \trianglelefteq G$, then either $H = K$ or $K = G$.

Lemma 2.2. *If H is a maximal normal subgroup of a group G, then the quotient G/H is a simple group.*

Proof. If $K/H \trianglelefteq G/H$, then from Theorem 2.5 we have $K \trianglelefteq G$, so that if $H < K$ then $K = G$. Hence, either $K = H$, in which case $K/H = H$ is the identity subgroup of G/H, or $K = G$, and K/H is the whole group G/H. ♢

Consider now a group G and a maximal normal subgroup G_1, if any; the quotient G/G_1 is simple. Let G_2 be a maximal normal subgroup of G_1 (we stress the fact that G_2 need not be normal in G, but only in G_1); the quotient G_1/G_2 is simple. Proceeding in this way, we construct a chain of subgroups $G \supset G_1 \supset G_2 \supset \ldots$ each normal in the preceding one, such that the quotient between two consecutive subgroups of the chain is a simple group (if G_i has no maximal subgroups, then $G_{i+1} = G_i$ and the chain stops at G_i). If G is finite, such a chain stops at $\{1\}$ (if G is simple, then the chain is $G \supset \{1\}$).

Definition 2.16. A chain of subgroups of a group G:

$$G \supset G_1 \supset G_2 \supset \ldots \supset G_{l-1} \supset G_l \tag{2.18}$$

in which each subgroup G_i is normal in the preceding one is a *normal chain.* The integer l is the *length* of the chain. A subgroup appearing in a normal series is a *subnormal* subgroup. If $G_l = \{1\}$, the chain is a *series.* If each G_i is maximal normal in G_{i-1}, in which case the quotients G_{i-1}/G_i are simple groups, then the series is a *composition series*, the quotients G_i/G_{i+1} are the *composition factors* of G. Note that in a finite group the product of the orders of the composition factors G_i/G_{i+1} is the order of the group.

Examples 2.7. 1. An infinite group may not admit a composition series. For example, in the group of integers $\mathbf{Z}$, a chain in which each subgroup is maximal in the preceding one has the form $\mathbf{Z} \supset \langle p \rangle \supset \langle pq \rangle \supset \langle pqr \rangle \supset \ldots$, where $p, q, r, \ldots$ are prime numbers. Such a chain never stops.

2. The orders of the composition factors of an abelian group are prime numbers, and therefore an abelian group admits a composition series if and only if is finite. Indeed, in (2.18) with $G_l = \{1\}$, G_{l-1} is simple abelian and therefore of prime order p; similarly, G_{l-2}/G_{l-1} is simple abelian, hence of prime order q, so $|G_{l-2}| = pq$, and so on, and the order of G is the product of these primes.

3. The group S^3 has the (unique) composition series:

$$S^3 \supset C_3 \supset \{1\},$$

with composition quotients C_2 and C_3. The group C_6 admits two composition series:

$$C_6 \supset C_3 \supset \{1\} \text{ and } C_6 \supset C_2 \supset \{1\}$$

with quotient groups C_2 and C_3, and C_3 and C_2, respectively.

These examples show that factor groups, their orders and the length of a composition series do not characterize a group.

Although a group may have several composition series, there are however invariants that we now illustrate (in the finite case) through an analogy to the natural numbers. For every natural number n there exists a sequence:

$$n > n_1 > n_2 > \ldots > n_{l-1} > n_l = 1$$

such that each n_i divides the previous one, and the quotient is a prime number. For instance, with $n = 120$, we have:

$$120 > 40 > 20 > 10 > 2 > 1$$

with quotients 3, 2, 2, 5 and 2. Such a sequence is not unique:

$$120 > 60 > 30 > 6 > 2 > 1$$

is another sequence, with quotients 2, 2, 5, 3 and 2. However, the prime numbers appearing as quotients are the same, up to the order. In particular, the two sequences have the same length, and the product of the primes is the number we started with. (All this is simply a way of expressing the fundamental theorem of arithmetic.)

In the case of a group, the simple groups of prime order obtained as composition factors play the role of the prime numbers obtained as quotients in the above sequences of integers.

Definition 2.17. Two composition series of a group G are said to be *isomorphic* if they have the same length and their composition factors are isomorphic, up to the order.

The following theorem is of fundamental importance. It originates in Galois theory, where the following result is proved: let $f_1(x)$ and $f_2(x)$ be two polynomials over a field with Galois groups G_1 and G_2, respectively. Adjoining the roots of $f_2(x)$ to the field, the group G_1 reduces to a normal subgroup G_1', while adjoining the roots of $f_1(x)$ the group G_2 reduces to a normal subgroup G_2'. Then (Jordan) $[G_1 : G_1'] = [G_2 : G_2']$ and (Hölder) the quotients G_1/G_1' and G_2/G_2' are isomorphic.

Theorem 2.27 (Jordan-Hölder). *Two composition series of a finite group are isomorphic.*

Proof. Let

$$\begin{array}{c} G \supset G_1 \supset G_2 \supset \ldots \supset G_{s-1} \supset G_s = \{1\} \\ G \supset H_1 \supset H_2 \supset \ldots \supset H_{t-1} \supset H_t = \{1\} \end{array} \tag{2.19}$$

be two composition series of the group. If $|G| = 1$ there is nothing to prove. (The smallest group with at least two composition series is the Klein group, on which the theorem can be verified directly.) If $G_1 = H_1 = L$, the two series of L:

$$L \supset G_2 \supset \ldots \supset G_{s-1} \supset G_s = \{1\}$$
$$L \supset H_2 \supset \ldots \supset H_{t-1} \supset H_t = \{1\}$$

are isomorphic by induction, and therefore the series (2.19) are also isomorphic (the first quotient is the same).

Then let $G_1 \neq H_1$. The subgroup G_1H_1 properly contains both G_1 and H_1, and is normal, being a product of two normal subgroups. By maximality, $G_1H_1 = G$. It follows, setting $K = G_1 \cap H_1$,

$$G/G_1 = G_1H_1/G_1 \simeq H_1/K, \quad G/H_1 = G_1H_1/H_1 \simeq G_1/K. \tag{2.20}$$

The factors H_1/K and G_1/K are simple groups because they are isomorphic to the simple groups G/G_1 and G/H_1, respectively. A composition series of K can then be extended to one of G_1 and to one of H_1, and these, in turn, to two series of G. We now have four series, the series (2.19) and two new ones:

$$\begin{array}{rl} i) & G \supset G_1 \supset G_2 \supset \ldots \supset \{1\}, \\ ii) & G \supset G_1 \supset K \supset \ldots \supset \{1\}, \\ iii) & G \supset H_1 \supset K \supset \ldots \supset \{1\}, \\ iv) & G \supset H_1 \supset H_2 \supset \ldots \supset \{1\}. \end{array}$$

The factors G/G_1 and G_1/K of *ii*) are isomorphic, by (2.20), to H_1/K and G/H_1, respectively, of *iii*); the remaining factors of *ii*) and *iii*) are those of the series chosen in K. Hence, *ii*) and *iii*) are isomorphic. Series *i*) and *ii*) have the same first factor; the others are isomorphic by induction (they are composition factors of G_1).

It follows that *i*) and *ii*) are isomorphic; the same holds for *iii*) and *iv*) by induction on $|H_1|$. This proves that *i*) and *iv*) are isomorphic. ◇

Hence, due to this theorem, a finite group uniquely determines a set of simple groups, its composition factors.

This motivates *Hölder's program* for the classification of all finite groups:

i) determine all finite simple groups;
ii) given two groups K and H, determine all groups G which contain a normal subgroup isomorphic to K and such that the factor group G/K is isomorphic to H. Such a group G is an *extension* of K *by* H.

Given a family of simple groups $S_1, S_2, \ldots, S_{l-1}$ (in this order), knowing how to solve *ii*) we can construct, at least in principle, all groups G whose composition factors are the S_i. Indeed, let $G_l = \{1\}$ and $G_{l-1} = S_{l-1}$, and let us determine the groups G_{l-2} which contain a normal subgroup isomorphic to S_{l-1} and such that the factor group is isomorphic to S_{l-2}. This will provide us with a certain number of groups G_{l-2}, and for each of these we will have a series:

$$G_{l-2} \supset G_{l-1} \supset G_l = \{1\}$$

whose factors are S_{l-2} and S_{l-1}. Proceeding in this way, we arrive at a certain number of series, all having as factors the given simple groups $S_1, S_2, \ldots, S_{l-1}$. However, the same group may be found several times among these. Let us give a few examples.

Examples 2.8. 1. With C_3, C_2 we obtain two groups, C_6 and S_3, and only one with C_2, C_3, that is C_6 (cf. *Ex.* 2.7, 3). This shows that the same group may be obtained via a different ordering of the simple groups.

2. Let us recover by the method above the groups of order 8 (cf. *Ex.* 1.8, 2). We assume that we know the groups of order 4. We know that a group of order 8 has a normal subgroup of order 4, and the latter a normal subgroup of order 2, so that a composition series has necessarily the form:

$$G \supset G_1 \supset G_2 \supset \{1\}$$

with quotients $\{C_2, C_2, C_2\}$. Starting from these three simple groups we have $G_2 = C_2$, and $G_1/C_2 \simeq C_2$. Then G_1 has order 4, and we have two possibilities: $G_1 \simeq C_4$ or $G_1 \simeq V$:

$$C_4 \supset C_2 \supset \{1\}, \quad V \supset C_2 \supset \{1\}.$$

Finally, $G/C_4 \simeq C_2$ implies $G \simeq C_8, D_4, \mathcal{Q}, U(15)$, whereas $G/V \simeq C_2$ implies $G \simeq D_4, \mathbf{Z}_2^{(3)}, U(15)$. Thus, we obtain the five groups of order 8 that we already know, with repetitions.

From the above, the importance is clear of the simple groups: they are the bricks with which all finite groups are built. As to the extension problem, it finds its natural place within the cohomology of groups as we shall see in Chapter 7. In the next two sections we consider two extension: the direct and the semidirect product. As to point *i*) of Hölder's program, we will see in Section 2.8 an infinite family of finite simple groups, the alternating groups, and in Chapter 3 another infinite family of simple groups, the projective special linear groups (cf. Section 3.7).

2.6 Direct Products

Given two groups H and K, the more immediate extension of one by the other is their *direct product*. The direct product of H and K, denoted $H \times K$, is the set of ordered pairs (h,k), $h \in H$, $k \in K$, with the componentwise product:

$$(h_1,k_1)(h_2,k_2) = (h_1h_2, k_1k_2).$$

The identity is the pair $(1,1)$, the inverse of (h,k) is (h^{-1},k^{-1}), and the operation is associative because so are those of the two groups (the *factors*). In additive notation, which is especially used for abelian groups, we have the *direct sum* $H \oplus K$.

The subsets

$$H^* = \{(h,1),\ h \in H,\ 1 \in K\} \text{ and } K^* = \{(1,k),\ 1 \in H,\ k \in K\}$$

form two subgroups isomorphic to H and K, respectively. Since

$$(h,1)(1,k) = (h,k) = (1,k)(h,1)$$

every element of G is the product of an element of H^* and one of K^*. Moreover, two such elements commute. In particular, $H^*, K^* \trianglelefteq G$. By identifying H^* with H and K^* with K we have:

$$\begin{aligned} &i)\ \ G = HK;\\ &ii)\ \ H,K \trianglelefteq G;\\ &iii)\ \ H \cap K = 1 \end{aligned} \tag{2.21}$$

(*internal* direct product). Conversely, if a group contains two subgroups H and K such that *i*), *ii*) and *iii*) hold, G is the internal direct product of H and K, and is isomorphic to the direct product defined above (*external*). Indeed, from *i*) and *iii*) it follows that an element $g \in G$ is a product of an element $h \in H$ and one of $k \in K$ in a unique way (cf. the observation following Theorem 2.2). From *ii*) and *iii*) we have that H and K commute elementwise (Theorem 2.5).

The notion of direct product extends immediately to a finite number of groups $G = H_1 \times H_2 \times \cdots \times H_n$ by considering n-tuples instead of pairs (but cf. Remark 2.3, 2 below) with the componentwise product. As h_i varies in H_i, the n-tuples $(1,1,\ldots,1,h_i,1,\ldots,1)$ form a subgroup isomorphic to H_i, still denoted H_i. The H_i commute elementwise, and their product is G. Moreover, $H_i \cap H_j = \{1\}$, $i \neq j$; but more is true, i.e. each H_i has trivial intersection not only with the H_j, $j \neq i$, but also with their product, because the product of n-tuples having 1 at place i still has 1 at place i. We say that G is the direct product of the subgroups $H_1, H_2, \ldots, H_n$ if, for $i = 1, 2, \ldots, n$,

$$\begin{aligned} &i)\ \ G = H_1H_2\cdots H_n;\\ &ii)\ \ H_i \trianglelefteq G;\\ &iii)\ \ H_i \cap H_1H_2\cdots H_{i-1}H_{i+1}\cdots H_n = \{1\}. \end{aligned} \tag{2.22}$$

Property (*iii*) ensures that the expression of an element $g \in G$ as a product $g = h_1h_2\cdots h_n$, $h_i \in H_i$ is unique. In particular, if G is finite, then $|G| = |H_1| \cdot |H_2| \cdots |H_n|$.

Remarks 2.3. **1.** A normal subgroup of a direct factor H is normal in the whole group. Indeed, H commutes elementwise with the other direct factors.

2. The n-tuples of elements belonging to the subgroups H_i of a direct product are the set of functions from the set $\{1, 2, \ldots, n\}$ to the union of the H_i such that $f(i) \in H_i$, and the componentwise product corresponds to the product of functions $fg(i) = f(i)g(i)$. The identity element is the function whose value is 1 at all i (the n-tuple $(1,1,\ldots,1)$, and f^{-1} is the function whose value at i is $f(i)^{-1}$. Then one can consider the direct product of an infinite family of groups $\{H_\lambda\}$, where λ varies in a set of indices Λ, i.e. the set of functions $f : \Lambda \to \bigcup_{\lambda \in \Lambda} H_\lambda$ such that $f(\lambda) \in H_\lambda$, and $f(\lambda) \neq 1$ only for a finite number of indices λ (functions with "finite support"), with the product $fg(\lambda) = f(\lambda)g(\lambda)$. Identity and inverse are defined as above. There is no need of the axiom of choice to affirm that the direct product of an infinite family of groups is nonempty. Since a group has a privileged element (the identity) there is at least the function f that chooses the identity in each of the groups. This function f is the identity of the direct product. If the set of all functions is considered, and not only those with finite support, then the product is the *cartesian product* of the groups H_λ.

Theorem 2.28. *Let G be an abelian p-group of order p^n in which all the elements have order p. Then G is a direct sum of n copies of $\mathbf{Z}_p$.*

Proof (Additive notation). Every element of G generates a subgroup of order p, and therefore G is a sum of subgroups isomorphic to $\mathbf{Z}_p$. Let k be minimum such that:

$$G = H_1 + H_2 + \cdots + H_k, \quad H_i \simeq \mathbf{Z}_p.$$

Then (*iii*) of (2.22) holds because if the intersection of one of the H_i with the sum of the others is not trivial, then it is the whole H_i, and G would be a sum of $k - 1$ subgroups H_j, against the minimality of k. It follows:

$$p^n = |G| = |\mathbf{Z}_p| \cdot |\mathbf{Z}_p| \cdots |\mathbf{Z}_p|, \quad k \text{ times},$$

and therefore $p^n = p^k$, $k = n$ and $G = \mathbf{Z}_p \oplus \mathbf{Z}_p \oplus \cdots \oplus \mathbf{Z}_p$, n times. ◇

An abelian p-group whose elements all have order p is called *elementary abelian.* In such a group the non-identity elements divide up $p - 1$ by $p - 1$ into subgroups of order p, so that if the group has order p^n the number of subgroups of order p is

$$\frac{p^n - 1}{p - 1} = p^{n-1} + p^{n-2} + \cdots + p + 1.$$

Such a group is the additive group of a vector space of dimension n over $\mathbf{Z}_p$. As already observed in *Ex.* 1.9, 3, over a prime field as $\mathbf{Z}_p$ the additive structure implies that of vector space because multiplication of a vector by a scalar reduces to a sum of the vector with itself. Hence, a direct sum of copies of $\mathbf{Z}_p$ is a vector space over $\mathbf{Z}_p$, and the subspaces are the subgroups of the additive group. In particular, in such a sum there are no characteristic subgroups (cf. *Ex.* 1.10, 4).

The groups $\mathbf{Z}_p$ are abelian simple groups, but direct products of isomorphic nonabelian simple groups also admit no characteristic subgroups, as the theorem below shows.

The dual notion to the notion of a maximal subgroup is that of a minimal subgroup.

Definition 2.18. A subgroup $H \neq \{1\}$ of a group G is *minimal* with respect to a property $\mathcal{P}$ if it contains no non trivial subgroup of G having property $\mathcal{P}$. In other words, if H has $\mathcal{P}$, $K \leq H$, then either $K = H$ or K does not have $\mathcal{P}$.

Examples 2.9. 1. If property $\mathcal{P}$ is simply that of being a subgroup, then a minimal subgroups has no proper subgroups, and hence is of prime order. A torsion free group (e.g. $\mathbf{Z}$) has no minimal subgroups.

2. If property $\mathcal{P}$ is that of being a normal subgroup, then a *minimal normal* subgroup properly contains no non trivial normal subgroups of the group. In particular, a minimal normal subgroup has no characteristic subgroups (such a group is said to be *characteristically simple*). A non simple finite group always has minimal normal subgroups.

Theorem 2.29. *A finite group has no characteristic subgroups if, and only if, it is either simple, or a direct product of isomorphic simple groups.*

Proof. If the group G is simple there is nothing to prove. If not, it admits a minimal normal subgroup H. Let $H = H_1, H_2, \ldots, H_m$ be the distinct images of H under the various automorphisms of G. These images are also minimal normal; their product is obviously characteristic and therefore equals G. Let n be minimum such that G is the product of n of the H_i, $G = H_1H_2 \cdots H_n$. Then for each i, $H_i \cap H_1H_2 \cdots H_{i-1}H_{i+1} \cdots H_n = \{1\}$. Indeed, this intersection is normal in G and is contained in H_i, and by the minimality of H_i is either $\{1\}$ or H_i; but if it equals H_i this subgroup is contained in the product of the other subgroups, against the minimality of n. Hence, conditions (2.22) are satisfied. Finally, the H_i are simple (cf. Remark 2.3, 1).

In order to prove the converse we need two lemmas.

Lemma 2.3. *Let* $G = H_1 \times H_2 \times \cdots \times H_n$, $\mathbf{Z}(H_1) = \{1\}$, $K \trianglelefteq G$ *and* $K \cap H_1 = \{1\}$. *Then* $K \subseteq H_2 \times \cdots \times H_n$.

Proof. Let $k \in K$ and $k = h_1h_2 \cdots h_n$, $h_i \in H_i$. Now K and H_1 are normal and their intersection is trivial, so they commute elementwise. If $h \in H_1$, then:

$$h \cdot h_1h_2 \cdots h_n = h_1h_2 \cdots h_n \cdot h = h_1h \cdot h_2 \cdots h_n,$$

where the second equality follows from the fact that H_1 commutes elementwise with all the H_i, $i \neq 1$. Hence $hh_1 = h_1h$, for all $h \in H_1$, and therefore $h_1 \in \mathbf{Z}(H_1) = \{1\}$, and $h_1 = 1$. Then $k = h_2h_3 \cdots h_n$, and the result. ◇

Lemma 2.4. *Let G be as in the previous lemma with the H_i nonabelian simple subgroups. Then the only non trivial normal subgroups of G are the H_i and their products two by two, three by three,..., $n-1$ by $n-1$.*

Proof. Obviously, the said subgroups are normal. Conversely, if $K \trianglelefteq G$, since by simplicity $\mathbf{Z}(H_1) = \{1\}$, by the previous lemma $K \subseteq H_2 \times \cdots \times H_n$. If equality holds, K is of the required type; if not, the result follows by induction on n. ♢

We return to the proof of Theorem 2.29.

Proof. By Theorem 2.28 we may assume that the H_i are simple and non abelian. A characteristic subgroup K is normal, so is one of those seen in the previous lemma. However, a permutation of the H_i yields an automorphism of G, and if $K = H_j H_l \cdots H_s$ and H_t does not appear in K, the transposition exchanging H_j and H_t and fixing all the other H_i moves K, unless all the H_i appear. But in the latter case $K = G$. ♢

Definition 2.19. A normal series

$$G \supset G_1 \supset G_2 \supset \ldots \supset G_{l-1} \supset G_l = \{1\}$$

(Definition 2.16) is said to be an *invariant series* if each G_i is a normal subgroup of G. An invariant series is said to be a *principal series*, or a *chief series* if each G_i is maximal among the normal subgroups of G which are properly contained in G_{i-1}, $i = 1, 2, \ldots, l$ ($G_0 = G$). The factors of a principal series are the *principal* or *chief factors* of G. Note that a principal factor G_{i-1}/G_i is a minimal normal subgroup of G/G_i.

It follows from this definition that principal series are to invariant ones as composition series are to normal ones. The maximality condition implies that the subgroups of a chief series are all distinct. Moreover, the composition factors are simple groups, but the principal factors may not be simple because there may exist normal subgroups of G_{i-1} properly containing G_i. However, they are characteristically simple because a characteristic subgroup H/G_i of a principal factor G_{i-1}/G_i would be normal in G/G_i, and H would be normal in G, against the maximality of G_i. Therefore, if a finite group determines a family of simple groups as composition factors, it also determines, as principal factors, a family of subgroups that are direct products of isomorphic simple groups (Theorem 2.29).

Theorem 2.30. *Let $G = H_1 \times H_2 \times \cdots \times H_n$, and let $K_i \trianglelefteq H_i$, $i = 1, 2, \ldots, n$. Then:*

i) $K_i \trianglelefteq G$;

ii) $G/K_1 K_2 \cdots K_n \simeq H_1/K_1 \times H_2/K_2 \times \cdots \times H_n/K_n$.

Proof. *i*) has already been seen in Remark 2.3, 1. As for *ii*), observe that $g \in G$ admits a unique expression as $g = h_1 h_2 \cdots h_n$, $h_i \in H_i$, so that the mapping:

$$G \to H_1/K_1 \times H_2/K_2 \times \cdots \times H_n/K_n,$$

given by $g \to (K_1h_1, K_2h_2, \ldots, K_nh_n)$, is well defined. It is clear that it is a surjective homomorphism, with kernel $K_1K_2 \cdots K_n$. ◊

Exercises

50. If $H \leq G$, and $\{G_i\}_{i=0}^{l}$ is a normal series of G, then $\{H_i\}_{i=0}^{l}$, where $H_i = H \cap G_i$, is a normal series of H, and a quotient H_{i-1}/H_i is isomorphic to a subgroup of the quotient G_i/G_{i+1}.

51. *i*) If H is subnormal and $K \leq G$, then $H \cap K$ is subnormal in K;
ii) the intersection of two subnormal subgroups is subnormal;
iii) the image of a subnormal subgroup under an automorphism of the group is again subnormal;
iv) the product of two subnormal subgroups is not necessarily a subgroup;
v) a subnormal subgroup of order coprime to the index is normal, and therefore unique of its order (cf. *ex.* 4).

52. The direct product is associative, $(H \times K) \times L \simeq (H \times (K \times L))$, and commutative, $H \times K \simeq K \times H$.

53. Prove that the product of a finite number of finite subgroups of a group that commute elementwise is a subgroup, and is a direct product if and only if its order is the product of the orders of the subgroups.

54. Let $G = H_1 \times H_2 \times \cdots \times H_n$ be a finite group, and let the orders of the H_i be pairwise coprime. Prove that:
i) if $K \leq G$, then K is the direct product of the subgroups $K \cap H_i$, $i = 1, 2, \ldots, n$;
ii) $\mathbf{Aut}(G)$ is isomorphic to the direct product of the groups $\mathbf{Aut}(H_i)$.

55. If $H, K \trianglelefteq G$, then $G/(H \cap K)$ is isomorphic to a subgroup of $G/H \times G/K$. If the indices of H and K are finite and coprime, then $G/(H \cap K)$ is isomorphic to $G/H \times G/K$. [*Hint*: consider the homomorphism $G \to G/H \times G/K$ given by $x \to (Hx, Kx)$.]

56. Prove that $S^m \times S^n$ is isomorphic to a subgroup of S^{m+n}.

57. The center of a direct product is the direct product of the centers of the factors.

58. In a finite nonabelian direct product in which every abelian subgroup is cyclic the direct factors are characteristic subgroups. The result is false if the group is abelian. [*Hint*: if a prime p divides the orders of two factors then there is a subgroup $C_p \times C_p$.]

59. A group G is said to be *complete* if its center is trivial and all the automorphisms are inner. If the center is trivial, then G is isomorphic to $\mathbf{I}(G)$, and if it is complete it is isomorphic to $\mathbf{Aut}(G)$. Prove that a complete group is a direct factor of every group in which it is contained as a normal subgroup.

60. Let n be an integer such that there exists only one group of order n. Prove that n is square-free (cf. Chapter 5, *ex.* 45). [*Hint*: if $p^2|n$, then there exist at least two groups of order n.]

61. Let $G = AH$, $A \trianglelefteq G$ abelian, $H \cap A = \{1\}$. Prove that A is minimal normal if and only if H is maximal.

2.7 Semidirect Products

We now consider a product that generalizes the direct product of two groups (but not of any number). This will provide a more interesting example of extension.

In the definition of direct product of two subgroups both subgroups are required to be normal. If only one of them is required to be normal we have the *semidirect product* of the two subgroups:

$$\begin{aligned} &i)\ \ G = HK;\\ &ii)\ \ K \trianglelefteq G;\\ &iii)\ \ H \cap K = 1. \end{aligned} \tag{2.23}$$

The subgroup H is a *complement of* K *in* G. Note that *i*) and *iii*) imply that the elements of H form a representative system of the cosets of K, and that an element of the group has a unique expression as a product of an element of H and one of K. Clearly, if the group is abelian, a semidirect product is direct.

In a semidirect product, an element $h \in H$ induces a mapping

$$\varphi_h : k \longrightarrow h^{-1}kh$$

of K in itself which is clearly an automorphism of K. Thus we have a mapping

$$\varphi : H \longrightarrow \mathbf{Aut}(K), \tag{2.24}$$

of H in the automorphism group of K. Note that, given $g_1, g_2 \in G$, we have $g_1 = h_1k_1$, $g_2 = h_2k_2$, and therefore:

$$g_1g_2 = h_1k_1 \cdot h_2k_2 = h_1h_2 \cdot h_2^{-1}k_1h_2 \cdot k_2 = h_1h_2 \cdot \varphi_{h_2}(k_1)k_2.$$

Conversely, given two groups H and K we may build a semidirect product by assigning a homomorphism φ as in (2.24) and defining in the cartesian product $H \times K$ the operation

$$(h_1, k_1)(h_2, k_2) = (h_1h_2, \varphi_{h_2}(k_1) \cdot k_2).$$

We obtain a group with identity (1,1) and inverse

$$(h, k)^{-1} = (h^{-1}, \varphi_{h^{-1}}(k^{-1})).$$

(Verification of associativity is routine.) As in the direct product, the pairs $(h, 1)$ and $(1, k)$ form two subgroups H^* and K^* isomorphic to H and K, respectively, and in H^*K^* we have

$$(h, 1)^{-1}(1, k)(h, 1) = (h^{-1}, k)(h, 1) = (1, \varphi_h(k)).$$

It follows that K^* is normal, and by identifying k with $(1, k)$ and h with $(h, 1)$, the image of k under φ_h is the conjugate of k by h.

From (2.23) we have $G/K \simeq H$, so that G is an extension of K by H. If in an extension $G/K \simeq H$ coset representatives may be so chosen as to form a subgroup isomorphic to H, the extension is a semidirect product (one also says that G *splits* over K). If in the homomorphism φ of (2.24) the image of every element of H is the identity automorphism of K, then the semidirect product is a direct product.

We will denote $H \times_\varphi K$ the semidirect product determined by a homomorphism φ as in (2.24).

Lemma 2.5. *If φ and ψ are two homomorphisms $H \to \mathbf{Aut}(K)$, and if there exist automorphisms α of K and β of H such that*

$$\varphi(h)\alpha = \alpha\psi(h^\beta)$$

for all $h \in H$, then the semidirect products relative to φ and ψ isomorphic.

Proof. The mapping $H \times_\varphi K \to H \times_\psi K$ given by $(h, k) \to (h^\beta, k^\alpha)$ is an isomorphism. ◇

Examples 2.10. 1. S^3 is the semidirect product of C_3 by C_2. More generally, D_n is the semidirect product of C_n by C_2. If $C_n = \langle r \rangle$ and $C_2 = \langle a \rangle$, where a is a flip with respect to an axis, then a induces by conjugation on C_n the automorphism $\psi : r^k \to a^{-1}r^k a = r^{-k}$. The homomorphism $\varphi : C_2 \to \mathbf{Aut}(C_n)$ associating a with ψ allows the construction of the semidirect product:

$$(a, 1)^{-1}(1, r^k)(a, 1) = (a^{-1}, r^k)(a, 1) = (1, (r^k)^{\varphi(a)}) = (1, (r^k)^\psi) = (1, r^{-k}).$$

The group D_4 is also the semidirect product of V by C_2: conjugation by the nonidentity element of C_2 exchanges two nonidentity elements of V and fixes the third. D_4 can also be obtained as an extension of $C_2 = \mathbf{Z}(D_4)$ by V, but this extension is not a semidirect product ($\mathbf{Z}(D_4)$ is contained in all the subgroups of order 4 of D_4, and therefore it cannot have trivial intersection with any of these).

2. *The affine group. i)* A transformation of the real line $\mathbf{R}$:

$$\varphi_{a,b} : x \longrightarrow ax + b, \; x \in \mathbf{R},$$

where $a, b \in \mathbf{R}$ and $a \neq 0$ is called an *affine transformation* or an *affinity*. The affinities form a group $\mathcal{A}$ under composition, the *affine group* of the real line.

$\mathcal{A}$ is a semidirect product. Indeed, if $a = 1$, the transformations $\varphi_b : x \to x+b$ (set $\varphi_{1,b} = \varphi_b$) form, as b varies, a subgroup $\mathcal{T}$ of $\mathcal{A}$ isomorphic to the additive group of the reals (with $\varphi_b^{-1} = \varphi_{-b}$), the subgroup of *translations*. The transformations $\psi_a : x \to ax$ also form a group, the subgroup of $\mathcal{H}$ of *homotheties* or *dilations* (with $\psi_a^{-1} = \psi_{a^{-1}}$), isomorphic to the multiplicative group of non zero elements of $\mathbf{R}$: $\psi_a\psi_b(x) = \psi_a(bx) = ab \cdot x$. We have:

$$\psi_{a^{-1}}\varphi_b\psi_a = x + a^{-1}b,$$

so that the conjugate of a translation is again a translation, i.e. $\mathcal{T} \lhd \mathcal{A}$. If $ax = x + b$, for some a and b and for all x, then $b = 0$ and $a = 1$; thus $\mathcal{H} \cap \mathcal{T} = \{1\}$. Moreover, every affinity is the product of a homothety and a translation, so that $\mathcal{A} = \mathcal{H}\mathcal{T}$. Hence, the three conditions (2.23) are satisfied. Finally, note that the conjugate of a nonidentity homothety

$$\varphi_{-b}\psi_a\varphi_b : x \longrightarrow ax + ab - b,$$

is no longer a homothety, i.e. $\mathcal{H}$ is not normal in $\mathcal{A}$, so that the product $\mathcal{H}\mathcal{T}$ is not direct. A nonidentity affinity has at most one fixed point. Indeed, if $ax + b = ay + b$, then $a(x - y) = 0$, and since $a \neq 0$ this implies $x - y = 0$ e $x = y$.

ii) Affinities may also be defined for the integers mod n:

$$\varphi : i \to hi + k \mod n, \; (h, n) = 1;$$

these form a group of order $n\varphi(n)$. This group is isomorphic to the normalizer N in S^n of the cyclic group generated by an $n-$cycle $c = (0, 1, \ldots, n-1)$. Indeed, let $\sigma \in N$, with $\sigma(0) = k$ and $\sigma c \sigma^{-1} = c^h$, $(h, n) = 1$. Then:

$$\sigma c^i(0) = \sigma c^i \sigma^{-1}(k) = c^{hi}(k) = c^{hi+k}(0) = hi + k,$$

and the mapping $\sigma \to \varphi$, where $\varphi : i \to hi + k$, with h and k determined by σ as above, is an isomorphism. Here too the group is a semidirect product: the two subgroups are the subgroup of the "homotheties" $i \to hi$ and the normal subgroup of the "translations" $i \to i + k$. As in the case of the reals, an affinity has at most one fixed point. Indeed, if $hi + k \equiv hj + k \mod n$, then $h(i - j) \equiv 0 \mod n$, and $(h, n) = 1$ implies $i - j \equiv 0 \mod n$, and $i \equiv j \mod n$ (see *ex.* 69).

3. If K is an abelian group, then the mapping $\sigma : k \to k^{-1}$ is an automorphism of K, so we may consider the semidirect product of $\langle\sigma\rangle$ by K. If K is a cyclic group of order n the result is the dihedral group D_n, and if $K = \mathbf{Z}$ is the group of integer then we have the *infinite dihedral group* D_∞. Every element of D_∞ not belonging to $\mathbf{Z}$ has order 2:

$$(\sigma, n)(\sigma, n) = (\sigma^2, \sigma(n) + n) = (1, -n + n) = (1, 0).$$

The group D_n is isomorphic to a quotient of D_∞, as we now show. In general, if L is a characteristic subgroup of a group K, an automorphism σ of K induces an automorphism $\overline{\sigma}$ of K/L, $\overline{\sigma}(Lx) = L\sigma(x)$. This is well defined because if lx is another representative of Lx, then $\overline{\sigma}(Llx) = L\sigma(lx) = L\sigma(l)\sigma(x) = L\sigma(x)$; L being characteristic, it contains $\sigma(l)$ for all $\sigma \in \mathbf{Aut}(L)$ and $l \in L$. Moreover, the mapping $\sigma \to \overline{\sigma}$ is a homomorphism $\mathbf{Aut}(K) \longrightarrow \mathbf{Aut}(K/L)$. It follows that a homomorphism $\varphi : H \longrightarrow \mathbf{Aut}(K)$, composed with the previous one, yields a homomorphism $\overline{\varphi} : H \longrightarrow \mathbf{Aut}(K/L)$:

$$\overline{\varphi}(h)(Lx) = L\varphi(h)(x),$$

using which one can construct the semidirect product of K/L by H. The mapping

$$H \times_\varphi K \longrightarrow H \times_{\overline{\varphi}} K/L,$$

given by $(h, k) \longrightarrow (h, Lk)$, is a surjective homomorphism with kernel $L^* = \{(1, x),\ x \in L\} \simeq L$. It follows:

$$H \times_{\overline{\varphi}} K/L \simeq (H \times_\varphi K)/L^*.$$

In the case of D_∞ we have:

$$D_\infty/\langle n\rangle \simeq \langle\sigma\rangle \times_{\overline{\varphi}} (\mathbf{Z}/\langle n\rangle) \simeq D_n,$$

(in $D_\infty/\langle n\rangle$ we have identified $\{(1, mn), m \in \mathbf{Z}\}$ with $\langle n\rangle$).

4. *Automorphisms of dihedral groups.* The elements of a dihedral group $D = \langle a, b\rangle$ (finite or infinite), with $a^2 = 1$ and b of finite ($b^n = 1$) or infinite order, are of the form $a^i b^j$, $i = 0, 1$ and j an integer. If $\alpha \in \mathbf{Aut}(D)$, then $b^\alpha \in \langle b\rangle$, since $\langle b\rangle$ is characteristic (in the finite case it is the unique cyclic subgroup of order n, in the infinite case it contains all the elements of infinite order). Hence, either $b^\alpha = b^k (k, n) = 1$ or $b^\alpha = b^{\pm 1}$. Moreover, $a^\alpha \notin \langle b\rangle$ (otherwise $a \in \langle b\rangle$), and therefore $a^\alpha = ab^j$. Consider the mapping

$$\gamma(ab^j) = ab^{j+1},\ \ \gamma(b^j) = b^j,$$

for all j (in the finite case, γ fixes the rotations and cyclically permutes the axial symmetries). It is easily seen that γ is an automorphism (for example, $\gamma(ab^j \cdot ab^k) = \gamma(a \cdot ab^{-j} \cdot b^k = b^{-j+k})$, $\gamma(ab^j)\gamma(ab^k) = ab^{j+1} \cdot ab^{k+1} = a \cdot ab^{-j-1} \cdot b^{k+1} = b^{-j+k}$ and $\gamma(ab^j) = \gamma(a) \cdot \gamma(b^j) = ab \cdot b^j = ab^{j+1}$; moreover, it is clear that γ is both injective and surjective), and that is of order n or infinite. Similarly, the mapping:

$$\beta_i(ab^s) = ab^{si},\ \ \beta_k(b) = b^k,\ (k, n) = 1,$$

for all i and s, is an automorphism. If $\alpha \in \mathbf{Aut}(D)$, then $b^\alpha = b^i, (i, n) = 1$ (finite case), or $b^\alpha = b^{\pm 1}$, and $a^\alpha = ab^j$; it follows $\alpha = \beta_i\gamma^j$, and setting $H = \{\beta_i,\ (i, n) = 1\}$, that $\mathbf{Aut}(D_n) = H\langle\gamma\rangle$. H and $\langle\gamma\rangle$ have trivial intersection, because if $\beta_i = \gamma^j$, for some i and j, then $b^{\beta_i} = b^{\gamma^j}$, $b^i = b^{\gamma^j} = b$, $i = 1$

and $\beta_i = \beta_1 = 1$. Moreover, $\beta_i^{-1}\gamma^j\beta_i = \gamma^{ji}$, so $\langle\gamma\rangle$ is normal in $\mathbf{Aut}(D_n)$; thus $\mathbf{Aut}(D_n)$ is the semidirect product of C_n by $\mathbf{Aut}(C_n)$, and $\mathbf{Aut}(D_\infty)$ is the semidirect product of $\mathbf{Z}$ by C_2, and as such is isomorphic to D_∞.

The semidirect product of a group by its automorphism group is the *holomorph* of the group. For example, we have seen above that D_∞ is the holomorph of $\mathbf{Z}$, that $\mathbf{Aut}(D_n)$ is the holomorph of C_n and $\mathbf{Aut}(D_\infty) \simeq D_\infty$ that of the integers. If G is a finite group of order n, and G_r is its image via the right regular representation, then the holomorph of G is isomorphic to the normalizer of G_r in S^n (cf. Theorem 3.28).

5. *Groups of order* 8. Taking advantage of the notion of semidirect product we now prove that there exist exactly five groups of order 8, those seen in *Ex.* 1.8, 2. The possible orders of the non identity elements are 2, 4 and 8. We distinguish various cases.

i) The group has only one element of order 2. If there is an element of order 8, the group is cyclic. So suppose that the group has six elements of order 4 and therefore three subgroups of order 4 that intersect in the unique subgroup of order 2. If $H = \langle x\rangle$ and $K = \langle y\rangle$ are two of these, the element $z = xy$ belongs neither to H nor to K, and therefore has order 4 and generates the third subgroup, $L = \langle z\rangle$. Now, $yz \notin L$ (otherwise $y \in L$) and $yz \notin K$ (otherwise $z \in K$), and therefore yz has order 4; hence $yz \in H$. If $yz = x^{-1}$, then $z^2 = xyz = xx^{-1} = 1$; hence $yz = x$, and similarly $zx = y$. Clearly, this group is the quaternion group.

ii) If all the elements have order 2, the group is elementary abelian, and therefore a direct product $\mathbf{Z}_2^{(3)}$ of three copies of $\mathbf{Z}_2$.

iii) We may assume that the group contains at least a subgroup C_4 and at least a C_2 not contained in C_4. Of the two possible homomorphisms $\varphi : C_2 \to \mathbf{Aut}(C_4) \simeq C_2$, the trivial one yields the direct product $C_2 \times C_4 \simeq U(15) \simeq U(20)$, and the isomorphism the group $C_2 \times_\varphi C_4$, where the image of φ is the automorphism exchanging the two generators of C_4. Clearly, the latter group is the dihedral group D_4.

6. Let $H = \mathbf{Z}_p \oplus \mathbf{Z}_p$, p an odd prime, and let u and v be generators of the two summands. The automorphism ψ given by $u \to u + v$, and $v \to v$, has order p, and thinking of H as a 2-dimensional vector space over $\mathbf{Z}_p$ with basis $\{u, v\}$, ψ is represented by the matrix $\begin{pmatrix} 1 & 1 \\ 0 & 1 \end{pmatrix}$. The semidirect product of H and $K = \langle\psi\rangle$ is a nonabelian group of order p^3 in which all the elements have order p. (The abelian group $\mathbf{Z}_p \oplus \mathbf{Z}_p \oplus \mathbf{Z}_p$ also has all the elements of order p; hence two groups may have the all the elements of the same order without being isomorphic.)

7. We now consider a few examples of groups that are not semidirect products:

i) the quaternion group $\mathcal{Q}$; its non identity subgroups have nontrivial intersections. Like D_4, it is an extension of its center by V;

ii) a cyclic p-group; its subgroups form a chain, so two nontrivial subgroups never have a nontrivial intersection;
iii) the integers; two subgroups $\langle m \rangle$ and $\langle n \rangle$ have in common the subgroup generated by $\mathrm{lcm}(m, n)$.

2.8 Symmetric and Alternating Groups

In this section we consider an infinite class of simple groups, the *alternating groups* A^n, $n \geq 5$.

A permutation of S^n splits into cycles, and if $(1, 2, \ldots, k)$ is a cycle, then

$$(1, 2, \ldots, k) = (1, 2)(1, 3) \cdots (1, k).$$

Hence:

Theorem 2.31. *Every permutation of S^n is a product of transpositions, and therefore S^n is generated by the transpositions.*

Every transposition is a product of transpositions that involve a fixed digit, 1 say:

$$(h, k) = (1, h)(1, k)(1, h),$$

((h, k) is expressed as the conjugate of $(1, k)$ by $(1, h)$).

Corollary 2.7. *S^n is generated by the $n - 1$ transpositions:*

$$(1, 2), \ldots, (1, 3), \ldots (1, n).$$

If we replace a product of disjoint cycles with a product of transpositions we gain something (we only have 2-cycles and not cycles of variable length) but we lose something (2-cycles do not commute in general and they are not uniquely determined):

$$(1, 2, 3) = (1, 2)(1, 3) \neq (1, 3)(1, 2) = (1, 3, 2),$$

and

$$(1, 2, 3) = (1, 2)(1, 3) = (2, 3)(1, 2) = (1, 3)(1, 2)(1, 3)(1, 2).$$

There is something invariant, however, in all these products: the parity of the number of transpositions that appear. This is the content of the next theorem.

Theorem 2.32. *If a permutation of S^n is a product of transpositions in two different ways, and if the number of transpositions in the first decomposition is even (odd), then it is even (odd) also in the second.*

Proof. If $\sigma \in S^n$ can be written as a product of both an even and an odd number of transpositions, the same holds for σ^{-1}. Choosing an even decomposition for σ and an odd one for σ^{-1}, we have that the identity $1 = \sigma\sigma^{-1}$ can be written as the product of an odd number of transpositions:

$$1 = \tau_1\tau_2\cdots\tau_{2m+1}.$$

Let i be a digit appearing in one of the τ_i. Observe that $(i,j)(k,l) = (k,l)(i,j)$ and $(i,j)(j,l) = (j,l)(i,l)$. Hence we can write the above product without changing the number of transposition as

$$1 = \sigma_1\sigma_2\cdots\sigma_r\sigma_{r+1}\cdots\sigma_{2m+1}, \tag{2.25}$$

where i does not appear in the first r transpositions and where $\sigma_j = (i,k_j)$, $j = r+1,\ldots,2m+1$. If (i,k_{r+1}) appears only once, then in the product (2.25) i goes to k_{r+1}, contrary to the product being the identity. Hence assume we have $(i,k_{r+1})\cdots(i,k_j)(i,k_{r+1})\cdots$; using $(i,k_j)(i,k_{r+1}) = (i,k_{r+1})(k_{r+1},k_j)$ we can move the second (i,k_{r+1}) near the first and cancel both of them, leaving the value of the product unaltered. By repeating this operation, the number of transpositions decreases by two each time, and the final result is a single transposition, contrary to the product being the identity. ♢

It follows that the *parity* of a permutation, is well defined, and we may give the following definition.

Definition 2.20. A permutation is *even* if is the product of an even number of transpositions, and *odd* in the other case[12].

By Theorem 2.32, the mapping $S^n \to \{1,-1\}$, $n > 1$, obtained by associating 1 with a permutation σ if this is even, and -1 if it is odd, is well defined. Moreover, since the parity of a product is even if and only if both permutations have the same parity, this mapping is a homomorphism, whose kernel is the set of even permutations.

Definition 2.21. The subset of even permutations of S^n is a subgroup, called the *alternating group*, denoted A^n.

Hence $S^n \to \{1,-1\} \simeq S^n/A^n$, and A^n has index 2 in S^n, so half of the permutations of S^n are even[13].

Theorem 2.33. *Let $n \geq 3$. Then:*
i) *A^n contains all the 3-cycles of S^n;*
ii) *for fixed i and j, A^n is generated by the following $n-2$ 3-cycles :*

$$(i,j,1),(i,j,2),\ldots,(i,j,k),\ldots,(i,j,n), \quad k \neq i,j.$$

[12] The parity of a permutation can also be defined in terms of the inversions it presents (Corollary 3.19). For another proof that parity is well defined see *ex.* 115 of Chapter 3.

[13] A^n has already been seen in its matrix representation (*ex.* 26 of Chapter 1).

Proof. *i*) A 3-cycle is even:

$$(i,j,k) = (i,j)(i,k), \tag{2.26}$$

so it belongs to A^n.

ii) For fixed i, every element of A^n is a product of transpositions containing i (Corollary 2.7), and these are even in number. Hence they can be paired off in the order they appear, and the product of each pair is a 3-cycle as in (2.26). This proves that an element of A^n is a product of 3-cycles. Moreover, for fixed i and j we have $(h,k,l) = (i,l,k)(i,l,k)(i,h,k)$, and $(i,h,k) = (i,j,h)(i,j,h)(i,j,k)$, as required. ◊

Remark 2.4. This theorem shows that 3-cycles play for A^n the role the transpositions, i.e. 2-cycles, do for S^n. Also note that a 2-cycle is a nonidentity element of S^n fixing the maximum possible number of digits. In A^n, such an element is a 3-cycle.

Lemma 2.6. *If a normal subgroup of A^n, $n \geq 3$, contains a 3-cycle, then it coincides with A^n.*

Proof. If $n = 3$, A^3 is cyclic of order 3 and there is nothing to prove. Let $n > 3$, $N \trianglelefteq A^n$, and let $(1,2,3) \in N$. A conjugate of this cycle by an even permutation still belongs to N. As $n > 3$, we have in A^n all the permutations of type $(1,2)(3,k)$, with $k = 4,5,\ldots,n$. But $(1,2,3)^{(1,2)(3,k)} = (2,1,k)$, and by Theorem 2.33 these 3-cycles generate A^n. ◊

Lemma 2.7. *Let $\{1\} \neq N \trianglelefteq A^n$, $n \neq 4$. Then N contains a 3-cycle.*

Proof. Let $\sigma \neq 1 \in N$. We split the proof into various parts.

1. Assume that in the decomposition into disjoint cycles σ has a cycle with more than three digits: $\sigma = (1,2,3,4,i,\ldots,j,k)\tau$, where τ is the product of the other cycles, if any. Conjugating σ by $(1,2,3) \in A^n$ we find $\sigma_1 = \sigma^{(1,2,3)} = (2,3,1,4,i,\ldots,j,k)\tau$, which is still an element of N. It follows $\sigma_1^{-1} = (1,3,2,k,j,\ldots,i,4)\tau^{-1}$, so $\sigma\sigma_1^{-1} = (1,k,3)(2)(4)(j)\ldots$ is a 3-cycle of N, as required. Therefore, we may assume that the cycles of σ have length at most 3.

2. If σ has more than one cycle, three cases are to be considered.

2*a*. σ has at least two cycles of length 3, $\sigma = (1,2,3)(4,5,6)\tau$. Conjugating by $(1,2,4)$ we have $\sigma_1 = (2,4,3)(1,5,6)\tau$, and $\sigma\sigma_1 = (1,4,6,3,5)(2)\tau^2$ has a cycle with more that three digits, and we are in case 1.

2*b*. σ has only one cycle of length 3. We may assume that the remaining cycles have length at most 2. But then σ^2 is a 3-cycle.

2*c*. σ is a product of disjoint transpositions: $\sigma = (1,2)(3,4)\tau$, where τ consists of disjoint transpositions or fixed points. By assumption, there exists a digit, 5 say, different from 1,2,3 and 4, appearing in τ either in a transposition, $\sigma = (1,2)(3,4)(5,6)\tau_1$ or as a fixed point, $\sigma = (1,2)(3,4)(5)\tau_2$. Conjugation by $(1,3)(2,5)$ yields, in the first case, $\sigma_1 = (3,5)(1,4)(2,6)\tau_1$, so $\sigma\sigma_1 =$

$(1,6,3)(2,4,5)\tau_1^2$ and we are in case $2a$. In the second, $\sigma_1 = (3,5)(1,4)(2)\tau_2$ and $\sigma\sigma_1 = (1,2,4,5,3)\tau_2^2$, and we are in case 1.

3. σ has only one cycle. If its length is greater than 3, we are in case 1; if its length is 2, it is a transposition, which is odd, so it cannot belong to A^n. Hence σ is the required 3-cycle. ♢

From the previous lemmas it follows:

Theorem 2.34. *If $n \neq 4$, A^n is a simple group.*

Theorem 2.35. *If $n \neq 4$, A^n is the unique nontrivial normal subgroup of the group S^n.*

Proof. If $n = 1,2,3$, the result is obvious. Let $n \geq 5$. With same proof of Lemma 2.7, except where it is excluded that σ is a transposition, we have that $N \trianglelefteq S^n$ contains a 3-cycle, and therefore all the 3-cycles because they are all conjugate. Hence $N \supseteq A^n$, and if $N \neq S^n$ then $N = A^n$. If N contains a transposition it contains all of them because they are all conjugate; but the transpositions generate S^n, and therefore $N = S^n$. ♢

Let us now consider the exceptional case $n = 4$. A normal subgroup is a union of conjugacy classes, so its order is the sum of the orders of these. But in S^4, the sums of class orders compatible with Lagrange's theorem are, apart from the trivial ones, $|C_1| + |C_3| + |C_4| = 12$ and $|C_1| + |C_4| = 4$ (*Ex.* 2.5). In the first case, the union of the three classes is a subgroup, i.e. A^4; in the second, the union of the two classes is a Klein group contained in A^4 and normal in S^4. In particular, A^4 is not simple.

A^4 provides a counterexample to the converse of Lagrange's theorem, and in fact the smallest counterexample.

Theorem 2.36. *A^4 has no subgroups of order* 6.

Proof. A subgroup H of order 6 should contain, by Cauchy's theorem, a subgroup of order 3, and therefore an element of order 3; being in A^4, this is a 3-cycle. However H is normal in A^4 (index 2), and therefore by Lemma 2.6, $H = A^4$.

Another proof is the following. A subgroup of order 6 being of index 2 should contain the squares of all the elements of A^4, i.e. the squares of the eight 3-cycles. But these are again the eight 3-cycles. ♢

Exercises

62. Prove that the transpositions $(1,2),(2,3),\ldots,(i,i+1),\ldots,(n-1,n)$ generate the whole group S^n.

63. *i)* S^n is generated by an n-cycle and a transposition exchanging two consecutive digits of the cycle.

ii) The 2-cycles generate S^n and the 3-cycles A^n. Which subgroups do the r-cycles, $r > 3$, generate?

64. A k-cycle is even if and only if k is odd.

65. The number of $k-$cycles of S^n is $\frac{1}{k}n(n-1)(n-2)\cdots(n-k+1)$.

66. In a permutation group, either all permutations are even or else exactly half of them are even.

67. Show that for the group S^4 Lagrange's theorem has an inverse. More precisely, S^4 admits:
i) a subgroup of order 12, and only one;
ii) three subgroups of order 8, dihedral;
iii) four subgroups of order 6 isomorphic to S^3;
iv) seven subgroups of order 4, three cyclic groups and four Klein groups;
v) four subgroups of order 3;
vi) nine subgroups of order 2.

68. Determine:
i) the conjugacy classes of S^5;
ii) the conjugacy classes of A^5;
iii) two elements of A^5 that are conjugate in S^5 but not in A^5 [*Hint:* consider a 5-cycle and its square.];
iv) use *ii)* to show that A^5 is simple.

69. *i)* Prove that the affine group of the integers mod 6 (*Ex.* 2.10, 2 *ii)*) is the dihedral group D_6.

ii) Consider $\mathbf{F}_4$, the 4-element field given by the four polynomials $0, 1, x, x+1$ with sum modulo 2 and product modulo the polynomial x^2+x+1. Prove that the affine group of this field is isomorphic to the alternating group A^4.

2.9 The Derived Group

In this section we consider a subgroup that in some sense provides a measure of how far the group is from being abelian.

Let a and b be two elements of a group, and consider the products ab and ba. There exists $x \in G$ such that $ab = ba \cdot x$, so

$$x = a^{-1}b^{-1}ab.$$

Definition 2.22. If a, b are two elements of a group G, the element $a^{-1}b^{-1}ab$ is called the *commutator* of a and b (in this order)[14]. It is denoted $[a, b]$. The subgroup generated by all commutators is the *derived group* or *commutator subgroup* G, and is denoted by G' or by $[G, G]$:

$$G' = \langle [a, b] \mid a, b \in G \rangle.$$

[14] Some authors define the commutator of a and b as the element $aba^{-1}b^{-1}$.

The identity is a commutator: $1 = [a, a]$, for all $a \in G$, and the inverse of a commutator is again a commutator:

$$[a, b]^{-1} = (a^{-1}b^{-1}ab)^{-1} = b^{-1}a^{-1}ba = [b, a].$$

However, the product of two commutators is not necessarily a commutator[15], so in order to have a subgroup it is necessary to consider the subgroup generated by the commutators.

Two elements commute if and only if their commutator is the identity, so a group is abelian if and only if the commutator subgroup is the identity subgroup.

Theorem 2.37. *i) G' is a characteristic subgroup;*
ii) the quotient G/G' is abelian;
iii) if $N \trianglelefteq G$ and G/N is abelian, then $G' \subseteq N$;
iv) if $H \leq G$ and $G' \subseteq H$, then H is normal.

Proof. i) If $\alpha \in \mathbf{Aut}(G)$, then

$$[a, b]^{\alpha} = (a^{-1}b^{-1}ab)^{\alpha} = (a^{\alpha})^{-1}(b^{\alpha})^{-1}a^{\alpha}b^{\alpha} = [a^{\alpha}, b^{\alpha}].$$

Hence an automorphism of G takes commutators to commutators and $(G')^{\alpha} \subseteq G'$. On the other hand, $[a, b] = [a^{\alpha^{-1}}, b^{\alpha^{-1}}]^{\alpha}$, and therefore $G' \subseteq (G')^{\alpha}$.

ii) By *i)*, G' is normal, and if $aG', bG' \in G/G'$ then:

$$aG'bG' = abG' = ba[a, b]G' = baG' = bG'aG',$$

where the third equality follows from $[a, b] \in G'$.

iii) If $aNbN = bNaN$, for all $a, b \in G$, then $abN = baN$ and $a^{-1}b^{-1}abN = N$, from which $[a, b] \in N$ and $G' \subseteq N$, and conversely.

iv) H/G' is normal in G/G' because G/G' is abelian. It follows (Theorem 2.5) $H \trianglelefteq G$. ◇

Point (*iii*) of this theorem can also be expressed by saying that G' is the smallest subgroup of G with respect to which the quotient is abelian.

Examples 2.11. 1. In S^3 we have the abelian quotient $S^3/C_3 \simeq C_2$, so $(S^3)' \subseteq C_3$, and either $(S^3)' = \{1\}$ or $(S^3)' = C_3$. Since S^3 is not abelian, $(S^3)' = C_3$. The two 3-cycles of C_3 may be expressed as commutators: $(1, 2, 3) = [(2, 3), (1, 3, 2)]$ and $(1, 3, 2) = [(1, 3, 2), (2, 3)]$.

2. The quotient $D_4/\mathbf{Z}(D_4)$ is the Klein group, hence the derived group D_4' is contained in the center and cannot be the identity subgroup. It follows $D_4' = \mathbf{Z}(D_4)$. Similarly, in the quaternion group the derived group coincides with the center.

---

[15] Cf. Carmichael, p. 39, *ex.* 30, or Kargapolov-Merzliakov, *ex.* 3.2.11.

3. Let $G = D_4 \times C_2$. D_4' is normal in G (Theorem 2.30, *i*)) and the quotient is isomorphic to $V \times C_2$ (same theorem, *ii*)), which is abelian. Hence $G' \simeq D_4' \simeq C_2$.

4. *A^n is the derived group of S^n for all n.* The 3-cycles are commutators: $(1,2,3) = (1,3)(1,2)(1,3)(1,2)$, and therefore $A^n \subseteq (S^n)'$. However, a commutator $\sigma^{-1}\tau^{-1}\sigma\tau$ is an even permutation, whatever the parity of σ and τ; hence $(S^n)' \subseteq A^n$ (this can also be seen by considering that S^n/A^n has order 2 and therefore is abelian).

5. Every element of A^5 is a commutator, and actually a commutator of elements of A^5[16]:

$$\begin{aligned}(1,2,3) &= [(2,3)(4,5),(1,3,2)],\\ (1,2)(3,4) &= [(1,3,4),(1,2,3)],\\ (1,2,3,4,5) &= [(2,5)(3,4),(1,4,2,5,3)].\end{aligned}$$

In A^3 or in A^4 a 3-cycle is not a commutator of elements of A^3 or A^4. For a 3-cycle to be a commutator of elements of A^n at least five digits are needed.

In order to generate the derived group of a group it is not necessary to take all the commutators. Indeed, let a and b belong to two cosets of the center of the group, and let x and y be two representatives of these cosets. Then $a = xz$ and $b = yz_1$, with z and z_1 in the center, so that:

$$[a,b] = [xz, yz_1] = z^{-1}x^{-1}z_1^{-1}y^{-1}xzyz_1 = x^{-1}y^{-1}xy = [x,y].$$

Hence the representatives of the cosets of the center suffice.

Theorem 2.38 (Schur). *If the center of a group G is of finite index then the derived group G' is finite.*

Proof. (The following proof is due to D. Ornstein[17]). If the center is of index m there are at most m^2 commutators, so G' is finitely generated. Let us show that an element of G' may be written as a product of at most m^3 commutators. If in a product of commutators a given commutator c appears m times, these copies of c may be brought together, possibly replacing a commutator by a conjugate. Now apply the following lemma.

Lemma 2.8. *If $a, b \in G$, and if, for some m, $(ab)^m \in \mathbf{Z}(G)$, then $[a,b]^m$ can be written as a product of $m-1$ commutators.*

Proof. For each r, $[a,b]^r$ can be written as a product of $(a^{-1}b^{-1})^r(ab)^r$ and $r-1$ commutators. This is clear for $r = 1$; assuming the result true for $r-1$ we have:

$$[a,b]^r = [a,b][a,b]^{r-1} = [a,b](a^{-1}b^{-1})^{r-1}(ab)^{r-1}c_{r-2}\cdots c_1,$$

[16] This holds for all A^n, $n \geq 5$ (Itô N.: Math. Japonicae **2** (1951), p. 59–60).

[17] Cf. Kaplansky I.: An introduction to differential algebra. Hermann, Paris (1957), p. 59.

where the c_i, $i = 1, 2, \ldots, r-2$ are commutators. The last expression equals

$$a^{-1}b^{-1}(a^{-1}b^{-1})^{r-1}(ab)[ab, (a^{-1}b^{-1})^{r-1}](ab)^{r-1}c_{r-2}\cdots c_1 =$$
$$(a^{-1}b^{-1})^r ab(ab)^{r-1}(ab)^{1-r}[ab, (a^{-1}b^{-1})^{r-1}](ab)^{r-1}c_{r-2}\cdots c_1$$

The element preceding c_{r-2} is the conjugate of a commutator, so again a commutator (Theorem 2.37, proof of *i*)); if we denote it by c_{r-1} we have:

$$[a, b]^r = (a^{-1}b^{-1}ab)^r = (a^{-1}b^{-1})^r(ab)^r c_{r-1}c_{r-2}\cdots c_1,$$

as required. Now $(ab)^m$ and $(ba)^m$ are conjugate, and if $(ab)^m \in \mathbf{Z}(G)$, then they are equal; but $(a^{-1}b^{-1})^m = (ba)^{-m} = (ab)^{-m}$, and the previous equality becomes:

$$[a, b]^m = c_{m-1}c_{m-2}\cdots c_1,$$

i.e. the result. 

Going back to the proof of the theorem, the element c^m may be written as a product of $m-1$ commutators. Since there are at most m^2 commutators, each element of G' may be written as a product of at most m^3 commutators. This shows that the length of the elements of G' is uniformly bounded, so G' is finite.

Example 2.12. The integers are a torsion free group with automorphism group of order 2. More generally, if G is a torsion free group and $\mathbf{Aut}(G)$ is finite, then G is abelian. Indeed, $\mathbf{I}(G)$ is finite, so $G/\mathbf{Z}(G)$ is finite. By Schur's theorem, G' is finite; G being torsion free this implies $G' = \{1\}$ and G abelian.

Exercises

70. Show that the derived group of D_n is generated by the square of the rotation of $2\pi/n$.

71. If $H \trianglelefteq G$ and $H \cap G' = \{1\}$ then H is contained in the center of G. [*Hint*: consider $(g^{-1}h^{-1}g)h$.]

72. If $a \sim a^2$ then a is a commutator. More generally, if $a \sim a^n$ then a^{n-1} is a commutator.

73. Two conjugate elements belong to the the same coset of the derived group. Hence the order of a conjugacy class is at most that of the derived group. [*Hint*: $G'ab = G'ba$.]

74. In a group of odd order, the product of all the elements, in any order, belongs to the derived group.

75. If $a^n = b^n = 1$, then:

$$(ab)^n = (a, b)(b, a^2)(a^2, b^2)(b^2, a^3)\cdots(b^{n-2}, a^{n-1})(a^{n-1}, b^{n-1}),$$

where $(x, y) = [x^{-1}, y^{-1}]$.

76. Let $a, b, c \in G$ be such that $a^n b^n = c^n$, where n is coprime to $|G|$. Prove that ab and c belong to the same coset of G'.

77. A commutator is a product of squares, and in fact of at most three squares.

78. If G is a p-group of order p^n, prove that $|\mathbf{Z}(G)| \neq p^{n-1}$, and if $n \geq 2$ then $|G'| \leq p^{n-2}$.

79. Let $\Delta(G)$ be the set of elements of G having a finite number of conjugates. Prove that:
i) $\Delta(G)$ is a characteristic subgroup;
ii) if H is a finitely generated subgroup of $\Delta(G)$, then the index of $\mathbf{Z}(H)$ in H is finite, and therefore H' is finite;
iii) G has a finite normal subgroup of order divisible by a given prime number p if, and only if, $\Delta(G)$ contains an element of order p.

80. Prove that if $(ab)^2 = (ba)^2$ for all pairs of elements $a, b \in G$, then G' is an elementary abelian 2-group. [*Hint*: use *ex.* 18, 17 and 77.]

81. If $[G : \mathbf{Z}(G)]^2 < |G'|$, then there are elements of G' that are not commutators. [*Hint*: a commutator is of the form $[a, b]$, where a and b are representatives of the cosets of the center.]

82. Let $|G'| \leq 2$. Prove that for all triples of elements x_1, x_2, x_3, there exists a nonidentity permutation of the x_i such that $x_1 x_2 x_3 = x_i x_j x_k$. Conversely, if this property holds, then the square of every element belongs to the center of the group.

83. Let $H \leq G$, and let T be a transversal of the right cosets of H. Prove that:
i) the subgroups $t^{-1}Ht$, $t \in T$, are all the conjugates of H;
ii) if $H = N_G(H)$, then the conjugates of H are all distinct;
iii) if G is simple, then for all $h \in H$ there exists $t \in T$ such that $t^{-1}ht \notin H$.

84. The derived group of a direct product of groups is the direct product of the derived groups of the factors.

85. A complete group H, with $H' < H$, cannot be the derived group of a group G. [*Hint*: $G = H \times \mathbf{C}_G(H)$ (cf. *ex.* 59); apply the previous exercise.]

3

Group Actions and Permutation Groups

3.1 Group actions

Group actions on sets provide a powerful way of obtaining information about a group. Moreover, a number of results seen in the preceding chapters whose proofs have a similar flavor may all be proved using the technique of group action.

Definition 3.1. Given a set $\Omega = \{\alpha, \beta, \gamma, \ldots\}$, a group G *acts* on Ω (or G *operates* on Ω, or Ω is a G-set) when a function $\Omega \times G \to \Omega$ is assigned such that, denoting α^g the image of the pair (α, g),

i) $\alpha^{gh} = (\alpha^g)^h$, $\alpha \in \Omega$, $g, h \in G$;
ii) $\alpha^1 = \alpha$, where 1 is the identity of G.

The assigned function is called an *action* of G on Ω[1], and the cardinality of Ω the *degree* of the group. Using a geometric language, we shall often call *points* the elements of Ω; *letters* and *digits* are also names for these elements. Note that if H is a subgroup of G and G acts on Ω then H also acts on Ω; this action of H is the *restriction* to H of the action of G. The notion of a group acting on a set generalizes that of a group of permutations, in the sense that it may well happen that there exist nonidentity elements g of G such that $\alpha^g = \alpha$ for all $\alpha \in \Omega$ (if G is a permutation group only the identity fixes everything).

Theorem 3.1. *If a group G acts on a set Ω, every element of G gives rise to a permutation of Ω. More precisely, the mapping $\Omega \to \Omega$ given by $\varphi_g : \alpha \to \alpha^g$ is, for every fixed element $g \in G$, a permutation of Ω.*

Proof. φ_g is injective: $\alpha^g = \beta^g \Rightarrow (\alpha^g)^{g^{-1}} = (\beta^g)^{g^{-1}} \Rightarrow \alpha^{gg^{-1}} = \beta^{gg^{-1}} \Rightarrow \alpha^1 = \beta^1$, and $\alpha = \beta$. Note that we have used both *i)* and *ii)* of Definition 3.1. It is surjective: if $\alpha \in \Omega$, let $\beta = \alpha^{g^{-1}}$; then $\beta^g = (\alpha^{g^{-1}})^g = \alpha^{g^{-1}g} = \alpha^1 = \alpha$, and α is the image of β. ◇

[1] *Right* action; see Remark 1.2.

Machì A.: Groups. An Introduction to Ideas and Methods of the Theory of Groups.
DOI 10.1007/978-88-470-2421-2_3, Springer-Verlag Italia 2012

Let S^Ω be the symmetric group on Ω, and let G be a group acting on Ω. By the above theorem we can consider the mapping $\varphi : G \to S^\Omega$ obtained by associating with $g \in G$ the permutation φ_g of Ω it induces: $g \to \varphi_g$. This mapping is a homomorphism, as is easily seen. In this way, we obtain a representation of the elements of G as permutations. To the product of two elements of G there corresponds the product of the two permutations representing them. The kernel of φ is given by

$$K = \{g \in G \mid \alpha^g = \alpha,\ \forall \alpha \in \Omega\},$$

i.e. K is the set of elements of G fixing all the points of Ω, and is called the *kernel of the action.* If $K = \{1\}$ the action is *faithful*, and in this case G is isomorphic to a subgroup of S^Ω; we shall then say that G is a group of permutations of Ω. In any case, the quotient G/K is a group of permutations of Ω, with the action defined by:

$$(\alpha, Kg) \to \alpha^g, \text{ i.e. } \alpha^{Kg} = \alpha^g,$$

which is well defined because if h is another representative of the coset Kg, then $h = kg$, $k \in K$, and therefore $\alpha^{kg} = (\alpha^k)^g = \alpha^g$, since $\alpha^k = \alpha$, for all $k \in K$ and $\alpha \in \Omega$. If $K = G$, the elements of G leave all the points of Ω fixed: the action is *trivial.*

An element α of Ω determines two subsets, one in Ω (the orbit of α), the other one in G (the stabilizer of α).

Definition 3.2. The *orbit* of α under the action of G is the subset of Ω:

$$\alpha^G = \{\alpha^g,\ g \in G\},$$

i.e. the set of points of Ω to which α is taken by the various elements of G.

Two orbits either coincide or are disjoint. Indeed, the orbits are the equivalence classes of the following relation ρ:

$$\alpha\rho\beta \text{ if there exists } g \in G \text{ such that } \alpha^g = \beta.$$

In particular, Ω is a disjoint union of orbits:

$$\Omega = \bigcup_{\alpha \in T} \alpha^G,$$

where α varies in a set T of representatives of the orbits. If Ω is finite, then:

$$|\Omega| = \sum_{\alpha \in T} |\alpha^G|.$$

If $H \leq G$, the orbits of G are unions of orbits of H. If g is a permutation, the orbits of the subgroup generated by g are the subsets of Ω on which g is a cycle. The orbits are also called *transitivity systems.*

Definition 3.3. G is *transitive* if there is only one orbit, *intransitive* in the other case.

In other words, G is transitive if, given any two points $\alpha, \beta \in \Omega$, there exists at least one element of G taking α to β.

Definition 3.4. The *stabilizer* of α is the subset of G:

$$G_\alpha = \{g \in G \mid \alpha^g = \alpha\},$$

i.e. the set of elements of G fixing α.

G_α is a subgroup of G. Indeed,
i) if $x, y \in G_\alpha$, then $\alpha^{xy} = (\alpha^x)^y = \alpha^y = \alpha$, so $xy \in G_\alpha$;
ii) $1 \in G_\alpha$;
iii) if $x \in G_\alpha$, $\alpha = \alpha^1 = \alpha^{xx^{-1}} = (\alpha^x)^{x^{-1}} = \alpha^{x^{-1}}$, so that $x^{-1} \in G_\alpha$.
The stabilizer of a point is also called the *isotropy group* of the point.

If an element of G belongs to the stabilizer of all the points of Ω, then it belongs to the kernel of the action, and conversely. Therefore, the kernel of the action is the intersection of the stabilizers of the points of Ω:

$$K = \bigcap_{\alpha \in \Omega} G_\alpha.$$

The relation between orbits and stabilizers is shown in the following theorem.

Theorem 3.2. *i) The cardinality of the orbit of a point α equals the index of the stabilizer of α:*

$$|\alpha^G| = [G : G_\alpha]; \tag{3.1}$$

in particular, if G is finite,

$$|G| = |G_\alpha||\alpha^G|, \tag{3.2}$$

and hence the cardinality of an orbit divides the order of the group.
ii) If β belongs to the orbit of α, then the stabilizers of β and α are conjugate. More precisely, if $\beta = \alpha^g$ then $G_\beta = (G_\alpha)^g$, i.e.

$$G_{\alpha^g} = (G_\alpha)^g.$$

Proof. i) It may very well happen that for two distinct elements $g, h \in G$ one has $\alpha^g = \alpha^h$. If we want to know how many distinct points α^g are obtained as g varies in G, we should be able to determine how many times one point is repeated. Now $\alpha^g = \alpha^h \Leftrightarrow \alpha^{gh^{-1}} = \alpha \Leftrightarrow gh^{-1} \in G_\alpha$, so that α^g and α^h are the same point if, and only if, g and h belong to the same (right) coset of the stabilizer of α, i.e. if, and only if, $g = xh$ with $x \in G_\alpha$. In other words, a point α^g is repeated as many times as there are elements in the coset $G_\alpha g$ (for $x \in G_\alpha$ we have $\alpha^{xg} = (\alpha^x)^g = \alpha^g$). Two elements g and h belonging to distinct cosets give rise to distinct α^g and α^h, and therefore there are as many points in the orbit α^G as there are cosets of G_α in G[2].

ii) If $x \in G_\alpha$, then $(\alpha^g)^{g^{-1}xg} = (\alpha^x)^g = \alpha^g$, i.e. $g^{-1}xg$ stabilizes $\alpha^g = \beta$. Thus $(G_\alpha)^g \subseteq G_{\alpha^g} = G_\beta$. Conversely, if $x \in G_{\alpha^g}$, then $(\alpha^g)^x = \alpha^g$, i.e. $gxg^{-1} \in G_\alpha$ and therefore $x \in (G_\alpha)^g$ and $G_{\alpha^g} = G_\beta \subseteq (G_\alpha)^g$. ♢

If α and β belong to the same orbit, then $\alpha^G = \beta^G$, and therefore, by *i*) of the previous theorem, $[G : G_\alpha] = |\alpha^G| = |\beta^G| = [G : G_\beta]$, i.e. the two stabilizers have the same index. Part *ii*) of the theorem says that, in addition, they are conjugate (by Theorem 2.21 two conjugate subgroups have the same index).

Corollary 3.1. *If G is transitive, $|\Omega| = [G : G_\alpha]$. If G is finite and transitive, then Ω is also finite and $|\Omega|$ divides $|G|$.*

To be able to exploit the group action to discover properties of a group one must find a convenient set on which have the group act. This set will be suggested by the nature of the problem, and will often be found inside the group itself. Some of the results previously found can be obtained by assuming the point of view of the action.

Examples 3.1. 1. Take as Ω the set underlying the group G, and let G act by right multiplication: $a^x = ax$. It is an action since $a \cdot 1 = a$, and the property $a^{xy} = (a^x)^y$ is simply the associative law of G: $a^{xy} = a(xy) = (ax)y = (a^x)^y$. G is transitive: given $a, b \in G$, there exists $x \in G$ such that $ax = b$ (i.e. $x = a^{-1}b$). The stabilizer of an element is the identity: if $ax = a$, then $x = 1$; a fortiori, the kernel of the action is the identity, so that the homomorphism $G \to S^\Omega$ is an isomorphism between G and a subgroup of S^Ω. In this way, one obtains the right regular representation of G (Definition 1.9). By defining $a^x = x^{-1}a$ one has the left regular representation.

If G is finite, $G = \{x_1, x_2, \ldots, x_n\}$, the image of an element $x \in G$ in the (right) regular representation is the permutation:

$$\begin{pmatrix} x_1 & x_2 & \ldots & x_n \\ x_1 x & x_2 x & \ldots & x_n x \end{pmatrix}.$$

Now $x_1 x = x_i$, some i, so that $x_i x = x_1 x^2$; similarly, $x_1 x^2 = x_j$ and $x_j x = x_1 x^3$, etc. If k is the order of x, then $x_1 x^{k-1} x = x_1 x^k = x_1$; hence the element x_1 belongs to the cycle $(x_1, x_1 x, x_1 x^2, \ldots, x_1 x^{k-1})$. (If $x_1 x^h = x_1$ with $h < k$, then $x^h = 1$, contrary to $o(x) = k$.) The same happens for the other cycles. Thus, the image of an element x of G in the regular representation is a permutation whose cycles all have the same length (permutations of this kind are said to be *regular*[3]), their length being the order of x.

[2] This argument is the same as that leading to (2.12); see *Ex.* 3.2, 2.
[3] See below, Section 3.5.

2. A group of linear transformations of a vector space V defines an action on the set underlying V and on the set of subspaces having the same dimension. The group $G = GL(V)$ is transitive on the subspaces having the same dimension. Indeed, if W_1 and W_2 have the same dimension with bases B_1 and B_2, respectively, let φ be a one-to-one correspondence between them. B_1 and B_2 can be extended to two bases of V, and φ to an invertible linear transformation of V taking W_1 onto W_2. The stabilizer G_v of a vector $v \neq 0$ is transitive on the set of vectors that are not multiples of v. Indeed, if u and w do not belong to the subspace generated by v, then v, u and v, w are two pairs of independent vectors that can be extended to two bases of V. A one-to-one correspondence between these two bases taking v to v and u to w extends to an invertible linear transformation of V fixing v, which therefore belongs to G_v and takes u to w.

Let us now show how the representation of *Ex.* 3.1, 1, can be used to prove the existence of subgroups of a given order in certain groups.

Corollary 3.2. *Let G be a group of order $2m$, m odd. Then there exists in G a subgroup of order m. In particular, G is not simple.*

Proof. In the regular representation, the image in S^{2m} of an element $x \in G$ of order 2, which exists by Cauchy, is a permutation that is a product of transpositions:

$$(x_1, x_1x)(x_2, x_2x) \cdots (x_k, x_kx).$$

All the elements of G appear in the above cycles, so $2k = 2m$, and $k = m$, an odd number. Hence the image $\overline{G}$ of G in S^{2m} contains an odd permutation, and therefore is not contained in A^{2m}. But

$$2 = |S^{2m}/A^{2m}| = |\overline{G}A^{2m}/A^{2m}| = |\overline{G}/\overline{G} \cap A^{2m}|, \tag{3.3}$$

so that $\overline{G} \cap A^{2m}$ is a subgroup of $\overline{G}$ of order m. Since $G \simeq \overline{G}$, such a subgroup exists in G, and having index 2 is normal. ◇

Two isomorphic subgroups of a group G having the same index become conjugate in the group S^G.

Theorem 3.3. *Let H and K be two isomorphic subgroups of a group G, $\varphi: H \to K$ an isomorphism. Then there exists $\tau \in S^G$ such that:*

$$\varphi'(\sigma(h)) = \tau^{-1}\sigma(h)\tau, \ h \in H,$$

where $\sigma(h)$ is the image of h in the regular representation, and φ' the isomorphism between $\sigma(H)$ and $\sigma(K)$ induced by φ.

Proof. Let $T = \{x_i\}, T' = \{y_i\}$ be two transversals of the left cosets of H and K, respectively. If $x \in G$, then $x = x_ih = y_jk$, for certain $h \in H$ and $k \in K$; define

$$\tau(x) = \tau(x_ih) = y_j\varphi(h).$$

It is immediate that τ is a permutation of the elements of G, and therefore $\tau \in S^G$. Clearly, $\sigma(\varphi(h)) = \varphi'(\sigma(h))$ for all $h \in H$. Moreover, for $h' \in H$,

$$\begin{aligned}(y_i\varphi(h))^{\tau^{-1}\sigma(h')\tau} &= x^{\sigma(h')\tau} = (x_i hh')^{\tau} = y_j\varphi(hh') = y_j\varphi(h)\varphi(h') \\ &= (y_i\varphi(h))^{\sigma(\varphi(h'))} = (y_i\varphi(h))^{\varphi'(\sigma(h'))}.\end{aligned}$$

For all $h' \in H$ we have $\varphi'(\sigma(h')) = \tau^{-1}\sigma(h')\tau$, as required. $\diamondsuit$

The last result admits a converse: if two subgroups H and K of a group G are conjugate in G, $K = H^g$, their images in S^G are also conjugate, $\sigma(K) = \sigma(H)^{\sigma(g)}$, and therefore these images have the same index (Theorem 2.21).

Examples 3.2. 1. *Ex.* 3.1, 1 can be generalized (without transitivity) considering subsets instead of elements. Let Ω be the family of nonempty subsets of the group G, $\Omega = \{\emptyset \neq \alpha \mid \alpha \subseteq G\}$, and let G act on Ω by right multiplication, $\alpha^g = \alpha g$ (if $\alpha = \{x, y, \ldots\}$, then $\alpha g = \{xg, yg, \ldots\}$). As in the quoted example, the fact that it is actually an action follows from the associative law of G.

If $\alpha = H$ is a subgroup, then $Hg = H$ if, and only if, $g \in H$, and therefore H coincides with its stabilizer; the orbit of H is the set of its right cosets. If G is finite, (3.2) becomes the well known equality

$$|G| = |H|[G : H].$$

Conversely, if α coincides with its stabilizer, $\alpha = G_\alpha$, then α is a subgroup (stabilizers are subgroups). Hence, the subgroups of a group may be characterized as the subsets which, in the above action, coincide with their stabilizers.

2. Let Ω be the set underlying G, and let G act on Ω by conjugation, $x^a = a^{-1}xa$. Since

$$x^{ab} = (ab)^{-1}x(ab) = b^{-1}(a^{-1}xa)b = b^{-1}x^a b = (x^a)^b,$$

conjugation is an action (as a left action, conjugation must be defined as $x^a = axa^{-1}$). The orbit of an element x of G is its conjugacy class, and the stabilizer is its centralizer. Then (3.1) becomes equality (2.12). Moreover, the kernel of this action is the center of G.

3. Now let Ω be the family of all subgroups of G, the action being again conjugation, $H^x = x^{-1}Hx$. The orbit of a subgroup H is its conjugacy class, and the stabilizer its normalizer. We recover the equality of Theorem 2.22.

4. Let H be a subgroup of G, and let Ω be the set of right cosets of H, $\Omega = \{Hx,\ x \in G\}$. G acts on Ω as follows: $(Hx)^g = Hxg$. (Note that *Ex.* 3.1, 1 is a special case of this: take $H = \{1\}$ and identify the elements of G with the cosets of $\{1\}$.) Given two cosets Hx and Hy, consider the element $g = x^{-1}y$; then $(Hx)^g = Hy$, i.e. any two right cosets of H belong to the same orbit. Thus, there is only one orbit, and the group is transitive. The stabilizer of a coset Hx is

$$G_{Hx} = \{g \in G \,|Hxg = Hx\};$$

but

$$Hxg = Hx \Leftrightarrow Hxgx^{-1} = H \Leftrightarrow xgx^{-1} \in H \Leftrightarrow g \in x^{-1}Hx,$$

i.e.

$$G_{Hx} = H^x.$$

In words, the stabilizer of a coset of H is the conjugate of H by a representative of that coset (changing representative: $Hy = Hx$, with $y = hx$, so that $H^y = H^{hx} = H^x$). In particular, the stabilizer G_H of H is H itself. Hence the kernel of the action is

$$K = \bigcap_{x \in G} H^x,$$

i.e. the intersection of all conjugates of H. In this sense, K is the largest normal subgroup of G contained in H. In particular, the action on the cosets of a subgroup $H \neq G$ is never trivial. Indeed, the kernel being contained in H it can never be the whole group G.

5. Using the action on the cosets, it can be shown that *if a simple group G has a subgroup of index n it embeds in A^n*. Indeed, G being simple, the kernel of the homomorphism $G \to S^n$ induced by the action of G on the cosets of the subgroup of index n is necessarily $\{1\}$, so that G embeds in S^n, and if the image is not contained in A^n then it has index 2 (see (3.3)), and G would not be simple.

Since the kernel of the action on the cosets of a subgroup is never the whole group, this action can be used to produce proper normal subgroups, in particular when one wants to prove that certain groups cannot be simple. An example is given in the following theorem. More examples in Section 3.2.

Theorem 3.4. *In a finite group G, a subgroup H of index p, where p is the smallest divisor of the order of the group, is normal.*

Proof. (Note that p is a prime.) Let G act on the cosets of H; we show that H is the kernel of the action. We have a homomorphism $G \to S^p$. If K is the kernel, then G/K is isomorphic to a subgroup of S^p, so that $|G/K|$ divides S^p. Hence, a prime q dividing $|G/K|$ also divides $p! = p(p-1)\cdots 2 \cdot 1$, and therefore $q \leq p$. But q, a divisor of $|G/K|$, also divides $|G|$; p being the smallest such, we have $q \geq p$. Hence $q = p$, and G/K is a p-group. Since $|G/K|$ divides $p!$, and p^2 does not, p^2 does not divide $|G/K|$, so that G/K has order p. Hence K has index p like H, and therefore the same order as H, and from $K \subseteq H$ it follows $H = K$. ◊

The two following facts, apparently unrelated, are both consequences of this theorem.

Corollary 3.3. *i) In a finite group, a subgroup of index* 2 *is normal (this also holds in an infinite group);*

ii) in a finite p-group, a subgroup of index p is normal.

Proof. In both cases, the index of the the subgroup is the smallest divisor of the order of the group. ◇

Let us now consider an application to the case of an infinite group.

Theorem 3.5. *Let G be a f.g. group. Then:*

i) for all n, the number of subgroups of G of index n is finite;

ii) if G has a subgroup of finite index, then it also has a characteristic subgroup of finite index.

Proof. i) The homomorphisms of a f.g. group G in a finite group are finite in number because they are determined once the images of the generators are assigned, and these images belong to a finite set. Hence, there are only a finite numbers of kernels. Now, a subgroup H of G of index n gives rise to a homomorphism of G in S^n whose kernel K is contained in H; the image of H is H/K. The subgroup K may also be the kernel of the action on the cosets of other subgroups of G of index n, but only of a finite number of them because the number of possible images is finite (these are subgroups of S^n, and if $H_1/K = H_2/K$ then $H_1 = H_2$). The number of K's being finite, we conclude that the number of subgroups of index n is also finite.

ii) Let $H \leq G$, $[G : H] = n$. The image of H under an automorphism ϕ of G has the same index n of H. By *i)*, as ϕ varies in the automorphism group of G we have only a finite number of these images. Therefore, their intersection has finite index, and is the required characteristic subgroup. ◇

In the case *ii)* of the theorem we know that there always exists a normal subgroup of finite index (Theorem 2.25) without the hypothesis of finite generation.

We have seen that the action on the cosets of a subgroup is transitive; we will see in a moment that this is essentially the only way in which a group can act transitively.

Definition 3.5. The actions of two groups G and G_1 on two sets Ω and Ω_1 are *similar* if there exists a one-to-one correspondence $\varphi : \Omega \to \Omega_1$ and an isomorphism $\theta : G \to G_1$ such that:

$$\varphi(\alpha^g) = \varphi(\alpha)^{\theta(g)}, \tag{3.4}$$

for every $\alpha \in \Omega$ and every $g \in G$. It is clear that the relation of similarity is an equivalence relation. For this reason, one also speaks of *equivalent* actions.

If $G = G_1$, θ will be an automorphism of G; if θ is the identity automorphism, (3.4) can be expressed by saying that φ *commutes with the action of G* or that φ *preserves the action of G.*

Examples 3.3. 1. The actions $a^x = ax$ and $a^x = x^{-1}a$, which give the right and left regular representations, are similar. Indeed, by taking as φ the mapping $a \to a^{-1}$, we have:

$$\begin{aligned} \varphi(a^x) &= \varphi(ax) &= x^{-1}a^{-1}, \\ \varphi(a)^x &= (a^{-1})^x &= x^{-1}a^{-1}. \end{aligned}$$

2. In case G and G_1 are two permutation groups of the same set Ω, i.e. $G, G_1 \subseteq S^{\Omega}$, then $\varphi \in S^{\Omega}$ and similarity becomes conjugation in S^{Ω}: (3.4) expresses the fact that G_1 is the conjugate of G by φ, i.e. that θ is the conjugation by φ. Indeed, write (3.4) as $(\alpha^g)^{\varphi} = (\alpha^{\varphi})^{\theta(g)}$, i.e. as $\alpha^{g\varphi} = \alpha^{\varphi\theta(g)}$. This equality holding for all α we have $g\varphi = \varphi\theta(g)$, i.e. $\theta(g) = \varphi^{-1}g\varphi$.

3. If $G = \{S_i\}$ and $G_1 = \{T_i\}$ are two groups of linear transformations (matrices) of a vector space V, the actions of the two groups are similar if there exist an isomorphism $\theta : S_i \to T_i$ and a bijection $\varphi : V \to V$ such that $vS_i\varphi = v\varphi T_i$, for all $v \in V$.

Theorem 3.6. *The actions of a group on the cosets of two subgroups are similar if, and only if, the two subgroups are conjugate.*

Proof. Let H and H^x be the two subgroups. The mapping:

$$\varphi : H^x y \to Hxy,$$

is well defined and bijective (see the proof of Theorem 2.21), and commutes with the action of G:

$$\begin{aligned} \varphi((H^x y)^g) &= \varphi(H^x yg) = Hx(yg), \\ (\varphi(H^x y))^g &= (Hxy)^g \quad = H(xy)g. \end{aligned}$$

Conversely, let H and K be two subgroups of a group G, and let the actions of G on the cosets of H and K be similar. The image under φ of the coset H is a coset of K, Ky say. For all $h \in H$, we have $\varphi(H) = \varphi(Hh) = \varphi(H)^h$, where the second equality follows from the fact that φ commutes with the action of G. Hence $Ky = (Ky)^h = Kyh$, so that $yhy^{-1} \in K$ for all $h \in H$, and therefore $H \subseteq K^y$. On the other hand, from $\varphi(H) = Ky$ it follows $\varphi(Hy^{-1}) = \varphi(H)^{y^{-1}} = (Ky)^{y^{-1}} = Kyy^{-1} = K$, so that $\varphi(Hy^{-1}k) = \varphi(Hy^{-1})^k = Kk = K = \varphi(Hy^{-1})$, for all $k \in K$; φ being injective, $Hy^{-1}k = Hy^{-1}$, and $K^y \subseteq H$. Thus $H = K^y$. ♢

Theorem 3.7. *Let G be a group acting transitively on a set Ω, and let $H = G_\alpha$ be the stabilizer of a point $\alpha \in \Omega$. Then the action of G on Ω is similar to that of G on the cosets of H seen in Ex.* 3.2, 4.

Proof. Let $\Omega_1 = \{Hx,\ x \in G\}$. Fix $\alpha \in \Omega$; by transitivity, every point of Ω has the form α^x, $x \in G$. Define $\varphi : \Omega \to \Omega_1$ by $\varphi : \alpha^x \to Hx$; φ is well defined and injective:

$$\alpha^x = \alpha^y \Leftrightarrow xy^{-1} \in G_\alpha = H \Leftrightarrow Hx = Hy,$$

and is also surjective, because α^x runs over all the points of Ω as x varies in G. Moreover, φ commutes with the action of G: if $\beta \in \Omega$, then $\beta = \alpha^x$ for some x, and we have $\varphi(\beta^g) = \varphi(\alpha^{xg}) = Hxg = (Hx)^g = \varphi(\alpha^x)^g = \varphi(\beta)^g$. ◇

In other words, given a transitive group and a fixed $\alpha \in \Omega$, we may identify α with G_α, i.e. with the set of elements of G taking α to α, and $\beta \in \Omega$ with the set of elements of G taking α to β, i.e. with the coset $G_\alpha x$, where $\alpha^x = \beta$ (cf. *Ex.* 3 of 1.46). Moreover, by transitivity, the stabilizers of any two elements are conjugate, and therefore, by Theorem 3.6, the actions of G on their cosets are similar.

From the previous two theorems we have:

Corollary 3.4. *Up to similarities, the number of transitive actions of a group G equals the number of conjugacy classes of subgroups of G (including $\{1\}$ and G).*

Remark 3.1. The actions on $\{1\}$ and G are similar to the regular action and to the trivial one on a singleton, respectively.

We close this section with two theorems: the first shows the relation existing between transitivity and normality, the second what is implied by a subgroup being already transitive.

Theorem 3.8.[4] *Let G be transitive on Ω, $|\Omega| = n$, and let N be a normal subgroup of G. Then G acts on the set of orbits of N, and this action is transitive. In particular, the orbits of N all have the same cardinality n/k, where k is their number. Moreover, the actions of N on these orbits are all similar. If n is prime, N is transitive (or trivial on Ω).*

Proof. We have to show that if Δ is an orbit of N, then for all $g \in G$, $\Delta^g = \{\alpha^g,\ \alpha \in \Delta\}$ is again an orbit of N, and that, if Δ_1 and Δ_2 are two orbits of N, there exists $g \in G$ such that $\Delta_1^g = \Delta_2$. Let $x \in N$; then $(\Delta^g)^x = \Delta^{gx} = \Delta^{x'g} = \Delta^g$, where $x' \in N$, and therefore Δ^g is an orbit of N. Then let $\alpha \in \Delta_1$ and $\beta \in \Delta_2$. By the transitivity of G on Ω there exists $g \in G$ such that $\alpha^g = \beta$, so that $\Delta_1^g \cap \Delta_2 \neq \emptyset$. Since we are dealing with orbits, this implies $\Delta_1^g = \Delta_2$.

As to the similarity of the action of N on its orbits, let Δ_1 and $\Delta_2 = \Delta_1^g$ be two of these orbits, φ the mapping $\Delta_1 \to \Delta_2$ given by $\varphi : \alpha \to \alpha^g$, and θ the automorphism of N induced by the conjugacy determined by g: $\theta(y) = g^{-1}yg$, $y \in N$. With these φ and θ we have the similitude we seek:

$$\varphi(\alpha^y) = \alpha^{yg} = (\alpha^g)^{g^{-1}yg} = \varphi(\alpha)^{\theta(y)}.$$

If n is prime, then $k = 1$ or $k = n$, and therefore N has one or n orbits. ◇

[4] This is the permutation group version of a theorem of Clifford in representation theory.

Theorem 3.9. *Let $H \leq G$, H transitive, and let $\alpha \in \Omega$. Then $G = HG_\alpha$.*

Proof. Let $g \in G$; given α^g and α, there exists $h \in H$ such that $\alpha^{gh} = \alpha$. Then $gh \in G_\alpha$, $g = gh \cdot h^{-1} \in G_\alpha H$, so $G = G_\alpha H = HG_\alpha$. ◊

As an illustration of this theorem, consider the group A^4, with $H = V$ the Klein group, and $G_\alpha = C_3$ the stabilizer of a point in $\Omega = \{1, 2, 3, 4\}$. Then $A^4 = HG_\alpha$.

Exercises

The actions of a group G on his subsets and on the cosets of a subgroup are those seen in the examples given in the text proper.

1. Consider the action of G on its subsets. Show that $|G_\alpha| \leq |\alpha|$, for all $\alpha \in \Omega$.

2. Give a new proof of the equality:

$$|HK| = \frac{|H||K|}{|H \cap K|}$$

for G finite and $H, K \leq G$, by letting K act on the subsets of G and considering the orbit and the stabilizer of H.

3. *i*) Let $H, K \leq G$. Show that the number of cosets Hx contained in HK equals the index of $H \cap K$ in K. [*Hint*: let K act on the subsets of G.]

ii) If G is finite, show that

$$[\langle H \cup K \rangle : H] \geq [K : H \cap K],$$

and that equality holds if, and only if, $\langle H \cup K \rangle = HK$.

4. If G is finite, $x \in G$ and $H \leq G$, show that

$$|\mathbf{C}_G(x)| \leq [G : H]|\mathbf{C}_H(x)|$$

and that equality holds if, and only if, $\mathbf{C}_G(x)H = G$. ($\mathbf{C}_H(x) = \mathbf{C}_G(x) \cap H$, the subgroup of the elements of H commuting with x.) [*Hint*: see *ex.* 3.]

5. Show that there are only two groups of order 6 by letting the group act on the cosets of a subgroup of order 2.

6. Prove Poincaré's Theorem 2.25 by letting the group act on the cosets of the subgroup of finite index.

7. For $H, K \leq G$, by a *double coset* of H and K we mean a subset of the form HaK, $a \in G$. Show that:

i) G is a disjoint union of double cosets, and therefore these cosets are the classes of an equivalence relation. Which one?

Assume now G finite.

ii) The number of right cosets of H contained in HaK equals the index of $H^a \cap K$ in K. Conclude that the number of elements of G contained in HaK equals $|H|[K : H^a \cap K]$, i.e.:

$$|HaK| = \frac{|H||K|}{[H^a \cap K]}.$$

[*Hint*: let K act on the cosets of H; the case $a = 1$ is in *ex.* 2.]

iii) If Ha_iK, $i = 1, 2, \ldots, m$, are the distinct double cosets of H and K, we have (Frobenius):

$$|G| = \sum_{i=1}^{m} \frac{|H||K|}{|H^{a_i} \cap K|};$$

it follows that the index of H in G is $[G : H] = \sum_{i=1}^{m}[K : H^{a_i} \cap K]$.

iv) If $k \in K$, then $[K : H^x \cap K] = [K : H^{xk} \cap K]$.

v) If $x_1, x_2, \ldots, x_m$ are representatives of the right cosets of H, then the number of conjugates H^{x_i} of H for which $[K : H^{x_i} \cap K] = t$ is a multiple of t.
[*Hint*: the number of cosets of the form Hxk is $t = [K : H^x \cap K]$, because $Hxk = Hxk_1 \Leftrightarrow Hxkk_1^{-1}x^{-1} = H \Leftrightarrow kk_1 \in H^x$. Thus the cosets $Hx, Hxk_2, \ldots, Hxk_t$ are all distinct.]

vi) Double cosets may have different cardinalities that do not divide the order of the group.

vii) What is the number of left cosets of K contained in HaK?

8. Prove that if the actions of two groups G and G_1 are similar (Definition 3.5), and Δ is an orbit of G on Ω, then $\varphi(\Delta)$ is an orbit of G_1 on Ω_1. Hence two similar actions have the same number of orbits of the same length, and in particular they are either both transitive or both intransitive.

9. Give an example of two non similar actions of the Klein group on a set of four elements.

10. The action of G on the left cosets of a subgroup H defined by $(aH)^g = (g^{-1}a)H$ is similar to that seen on the right cosets.

11. If $H \leq G$, the action of G by conjugation on the set of conjugates of H is similar to that of G on the cosets of $\mathbf{N}_G(H)$.

12. Let $\Gamma = \{\alpha_1, \alpha_2, \ldots, \alpha_t\}$ be an orbit of the action of G on its subsets. Show that:

i) $G = \bigcup_{i=1}^{t} \alpha_i$;

ii) if the above union is disjoint, then one of the α_i is a subgroup and the other ones are its right cosets.

13. *i)* Let $H \leq G$, and let G act on Ω. Show that the normalizer of H acts on the set Δ of points fixed by elements of H.

ii) Let $H \leq G$ and let α be the unique point of Ω fixed by every element of H. Show that the normalizer of H is contained in G_α.

14. The automorphisms of a group G fixing all the conjugacy classes of G form a normal subgroup of $\mathbf{Aut}(G)$. (This subgroup contains $\mathbf{I}(G)$, but in general it does not coincide with it; see footnote 6 of Chapter 2.)

15. If $H \trianglelefteq G$, then G/H acts on H/H'.

16. In a transitive group of prime degree the subgroups of a normal series are transitive. [*Hint*: Theorem 3.8.]

3.2 The Sylow Theorem

If m is an integer dividing the order of a group G, we know that there does not necessarily exist a subgroup of order m in G; that is, the converse of Lagrange's theorem does not hold in general (the alternating group A^4 providing the smallest counterexample). If $m = p$, a prime, Cauchy's theorem ensures the existence of subgroups of order p. Sylow's theorem, that we now prove, says that, if $|G| = p_1^{h_1} p_2^{h_2} \cdots p_t^{h_t}$ is the prime factorization of the order of a group G, then there exist in G at least one subgroup of order $p_1^{h_1}$, one of order $p_2^{h_2}$, ..., one of order $p_t^{h_t}$. The main ingredient of the proof will be the action of a group on a set whose cardinality is coprime to the order of the group (*coprime* action).

Lemma 3.1. *Let P be a finite p-group acting on a set Ω such that $p \nmid |\Omega|$. Then there exists a point of Ω fixed by every element of P.*

Proof. The orbits of P on Ω have cardinality a power of p. However, not all can have cardinality p^k with $k > 0$, otherwise $|\Omega|$ would be divisible by p. Thus, for at least one orbit, we have $k = 0$; this orbit has only one point, which is obviously fixed by every element of P. ◊

We already know that a finite p-group has a non trivial center. This result can be proved, together with a generalization, using the above lemma.

Corollary 3.5. *Let P be a finite p-group, and $H \neq \{1\}$ a normal subgroup of P. Then $H \cap Z(P) \neq 1$. In particular, the center of P is non trivial.*

Proof. Take as Ω the set $H \setminus \{1\}$ and let P act on Ω by conjugation (it is an action because the image of an element of Ω still belongs to Ω). If H has order p^h, Ω has order $p^h - 1$, and therefore p does not divide $|\Omega|$. By Lemma 3.1, there exists $x \in \Omega$ fixed by every element of $g \in P$, $x^g = x$. The action being conjugation, this means $g^{-1}xg = x$, i.e. x commutes with every element of P. Hence $x \in Z(P)$, and $x \neq 1$ because $x \in \Omega$. ◊

Corollary 3.6. *Let P be a finite p-group, H a proper subgroup of P. Then H is properly contained in its normalizer, $H < P \Rightarrow H < N_P(H)$.*

Proof. If $H \triangleleft P$ there is nothing to prove. Assume H is not normal, and let $\Omega = \{H^a, H^b, \ldots\}$ be the set of conjugates of H other than H. The subgroup H acts on Ω by conjugation (conjugation by elements of H keeps an element of Ω again in Ω, because if $H^g \in \Omega$, and $(H^g)^h \notin \Omega$, then $(H^g)^h = H$,

$gh \in N_P(H)$, $g \in N_P(H)$ and $H^g = H \notin \Omega$). The conjugacy class of H has cardinality p^k, some k ($k > 0$ since H is not normal), and therefore Ω has cardinality $p^k - 1$, a number not divisible by p. By Lemma 3.1 there exists an element H^g of Ω fixed by every element of H: $(H^g)^h = H^g$, from which $gHg^{-1} \subseteq N_P(H)$. Since it cannot be $gHg^{-1} \subseteq H$, otherwise $H^g = H$, there exists h such that $ghg^{-1} \in N_P(H)$ and $ghg^{-1} \notin H$, as required. ◇

Corollary 3.7. *In a finite p-group of order p^n a maximal subgroup is normal, and therefore is of index p.*

Definition 3.6. Let p be a prime, and let p^n be the highest power of p dividing the order of a finite group G. A subgroup S of order p^n is called a *Sylow p-subgroup of G*, or simply a *p-Sylow*. The set of Sylow p-subgroups of G is denoted by $Syl_p(G)$, and their number by $n_p(G)$, or simply n_p.

Theorem 3.10 (Sylow). *Let p be a prime dividing the order of a finite group G. Then:*

i) there exists in G at least one p-Sylow S;

ii) if P is a p-subgroup of G and S is a p-Sylow, then P is contained in a conjugate of S. It follows:

  a) every p-subgroup, and in particular every p-element, is contained in some p-Sylow;

  b) all p-Sylows are conjugate; hence, if a p-Sylow is normal, it is the unique p-Sylow of G.

iii) a) $n_p \equiv 1 \bmod p$;

  b) $n_p = [G : \mathbf{N}_G(S)]$, where S is any p-Sylow. In particular, n_p divides the index of S[5].

Before proving the theorem, we establish a preliminary lemma of an arithmetic character.

Lemma 3.2. *If p is a prime and $p \nmid m$, then $p \nmid \binom{p^n m}{p^n}$.*

Proof. We have: $(p^n m)!/p^n!(p^n m - p^n)!$, i.e.

$$\binom{p^n m}{p^n} = \frac{p^n m(p^n m - 1)\dots(p^n m - i)\dots(p^n m - p^n + 1)}{p^n(p^n - 1)\dots(p^n - i)\dots(p^n - p^n + 1)}.$$

Since $i < p^n$, the power of p dividing $p^n m - i$ is the power of p dividing i, and the same holds for $p^n - i$. Therefore, the corresponding terms on the numerator and the denominator are divisible by the same power of p. The result follows. ◇

[5] The three parts of the theorem are sometimes considered as three distinct theorems.

We now come to the proof of the theorem.

Proof. *i)* Let p^n be the highest power of p dividing $|G|$, and let G act by right multiplication on the set of subsets of G of cardinality p^n:

$$\Omega = \{\alpha \subseteq G \mid |\alpha| = p^n\}.$$

We have $|G| = p^n m$, and therefore $|\Omega| = \binom{p^n m}{p^n}$, with $p \nmid m$. By Lemma 3.2, $p \nmid |\Omega|$. Hence there is an orbit of the action whose cardinality is not divisible by p. Let Δ be such an orbit, and let $\alpha \in \Delta$; then $|G| = |G_\alpha||\Delta|$, and therefore $p^n||G_\alpha|$. On the other hand, if $x \in \alpha$ and $g \in G_\alpha$ then $xg \in \alpha$, and therefore $xG_\alpha \subseteq \alpha$. Thus,

$$|G_\alpha| = |xG_\alpha| \leq |\alpha| = p^n,$$

and hence $|G_\alpha| = p^n$. We have found a p-Sylow as the stabilizer of a subset belonging to an orbit of cardinality not divisible by p.

ii) P acts by multiplication on the set of right cosets of S, a set of cardinality $[G : S]$ and hence not divisible by p. Then there exists Sg fixed by every element of P: $Sgx = Sg$, for all $x \in P$, from which $gxg^{-1} \in S$, $x \in S^g$ and hence $P \subseteq S^g$:

a) S^g has the same order as S, and therefore is also a p-Sylow. Moreover, if x is a p-element, the subgroup P generated by x is contained in a p-Sylow;

b) if P is also Sylow, then $P = S^g$ because they have the same order. Any two p-Sylows are therefore conjugate, so that the p-Sylows make up a unique conjugacy class. It follows that if a p-Sylow is normal, then it is the unique p-Sylow of the group. In particular, S is the unique p-Sylow of its normalizer. If P is a normal p-subgroup of G, then $P \subseteq S$ implies $P^g = P \subseteq S^g$, all $g \in G$, and hence P is contained in all the p-Sylows of the group.

iii) *a)* Let $S_0, S_1, \ldots, S_r$, $r = n_p - 1$, be all the p-Sylows of G. As in the proof of Corollary 3.6 (with S_0 playing the role of H) let S_0 act by conjugation on the set $\{S_1, S_2, \ldots, S_r\}$. If $g \in S_0$ stabilizes S_i, then $S_i^g = S_i$ and $g \in \mathbf{N}_G(S_i)$. But since S_i is the unique p-Sylow of $\mathbf{N}_G(S_i)$ and g is a p-element, by *ii, a)* $g \in S_i$. This shows that the stabilizer of S_i in S_0 is $S_0 \cap S_i$, so that the orbit of S_i has cardinality $[S_0 : S_0 \cap S_i]$. No orbit can have cardinality 1: this would imply $S_0 \cap S_i = S_0$, i.e. $S_0 = S_i$. It follows $r = kp$. Adding S_0, we have $n_p = 1 + kp$. *b)* follows from the fact that the p-Sylows are all conjugate, and therefore the number of p-Sylows equals the index of the normalizer of any of them. Moreover, since $\mathbf{N}_G(S) \supseteq S$, we also have that n_p divides the index of S. ♢

The action of G by conjugation on the set of its p-Sylows (or on the cosets of the normalizer of one of them) yields a homomorphism of G in S^{n_p}. This remark will be useful to prove that groups of a certain order cannot be simple.

Corollary 3.8. *If p^h, p a prime, divides the order of a finite group G, then there exists in G a subgroup of order p^h. In particular, for $h = 1$ we have Cauchy's theorem.*

Proof. If $S \in Syl_p(G)$, with $|S| = p^n$, then $h \leq n$ and therefore $p^h | p^n$. Thus there exists in S, and so in G, a subgroup of order p^h (for p-groups, Lagrange's theorem has an inverse; see *Ex.* 2.3, 2). ◊

Corollary 3.9. *Let G be a finite abelian group. Then:*

i) *G has only one p-Sylow for all p dividing $|G|$;*

ii) *G is the direct product of its p-Sylows, for the various primes p dividing $|G|$;*

iii) *G is cyclic if, and only if, its Sylow subgroups are cyclic.*

Proof. i) Like all the subgroups of G, a p-Sylow is normal, and therefore, by *ii, b)* of Sylow's theorem, unique.

ii) The Sylow subgroups being normal, their product $S_1 S_2 \cdots S_i \cdots S_t$ is a subgroup. If $1 \neq x = x_1 x_2 \cdots x_{i-1} x_{i+1} \cdots x_t$, $x_k \in S_k$, then $o(x) = \text{lcm}(o(x_k))$, and therefore $x \notin S_i$. Then S_i has a trivial intersection with the product of the other Sylow subgroups. This shows that the product of the S_i is direct.

iii) If G is cyclic, every subgroup is cyclic. Conversely, take a generator in each p-Sylow for the various primes p dividing $|G|$. The product of these generators is an element of order $|G|$. ◊

Note that, even if the group is not abelian, *i)* and *ii)* of this corollary are equivalent; the proof is the same as that given above. If these conditions hold, the group is said to be nilpotent (we shall be considering nilpotent groups in Chapter 5).

Here is another proof of Sylow's theorem, based on Cauchy's theorem. Let $\mathcal{S}$ be the set of the maximal p-subgroups of G, and let G act on $\mathcal{S}$ by conjugation. In the action of a subgroup S of $\mathcal{S}$ no element P of $\mathcal{S}$ is left fixed, otherwise S normalizes P and SP would be a subgroup properly containing S, against the maximality of S. Hence S itself is the only fixed point of the action of S. Let Ω be the orbit of G to which S belongs. Since S only fixes itself in this orbit, and since the other orbits of S in Ω have cardinality a non trivial power of p, we have $|\Omega| \equiv 1 \bmod p$. Let $P \notin \Omega$. Since P only fixes P, it fixes no point in its action on Ω; it follows $|\Omega| \equiv 0 \bmod p$, a contradiction. Hence there are no points outside Ω, that is $\Omega = \mathcal{S}$, and G is transitive on $\mathcal{S}$, so $\mathcal{S} = [G : \mathbf{N}_G(S)]$, and this index is congruent to 1 mod p. If p divides $[\mathbf{N}_G(S) : S]$, by Cauchy's theorem $\mathbf{N}_G(S)/S$ has a subgroup H/S of order p, so H contains S and has order $p|S|$, against the maximality of S. Since p does not divide either $[G : \mathbf{N}_G(S)]$ or $[\mathbf{N}_G(S) : S]$, it does not divide their product $[G : S]$. Hence S is Sylow. This proves existence; the above discussion proves *i)*, *ii)* and *iii)* of the theorem (part *ii)a)* follows because every p-subgroup is contained in a maximal p-subgroup). For other proofs of various parts of the theorem see *ex.* 22, 59 and 60 below.

Examples 3.4. **1.** Consider the upper unitriangular $n \times n$ matrices with coefficients in $\mathbf{F}_q$, $q = p^f$, i.e. the subgroup of $GL(n,q)$ of matrices ($*$ denotes any elements of $\mathbf{F}_q$):

$$\begin{pmatrix} 1 & & & \\ & 1 & & * \\ & & \ddots & \\ 0 & & & 1 \end{pmatrix}.$$

There are q^{n-1} choices for the elements of the first row, q^{n-2} for those of the second,..., q for those of the second to the last, and only one for the last. This gives a total of

$$q^{n-1} \cdot q^{n-2} \cdots q \cdot 1 = q^{\frac{n(n-1)}{2}}$$

matrices. The group $GL(n,q)$ has

$$(q^n - 1)(q^n - q) \cdots (q^n - q^{n-1}) =$$
$$q^{\frac{n(n-1)}{2}}(q-1)(q^2-1)\cdots(q^n-1) = q^{\frac{n(n-1)}{2}} r$$

elements, where $p \nmid r$ (*Ex.* 1.3, 2). Then $p^{\frac{fn(n-1)}{2}}$ is the highest power of p dividing $|GL(n,q)|$, so that the upper unitriangular matrices form a Sylow p-subgroup of $GL(n,q)$. These matrices have determinant 1, and therefore belong to $SL(n,q)$), and this holds for the elements of any other p-Sylow because the p-Sylows are all conjugate, and conjugation does not alter the value of the determinant. (This can also be seen directly by observing that if A is a p-element, then 1=det(I)=det(A^{p^k})=det$(A)^{p^k}$, where I is the identity matrix; the multiplicative group of $\mathbf{F}_q$ being of order $q-1$, and therefore not divisible by p, we have $\det(A) = 1$.)

The existence of a Sylow p-subgroup in $GL(n,q)$ can be used to prove the existence of a Sylow p-subgroup in any finite group. Let G be any finite group. By Cayley's theorem, G is isomorphic to a subgroup of S^n, $n = |G|$, and S^n is isomorphic to the subgroup of $GL(n,q)$ given by the permutation matrices. Hence G is isomorphic to a subgroup of $GL(n,q)$. Let G act by multiplication on the set Ω of right cosets of the Sylow p-subgroup S of $GL(n,q)$. Since $p \nmid |\Omega|$, there is an orbit of the action of G whose order is not divisible by p. If H is the stabilizer of a point Sg in this orbit, then $Sgh = Sg$, for all $h \in H$, i.e. $H \subseteq S^g$, so that H is a p-group, and since the index of H in G is $|\Omega|$ which is not divisible by p, H is a Sylow p-subgroup of G.

2. The dihedral group D_4 induces a permutation of the four vertices and therefore embeds in S^4. We know (*Ex.* 1.7, 6) that there are essentially three ways of numbering the vertices of a square, each of which corresponds to a dihedral group. These three subgroups have in common the Klein group $\{I, (1,2)(3,4), (1,3)(2,4), (1,4)(2,3)\}$, and since $|S^4| = 24 = 2^3 \cdot 3$, and $n_2 = 3$, they are the Sylow 2-subgroups of S^4. The eight elements of order 3 of S^4 are

3-cycles and distribute in pairs in four subgroups of order 3. Thus $n_3 = 4$, and these also are the 3-Sylows of A^4 because 3-cycles are even permutations. The only 2-elements of S^4 that are even permutations are those of the Klein group seen above; therefore this is the unique 2-Sylow of A^4.

The following examples illustrate the use of Sylow theorem to study groups for which the order has a special decomposition into prime factors, and in particular to show that such groups are not simple.

3. *A group G of order the product of two primes pq, $p < q$, $p \nmid (q-1)$, is cyclic.* We have $n_q | p$, and therefore $n_q = 1$ or p. But $n_q \equiv 1 \bmod q$, and if $n_q = p$ then $q|(p-1)$, contrary to $p < q$. Then $n_q = 1$. As for n_p, we have $n_p | q$, $n_p = 1, q$. If $n_p = q$, then $n_p \equiv 1 \bmod p$ would imply $p|(q-1)$, contrary to $p \nmid (q-1)$. Hence G has only one p- and only one q-Sylow subgroups, that therefore are normal, and having trivial intersection they commute elementwise. The product of an element of order p by one of order q has order pq and generates the group. Note that the assumption $p \nmid (q-1)$ is necessary as the example of the group S^3 shows[6].

If $p|(q-1)$, besides the cyclic group there exists one more group, the semidirect product of C_p by C_q (we have $C_p \leq \mathbf{Aut}(C_q)$). This is a *Frobenius group* (see Definition 5.12).

4. *A group G of order pqr, p, q, r primes and $p < q < r$ has a normal Sylow subgroup.* If for no prime there is only one Sylow subgroup, then $n_r = p, q$ or pq. Now $n_r = p$ is impossible because $p \equiv 1 \bmod r$ means $r|(p-1)$, contrary to $r > p$. Similarly if $n_r = q$. Thus $n_r = pq$, and G has $pq(r-1)$ r-elements. The same argument shows that $n_p \geq q$ and $n_q \geq r$, so that G has at least $q(p-1)$ p-elements and $r(q-1)$ q-elements. It follows $|G| \geq pq(r-1)+r(q-1)+q(p-1)+1$, and being $r(q-1)+q(p-1)+1 > pq$ because $rq > q+r-1$, we have the contradiction $|G| > (r-1)pq + pq = pqr = |G|$.

5. *A group of order p^2q, p and q primes and $p < q$, has a normal p- or q-Sylow subgroup.* If $n_q \neq 1$, then $n_q = p$ or p^2. The first possibility is excluded because $q|(p-1)$ is impossible. As for the second, $q|(p^2-1) = (p-1)(p+1)$ implies $q|(p+1)$, i.e. $q = 3$ $p = 2$, and the group has order 12. From $n_3 = 4$ it follows that the group has eight 3-elements, and there is room left for only one subgroup of order 4.

6. *If a group G has a subgroup H of index* 2, 3 *or* 4, *then it is not simple* (*or else has order* 2 *or* 3). In the three cases, letting G act on the cosets of H we have a homomorphism $G \to S^i$, $i = 2, 3, 4$. If the kernel is not $\{1\}$, G is not simple; otherwise, G embeds in S^i, and therefore has order $2, 3, 2^2, 2 \cdot 3, 2^3, 2^2 3, 2^3 3$. Thus either $|G|$ has order a prime, the power of a prime, $2^2 3$, or $G \simeq S^3$ or $G \simeq S^4$: the latter groups have A^3 or A^4 as normal subgroups, respectively.

---

[6] More generally, a group of order n is cyclic if $(n, \phi(n)) = 1$ (cf. *ex.* 45 of Chapter 5).

7. *A group G of order* 30 *has normal* 3- *and* 5-*Sylow subgroups.* The order $2\cdot 3\cdot 5$ of G is of the form $2m$, with m odd. Therefore there exists a subgroup H of order $m=15=3\cdot 5$ (Corollary 3.2), and since $3 \nmid (5-1)$, H is cyclic and admits only one 3- and only one 5-Sylow subgroup, that are therefore characteristic. But $H \trianglelefteq G$ (index 2) then implies that these two Sylow subgroups are normal in the whole group G.

8. *A group of order* 120 *cannot be simple.* We have $120=2^3\cdot 3\cdot 5$, and therefore $n_5=1$ or 6. If $n_5=6$ and G is simple, then G embeds in A^6 (*Ex.* 3.2, 5); but then A^6, which is of order 360, would have a subgroup of index 3 and it would not be simple (*Ex.* 6 above).

9. *In a group of order* 36 *a* 2- *or a* 3-*Sylow is normal.* First, G is not simple because it has a subgroup of index 4 (*Ex.* 6 above), i.e a 3-Sylow S. If the kernel of $G \to S^4$ is the whole of S, S is normal. Otherwise the kernel is a C_3 (it cannot be $\{1\}$ because we would have $G \leq S^4$, which is absurd), and G has four 3-Sylows, which intersect in this C_3, and $6\cdot 4+2=26$ 3-elements. A group of order 9 being abelian, the four 3-Sylows centralize C_3, so that $\mathbf{C}_G(C_3)$ has order divisible by 9 and by 4, and therefore is the whole group G. In other words, $C_3 \subseteq \mathbf{Z}(G)$. If a 2-Sylow is a $C_4=\langle y\rangle$, the product of y by a generator x of C_3 yields an element of order 12. But a C_{12} has four elements of order 12 and two of order 6; together with the 26 3-elements we have a total of 32 elements, and there is room left for only one 2-Sylow. If a 2-Sylow is a Klein group, the three elements of order 2 multiplied by x and x^{-1} yield six elements of order 6, and again there is room left for only one 2-Sylow.

10. *A group of order* 1056 *is not simple.* We have $|G|=2^5\cdot 3\cdot 11$. If G is simple, then $n_{11}\neq 1$, $n_{11}=12$, and therefore $[G:\mathbf{N}_G(C_{11})]=12$, $\mathbf{N}_G(C_{11})=88=2^3\cdot 11$. But $\mathbf{N}_G(C_{11})/\mathbf{C}_G(C_{11})$ is isomorphic to a subgroup of $\mathbf{Aut}(C_{11})$ which has order 10. It follows that 2^2 divides $|\mathbf{C}_G(C_{11})|$, so that there is an element of order 2 that centralizes C_{11}, and this element, multiplied by a nonidentity element of C_{11}, gives an element of order 22. However, by assumption G is simple, and $n_{11}=12$ implies that it embeds in A^{12}, a group that does not admit elements of order 22 (such an element must have at least one 2-cycle and one 11-cycle, so that at least 13 digits are needed).

We know that the alternating group A^5 is simple. We now prove this using Sylow's theory. Actually, we will prove something more general, namely, that a group of order 60 in which the maximum order of an element is 5 is a simple group.

11. A^5 *is simple.* From the possible cycle structures we see that the maximum order of an element of A^5 is 5. The proper divisors of $|A^5|=60$ are 2,3,4,5,6,10,12,15,20 and 30. A^5 has no subgroups of order 15 or 30 because these groups have elements of order 15 (*Ex.* 2 and 7), nor has normal subgroups of order 3 or 5, because multiplication by an element of order 5 or 3 (that exist by Sylow or Cauchy) would yield a subgroup of order 15. For the same reason there are no normal subgroups of order 2 or 4: the quotients would be of order 30 or 15 and therefore would have elements of order 15, and the corresponding group elements would be of order a multiple of 15. A group H of order 10 has a unique 5-Sylow (index 2, therefore normal), and therefore characteristic in H; if $H \lhd G$ then $S \lhd G$, already excluded (or also because the product of H by a C_3 is a subgroup of order 30). Similarly, if $|H| = 20$, a 5-Sylow is characteristic in H. If $|H| = 6$, a 3-Sylow is characteristic, and if $|H| = 12$ the same happens for a 2- or a 3-Sylow (*Ex.* 5). (That A^5 is the unique simple group of order 60 will be shown in *Ex.* 3.5, 3.)

12. Consider again A^5. Since 4 divides 12, which is the order of A^4, a 2-Sylow of A^5 (which is a Klein group because there are no elements of order 4, an element of order 4 in S^5 being a 4-cycle and therefore odd) is contained in A^4, and is normal there. Then its normalizer in A^5 has order at least 12, and since it cannot be equal to A^5 because of the simplicity of A^5, exactly 12, and therefore is A^4. It follows that $n_2 = [A^5 : \mathbf{N}_{A^5}(V)] = 5$. A^5 acts by conjugation on the set of these 2-Sylows, and the action is transitive, and therefore similar to that of the group on the cosets of the stabilizer of a 2-Sylow, that is, on the cosets of its normalizer, which is A^4. As the set of five elements on which A^5 acts one can take the five 2-Sylows of A^5.

As for the Sylow 3-subgroups we have, again because A^5 is simple, that $n_3 = 10$. The index of the normalizer of a 3-Sylow is therefore 10, its order is 6, and since there are no elements of order 6, this normalizer is S^3. Finally, $n_5 = 6$, the normalizer of a C_5 has order 10, and since there are no elements of order 10, it is the dihedral group D_5 (an element of order 2 inverts a generator of C_5).

A^5 provides an example of the fact that the product of a set of Sylow subgroups, one for each prime dividing the order of a group, is not necessarily the whole group. If $|G| = p_1^{h_1} p_2^{h_2} \cdots p_t^{h_t}$ e $|S_i| = p_i^{h_i}$, in order that a product $S_{i_1} S_{i_2} \cdots S_{i_t}$ be equal to G we must have:

$$|S_{i_1} S_{i_2} \cdots S_{i_t}| = |G| = p_1^{h_1} p_2^{h_2} \cdots p_t^{h_t} = |S_{i_1}||S_{i_2}| \cdots |S_{i_t}|.$$

Thus the cardinality of G must be equal to that of the cartesian product set of the S_i, and this happens if, and only if, an element $g \in G$ can be expressed in a unique way as a product $g = x_{i_1} x_{i_2} \cdots x_{i_t}$, $x_{i_j} \in S_{i_j}$. Uniqueness of the expression may depend on the choice of the Sylow subgroups and on the order in which the product is performed. In A^5, by taking:

$$\begin{aligned} V &= \{I, (1,2)(3,4), (1,3)(2,4), (1,4)(2,3)\}, \\ C_3 &= \{I, (1,3,5), (1,5,3)\}, \\ C_5 &= \{I, (1,2,3,4,5), (1,3,5,2,4), (1,4,2,5,3), (1,5,4,3,2)\}, \end{aligned}$$

we have $VC_3C_5 \neq A^5$. Indeed, there are for example two distinct expressions for the cycle (1,2,3,4,5):

$$(1,2,3,4,5) = I \cdot I \cdot (1,2,3,4,5) = (1,2)(3,4) \cdot (1,3,5) \cdot I.$$

It can be proved that no permutation of the chosen three Sylows gives A^5. If instead we choose $C_3 = \{I, (1,2,3), (1,3,2)\}$, then with V and C_5 as above, we have $VC_3 = C_3V = A^4$ and therefore $VC_3C_5 = C_3VC_5 = A^5$. However, for no choice of a 2-, a 3- and a 5-Sylow $VC_5C_3 = A^5$ (*ex.* 47)[7].

Let us consider the relation between the Sylow subgroups of a group and those of its subgroups and quotients. There are no surprises.

Theorem 3.11. *The Sylow subgroups of a subgroup H of a group G are obtained by intersecting H with the Sylow subgroups of G. Moreover two distinct Sylows p-subgroups of H come from two different Sylow p-subgroups of G.*

Proof. If $P \in Syl_p(H)$, by *ii*) of the Sylow theorem there exists $S \in Syl_p(G)$ such that $P \subseteq S$. Therefore $P \subseteq H \cap S$; $H \cap S$ being a p-subgroup of H and P a Sylow p-subgroup of H we have equality: $P = H \cap S$. Let $P_1 = H \cap S_1$ and $P_2 = H \cap S_2$. If $S_1 = S_2$ we have $P_1 = P_2$. ♢

Corollary 3.10. *The number of Sylow p-subgroups of a subgroup does not exceed that of the Sylow p-subgroups of the group.*

It should be noted that the theorem just proved does not say that the intersection of a p-Sylow of a group G with a subgroup H is a p-Sylow of H. It only says that there exists a p-Sylow of G whose intersection with H is a p-Sylow of H. However, if H is normal, and S is any Sylow subgroup, then $H \cap S$ is Sylow in H as the following theorem shows.

Theorem 3.12. *Let $H \trianglelefteq G$, and let $S \in Syl_p(G)$. Then $H \cap S \in Syl_p(H)$.*

Proof. There exists S_1 such that $H \cap S_1 \in Syl_p(H)$. But $S_1 = S^g$, some $g \in G$, so $H \cap S_1 = H \cap S^g = H^g \cap S^g = (H \cap S)^g$. Then $(H \cap S)^g \in Syl_p(H)$, and so also $H \cap S \in Syl_p(H)$ because it is contained in $H^{g^{-1}} = H$ and has the same order as $(H \cap S)^g$. ♢

Theorem 3.13. *Let $H \leq G$, $p \nmid [G : H]$. Then:*

i) *H contains a p-Sylow of G;*

ii) *if H is normal, then it contains all the p-Sylows of G.*

Proof. i) From $|G| = |H|[G : H]$ and the assumption, we have that the highest power of p dividing $|G|$ also divides $|H|$. Hence a p-Sylow of H has the same order as a p-Sylow of G.

ii) The p-Sylows being all conjugate and H containing one of them, by normality it contains all of them. ♢

---

[7] Groups the product gives the whole group independently of the choice of the Sylow subgroups and of the order in which they appear are the solvable groups (cf. Chapter 5, *ex.* 72).

Let us now see what happens with the quotients.

Theorem 3.14. *The p-Sylows of a quotient group G/N are the images of the p-Sylows of G in the canonical homomorphism $G \to G/N$.*

Proof. If $S \in Syl_p(G)$, its image SN/N is a p-subgroup of G/N of index $[G/N : SN/N] = [G : SN] = [G : S]/[SN : S]$, a divisor of $[G : S]$, which is not divisible by p. But a p-subgroup whose index is not divisible by p has order the highest power of p that divides $|G|$, and therefore is a p-Sylow. Hence, the image of a p-Sylow is a p-Sylow of the quotient.

Conversely, if K/N is a p-Sylow of G/N, then p does not divide the index $[G/N : K/N] = [G : K]$, and therefore K contains a p-Sylow S of G. It follows $SN/N \subseteq K/N$; but, by what we have just seen, SN/N is Sylow, and since, by assumption, K/N is also Sylow, we have $K/N = SN/N$. Therefore, the p-Sylows of a quotient are obtained as images of the p-Sylows of the group.◇

Corollary 3.11. *The number of p-Sylows of a quotient does not exceed the number of p-Sylows of the group.*

The intersection of two Sylow p-subgroups S_1 and S_2 is a *maximal Sylow intersection* (for the prime p) if the intersection $S_1 \cap S_2$ has maximal cardinality among all possible intersections between two Sylow p-subgroups. The intersection of two Sylow p-subgroups S_1 and S_2 is a *tame intersection* if $\mathbf{N}_{S_1}(S_1 \cap S_2)$ and $\mathbf{N}_{S_2}(S_1 \cap S_2)$ are both Sylow p-subgroups of $\mathbf{N}_G(S_1 \cap S_2)$. Trivial examples of tame intersections are $S_1 \cap S_2 = \{1\}$ and, if $S_1 = S_2$, $S_1 \cap S_2 = S_1$. More interesting is the following result.

Lemma 3.3. *A maximal Sylow intersection is a tame intersection.*

Proof. Let $N = N_G(S_1 \cap S_2)$. If $N \cap S_1$ and $N \cap S_2$ coincide, then $N \cap S_1 = (N \cap S_1) \cap (N \cap S_1) = (N \cap S_1) \cap (N \cap S_2) = N \cap S_1 \cap S_2 = S_1 \cap S_2$. But $N \cap S_1 = \mathbf{N}_{S_1}(S_1 \cap S_2)$ and this properly contains $S_1 \cap S_2$ (Corollary 3.6). This proves that $N \cap S_1$ and $N \cap S_2$ are distinct. If $N \cap S_1$ is not Sylow in N, there exists a p-Sylow of N properly containing it, and this is obtained as the intersection of N with a p-Sylow S of G: $N \cap S_1 < N \cap S$. Therefore, $S_1 \cap S_2 < N \cap S_1 < N \cap S$, and in particular $S_1 \cap S_2 < S$. Since $S_1 \cap S_2 < S_1$, it follows that $S_1 \cap S_2 < S_1 \cap S$; by the maximality of $S_1 \cap S_2$ this is only possible if $S_1 = S$. Hence, $N \cap S_1 = N \cap S$, and $N \cap S_1$ is Sylow in N. Similarly for S_2. ◇

Since for no prime p is $2 \equiv 1 \bmod p$, besides $N \cap S_1$ and $N \cap S_2$ the normalizer $\mathbf{N}_G(S_1 \cap S_2)$ contains at least one more p-Sylow, and hence at least three. Lemma 3.3 will be used in some of the examples below to prove that certain groups cannot be simple.

Examples 3.5. 1. *A group of order $p^n q$, p and q primes, is not simple* (see *Ex.* 3.4, 5). We have $n_p = 1, q$. If $n_p = 1$, there is nothing to prove. Let $n_p = q$;

if the maximal p-Sylow intersection is $\{1\}$, since each of them contains $p^n - 1$ nonidentity elements, the number of p-elements is $(p^n - 1)q = p^n q - q = |G| - q$. There is room left for q elements, and since a q-Sylow has q elements, it is unique and therefore normal. Now let $S_1 \cap S_2 \neq \{1\}$ be a maximal p-Sylow intersection, and N its normalizer. By the above lemma, N has more than one p-Sylow, and therefore exactly q (its order is of the form $p^k q$). These are obtained intersecting N with q distinct p-Sylow of G, i.e. with all the p-Sylow of G, and such intersections all contain $S_1 \cap S_2$ which is normal in N. Then all p-Sylows of G also contain the nonidentity subgroup $S_1 \cap S_2$, and therefore so does their intersection, which is normal, and is the required normal subgroup.

2. *A group of order* 144 *is not simple.* We have $144 = 2^4 \cdot 3^2$; the only case to deal with is $n_3 = 16$. If any two 3-Sylows have trivial intersection, they contain a total of $16 \times 8 = 128$ 3-elements. There is room left for 16 elements, that make up a unique 2-Sylow. Assume now that S_1 and S_2 are two 3-Sylows such that $|S_1 \cap S_2| = 3$. The normalizer N of this intersection contains S_1 and S_2 (they are abelian) and at least one more 3-Sylow, and since $n_3 \equiv 1 \bmod 3$, also another one. Since the product $n_3 \cdot 3^2$ divides $|N|$, $|N| \geq n_3 \cdot 9 \geq 4 \cdot 9$, and therefore $|N| = 36, 72$ or 144, of indices 4, 2 and 1, and the group is not simple.

3. A^5 *is the only simple group of order* 60. We show that if G is simple with $|G| = 60 = 2^2 \cdot 3 \cdot 5$ then $n_2 = 5$, and G embeds in A^5; their orders being the same, they are equal. From $n_5 = 6$ and $n_3 = 10$ we have 24 5-elements and 20 3-elements. Now, $n_2 = 5$ or 15. If $n_2 = 15$, and the maximal intersection of two 2-Sylows is $\{1\}$, then we have 45 2-elements which, added to the previous ones, give a total 89 elements: too many. Now let $|S_1 \cap S_2| = 2$; the normalizer N of this intersection contains the two 2-Sylows and at least one more; thus $|N| \geq 12$, of index ≤ 5. By simplicity, $|N| = 12$. However, since a 2-Sylow is not normal in N, a 3-Sylow must be normal, and this is also a 3-Sylow of G. Hence, the normalizer of a 3-Sylow would have at least 12 elements, but since $n_3 = 10$, the normalizer of a 3-Sylow has order 6. In all cases, the assumption $n_2 = 15$ leads to a contradiction. Thus $n_2 = 5$, as required.

4. A^5 *is the only group of order* 60 *that has six Sylow* 5*-subgroups.* Let us show that such a group G must be simple. Let $\{1\} \neq N \lhd G$; if $5 \mid |N|$, then N contains a 5-Sylow, and being normal it contains all of them. It follows $6 \mid |N|$ and therefore also $30 \mid |N|$. But $|N| > 30$, because a group of order 30 has only one Sylow 5-subgroup (*Ex.* 3.4, 7); hence $N = G$. If $5 \nmid |N|$, then $|N| = 2, 3, 4, 6, 12$, and in each case G/N is a group having a unique 5-Sylow subgroup C_5N/N and therefore normal in G/N. Hence $C_5N \trianglelefteq G$, 5 divides $|C_5N|$, and if $C_5N < G$ we have again a normal subgroup whose order is divisible by 5. If $C_5N = G$, then $|G| = 5 \cdot |N|$, so $|N| = 12$, and a 2- or a 3-Sylow subgroup S is normal in N. Then G/S has order 15 or 20, and again a 5-Sylow is normal in G/S. Here C_5S is normal in G and has order 20 or 15, so it is not the whole of G.

5. *A group of order* 396 *is not simple.* $396 = 2^2 \cdot 3^2 \cdot 11$; we should only consider $n_{11} = 12$. The normalizer of C_{11} has order 33, it is cyclic ($3 \nmid (11-1)$), and therefore it also normalizes C_3. However, this C_3 is contained in a 3-Sylow of order 3^2, and therefore abelian. Thus the normalizer of C_3 has order divisible by 11 and by 9, so at least 99, and index at most 4.

6. *Groups of order* 12. We show that, up to isomorphisms, there are five such groups, two abelian and three nonabelian. By *Ex.* 3.4, 5, such a group has a normal Sylow subgroup, and therefore is a semidirect product G of a 2- and a 3-Sylow. There are two possibilities for a 2-Sylow: V and C_4, and only one for a 3-Sylow, C_3. Let us consider various cases.

i) $V \lhd G$, $G = C_3 \times_\varphi V$. Since $\mathbf{Aut}(V) \simeq S^3$, setting $C_3 = \{1, x, x^2\}$ we have three possibilities for a homomorphism $C_3 \to S^3$:

$$\varphi_1 : x \to 1,\ \varphi_2 : x \to (1,2,3),\ \varphi_3 : x \to (1,3,2).$$

In the first case, the product is direct: $C_3 \times V$, the group $U(21)$. The second and third case give isomorphic groups (Lemma 2.5), i.e. $C_3 \times_\varphi V$, which in turn are isomorphic to A^4. This can be verified directly, or else by observing that the action of G on the cosets of C_3 induces a homomorphism of the group in S^4 of kernel $\{1\}$ (because C_3 is not normal in G) and taking into account that A^4 is the unique subgroup of order 12 of S^4.

ii) $C_4 \lhd G$ and $G = C_3 \times_\varphi C_4$. There is only one homomorphism of C_3 in $\mathbf{Aut}(C_4) \simeq C_2$, the trivial one, and therefore only one group, the direct product $C_3 \times C_4 \simeq C_{12}$.

iii) $C_3 \lhd G$ and $G = V \times_\varphi C_3$. Since $\mathbf{Aut}(C_3) \simeq C_2$, there are two possible images for V in C_2: the identity and C_2. In the first case we find again the group $V \times C_3$. In the second, there are three homomorphisms of $V = \{1, a, b, c\}$ in C_2, obtained by sending 1 and an element different from 1 to the identity automorphism, and the other two to the inversion $x \to x^{-1}$. However, these groups are all isomorphic (Lemma 2.5, taking the three automorphisms of V that interchange two nonidentity elements and fix the third one, and $\alpha = 1$ in the notation of the lemma). Thus we have the group $V \times_\varphi C_3$. The kernel of φ is a C_2, $\{1, a\}$, say, and being normal is contained in all three 2-Sylows, and being a normal C_2 is also contained in the center of the group (this can also be seen by the fact that the kernel induces the identity on C_3). The product of C_2 with a C_3 yields a C_6, and the automorphism $\varphi(b)(= \varphi(c))$ that interchanges x with x^{-1} also interchanges ax, which has order 6, and its inverse. Now it is clear that $V \times_\varphi C_3 \simeq C_2 \times_{\varphi'} C_6$ that we know to be isomorphic to D_6.

iv) $C_3 \lhd G$ and $G = C_4 \times_\varphi C_3$. As above, the three C_4 intersect in a C_2, which is normal, so the group contains a C_6. The elements of order 4 invert the generators of C_6, and therefore their squares centralize them. We are in presence of a new group, the last of the series, which is denoted by T. Let $C_4 = \{1, a, a^2, a^3\}$, $C_3 = \{1, b, b^2\}$. C_3 being normal, we have $C_3 a = a C_3$, and therefore $ba \in aC_3 = a\{1, b, b^2\} = \{a, ab, ab^2\}$, and the only possibility is $ba = ab^2$. Thus $T = \{1, a, a^2, a^3, b, b^2, ab, ab^2, a^2b, a^2b^2, a^3b, a^3b^2\}$, with the product determined by the relations $a^4 = b^3 = 1, ba = ab^2$ (see *ex.* 53).

The following lemma may be useful to prove the non simplicity of certain groups when the information contained in Lemma 3.3 is not sufficient.

Lemma 3.4. *If $n_p \not\equiv 1 \bmod p^2$ and the order of a p-Sylow is p^n, then for every p-Sylow S there is another p-Sylow S_1 such that $S \cap S_1$ has order p^{n-1}. Hence the normalizer of $S \cap S_1$ contains both S and S_1.*

Proof. As in the proof of the Sylow theorem, *iii*), with $S_0 = S$, if $[S_0 : S_0 \cap S_i] \geq p^2$ for all i, then $n_p \equiv 1 \bmod p^2$. Moreover, $S \cap S_1$ is maximal in S and S_1 and therefore normal in both of them. ♢

Example 3.6. *A group of order* 432 *cannot be simple.* We have $432 = 2^4 \cdot 3^3$, so that $n_3 = 1, 4$ or 16. In the first two cases G is not simple. In the third, we have $16 \not\equiv 1 \bmod 9$. If $H = S \cap S_1$ is an intersection of order 9 of two 3-Sylows (Lemma 3.4), then $N = \mathbf{N}_G(H)$ contains both S and S_1, and therefore two more 3-Sylows. Then $|N|$ is divisible by 27 and by 4 or 16, and N it has order at least 108 and index at most 4.

Theorem 3.15 (Frattini argument). *Let $H \trianglelefteq G$ and P a p-Sylow of H. Then $G = H\mathbf{N}_G(P)$.*

Proof. Recall (Theorem 3.9) that if H is a transitive subgroup of a group G, then G is a product of H and the stabilizer of a point. In our case, G acts by conjugation on the set Ω of the p-Sylows of H, because if $P \in Syl_p(H)$, then $P^g \subseteq H$ (H is normal), and $P^g \in Syl_p(H)$ because it has the same order as P. On the other hand, by the Sylow theorem, *ii*), *b*), H is transitive on Ω, and therefore, by the quoted theorem, $G = HG_P$. But $G_P = \mathbf{N}_G(P)$, and the claim follows. ♢

Theorem 3.16. *Let S be a p-Sylow of G and K a subgroup of G containing the normalizer of S. Then $\mathbf{N}_G(K) = K$. In particular, for $K = \mathbf{N}_G(S)$, the normalizer of the normalizer of S coincides with the normalizer of S:*

$$\mathbf{N}_G(\mathbf{N}_G(S)) = \mathbf{N}_G(S).$$

Proof. By the Frattini argument with $\mathbf{N}_G(K)$ playing the role of G we have, being $\mathbf{N}_G(S) \subseteq K$, that $\mathbf{N}_G(K) = K\mathbf{N}_G(S) = K$. ♢

If $K = \mathbf{N}_G(S)$ the above result can also be proved as follows. Let $g \in \mathbf{N}_G(\mathbf{N}_G(S))$. Then $S^g \subseteq \mathbf{N}_G(S)$, and since $S \trianglelefteq \mathbf{N}_G(S)$ is the unique p-Sylow of $\mathbf{N}_G(S)$ we have $S^g = S$ and $g \in \mathbf{N}_G(S)$, and equality follows.

Theorem 3.17. *Let P be a p-group of order p^n. Then the number n_s of subgroups of order p^s, $0 \leq s \leq n$, is congruent to* 1 mod p.

Proof. If $s = n$ there is nothing to prove. Let us first consider the two extreme cases $s = 1$ and $s = n - 1$, and then the intermediate ones $1 < s < n - 1$.

i) $s = 1$. The idea of the proof is to fix a subgroup of order p and to show that the other ones, if any, distribute p by p. Let $H \subseteq \mathbf{Z}(P)$ of order p; if it is the unique subgroup of order p there is nothing else to prove. If not, let $H_1 \neq H$ of order p. The product HH_1 has order p^2 and is of type $\mathbf{Z}_p \times \mathbf{Z}_p$, and contains $p^2 - 1$ elements of order p that come $p - 1$ by $p - 1$ in $p + 1$ subgroups of order p. If H_2 is not one of these, the group HH_2 contains p subgroups of order p different from those of HH_1, and in general if $H_{k+1} \not\subseteq HH_k$, the subgroup HH_{k+1} contains p subgroups of order p different from those of HH_k. Thus at each step we obtain p new subgroups; together with H we have $1 + hp$ subgroups of order p.

ii) $s = n - 1$. Fix a subgroup M of order p^{n-1} and let us show that the remaining ones distribute p by p. Let $M_1 \neq M$ of order p^{n-1}; then the intersection $M \cap M_1$ has index p^2 in P (in case *i)* the product had order p^2) because

$$p = |P/M| = |MM_1/M| = |M_1/M \cap M_1| = \frac{p^{n-1}}{|M \cap M_1|}, \tag{3.5}$$

from which $|M \cap M_1| = p^{n-2}$ and $[P : M \cap M_1] = p^2$. Setting $K_1 = M \cap M_1$, the group P/K_1 is not cyclic because it contains at least two subgroups of order p, i.e. M/K_1 and M_1/K_1; therefore, it is the group $C_p \times C_p$. The $p + 1$ subgroups of order p of this group come from subgroups of G containing K_1 and have order p^{n-1}. Thus besides M we have p subgroups of order p^{n-1}, i.e. a total of $1 + p$ such subgroups. If M_2 is a subgroup of order p^{n-1} that does not contain K_1, let $K_2 = M \cap M_2$; as above, we obtain $p + 1$ subgroups of order p^{n-1} containing K_2, one of which is M ($M = K_1K_2$, and is the unique subgroup which appears also this time). Therefore, there are p new subgroups, and at this point we have a total of $1 + 2p$ subgroups of order p^{n-1}. Iterating this procedure, at step k we have $1 + kp$ subgroups of order p^{n-1}, as required. Observe that this proof is dual to the previous one, according to the dualities maximal–minimal (a subgroup of order p is minimal), order–index, union (product)–intersection.

iii) $1 < s < n-1$. Let $|H| = p^s$, and consider $\mathbf{N}_P(H)/H$. By *i)*, this group contains a number of subgroups of order p congruent to 1 mod p, and these come from subgroups $H_1 \supset H$ of $\mathbf{N}_P(H)$ of order p^{s+1}. But all the subgroups of order p^{s+1} containing H normalize H, and therefore those of $\mathbf{N}_P(H)$ are all the subgroups of P containing H. It follows that the number of subgroups of order p^{s+1} containing H is congruent to 1 mod p. Next, let $H_1, H_2, \ldots, H_{n_s}$ be the subgroups of P of order p^s, and $K_1, K_2, \ldots, K_{n_{s+1}}$ those of order p^{s+1}.

In a grid like the following:

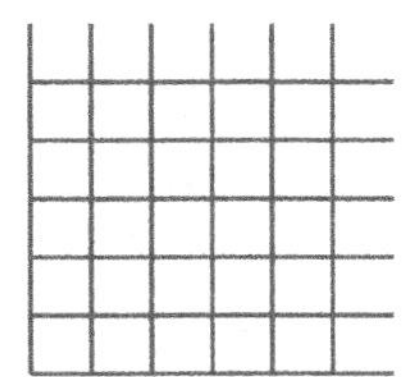

put the H_i in abscissa and the K_j in ordinate, and mark a point at the intersection of row K_j with column H_i if H_i is contained in K_j. If there are a_i points marked on the column H_j, a_i is the number of subgroups of order p^{s+1} containing H_i. For what we have seen above, we have $a_i \equiv 1 \bmod p$. Then the total number of points marked on the columns is:

$$\sum_{i=1}^{n_s} a_i \equiv 1 + 1 + \cdots + 1 = n_s \bmod p.$$

Let b_j be the number of points marked on the row K_j, i.e. the number of subgroups of order p^s contained in K_j; by *ii)* we have $b_j \equiv 1 \bmod p$ and therefore

$$\sum_{j=1}^{n_{s+1}} b_j \equiv 1 + 1 + \cdots + 1 = n_{s+1} \bmod p.$$

But the number of points on the rows is the same as that on the columns, and therefore $n_s \equiv n_{s+1} \bmod p$. Since $n_1 \equiv 1 \bmod p$ we have $n_2 \equiv 1 \bmod p$, $n_3 \equiv n_2 \equiv 1 \bmod p$, etc. ◇

Exercises

17. Give a new proof of Corollary 3.6 by induction on $|G|$, distinguishing the two cases $\mathbf{Z}(G) \subseteq H$ and $\mathbf{Z}(G) \not\subseteq H$. [*Hint*: *ex.* 40 of Chapter 2.]

18. A non cyclic finite p-group cannot be generated by conjugate elements. [*Hint*: if $\langle x \rangle \neq G$, then $\langle x \rangle \subseteq M$, M maximal and therefore normal (Corollary 3.6).]

19. *i)* Let G be a finite p-group, $H \trianglelefteq G$. Then H contains a subgroup of index p and normal in G. [*Hint*: consider $G/(H \cap \mathbf{Z}(G))$ and induct.]

ii) If H is as in *i)* and contains an abelian subgroup A of index p, then it also contains an abelian subgroup of index p and normal in G. [*Hint*: if A is not unique, consider A_1, $H = AA_1$, from which $A \cap A_1 \subseteq \mathbf{Z}(H)$; distinguish the two cases $A \cap A_1 \subset \mathbf{Z}(H)$ and $A \cap A_1 = \mathbf{Z}(H)$.]

20. Show that a finite group G has a quotient of order p if, and only if, for every p-Sylow S of G one has $S \cap G' < S$. [*Hint*: $|G/H| = p$ implies $G' \subseteq H$, and if $S \cap G' = S$ then $S \subseteq G' \subseteq H$ so that $p \nmid |G/H|$.]

21. Prove that if P is a p-subgroup of a group G and S is a p-Sylow of G, then the normalizer in P of S, $\mathbf{N}_P(S) = P \cap \mathbf{N}_G(S)$, equals $P \cap S$. [*Hint*: S is the unique p-Sylow of its normalizer; apply Theorem 3.10, *ii) a)*.]

22. Give a new proof of the existence of a p-Sylow in a group G as follows:
i) if p does not divide the index of a proper subgroup, the result follows by induction;
ii) if p divides the index of every proper subgroup, from the class equation it follows that p divides the order of the center of G;
iii) if H is a subgroup of order p of the center (which exists by Cauchy theorem), apply induction to G/H.

23. Let G be a finite p-group acting on a finite set Ω, and let $\Gamma = \{\alpha \in \Omega \mid \alpha^g = \alpha, \forall g \in G\}$. Show that $|\Omega| \equiv |\Gamma| \bmod p$.

24. (Gleason) Let G and Ω be finite, and assume that for each $\alpha \in \Omega$ there exists a p-element $x \in G$ (a p-subgroup H) such that α is the unique point of Ω fixed by x (by every element of H). Show that G is transitive on Ω. [*Hint*: let $\Delta_1 \neq \Delta_2$ be two orbits, $\gamma \in \Delta_1$, $x \in G$ a p-element such that γ is the unique point fixed by x. Apply the previous exercise.]

25. Use *ex.* 23 and 24 to prove that $n_p \equiv 1 \bmod p$ and that the Sylow p-subgroups form a unique conjugacy class.

26. If G/N is a p-group and $S \in Syl_p(G)$, then $G = NS$. [*Hint*: SN/N is Sylow in G/N.]

27. If G is finite and $x \in G$, the *p-components* of x are the elements appearing in the decomposition of x given in *ex.* 15, *i)* of Chapter 1. Show that if x and y are two conjugate elements of G, their p-components are conjugate. Moreover, if $x_i \in S_1$ and $y_i \in S_2$ are the p_i-components of x and y, and $S_1, S_2 \in Syl_p(G)$, then a conjugate of y_i belongs to S_1. (The conjugacy of x and y is thus reduced to that between elements of the same Sylow p-subgroup.)

28. If G is not abelian and, for all p, the p-Sylows are maximal subgroups then the center of G is the identity. [*Hint*: if p divides $|\mathbf{Z}(G)|$ and Q is a Sylow q-subgroup, then $G = QC_p$, and C_p is maximal and central.]

29. Let $H \trianglelefteq G$, $P \in Syl_p(H)$. Show that there exists $S \in Syl_p(G)$ such that $S \subseteq \mathbf{N}_G(P)$. [*Hint*: $P = S \cap H \trianglelefteq S$.]

30. Determine the number of Sylow p-subgroups of S^p, and use the result to prove *Wilson's theorem*:

$$(p-1)! \equiv -1 \bmod p.$$

[*Hint*: p is the highest power of p dividing $p! = |S^p|$, and the elements of order p of S^p are p-cycles.]

31. Let $p||G|$. Define $O_p(G)$ to be the maximal normal p-subgroup of G , i.e. the product of all normal p-subgroup of G. Show that:
i) $O_p(G)$ coincides with the intersection of all the p-Sylows and is characteristic in G;
ii) if H is a subgroup of index p, then $O_p(H)$ is normal in G.

32. Define $O^p(G)$ to be the subgroup generated by the elements of order not divisible by the prime p. Show that:
i) $O^p(G)$ is characteristic in G;
ii) p does not divide the index in $O^p(G)$ of the commutator subgroup $O^p(G)'$.

33. If $p > 2$ is the smallest divisor of the order of a group G, and a p-Sylow S has order p^2 and is normal, then S is contained in the center of G. [*Hint*: $S = C_{p^2}$ or $C_p \times C_p$; Theorem N/C.]

34. Let G act on Ω, and let S be a p-Sylow of G. If p^k divides $|\alpha^G|$, some $\alpha \in \Omega$, then p^k already divides $|\alpha^S|$.

35. Let G be a finite group in which any two 2-Sylows have trivial intersection. Show that if S is a 2-Sylow, no coset of S, except S itself, contains more than one involution. [*Hint*: let $x \in Sx$ and sx be involution; $sxsx = 1$, etc.]

36. Let G be a finite group in which any two p-Sylows, for a given p, have trivial intersection. Show that if S is a p-Sylow, no coset of the normalizer of S, except the normalizer itself, contains more than $n_p - 1$ p-elements. [*Hint*: show that two p-elements of a coset of $\mathbf{N}_G(S)$ belong to different Sylow subgroups.]

37. Let $S \in Syl_2(G)$, and let M be a maximal subgroup of S. Let $x \in S$ be an involution not conjugate in G to any element of M, and let Ω be the set of cosets of M. Show that:

i) $|\Omega| = 2m$, m odd;
ii) in the action of G on Ω, x fixes no element;
iii) in the homomorphism of G in S^{2m} induced by the action the image of x is an odd permutation;
iv) G has a subgroup of index 2 not containing x.

Conclude that x does not belong to the derived group G' of G.

38. A group of order p^2q^2, $p < q$, has a normal Sylow subgroup. [*Hint*: show that if $n_q \neq 1$, then G has order 36.]

39. Let S be a non normal 2-Sylow of a group G such that $S \cap S^g = \{1\}$ if $S \neq S^g$. Show that G has a unique conjugacy class of involutions. [*Hint*: Lemma 2.1.]

40. Let $S \in Syl_p(G)$ be normal, and let $x \in G$ be an element of odd order. Show that x is not conjugate to its inverse. [*Hint*: if $y^{-1}xy = y^{-1}$ distinguish the cases $o(y)$ odd or even, and recall that if an element y inverts x then y^2 centralizes x (Chapter 2, *ex.* 14); if $o(y)$ is even, write $y = tu$, with $o(t) = 2^k$ and $o(u)$ odd.]

41. A group of order 108 has a normal subgroup of order 9 or 27.

42. If $S \in Syl_p(G)$ then $G = G'\mathbf{N}_G(S)$. [*Hint*: $G'S \trianglelefteq G$.]

43. A group of order $12p$, $p > 5$, is not simple. [*Hint*: there are three cases, $p > 11$, $p = 11$ and $p = 7$.]

44. G is not simple if it has order 180, 288, 315, 400 or 900.

45. Let $G = HK$, $H, K \trianglelefteq G$. If $S \in Syl_p(G)$ then $S = (S \cap H)(S \cap K)$.

46. *i)* Prove that in the group A^8 the normalizer of a 7-Sylow has order 21, and use this fact to prove that a group of order 336 is not simple.

ii) Prove that in the group A^{12} the normalizer of an 11-Sylow has order 55, and use this fact to prove that a group of order 264 or 396 is not simple.

47. List the subgroups of A^5, and show that for no choice of a 2-, 3- and 5-Sylow one has $S_2S_5S_3 = A^5$.

48. A group of order 24 in which the normalizer N of a 3-Sylow has order 6 is isomorphic to S^4. [*Hint*: the kernel of $G \to S^4$ is contained in N.]

49. Prove that there are five groups of order 18 and two groups of order 21. [*Hint*: for $|G| = 18$, there are three direct products and two semidirect products.]

50. Show that a normal Sylow p-subgroup is the unique Sylow p-subgroup in the two following ways:
i) using *ex.* 4 of Chapter 2;
ii) assuming that there is another p-Sylow and considering the product of the two.

51. Show that the automorphisms of S^4 are all inner. [*Hint*: S^4 has four 3-Sylows S_i, $i = 1, 2, 3, 4$, each generated by a 3-cycle fixing the digit i, and $\alpha \in \mathbf{Aut}(S^4)$ permutes them; if $\sigma \in S^4$ is such that $\sigma(i) = j$, then on the 3-Sylows $\sigma^{-1}S_i\sigma = S_{\sigma(i)} = S_j$.]

52. Prove that the two matrices $\begin{pmatrix} 0 & i \\ i & 0 \end{pmatrix}$ and $\begin{pmatrix} \omega & 0 \\ 0 & \omega^2 \end{pmatrix}$, where i and ω are fourth and third primitive roots of unity, respectively, generate a subgroup of $GL(2, \mathbf{C})$ isomorphic to the group T of *Ex.* 3.5, 6, *iv*).

53. A *dicyclic* group is a group generated by an element s of order $2n$ and an element a of order 4 such that $a^2 \in \langle s \rangle$ (hence $a^2 = s^n$) and $a^{-1}sa = s^{-1}$. Prove that a dicyclic group has order $4n$, that for $n = 2$ it is the quaternion group, and that for $n = 3$ it is the group T of *Ex.* 3.5, 6 (as a product of two cyclic groups, not a semidirect product this time). The quotient of the group by the subgroup $\langle a \rangle$ is the dihedral group D_n. If n is a power of 2, the group is also called a *generalized quaternion* group.

54. (Bertrand's Theorem) Let $n \geq 5$. Show that:

i) S^n has no subgroups of index k, $2 < k < n$, but has subgroups of index 2 and n;
ii) A^n has no subgroups of index k, $1 < k < n$, but has subgroups of index n.

In the language of the classical theory of equations this result is stated by saying that permuting in all possible ways the variables of a polynomial function of $n \neq 4$ variables only 1, 2 or n functions can be obtained. For $n = 4$ we have seen (*Ex.* 1.7, 6) that permuting the variables in the function $x_1x_2 + x_3x_4$ three functions are obtained, and indeed S^4 contains three subgroups of index 3 (dihedral of order 8).

55. Let φ be a function of n variables, $\varphi = \varphi_1, \varphi_2, \ldots, \varphi_k$ those obtained from φ by permuting the variables according to the elements of S^n, and let φ_i belong to the subgroup G_i (*Ex.* 1.7, 6). Show that $K = \bigcap_i G_i$ is normal in S^n, and therefore, if $n \neq 4$, then $K = A^n$ or $K = S^n$. (This means that for $n \neq 4$, there exist no functions, besides the symmetric ones ("one valued") and the alternating ones ("two valued"), such that all the values taken remain unchanged under one and the same nonidentity permutation. In Galois theory, this corresponds to the fact that for $n > 4$ there is no Galois resolvent of degree k, $2 < k < n$, for the general equation of degree n.) For $n = 4$, the function given in the previous exercise has three values, and the intersection of the three dihedral groups fixing each value is the normal Klein subgroup of S^4.

56. Show that there are four groups of order 30. [*Hint*: from *Ex.* 3.4, 7 there is a subgroup H of order 15. Consider the possible semidirect products of a C_2 with H, and show that, besides the cyclic one, the groups obtained have centers of order 5,3 and 1, respectively.]

57. Prove that the Sylow p-subgroup of $GL(n,q)$ given by the upper unitriangular matrices is normal in the subgroup B of all upper triangular matrices.

58. The number of p-Sylows of a subgroup is not in general a divisor of the order of the number of the p-Sylows of the group.

59. Prove that any two Sylow p-subgroups are conjugate using *ex.* 7. [*Hint*: with H and K two p-Sylows, from Frobenius formula we have $|G| = p^r m = \sum_{i=1}^{m} = |H||K|/d_i$, where $d_i = |K \cap a_i^{-1} H a_i|$.]

60. Prove that the number of p-Sylows containing a given p-subgroup P is congruent to 1 mod$p$. (This generalizes part *iii*) of Sylow theorem.) [From *ex.* 7, *iii*), with $H = N_G(S)$, S a p-Sylow, and $K = P$, each term of the sum $\sum_i [P : N_G(S)^{a_i} \cap P]$ is a power of p; the sum being equal to $|G : N_G(S)|$, it is coprime to p so that some of the summands are equal to 1, i.e., $P \subseteq N_G(S)^a$, $a = a_i$, some i. But then $P \subseteq S^a$. Conversely, if $P \subseteq S^a$, then $[P : N_G(S)^{a_i} \cap P] = 1$. If n_P is the number of p-Sylows containing P, since $S^x = S^y$ implies $N_G(S)xP = N_G(S)yP$, the number of summands equal to 1 in the above sum is n_P. Therefore, $[G : N_G(S)] \equiv n_P \bmod p$. The left hand side being independent of P, setting $P = S$, we have $[G : H] = 1$ because $n_S = 1$, and $n_P \equiv 1 \bmod p$.]

3.3 Burnside's Formula and Permutation Characters

Formula (3.1) says how many points an orbit contains. Let us now see how many orbits there are. Throughout this section G will be a finite group.

Definition 3.7. The *character* of an action is the function χ defined on G by $\chi(g)$ =number of points of Ω fixed by g.

Remark 3.2. If P is the permutation matrix representing an element σ of S^n, the number of 1's on the diagonal of P is the number of points fixed by σ. Then $\chi(\sigma)$ is the trace of P.

The following theorem relates the number of orbits to that of the fixed points of the elements of G.

Theorem 3.18 (Burnside's counting formula). *The number N of orbits of the group G acting on a set Ω is:*

$$N = \frac{1}{|G|} \sum_{g \in G} \chi(g). \tag{3.6}$$

Proof. In a grid as that of page 113, put in abscissa the points of Ω and in ordinate those of G, marking a point at the intersection of the column of $\alpha \in \Omega$ and the row of $g \in G$ if g fixes α. Let us count the marked points.

On a row there are as many marked points as there are points of Ω fixed by the corresponding element g, i.e. $\chi(g)$. The total number of marked points on the rows is therefore $\sum_{g \in G} \chi(g)$. On a column there are as many marked points as there are elements of G fixing the corresponding point of Ω. The total number of points on the rows is therefore $\sum_{\alpha \in \Omega} |G_\alpha|$. Since the number of points counted in the two ways is the same, we have:

$$\sum_{g \in G} \chi(g) = \sum_{\alpha \in \Omega} |G_\alpha|. \tag{3.7}$$

Now, two elements belonging to the same orbit have stabilizers of the same order; the contribution of each orbit to the sum on the right hand side is therefore $|\alpha^G||G_\alpha|$, i.e. $|G|$. If there are N orbits, the sum on the right side equals $N|G|$. The result follows. ◇

Note that in the case of a transitive group $N = 1$ and (3.6) becomes $|G| = \sum_{g \in G} \chi(g)$.

The number of orbits is therefore obtained as an average: the total number of points fixed by the elements of G divided by the number of elements of G. In other words, an element of G fixes, in the average, a number of points equal to the number of orbits. In particular, in a transitive group an element fixes, in the average, one point. However,

Corollary 3.12. *In a transitive group there exist elements fixing no point.*

Proof. By transitivity, $|G| = \sum_{g \in G} \chi(g)$. If $\chi(g) \geq 1$ for all g, then $|G| = \chi(1) + \sum_{1 \neq g \in G} \chi(g) \geq \chi(1) + |G| - 1$, so that $|\Omega| = \chi(1) = 1$ (of course we assume $|\Omega| > 1$). ◇

Remark 3.3. A much deeper result asserts the existence of an element of prime power order fixing no point[8]. Here "prime power" cannot be replaced with "prime".

Corollary 3.13. *i) A finite group cannot be the set-theoretic union of proper conjugate subgroups;*

ii) if $H < G$, there exists a conjugacy class of G that has trivial intersection with H.

Proof. *i)* Let $G = \bigcup_{x \in G} H^x$, $H \neq G$. G acts transitively on the cosets of H, but if $g \in G$ then $g = x^{-1}hx$, some x and h, so that g fixes the coset Hx. Therefore, every element of the group fixes some point, contrary to the above corollary (see *ex.* 70 below).

ii) Otherwise, given $g \in G$, there exists $a \in G$ such that $aga^{-1} \in H$. Then g fixes the coset Ha in the action of *i)*. ◇

[8] Fein–Kantor–Schaker (1981).

Let a_i be the number of elements $g \in G$ such that $\chi(g) = i$. Then $\sum_{i=0}^{n} ia_i = \sum_{g\in G} \chi(g)$, where $n = |\Omega|$. The number of elements of G being $\sum_{i=0}^{n} a_i$, (3.6) can be written as:

$$\sum_{i=0}^{n} ia_i = N \sum_{i=0}^{n} a_i. \tag{3.8}$$

Examples 3.7. 1. If we color the sides of a square with two colors in all possible ways we obtain 16 colored squares. The dihedral group D_4 acts on the set of these squares; if $g \in D_4$ we have:

$g \in D_4$	$\chi(g)$	*squares left fixed*
id.	16	all
rot. of $\frac{\pi}{2}$	2	all sides of the same color
rot. of $\frac{3}{2}\pi$	2	all sides of the same color
rot. of π	4	opposite sides of the same color
diag. symm.	4	concurrent sides of the same color
other diag. symm.	4	concurrent sides of the same color
axial symm.	8	two opposite sides of the same color
other axial symm.	8	two opposite sides of the same color

By (3.6), $N = \frac{1}{|D_4|} \sum_{g\in D_4} \chi(g) = \frac{1}{8} \cdot 48 = 6$, and if the colors are W and B, the six orbits have, respectively:

- 1 square with all sides W;
- 1 square with all sides B;
- 4 squares with one side W and three B;
- 4 squares with one side B and three W;
- 4 squares with two concurrent sides W and the other two B;
- 2 squares with two opposite sides W and the other two B.

This means that there are essentially six ways of coloring the sides of a square using two colors. The criterion according to which two squares are to be considered distinct reflects the choice of the group acting on the set of squares. If we take the subgroup H of D_4 consisting of the identity and an axial symmetry we obtain 11 orbits, i.e. according to the group H there are eleven essentially different ways of coloring a square with two colors. In general, if $H \leq G$, since the orbits of G are unions of orbits of H, the smaller H, the larger the number of different objects. The extreme case is that of the identity subgroup, for which all the object of the set on which it acts are different.

2. Let us now count the ways in which the vertices of a square can be colored using four colors with all vertices of different colors, in other words, in how many essentially different ways the vertices of a square can be numbered.

There are 4 choices for the first vertex, 3 for the second, 2 for the third and 1 for the last one. Letting D_4 act, a nonidentity symmetry takes a square to a square different from it; thus $\chi(g) = 0$ for $g \neq 1$, and being $\chi(1) = 24$ we have $N = \frac{24}{8} = 3$. The three orbits may be represented by the squares numbered 1234, 1243 and 1324.

We now state a few properties of the function χ.

Theorem 3.19. *i) If x and y are conjugate, then $\chi(x) = \chi(y)$;*
ii) if x and y generate the same subgroup, then $\chi(x) = \chi(y)$;
iii) two similar actions have the same character.

Proof. i) x fixes α if and only if $g^{-1}xg$ fixes α^g;
ii) if an element fixes a point, the same holds for every power, and since x and y are powers of each other we have the result;
iii) from $\varphi(\alpha^g) = \varphi(\alpha)^{\theta(g)}$ it follows that g fixes α if, and only if, $\theta(g)$ fixes $\varphi(\alpha)$. ◇

In the symmetric group, if two elements generate the same subgroup they are conjugate. Indeed, if $o(x) = m$ and $\langle x \rangle = \langle y \rangle$, then $y = x^k$ with $(m, k) = 1$; if c is a cycle of x, then $o(c^k) = o(c)$ so that c^k and c have the same length. Then the partitions of n induced by the cycles of x and y are the same, and x and y are conjugate. Hence, in the symmetric group, *ii)* follows from *i)*.

By *i)* of the preceding theorem, χ has the same value on the elements of a conjugacy class. For this reason, χ is called a *class function* (also called a *central* function).

Statement *iii)* of Theorem 3.19 cannot be inverted, i.e two actions with the same character are not necessarily similar, as *Ex.* 3.8 below shows. First a theorem.

Theorem 3.20. *In the action of a group G on the cosets of a subgroup H we have, for $g \in G$,*

$$\chi(g) = \frac{|\mathbf{C}_G(g)|}{|H|}|\mathbf{cl}(g) \cap H|. \tag{3.9}$$

In particular, the action of G on the cosets of two subgroups H and K have the same character if, an only if,

$$|\mathbf{cl}(g) \cap H| = |\mathbf{cl}(g) \cap K|, \tag{3.10}$$

for all $g \in G$.

Proof. A coset Hx is fixed by the elements of H^x and only by them; therefore, the conjugates of g fixing it are the elements of $\mathbf{cl}(g) \cap H^x$. But the correspondence $\mathbf{cl}(g) \cap H \to \mathbf{cl}(g) \cap H^x$ given by $a^{-1}ga = h \to (ax)^{-1}g(ax) = h^x$ is one-to-one, and therefore $|\mathbf{cl}(g) \cap H^x| = |\mathbf{cl}(g) \cap H|$. In a grid like that of page 113 put in abscissa the conjugates $g_1, g_2, \ldots$ of g, and in ordinate the cosets $Hx_1, Hx_2, \ldots$ of H, and mark a point at the intersection of the column g_i and the Hx_j row if g_i fixes Hx_j. On the row of Hx_j there are $|\mathbf{cl}(g) \cap H^{x_j}| = |\mathbf{cl}(g) \cap H|$ marked points, for a total of $|\mathbf{cl}(g) \cap H^x| \cdot [G : H]$ points on the rows. On the g_i column there are $\chi(g_i)$ marked points, and since two conjugate elements fix the same number of points on the columns we have a total of $|\mathbf{cl}(g)|\chi(g)$ marked points. Equality (3.9) follows. ◇

Theorem 3.21. *Let H and K be two subgroups of a finite group G such that (3.10) holds. Then the two subgroups have the same number of elements of each order. Conversely, if the latter property holds, then the two subgroups embed in S^n, $n = |H| = |K|$, and (3.10) holds for all $g \in S^n$.*

Proof. Assume (3.10) holds. The set of elements of a given order decomposes into a disjoint union of conjugacy classes, $U = \bigcup \mathbf{cl}(g)$, so that $|U \bigcap H| = \sum |\mathbf{cl}(g) \bigcap H|$. By assumption, the elements of the summation are the same for K, and we have the result. Conversely, the two subgroups having the same order n embed in S^n via the regular representation. Let $g \in S^n$; if both intersections in (3.10) are empty, there is nothing to prove. So assume there is an element $h \in \mathbf{cl}(g) \bigcap H$; then there exists $k \in K$ having the same order as h and conjugate to h in S^n (they have the same cycle structure). It follows that $\mathbf{cl}(g) \cap K$ is also nonempty. Now $\mathbf{cl}(g) \bigcap H$ contains the elements of H having order equal to that of g, and so does $\mathbf{cl}(g) \bigcap K$; by assumption, the cardinalities of these two sets are the same. ◇

Condition (3.10) is equivalent to saying that for every element h of H there is an element of K that is conjugate in G to h. We shall say that two subgroups satisfying (3.10) are *almost conjugate.*

Example 3.8. Consider the following subgroups of S^6:

$$V_1 = \{I, (1,2)(3,4)(5)(6), (1,3)(2,4)(5)(6), (1,4)(2,3)(5)(6)\},$$
$$V_2 = \{I, (1,2)(3,4)(5)(6), (1,2)(5,6)(3)(4), (3,4)(5,6)(1)(2)\}.$$

V_1 and V_2 are both isomorphic to the Klein group. The orbits of V_1 are $\{1,2,3,4\}, \{5\}, \{6\}$, those of V_2 are $\{1,2\}, \{3,4\}, \{5,6\}$, so that V_1 and V_2 are not conjugate. Let $g \in S^6$; if $g = 1$, $\mathbf{cl}(1) = \{1\}$, and (3.10) is satisfied. If $g \neq 1$, and the cycle structure of g is $(i,j)(h,k)(l)(m)$, then all the nonidentity elements of V_1 and V_2 belong to the conjugacy class of g, so that $|\mathbf{cl}(g) \cap V_1| = |\mathbf{cl}(g) \cap V_2| = 3$. If g has a different cycle structure, then g cannot be conjugate to an element of V_1 or V_2, and therefore both intersections are empty. Then (3.10) is satisfied for every $g \in S^6$. By Theorem 3.20, the actions of S^6 on the cosets of V_1 and V_2 have the same character, but they are not similar because V_1 and V_2 are not conjugate[9]. (Further examples in *ex.* 78, 79 and 132.)

---

[9] See Theorem 6.7 and Remark 6.1. This example goes back to Gaßmann F.: Math. Z. **25** (1926), p. 665–675.

Remark 3.4. A theorem of Sunada states that if G is a finite group acting on a Riemannian manifold by isometries, and H and K are two subgroups of G acting without fixed points and satisfying (3.10), then the quotient manifolds with respect to H and K are isospectral (the spectra of the Laplace operator that acts on smooth functions on the manifolds are the same). If H and K are conjugate the manifolds are isometric[10].

Characters are additive functions: if G acts on Ω_1 and Ω_2 with characters χ_1 and χ_2, then it also acts on $\Omega_1 \cup \Omega_2$, and the number of points fixed by $g \in G$ in this action is the sum of the number of points fixed on Ω_1 and Ω_2: $\chi(g) = \chi_1(g) + \chi_2(g)$. It follows that one can define the sum $\chi = \chi_1 + \chi_2$ of two characters χ_1 and χ_2 as the character χ of the action of G on $\Omega_1 \cup \Omega_2$.

It is also possible to define the product of two characters. Consider the direct product $G \times G$ and its action on $\Omega_1 \times \Omega_2$ given by $(\alpha, \beta)^{(g,h)} = (\alpha^g, \beta^h)$. Then $\chi((g,h)) = \chi_1(g)\chi_2(h)$. The "diagonal" subgroup $\{(g,g), g \in G\}$ of $G \times G$ is isomorphic to G, and this allows one to define an action of G on $\Omega_1 \times \Omega_2$: $(\alpha, \beta)^g = (\alpha^g, \beta^g)$. If χ is the character of this action, then $\chi(g) = \chi_1(g)\chi_2(g)$; this equality defines the product of the two characters.

If A and B are two permutation matrices representing g then, in the case of the sum, $\chi(g)$ is the trace of the permutation matrix having on the main diagonal two blocks equal to A and B and zero elsewhere. In the case of the product, $\chi(g)$ is the trace of the tensor product of A by B (the matrix obtained by replacing the 1's of A with the matrix B, and zero elsewhere).

Exercises

61. How many essentially different squares are obtained coloring the sides using n colors in all possible ways? What if it is required that each side has a different color? [*Hint*: following the list of *Ex.* 3.7, 1, the elements of D_4 fix, respectively, $n^4, n, n, n^2, n^2, n^2, n^3, n^3$ squares.]

62. *i*) Let G be cyclic of order n, $G = \langle g \rangle$. Show that (3.6) may be written as $N = \frac{1}{n}\sum_{d|n} \chi(g^d)\varphi(\frac{n}{d})$, where φ is Euler's function.

ii) Using n pearls of a different colors, how many necklaces are obtained?

63. If $H \leq G$, show that $\sum_{x \in G} |\mathbf{C}_H(x)| = \sum_{y \in H} |\mathbf{C}_G(y)|$. [*Hint*: let H act on G by conjugation, and apply (3.7).]

64. Let $c(G)$ be the number of conjugacy classes of G. Show that if $H < G$, then *i*) $c(H) < c(G)[G:H]$ and *ii*) $c(G) \leq c(H)[G:H]$, and if in *ii*) equality holds, then H is normal, but the converse is not true. [*Hint*: in the action by conjugation of G on itself $\chi(g) = |\mathbf{C}_G(g)|$, and if $c(G)$ is the number of conjugacy classes we have, by (3.6), $c(G) = (1/|G|)\sum_{g \in G} |\mathbf{C}_G(g)|$, and similarly for H. As for *ii*) see *ex.* 4.]

---

[10] Gordon C.S., Webb D.L., Wolpert S.: One cannot hear the shape of a drum. Bull. AMS **27** (1992), 134–138.

65. If G acts non trivially with N orbits, there exists an element $g \in G$ that fixes less than N elements: $\chi(g) < N$. [If G is transitive, $N = 1$, we obtain Corollary 3.12.]

66 (Jordan[11]). Let G be a transitive group. Show that there are at least $|\Omega| - 1$ elements of G such that $\chi(g) = 0$.

67. Let G be a transitive permutation group. Show that:

i) if G has degree n, $\chi(g) \leq 1$ for all $1 \neq g \in G$ if, and only if, $\chi(g) = 0$ for $n - 1$ elements of G;

ii) if G has degree p, a prime, and $\chi(g) \leq 1$ for every $1 \neq g \in G$, then G has a unique subgroup of order p;

iii) if G is as in *ii)*, then G is isomorphic to a subgroup of the affine group over the integers mod p (*Ex.* 2.10, 2, *ii)*)[12], and the two actions are similar.

68. Prove that Corollary 3.13 implies Corollary 3.12.

69. Let $\alpha \in \mathbf{Aut}(G)$, $p | o(\alpha)$, $p \nmid |G|$. Show that α cannot fix all the conjugacy classes of G. [*Hint*: if $H = \{g \in G \mid g^\alpha = g\}$, there exists $g \in G$ such that $\mathbf{cl}(g) \cap H = \emptyset$, otherwise $G = \bigcup_{x \in G} H^x$ (Corollary 3.13).]

70. An infinite group may be the union of the conjugates of a proper subgroup. [*Hint*: consider $G = GL(V)$, V of dimension greater that 1, a vector $v \neq 0$, H the set of the $x \in G$ having v has an eigenvector; then $H < G$. Over the complex numbers, every $x \in G$ has an eigenvector u and there exists $g \in G$ such that $v^g = u$; it follows that G is a union of conjugates of H.]

Definition 3.8. The *rank* of a transitive group G is the number of orbits on Ω of the stabilizer G_α of a point α. (By transitivity, this number does not depend on α.)

71. Show that the rank of a group G equals the number of double cosets of G_α. [*Hint*: if $G = \bigcup_{i=1}^t G_\alpha x_i G_\alpha$, then the sets $\Omega_i = \{\alpha^g,\ g \in G_\alpha x_i G_\alpha,\ i = 1, 2, \ldots, t\}$ are the orbits of G_α.]

72. If r is the rank of G show that $r|G| = \sum_{g \in G} \chi(g)^2$. [*Hint*: let $\Omega = \{1, 2, \ldots, n\}$, $G = \bigcup_{i=1}^t G_1 t_i$, $t_1 = 1$; $g \in G$ fixes i if, and only if, $g \in t_i^{-1} G_1 t_i$, and if $\chi(g) = r$, g belongs to exactly r conjugates of G_1. Summing over the t conjugates of G_1, the contribution of g is $\chi(g)^2$. If a subgroup is not conjugate of G_1, its contribution to the sum is zero.]

Definition 3.9. G is 2-*transitive* if given two ordered pairs (α, β) and (γ, δ) of distinct points of Ω there exists $g \in G$ such that $\alpha^g = \gamma$ and $\beta^g = \delta$. More generally, G is k-*transitive* if, given two ordered k-tuples $(\alpha_1, \alpha_2, \ldots, \alpha_k)$ and $(\beta_1, \beta_2, \ldots, \beta_k)$ of distinct points of Ω, there exists $g \in G$ such that $\alpha_i^g = \beta_i$, $i = 1, 2, \ldots, k$. In other words, G acts on the set Ω_k of ordered k-tuples of distinct points of Ω as $(\alpha_1, \alpha_2, \ldots, \alpha_k)^g = (\alpha_1^g, \alpha_2^g, \ldots, \alpha_k^g)$, and this action is transitive. If, given two k-tuples, there is only one element g taking one k-tuple to the other, G is *sharply* k-*transitive*. If $|\Omega| = n$, then $|\Omega_k| = n(n-1) \cdots (n-k+1)$. Note that 1-transitive simply means transitive.

---

[11] A translation into number theory and topology of this result may be found in Serre J.-P.: On a theorem of Jordan. Bull. AMS **40**, N. 4 (2003), 429–440.

[12] For this reason, the action of a transitive group in which every element fixes at most one point is said to be *affine*.

73. Prove the following:

i) G is 2-transitive if, and only if, is of rank 2, i.e. G_α is transitive on $\Omega \setminus \{\alpha\}$. (Thus, by *ex.* 71, G is 2-transitive if, and only if, $G = G_\alpha \cup G_\alpha x G_\alpha$.) [*Hint: (i)* If G_α is transitive on $\Omega \setminus \{\alpha\}$, let $(\beta, \gamma), (\eta, \epsilon)$ be two pairs of elements of Ω. Let $\beta^g = \alpha$, $\alpha^h = \delta$, and $(\gamma^g)^k = \epsilon^{h^{-1}}$, $g, h \in G$, $k \in G_\alpha$.]

ii) G is k-transitive if, and only if, is of rank k, i.e. G_α is transitive on $\Omega \setminus \{\alpha\}$.

iii) G is k-transitive if, and only if, the stabilizer $G_{\alpha_1} \cap G_{\alpha_2} \cap \ldots \cap G_{\alpha_k}$ of any k points is transitive on the remaining ones.

74. A^n is $n-2$ transitive, but not $n-1$ transitive. [*Hint:* consider the stabilizer H of a point, n say, and let it act on the remaining points; induct on n.]

75. *i)* Prove that if G is k-transitive, then

$$|G| = n(n-1)\cdots(n-k+1)|G_{\alpha_1} \cap G_{\alpha_2} \cap \ldots \cap G_{\alpha_k}|;$$

ii) if G is sharply k-transitive, then $|G| = n(n-1)\cdots(n-k+1)$ (hence sharply k-transitive means that G is regular as a group acting on the set of ordered k-tuples).

iii) A^n is sharply $(n-2)$-transitive.

76. Let G be of degree n and k-transitive, $k \geq 2$. Prove that:

i) if G contains a transposition, then $G = S^n$;

ii) if G contains a 3-cycle, then $G \supseteq A^n$.

77. If χ_k is the character of G on Ω_k (see the above definition) prove that $\chi_k(g) = \chi(g)(\chi(g)-1)\cdots(\chi(g)-k+1)$. Conclude that G is k-transitive if, and only if, $|G| = \sum_{g \in G} \chi_k(g)$.

78. Embed the two groups of order p^3 of *Ex.* 2.10, 6 in S^n, $n = p^3$, via the regular representation, and prove directly that their images in S^n are almost conjugate.

79. Determine two actions of the Klein group on a 12-element set with the same character but not equivalent.

Definition 3.10. An element of $\sigma \in S^n$ has a *descent* at i if $\sigma(i) \leq i$. The descent is *strict* if $\sigma(i) < i$.

80. *i)* Let d_σ be the number of descents of σ. Show that for a subgroup G of S^n we have $\frac{1}{|G|}\sum_{\sigma \in G} d_\sigma = \frac{n+N}{2}$, where N is the number of orbits of G. [*Hint*: mimic the proof of Theorem 3.18 by marking a point on the grid of page 113 if $\sigma(i) \leq i$. The total number of marked points on the rows is $\sum_{\sigma \in G} d_\sigma$, and on the columns is $\sum_{i=1}^n |\{\sigma | \sigma(i) \leq i\}|$. Split the latter sum into the contributions of the single orbits, each orbit giving a contribution equal to $|G|(n_k+1)/2$, where n_k is the cardinality of the k-th orbit.]

ii) Let d'_σ be the number of strict descent of σ. Show that $\frac{1}{|G|}\sum_{\sigma \in G} d'_\sigma = \frac{n-N}{2}$.

iii) The difference $d_\sigma - d'_\sigma$ equals $\chi(g)$. One obtains Burnside's formula (3.6) subtracting sidewise the equality of *ii)* from that of *i)*.

81. Let $G \leq S^n$, $\sigma \in G$, and let $z_k(\sigma)$ be the number of k-cycles of σ. If C_k is the set of k-cycles appearing in the cycle decomposition of the elements of G, the group G acts by conjugation on C_k: if $c = (1, 2, \ldots, k)$ is a k-cycle of $\tau \in G$, $c^\sigma = (1^\sigma, 2^\sigma, \ldots, k^\sigma)$ is a k-cycle of $\tau^\sigma \in G$. Show that:

i) the elements of G containing a given cycle $c \in \sigma$ are those belonging to the coset $H\sigma$, where H is the subgroup of the elements of G fixing the points appearing in the cycle;
ii) the stabilizer of c is $G_c = \bigcup_{i=0}^{k-1} H\sigma^i$, where σ contains c (disjoint union);
iii) the orbit of c has cardinality $|c^G| = |G|/k|H|$;
iv) the total number of orbits of G on C_k is $P_k = \frac{1}{|G|}\sum_{\sigma \in G} kz_k(\sigma)$ (P_1 is the number of orbits of G on the n points);
v) for $k = 1$, Burnside's formula (3.6) is obtained (the action is on the cycles of length 1, i.e. on the n digits);
vi) if $C = \bigcup_{k=1}^{n} C_k$, in the action of G by conjugation on C the number of orbits is n;
(The n-tuple $P = (P_1, P_2, \ldots, P_n)$ is the *Parker vector* of the group G.)
vii) the Parker vector of S^n is (1,1,...,1), n times.

3.4 Induced Actions

In this section we see how an action of a subgroup can be extended to an action of the group.

Let $H \leq G$ be of finite index, and let an action of H on a set Ω be given. Let T be a transversal of the right cosets of H. If $g \in G$ and $x \in T$, the element xg belongs to a certain coset Hy, $y \in T$; let h be the unique element of H such that $xg = hy$. Then an action of G on the product set $\Omega \times T$ can be defined as follows: $(\alpha, x)^g = (\alpha^h, y)$. It is an action: if $g, g_1 \in G$, and $yg_1 = h_1 z$, $z \in T$, then $((\alpha, x)^g)^{g_1} = (\alpha^h, y)^{g_1} = (\alpha^{hh_1}, z)$, and since $x(gg_1) = (xg)g_1 = (hy)g_1 = h(yg_1) = hh_1 z$ we also have $(\alpha, x)^{gg_1} = (\alpha^{hh_1}, z)$.

Definition 3.11. The above defined action of G on the product set $\Omega \times T$ is said to be *induced* by the action of H on Ω.

We shall see in a moment that a change of coset representatives leads to similar actions. Note that if the action of H is transitive, that of G on $\Omega \times T$ is also transitive. Indeed, given (α, x) and (β, y), there exists $h \in H$ such that $\alpha^h = \beta$; then with $g = x^{-1}hy$ we have $(\alpha, x)^g = (\beta, y)$.

Examples 3.9. 1. Let $|\Omega| = |H| = 1$. Then the elements of T are the elements of G, and the set $\Omega \times T$ is that of the pairs (α, x), where $x \in G$ and α is the unique element of Ω. Thus the degree of the action is $|G|$. If $xg = y$ (i.e. $xg = 1 \cdot y$) then $(\alpha, x)^g = (\alpha, y) = (\alpha, xg)$. It is clear that this action is similar to the one that gives rise to the regular representation of G: $x^g = xg$ (with $\varphi : x \to (\alpha, x)$). Thus, *the regular representation of a group G is induced by the representation of degree* 1 *of the identity subgroup.*

2. Let again $|\Omega| = 1$, and let H be any subgroup of G. It is clear that the induced action is now that of G on the cosets of H. If H is normal we have the regular representation of G/H.

3. Consider now the regular representation of a subgroup H. In this case $\Omega = H$. The action of G on the pairs (h, x) is given by $(h, x)^g = (h^{h_1}, y) = (hh_1, y)$, where $xg = h_1 y$. If $a \in G$, a has a unique expression as a product $a = hx$. Therefore, the correspondence $\varphi : a \to (h, x)$ is one-to-one: $\varphi(a)^g = (h, x)^g = (hh_1, y)$, and since $ag = hxg = hh_1 y$, $\varphi(a^g) = \varphi(ag) = (hh_1, y)$. Hence, *the regular representation of a group is induced by the regular representation of any of its subgroups.*

Let us now show that the induced action does not depend on the choice of T. Let $T = \{t_1, t_2, \ldots, t_m\}$ and $T' = \{y_1, y_2, \ldots, y_m\}$. The idea of the proof is to go from T to T' by replacing the t_i with the y_i one at a time. To fix ideas, pick t_1 and consider $T_1 = \{y_1, x_2, \ldots, x_m\}$. The action of G on $\Omega \times T$ is similar to that on $\Omega \times T_1$. Indeed, define $\varphi : \Omega \times T \to \Omega \times T_1$ as follows. If $y_1 = h_1 t_1$, for the pairs whose second element is t_1 set $\varphi(\alpha, t_1) = (\alpha^{h_1^{-1}}, y_1)$ (the mapping $\alpha \to \alpha^{h_1^{-1}}$ is a permutation of Ω); for the other pairs take for φ the identity mapping. For the first pairs we have, if $t_1 g = ht_j$ with $j \neq 1$, that $\varphi((\alpha, t_1)^g) = \varphi(\alpha^h, t_j) = (\alpha^h, t_j) = (\alpha^{h_1^{-1} h_1 h}, t_j) = (\alpha^{h_1^{-1}}, y_1)^g = (\varphi(\alpha, t_1))^g$, because $y_1 g = h_1 t_1 g = h_1 h t_j$. If $j = 1$, then $\varphi((\alpha, t_1)^g) = \varphi(\alpha^h, t_1) = (\alpha^{hh_1^{-1}}, y_1)$; on the other hand, $(\varphi(\alpha, t_1))^g = (\alpha^{h_1^{-1}}, y_1)^g = (\alpha^{hh_1^{-1}}, y_1)$ since $y_1 g = h_1 t_1 g = h_1 h t_1 = h_1 h h_1^{-1} y_1$. For the other pairs, if $t_i g = h' t_k$, and $k \neq 1$, then $\varphi((\alpha, t_i)^g) = \varphi(\alpha^{h'}, t_k) = (\alpha^{h'}, t_k) = (\alpha, t_i)^g = \varphi(\alpha, t_i)^g$. If $k = 1$, then $\varphi(\alpha, t_i)^g) = \varphi(\alpha^{h'}, t_1) = (\alpha^{h' h_1^{-1}}, y_1) = (\alpha, t_i)^g = (\varphi(\alpha, t_i))^g$.

Hence, the actions of G on $\Omega \times T$ and $\Omega \times T_1$ are similar. In the same way, one sees that the action of G on $\Omega \times T_1$ is similar to that on $\Omega \times T_2$, where T_2 is obtained by replacing an element of T_1, t_2 say: $T_2 = \{y_1, y_2, t_3, \ldots, x_m\}$. Proceeding in this way, the action of G on $\Omega \times T$ proves to be similar to that on $\Omega \times T'$.

Now let H and K be two subgroups of G, with $H \subseteq K$. If H acts on Ω, consider the action of K induced by that of H, and then that of G induced by that of K. As we shall see in the following theorem the result is the same as that obtained by inducing directly from H to G. Recall that if $T_1 = \{t_1, t_2, \ldots\}$ is a transversal for H in K and $T_2 = \{y_1, y_2, \ldots\}$ one for K in G, then $T = \{t_i y_j, t_i \in T_1, y_j \in T_2\}$ is a transversal for H in G.

Theorem 3.22 (Transitivity of induction). *Let $H \subseteq K$ be two subgroups of G with H acting on Ω. Then the action induced from H to G is similar to that obtained inducing first from H to K and then from K to G.*

Proof. With the above notation, the action induced from H to K on $\Omega \times T_1$ is given by $(\alpha, t_i)^k = (\alpha^h, t_j)$, where $t_i k = ht_j$, and that induced from K to G (on $(\Omega \times T_1) \times T_2$) is $((\alpha, t_i), y_j)^g = ((\alpha, t_i)^k, y_s) = ((\alpha^h, t_j), y_s)$, where $y_j g = ky_s$. But $t_i y_j g = ht_j y_s$, and therefore the action of G on $\Omega \times T$ is $(\alpha, t_i y_j)^g = (\alpha^h, t_j y_s)$. The correspondence $\varphi : (\alpha, t_i y_j) \to ((\alpha, t_i), y_j)$ is one-to-one, and, for what we have just proved, preserves the action of G. ♢

Let us now investigate how characters behave with respect to induction. Denote by χ^G the character of G induced by the character χ of a subgroup,

Theorem 3.23. *Let $H \leq G$, χ a character of H. Then:*
i) $\chi^G(g) = 0$, if g is not conjugate to an element of H;
ii) put $\chi(y) = 0$ if $y \notin H$; then:

$$\chi^G(g) = \frac{1}{|H|} \sum_{x \in G} \chi(xgx^{-1}). \tag{3.11}$$

Proof. *i)* If in the induced action $g \in G$ fixes a point $(\alpha, t_i) \in \Omega \times T$, then from $t_i g = ht_j$ we have $(\alpha, t_i)^g = (\alpha^h, t_j) = (\alpha, t_i)$, and in particular, $t_i = t_j$, and therefore $t_i g t_i^{-1} = h \in H$.

ii) From the proof of *i)* we have, if m is the index of H, $\chi^G(g) = \sum_{i=1}^m \chi(t_i g t_i^{-1})$. Now, if $h \in H$, by *i)* of Theorem 3.19 $\chi(ht_i g t_i^{-1} h^{-1}) = \chi(t_i g t_i^{-1})$, and therefore

$$\chi(t_i g t_i^{-1}) = \frac{1}{|H|} \sum_{h \in H} \chi(ht_i g t_i i^{-1} h^{-1}).$$

It follows

$$\sum_{i=1}^m \chi(t_i g t_i^{-1}) = \frac{1}{|H|} \sum_{i=1}^m \sum_{h \in H} \chi(ht_i g t_i^{-1} h^{-1}).$$

As h varies in H, and t_i in the transversal T for H, the products ht_i run over the elements of G. The result follows. ◇

Exercises

82. Let $H \leq G$, K be the kernel of an action of H, K_1 that of the action of G induced by the action of H. Show that if $K \trianglelefteq G$ then $K \subseteq K_1$.

83. Denote by τ_H the character of the restriction to the subgroup H of the character τ of G. Show that if χ is a character of H then $\chi^G \tau = (\chi \tau_H)^G$.

84. Let $G = HK$, $H, K \leq G$, and let an action of H on a set Ω be given. Prove that the following actions of K are equivalent:

i) the action induced to G by that of H and then restricted to K;
ii) the restriction of the action of H to $H \cap K$ and then extended to K.

85. (Mackey). Let $G = \bigcup_a HaK$ be the partition of G into double cosets of H and K. Let T be a transversal of the right cosets of H in G, and let T_a be the set of elements of T belonging to HaK. If H acts on Ω, show that:

i) $\Omega \times T = \bigcup_a (\Omega \times T_a)$, and K acts on each of the $\Omega \times T_a$;
ii) the action of K on $\Omega \times T_a$ is similar to that of K induced by the action of $a^{-1}Ha \cap K$ on Ω defined by $\alpha^x = \alpha^{axa^{-1}}$, for $x \in a^{-1}Ha \cap K$.

86. *i)* Prove that if $H \leq G$, and χ is a character of H, then (3.11) may be written as

$$\chi^G(g) = |\mathbf{C}_G(g)| \sum_{i=1}^r \frac{\chi(x_i)}{|\mathbf{C}_H(x_i)|}, \tag{3.12}$$

where the $h_i, i = 1, 2, \ldots, r$, are representatives of the conjugacy classes of H contained in that of g in G, and where $\chi^G(g) = 0$ if $r = 0$, i.e. $H \cap \mathbf{cl}(g) = \emptyset$. [*Hint*: if $r = 0$, there are no conjugates of g in H, and from (3.11) we have $\chi(xgx^{-1}) = 0$ and $\chi^G(g) = 0$. If $r > 0$, let xgx^{-1} be conjugate in H to one of the h_i; in this case, $\chi(xgx^{-1})$ equals $\chi(h_i)$, and this happens $|\mathbf{C}_G(g)|$ times; there are $[H : \mathbf{C}_H(h_i)]$ elements in the class of h_i.]

ii) Let k_i be the order of the class of h_i; then (3.12) can be written as

$$\chi^G(g) = \frac{|G|}{|H|} \sum_{i=1}^{r} \frac{k_i}{|\mathbf{cl}_G(g)|} \chi(h_i).$$

3.5 Permutations Commuting with an Action

Let A_G be the set of permutations of Ω commuting with the action of G:

$$A_G = \{\varphi \in S^\Omega \mid \varphi(\alpha^g) = \varphi(\alpha)^g\}$$

for all $\alpha \in \Omega$ and $g \in G$. A_G is a group, subgroup of S^Ω. Indeed, $1 \in A_G$; if $\varphi, \psi \in A_G$ then

$$(\varphi\psi)(\alpha^g) = \varphi(\psi(\alpha^g)) = \varphi(\psi(\alpha)^g) = (\varphi(\psi(\alpha)))^g = ((\varphi\psi)(\alpha))^g.$$

Moreover, with $\alpha = \varphi(\beta)$,

$$\varphi^{-1}(\alpha^g) = \varphi^{-1}(\varphi(\beta)^g) = \varphi^{-1}(\varphi(\beta^g)) = \beta^g = \varphi^{-1}(\alpha)^g,$$

and so also $\varphi^{-1} \in A_G$.

Definition 3.12. The group A_G is the *automorphism group of the group G acting on Ω*. If $G \leq S^n$, then A_G is the centralizer of G in S^n.

Remark 3.5. It goes without saying that the group A_G is not to be confused with the group $\mathbf{Aut}(G)$.

Our purpose now is to determine for a transitive group G the structure of A_G from that of G. The result will be that A_G is isomorphic to a quotient of a subgroup of G[13]. This is a further demonstration that transitivity allows us to work inside the group.

Throughout this section Ω will be finite, $|\Omega| = n$. Let us consider a few properties of A_G in the case of some special actions.

Definition 3.13. A group G acting on Ω is *semiregular* if $G_\alpha = \{1\}$ for every $\alpha \in \Omega$. It is *regular* if it is both semiregular and transitive.

[13] A quotient of a subgroup of a group G is also called a *section* of G.

Remarks 3.6. **1.** A semiregular group is necessarily a permutation group because $G_\alpha = \{1\}$ for every $\alpha \in \Omega$ implies that the kernel of the action is $\{1\}$.

2. In Galois theory, let $K' = K(\alpha_1, \alpha_2, \ldots, \alpha_n)$ be the extension of the field K obtained by adding the roots α_i of a polynomial $f(x)$. Then the Galois group G of $f(x)$, i.e. the group of permutations of the α_i that preserve all relations among them, is a semiregular group if, and only if, $f(x)$ is a normal polynomial, that is, if the α_i may all be expressed as a rational function (polynomial) of any one of them (each α_i is then a primitive element of the field K'). If in addition $f(x)$ is irreducible over K, then G is also transitive, and therefore regular. In general, if γ is a primitive element of K', i.e. $K' = K(\gamma)$, and G_1 is the Galois group of the minimal polynomial $g(x)$ of γ over K, then G_1 is isomorphic to G as an abstract group, and as a permutation group of the roots of $g(x)$ is regular (G_1 affords the regular representation of G).

Let G be semiregular, and let $\alpha^g = \beta$ and also $\alpha^h = \beta$. Then $\alpha^g = \alpha^h$, $\alpha^{gh^{-1}} = \alpha$ so that $gh^{-1} = 1$ and $g = h$. Thus, semiregular means that given α and β, there exists at most one element of G taking α to β. If G is transitive, such an element exists; therefore, regular means that, given α and β, there exists exactly one element taking α to β. Note also that a subgroup H of a semiregular group is itself semiregular.

Theorem 3.24. *Let G be semiregular on Ω. Then all the orbits of G have the same cardinality, namely $|G|$. In particular, if G is regular, $|G| = |\Omega|$, i.e. order and degree are equal.*

Proof. Under the given assumptions, (3.1) means $|\alpha^G| = [G : G_\alpha] = [G : \{1\}] = |G|$. If G is regular, then $\alpha^G = \Omega$, and $|\Omega| = |G|$. ◊

It follows that a proper subgroup H of a regular group cannot be regular (otherwise $|H| = |\Omega| = |G|$).

Corollary 3.14. *Let G be a semiregular group. Then the cycles of an element of G all have the same length.*

Proof. The subgroup generated by $g \in G$ is itself semiregular, and its orbits are the cycles of g. ◊

Definition 3.14. A permutation is *regular* if its cycles all have the same length.

Thus, a group is semiregular if its elements are regular permutations.

Theorem 3.25. *If G is transitive, then A_G is semiregular.*

Proof. It is enough to show that if $\varphi \in A_G$ fixes a point, then it fixes all the points. Let $\varphi(\alpha) = \alpha$, and let $\beta \in \Omega$. By transitivity, there exists $g \in G$ such that $\alpha^g = \beta$. Then $\varphi(\beta) = \varphi(\alpha^g) = \varphi(\alpha)^g = \alpha^g = \beta$. ◊

Theorem 3.26. *Let G be a transitive group acting on a set Ω. Then two elements α and β of Ω have the same stabilizer if, and only if, there exists $\varphi \in A_G$ such that $\varphi(\alpha) = \beta$.*

Proof. First we show that if $\varphi(\alpha) = \beta$ then $G_\alpha = G_\beta$. We have

$$g \in G_\alpha \Leftrightarrow \alpha^g = \alpha \Leftrightarrow \varphi(\alpha^g) = \varphi(\alpha) \Leftrightarrow \varphi(\alpha)^g = \varphi(\alpha) \Leftrightarrow g \in G_{\varphi(\alpha)} = G_\beta.$$

Suppose now $G_\alpha = G_\beta$; we construct φ as follows. By the transitivity of G, as g varies in G, α^g runs over all the elements of Ω; we define $\varphi : \alpha^g \to \beta^g$. φ is well defined and injective: $\alpha^g = \alpha^h \Leftrightarrow gh^{-1} \in G_\alpha = G_\beta \Leftrightarrow \beta^g = \beta^h$. It is surjective: if $\gamma \in \Omega$, there exists $g \in G$ such that $\beta^g = \gamma$, so that $\alpha^g \to \gamma$.

It remains to show that $\varphi \in A_G$, namely that φ commutes with the action of G. If $\gamma \in \Omega$, then $\gamma = \alpha^g$; it follows $\varphi(\gamma^h) = \varphi((\alpha^g)^h) = \varphi(\alpha^{gh}) = \beta^{gh} = (\beta^g)^h = \varphi(\alpha^g)^h = \varphi(\gamma)^h$ for all $h \in G$. ◊

Lemma 3.5. *Let G be transitive on Ω, G_α the stabilizer of a point, and let $\Delta = \{\beta \in \Omega \mid G_\beta = G_\alpha\}$. Then the normalizer of G_α is transitive on Δ.*

Proof. First let us show that $N = \mathbf{N}_G(G_\alpha)$ actually acts on Δ, namely, if $\beta \in \Delta$ and $g \in N$, then $\beta^g \in \Delta$. Indeed, $G_\alpha = (G_\alpha)^g = (G_\beta)^g = G_{\beta^g}$, so that the stabilizer of β^g is G_α, i.e. $\beta^g \in \Delta$.

We shall show more than the transitivity of N on Δ, namely that given $\gamma, \delta \in \Delta$ an element of G taking γ to δ, which exists by the transitivity of G, already belongs to N. Indeed, let $g \in G$ be such that $\gamma^g = \delta$. Then for all $x \in G_\alpha$ we have $\gamma^{gx} = \delta^x = \delta = \gamma^g$, so that $gxg^{-1} \in G_\gamma = G_\alpha$, $gG_\alpha g^{-1} \subseteq G_\alpha$ and $g^{-1}G_\alpha g = G_\alpha$. ◊

We now show that the structure of A_G is determined by that of G.

Theorem 3.27. *Let G be transitive. Then the group A_G is isomorphic to the quotient $\mathbf{N}_G(G_\alpha)/G_\alpha$.*

Proof. Set $H = G_\alpha$ and $N = \mathbf{N}_G(H)$. The proof consists in showing how an element of A_G determines an element of N/H. We have seen in the above lemma that N acts on Δ; the kernel of this action is H, so that it is the quotient N/H that actually acts on Δ, and the action is transitive. If $g \in N$ fixes an element of Δ, then $g \in H$ (the elements of Δ are those whose stabilizer is G_α). It follows that N/H moves all the points of Δ, so that given α and β in Δ there exists a unique element $Hg \in N/H$ such that $\alpha^{Hg} = \beta$. Now, if $\varphi \in A_G$, α and $\varphi(\alpha)$ have the same stabilizer (Theorem 3.26), and therefore, since $\alpha \in \Delta$, they both belong to Δ. Thus there exists $Hg \in N/H$ such that $\varphi(\alpha)^{Hg} = \alpha$, and such an element is unique. In this way, φ determines a unique element Hg, the one that takes $\varphi(\alpha)$ to α. Therefore, the mapping $A_G \to N/H$ defined by $\varphi \to Hg$ is well defined. It is injective, because if $\psi \to Hg$, then $\varphi(\alpha)^{Hg} = \psi(\alpha)^{Hg}$, and therefore $\varphi(\alpha) = \psi(\alpha)$, contradicting A_G being semiregular. Let us show that it is surjective. Given $Hg \in N/H$, α and α^{Hg} both belong to Δ, and therefore have the same stabilizer, i.e. H. Then by Theorem 3.26 there exists $\varphi \in A_G$ such that $\varphi(\alpha^{Hg}) = \alpha$. But $\alpha = \varphi(\alpha^{Hg}) = \varphi(\alpha^g) = \varphi(\alpha)^g = \varphi(\alpha)^{Hg}$(among other things, this sequence of equalities shows that φ commutes with the action of N/H on Δ). Thus the mapping is surjective. Finally, we show that it is a homomorphism. If $\varphi \to Hg$ and $\psi \to Hg_1$ then, by first letting act φ and then ψ, we have $\psi\varphi(\alpha) = \psi(\varphi(\alpha)) = \psi(\alpha^{Hg^{-1}}) = \psi(\alpha)^{Hg^{-1}} = \alpha^{Hg_1^{-1}Hg^{-1}} = \alpha^{H(gg_1)^{-1}}$, from which $\psi\varphi(\alpha)^{Hgg_1} = \alpha$. This shows that $\varphi\psi \to Hg \cdot Hg_1$, as required. ♢

Corollary 3.15. $|A_G| = |\Delta|$, *that is, the order of* A_G *equals the number of elements fixed by* G_α.

Proof. $\mathbf{N}_G(G_\alpha)/G_\alpha$ acts transitively on Δ, and the stabilizer of a point is the identity. By (3.1), $|\Delta| = |\mathbf{N}_G(G_\alpha)/G_\alpha|$ and the result follows. ♢

We recall that if G is a permutation group, $G \leq S^n$, then A_G is the centralizer of G in S^n.

Corollary 3.16. *Let* G *be a regular permutation group. Then* G *is isomorphic to its own centralizer.*

Proof. G being regular, $G_\alpha = 1$. It follows $G \simeq A_G$. ♢

Corollary 3.17. *Let* $|G| = n$, *and let* G_r *and* G_l *be the images of* G *in* S^n *under the right and left regular representations, respectively. Then these two subgroups centralize each other. In particular, if* G *is abelian, they coincide.*

Proof. Let $r_x \in G_r$ and $l_y \in G_l$. Then for all $a \in G$ we have $r_x l_y(a) = r_x(y^{-1}a) = (y^{-1}a)x = y^{-1}(ax) = l_y(r_x(a)) = l_y r_x(a)$, and this shows that G_r and G_l centralize each other; denoting C the centralizer we have $G_r \subseteq C(G_l)$ and $G_l \subseteq C(G_r)$. G_l being regular, it is isomorphic to its own centralizer (Corollary3.16), $G_r \subseteq C(G_l) \simeq G_l$, and since $|G_r| = |G_l|$ (they are both equal to $|G|$), we have $G_r = C(G_l)$. Similarly, $G_l = C(G_r)$. Note that the intersection $G_r \cap G_l$ is the image of the center of G in both representations, so that, if G is abelian, then $G_r = G_l$. ♢

Let A be the image of $\mathbf{Aut}(G)$ in S^n. An automorphism fixes 1, and $1^x = x$, so that $G_r \cap A = \{1\}$. Now, A normalizes G_r. Indeed, let $x \in G$, and let r_x be the image of x in S^n, and let $(x_1, x_1x, \ldots, x_1x^k)$ be a cycle of r_x. If σ_α is the image of the automorphism α in S^n, then $\sigma_\alpha(x_i) = x_i^\alpha$, and therefore, by conjugating the above cycle of r_x with σ_α, we obtain $(x_1^\alpha, (x_1x)^\alpha, \ldots, (x_1x^k)^\alpha) = (x_1^\alpha, x_1^\alpha x^\alpha, \ldots, x_1^\alpha(x^\alpha)^k)$, which is a cycle of r_{x^α}. Let us show that the semidirect product of G_r by A is the normalizer of G_r in S^n. Let $\tau \in \mathbf{N}_{S^n}(G_r)$; then τ induces, by conjugation, an automorphism φ of G_r, and this is given by an element of A. It follows $r_x^\tau = r_x^\varphi$, for all $r_x \in G_r$, so that $\tau\varphi^{-1}$ centralizes G_r and therefore belongs to G_l. But $G_l \subseteq G_rA$. Indeed, let $l_x \in G_l$ and let $p_x \in A$ be the conjugation by x; then, for all $g \in G$, $r_{x^{-1}}p_x(g) = r_{x^{-1}}(x^{-1}gx) = x^{-1}gx \cdot x^{-1} = x^{-1}g = l_x(g)$, so that $l_x = r_{x^{-1}}p_x \in G_rA$. Finally, since $\tau\varphi^{-1} \in G_l$ and $\varphi \in A$ we have the result.

Summing up (see *Ex.* 2.10, 6):

Theorem 3.28. *The normalizer of G_r in S^n is the holomorph of G.*

When G is not abelian, the correspondence $\varphi : g \rightarrow g^{-1}$ is not an automorphism of G, and therefore it does not belong to the holomorph of G. But $\varphi^{-1}r_x\varphi(g) = \varphi^{-1}r_x(g^{-1}) = \varphi^{-1}(g^{-1}x) = x^{-1}g = l_x(g)$, and therefore φ interchanges G_r and G_l. Moreover, φ centralizes A because, if $\alpha \in \mathbf{Aut}(G)$, then $\alpha\varphi(x) = \alpha(x^{-1}) = (\alpha(x))^{-1} = \varphi\alpha(x)$, $x \in G$. Thus φ normalizes $K = G_rA$, and in the semidirect product $\langle\varphi\rangle K$ the subgroups G_r and G_l are conjugate.

Exercises

87. Determine a transitive and faithful but not regular action of D_4.

88. Prove the equivalence of the following two statements:
i) the only faithful and transitive action of the group G is the regular one;
ii) every subgroup of G of prime order is normal.

From now on G will be a permutation group.

89. Let H be a regular subgroup of G; prove that $G = HG_\alpha$ with $H \cap G_\alpha = 1$. [*Hint*: Theorem 3.9.]

90. Let $H \leq G$, H regular. Given $\alpha \in \Omega$, we obtain a one-to-one correspondence $\Omega \rightarrow H$ if we associate with $\gamma \in \Omega$ the unique (H is regular) element of H taking α to γ; note that $\alpha \rightarrow 1$. Thus, one can make G act on H by defining h^g as the unique element of H that takes α to α^{hg}. Prove that:
i) if $k \in H$ then $h^k = hk$;
ii) if $g \in G_\alpha$ then $h^g = g^{-1}hg$.

91. If the centralizer of a subgroup is transitive, then the subgroup is semiregular.

92. A transitive abelian group is regular.

93. Let $H \leq G$, H transitive and abelian. Prove that H equals its own centralizer. [*Hint*: $\mathbf{C}_G(\mathbf{C}_G(H)) \supseteq H$. Then, $\mathbf{C}_G(\mathbf{C}_G(H)) = H$ (H is said to have the *double centralizer property*).]

94. If G is transitive, then $x \in \mathbf{Z}(G)$ is regular.

95. An element of S^n is regular if, and only if, it is a power of an n-cycle.

Definition 3.15. If G is transitive, a subset Δ of Ω is a *block* (*imprimitivity system*) if either $\Delta^g = \Delta$ or $\Delta^g \cap \Delta = \emptyset$, for all $g \in G$. The singletons, the whole set Ω and the empty set are *trivial* blocks; if these are the only blocks, G is *primitive*, otherwise G is *imprimitive*[14].

The blocks are the classes of an equivalence relation ρ on Ω *compatible with the action of* G, i.e. $\alpha\rho\beta \Rightarrow \alpha^g\rho\beta^g$, $g \in G$. In particular, the blocks all have the same cardinality, which therefore divides the degree of G.

[14] With reference to Remark 3.6, 2, the Galois group of $f(x)$ is primitive as a permutation group if, and only if, every element of the field K' is primitive.

A group can be imprimitive in more than one way. For instance, in S^6, the subgroup generated by the cycle $\sigma = (1,2,3,4,5,6)$ has the blocks $\Delta_1 = \{1,3,5\}, \Delta_2 = \{2,4,6\}$, but also the blocks $\Gamma_1 = \{1,4\}, \Gamma_2 = \{2,5\}, \Gamma_3 = \{3,6\}$ (corresponding to the cycles of σ^2 and σ^3, respectively).

96. The intersection of two blocks is a block.

97. If $H \leq G$, a block of G is also a block of H. If Δ is a block of H, Δ^g is a block of H^g.

98. A 2-transitive group is primitive.

99. Let $\alpha \in \Omega$. Prove that there exists a one-to-one correspondence between the set of blocks containing α, $\mathcal{D} = \{\Delta \mid \alpha \in \Delta\}$, and the set of subgroups H of G containing G_α: $\mathcal{S} = \{H \leq G \mid G_\alpha \subseteq H\}$, by showing that:
i) if $H \in \mathcal{S}$ e $\alpha \in \Omega$, the orbit α^H is a block;
ii) the mapping $\theta : \mathcal{S} \to \mathcal{D}$ given by $H \to \alpha^H$ is injective;
iii) if $\Delta \in \mathcal{D}$, the set $\theta'(\Delta) = \{x \in G \mid \alpha^x \in \Delta\}$ belongs to $\mathcal{S}$;
iv) θ and θ' are inverse of each other.

100. A transitive group is primitive if, and only if, the stabilizer of a point is a maximal subgroup.

101. A transitive group on p elements, p a prime, is primitive. [*Hint*: G_α has index p.]

102. If $N \trianglelefteq G$ and Δ is an orbit of N, then Δ is a block of G. Hence, a non trivial normal subgroup of a primitive group is transitive. [*Hint*: Δ^x is an orbit of $N^x = N$.]

103. If G is primitive, then either $\mathbf{Z}(G) = \{1\}$ or $|G| = p$, a prime.

104. The group D_4 acting on the vertices of a square is a transitive group which is not primitive. Show that there exists a non trivial equivalence relation compatible with the action of the group. [*Hint*: let the vertices be 1,2,3,4 in the circular order. Then {1,3} and {2,4} are non trivial blocks (corresponding to the orbits of the normal subgroup $\{I, (1,2)(3,4)\}$, the stabilizer of 4 is {1,(2,3)} which is not maximal, and $\{I, (1,3)(2,4)\}$ is normal but not transitive.]

105. If G is primitive and the stabilizer of a point is a simple group, then either G is simple, or it contains a regular normal subgroup.

106. Let G be the direct product of two regular subgroups H_1 and H_2. Prove that:
i) $H_1 \simeq H_2$;
ii) if the two subgroups are simple, then G is primitive.

107. Prove that A^n is simple, $n \geq 6$, assuming the simplicity of A^5 and using $n = 5$ as basis of induction, by showing that:
i) A^n is primitive;
ii) if A^n is not simple, then it contains a regular normal subgroup;
iii) a normal subgroup of A^n cannot be regular.

3.6 Automorphisms of Symmetric Groups

We devote this section to proving a theorem that describes the automorphisms of symmetric groups. We will see that the case $n = 6$ is special. First a lemma[15].

Lemma 3.6. *An automorphism γ of S^n sending transpositions to transpositions is inner.*

Proof. Let $T(i)$ be the set of transpositions containing the digit i. Let us show that, for a certain j, $T(i)^\gamma = T(j)$. Since two transpositions of $T(i)$ do not commute, two transpositions of $T(i)^\gamma$ do not commute either; therefore, they all have one and the same digit in common (if $(j, u), (j, v)$ and (u, v) belong to $T(i)^\gamma$, their product has order 2, whereas the product of three elements of $T(i)$ has order 4). It follows $T(i)^\gamma \subseteq T(j)$, for a certain j, and by applying to $T(j)$ the inverse of γ we have equality. Thus, with γ there is associated the permutation σ such that $\sigma(i) = j$ if $T(i)^\gamma = T(j)$, $i = 1, 2, \ldots, n$. We show that γ coincides with the inner automorphism induced by σ, and since every element of S^n is a product of transpositions, it suffices to show that this happens for the transpositions. Observing that $T(i) \cap T(j) = (i, j)$, we have $(i, j)^\gamma = T(i) \cap T(j) = T(i^\sigma) \cap T(j^\sigma) = (i^\sigma, j^\sigma) = \sigma^{-1}(i, j)\sigma$, as required. ◇

An automorphism γ preserves the order of the elements and sends conjugacy classes to conjugacy classes. It follows that γ sends the conjugacy class C_1 of the transpositions to a class C_k of k disjoint transpositions. For a given k, the class C_k contains:

$$t_k = \frac{1}{k!}\binom{n}{2}\binom{n-2}{2}\cdots\binom{n-2(k-1)}{2} = \frac{1}{k!2^k}n(n-1)\cdots(n-2k+1)$$

elements, so that, if $C_1^\gamma = C_k$ with $k > 1$, it must be $t_1 = t_k$, from which $(n-2)(n-3)\cdots(n-2k+1) = k!2^{k-1}$. The right hand side is positive, so $n \geq 2k$, and

$$(n-2)(n-3)\cdots(n-2k+1) \geq (2k-2)(2k-3)\cdots(2k-2k+1) = (2k-2)!$$

For $k = 2$ this equality becomes $(n-2)(n-3) = 4$, which is satisfied by no n. For $k = 3$, since $n \geq 6$, the equality is satisfied for $n = 6$, whereas if $n > 6$ the left hand side is at least $5 \cdot 4 \cdot 3 \cdot 2 = 120$, while the right hand side is $3!2^2 = 24$. If $k \geq 4$ one always has $(2k-2)! > k!2^{k-1}$ (induction on k:

$$(2(k+1)-2)! = (2k)! = 2k(2k-1)(2k-2)! > 2k(2k-1)k!2^{k-1} = k(2k-1)k!2^k$$

and since $k(2k-1) > k+1$ the last quantity is greater than $(k+1)!2^k$). It follows that for, $n \neq 6$, there are no classes of involutions that are not transpositions but that have the same cardinality of the class of the transpositions. From the above lemma it follows:

[15] We follow Kargapolov-Merzliakov, Theorem 5.3.1.

Theorem 3.29. *Let $n \neq 6$. Then:*

i) (Hölder) the automorphisms of S^n are all inner;

ii) the subgroups of S^n of index n are the stabilizers of the digits $1, 2, \ldots, n$.

Proof. i) follows from the lemma.

ii) Let $[S^n : H] = n$, and consider S^n in its action on the cosets of H by identifying the digits $1, 2, \ldots, n$ with the cosets of H (let $H = 1$). The homomorphism $\varphi : S^n \to S^n$ induced by this action has kernel $\{1\}$, and therefore is an automorphism of S^n. The image $\varphi(H)$ of H fixes the digit 1; it follows that $\varphi(H) \subseteq S^{n-1}$, and $\varphi(H) = S^{n-1}$. But the automorphisms of S^n are all inner: $\varphi = \gamma_\sigma$ for a certain σ, so that $\varphi(H) = \gamma_\sigma(H)$, $H = (S^{n-1})^{\sigma^{-1}}$. H being a conjugate of the stabilizer of a digit, it is itself the stabilizer of a digit: if γ fixes i, then $\sigma^{-1}\gamma\sigma$ fixes i^σ. ◇

Remark 3.7. This result has an interesting consequence in the theory of equations. Let $f(x) = x^n + a_1 x^{n-1} + a_2 x^{n-2} + \cdots + a_n$ be a polynomial over a field K, $\alpha_1, \alpha_2, \ldots, \alpha_n$ its distinct roots. Let $\varphi = \varphi_1, \varphi_2, \ldots, \varphi_n$ be n rational functions of the n roots that are transitively permuted by the $n!$ permutations of the α_i, and let H be the stabilizer of one of the functions φ_i. Then H has index n in S^n, and therefore, if $n \neq 6$, is the stabilizer of one of the roots, α_1 say. It follows that φ_1 is a symmetric function of the remaining roots $\alpha_2, \ldots, \alpha_n$, and therefore is a rational function of the elementary symmetric functions of these roots with coefficients in $K(\alpha_1)$. But the elementary symmetric functions of $\alpha_2, \ldots, \alpha_n$ are rational functions of the coefficients of the polynomial and of α_1, so that φ_1 is a rational function φ of α_1 with coefficients in K. It follows that it is possible to choose the indices of the φ_i so that the n functions φ_i are expressed as values of φ at the α_i: $\varphi_i = \varphi(\alpha_i)$, $i = 1, 2, \ldots, n$. For example, with $n = 3$, let $\alpha_1, \alpha_2, \alpha_3$ be the three roots of $x^3 + qx + r$, and consider the three functions $\varphi_1 = \alpha_1^2 + \alpha_2\alpha_3, \varphi_2 = \alpha_2^2 + \alpha_3\alpha_1, \varphi_3 = \alpha_3^2 + \alpha_1\alpha_2$. S^3 acts transitively on these three functions by permuting the three roots, and since $\alpha_1\alpha_2\alpha_3 = -r$, with $\varphi(x) = x - \frac{r}{x}$ we have $\varphi_i = \varphi(\alpha_i)$, $i = 1, 2, 3$[16].

S^6 admits outer automorphisms and subgroups of index n that are not stabilizers of digits. See *ex.* 108 below.

Exercise

108. *i)* Show that S^6 contains, besides the S^5 obtained fixing a point, and therefore not transitive, and conjugate to one another, six more subgroups isomorphic to S^5, also all conjugate to one another, but transitive. [*Hint*: letting S^5 act by conjugation on the set of its six 5-Sylows, S^5 embeds in S^6.]

ii) Let H be one of the transitive subgroups isomorphic to S^5. Show that H does not contain transpositions. [*Hint*: H contains a 5-cycle, and if contains a transposition then by transitivity it coincides with S^6.]

---

[16] See Todd J.A.: The 'odd' number six. Proc. Cambridge Phil. Soc. **41** (1945), 66–68. In this paper there is an example of six rational functions φ_i of six elements α_i on which S^6 acts transitively, but for which there exists no rational function $\varphi(x)$ such that $\varphi_i = \varphi(\alpha_i)$, $i = 1, 2, \ldots, 6$.

iii) Let H be as in *ii*), and let φ be the action of S^6 on the conjugates of H. Show that φ induces an automorphism of S^6.

iv) The automorphism of *iii*) is not inner. [*Hint*: if τ is the transposition $(1,2)$ and φ is inner, then $\varphi(\tau)$ would be a transposition, and therefore would have four fixed points and τ would transpose two conjugates of H fixing the other four.]

v) Prove that $\mathbf{Aut}(S^6)/\mathbf{I}(S^6) \simeq C_2$. [*Hint*: by the discussion preceding Theorem 3.29, an automorphism which is not inner interchanges the conjugacy class of the transpositions and that of the products of three disjoint transpositions.]

vi) Prove that $|\mathbf{Aut}(S^6)| = 1440$.

3.7 Permutations and Inversions

In the sequel $z(\gamma)$ denotes the number of cycles of the permutation γ, including cycles of length 1 (fixed points).

Lemma 3.7 (Serret). *Let σ be a permutation and $\tau = (i,j)$ a transposition of S^n. Then:*

$$z(\sigma\tau) = \begin{cases} z(\sigma)+1, \text{ if } i \text{ and } j \text{ belong to the same cycle of } \sigma, \\ z(\sigma)-1, \text{ otherwise.} \end{cases}$$

In the first case we say that τ disconnects *σ, in the second that τ* connects *σ.*

Proof. Let $\sigma = (i,\ldots,k,j,\ldots,s)\ldots$; then $\sigma\tau = (i,\ldots,k)(j,\ldots,s)\cdots$ and $\sigma\tau$ has one cycle more than σ. If $\sigma = (i,\ldots,k)(j,\ldots,l)\cdots$, then $\sigma\tau = (i,\ldots,t,$ $j,\ldots,s)\cdots$, and $\sigma\tau$ has one cycle less than σ. ◇

Corollary 3.18. *The minimum number of transpositions whose product is a cycle of length n is $n-1$.*

Proof. Let $\sigma = (1,2,\ldots,n) = \tau_1\tau_2\cdots\tau_k$. Multiply on the right by τ_k, then by τ_{k-1} up to τ_1: the permutation $\sigma\tau_k$ has two cycles, $\sigma\tau_k\tau_{k-1}$ at most three (exactly three if the two letters of τ_{k-1} belong to the same cycle of $\sigma\tau_k$),..., $\sigma\tau_k\tau_{k-1}\cdots\tau_1$ at most $k+1$. But the latter permutation is the identity, that has n cycles. Thus $n \leq k+1$, and $k \geq n-1$. ◇

The product of two n-cycles is an even permutation (they are both even or both odd), and the product of an n-cycle and an $(n-1)$-cycle is an odd permutation (one is even and the other one odd). Conversely:

Theorem 3.30. *i*) *An even permutation of S^n is the product of two n-cycles;* *ii*) *an odd permutation of S^n is the product of an n-cycle and an $(n-1)-$cycle.*

Proof. *i*) If $n=1$, $(1) = (1)\cdot(1)$ ((1) is even); if $n=2$, the only even permutation is the identity, and we have $I = (1,2)(1,2)$, a product of two 2-cycles. Let $n \geq 3$, and let σ be even. If $\sigma = I$, and c is an n-cycle, then

$I = cc^{-1}$. If $\sigma \neq I$, let $\sigma(1) = 2$. Let us consider the 3-cycle $(1, k, 2)$, $k \neq 1, 2$. Then $\sigma(1, k, 2) = (1)\sigma'$, with $\sigma' \in S^{n-1}$. By induction on n, $\sigma' = c'c''$ and therefore $\sigma = (1)\sigma'(1, 2, k) = (1)\sigma'(1, 2)(1, k) = (1)c' \cdot (1)c'' \cdot (1, 2)(1, k) = (1)c' \cdot (1, k)[(1)c''(1, 2)]^{(1,k)}$. Since $k \in c'$, the transposition $(1, k)$ connects $(1)c'$, and therefore $(1)c' \cdot (1, k)$ is an n-cycle. Similarly, $2 \in c''$ and therefore $(1)c''(1, 2)$ is an n-cycle, and then so is its conjugate by $(1, k)$.

ii) If $n = 1$ there are no odd permutations; if $n = 2$, $(1, 2) = (1, 2) \cdot (1)$. If $n \geq 3$ we have, as above, $\sigma(1, k, 2) = (1)\sigma'$ and $\sigma = (1)c' \cdot (1)(l)c'' \cdot (1, 2)(1, k)$. We distinguish two cases.

a) $l = 2$. Then (1,2) commutes with $(1)(2)c''$ and we have $\sigma = (1)c'(1, 2) \cdot (1)(2)c''(1, k)$, where the first factor is an n-cycle, and the second an $(n-1)$-cycle because $k \in c''$, and therefore $(1, k)$ connects $(1)c''$.

b) $l \neq 2$; as above, $\sigma = (1)c'(1, k) \cdot [(1)(l)c''(1, 2)]^{(1,k)}$, and here also the first factor is an n-cycle, and the second an $(n-1)$-cycle. ◇

Let now $\sigma = \begin{pmatrix} 1 & 2 & \dots & n \\ k_1 & k_2 & \dots & k_n \end{pmatrix}$ be an element of S^n, and consider the n-tuple $(k_1, k_2, \dots, k_n)$. An element $\alpha \in S^n$ can act on this n-tuple in two ways: either by permuting the digits, $(k_1, k_2, \dots, k_n)^\alpha = (\alpha(k_1), \alpha(k_2), \dots, \alpha(k_n))$, or by permuting the indices: $(k_1, k_2, \dots, k_n)^\alpha = (k_{\alpha(1)}, k_{\alpha(2)}, \dots, k_{\alpha(n)})$. These two actions correspond to the two permutations obtained from the first by multiplying σ by α on the right: $\sigma\alpha = \begin{pmatrix} 1 & 2 & \dots & n \\ \alpha(k_1) & \alpha(k_2) & \dots & \alpha(k_n) \end{pmatrix}$, the second on the left: $\alpha\sigma = \begin{pmatrix} 1 & 2 & \dots & n \\ k_{\alpha(1)} & k_{\alpha(2)} & \dots & k_{\alpha(n)} \end{pmatrix}$. In the multiplication by α on the right the digits $1, 2, \dots, n$ are permuted according to α^{-1}, while in the multiplication on the left the places $1, 2, \dots, n$ undergo the same permutation.

Remark 3.8. It is hardly necessary to point out that the result of the multiplication on the right or on the left (the permutation of the digits $i, j, \dots$ or of the digits at the places $i, j, \dots$) depends on how the action of a product $\sigma\tau$ is defined, i.e. if σ acts first and then τ, or viceversa.

Definition 3.16. Let $\sigma = \begin{pmatrix} 1 & 2 & \dots & n \\ k_1 & k_2 & \dots & k_n \end{pmatrix} \in S^n$. The number of digits to the left of k_i and that are greater than k_i of the sequence $k_1\ k_2 \dots k_n$[17], is the number of *inversions* of k_i. If b_j is the number of inversions of digit j, $j = 1, 2, \dots, n$, the table $[b_1, b_2, \dots, b_n]$ is the *inversion table* of the permutation σ, and the number of *inversions of the permutation* is the sum of the b_i. Observe that $b_i \leq n - i$, $i = 1, 2, \dots, n$.

Example 3.10. In the permutation 541632 digit 1 presents two inversions, 2 four, 3 three, 4 one, 5 and 6 zero. The table of inversions is therefore $[2, 4, 3, 1, 0, 0]$ and the number of inversions is 10. We now want to reduce

[17] In Combinatorics, a permutation is defined as a rearrangement of the sequence 1,2,...,n. In this section, we will use this definition interchangeably with ours.

541632 to 123456 by interchanging each time two consecutive digits. Digit 1 can be shifted to the first position by first interchanging it with 4: 514632, and then with 5: 154632, i.e. by first interchanging the digits that are at the second and third place and then those that are at the second and first ones. We have seen above that interchanging the digits that are at the i-th and j-th place amounts to multiplying σ on the left by the transposition (i,j), so that the two mentioned operations amount to $\sigma' = (1,2)(2,3)\sigma$. Proceeding in this way, to take 2 to the second place we multiply σ' on the left in turn by (5,6),(4,5),(3,4) and (2,3); to take 3 to the third place we multiply the result by (5,6),(4,5) and (3,4). Finally, we take 4 to the fourth place by multiplying by (4,5), and we arrive at the identity permutation. Hence, $(4,5)(3,4)(4,5)(5,6)(2,3)(3,4)(4,5)(5,6)(1,2)(2,3)\sigma = I$, from which we deduce the following expression of σ as a product of transpositions of consecutive digits: $\sigma = (2,3)(1,2)(5,6)(4,5)(3,4)(2,3)(5,6)(4,5)(3,4)(4,5)$. This is true in general. Thus, *the transpositions of the form* $(i, i+1)$ *generate* S^n[18]. The 10 transpositions necessary to write σ correspond, by construction, to the 10 inversions of σ. Moreover, 10 is the minimal number of such transpositions. This is the content of the following theorem.

Theorem 3.31. *The minimal number of transpositions of the form* $(i, i+1)$ *whose product is* σ *equals the number* $I(\sigma)$ *of inversions of* σ.

Proof. As seen above, with a product of transpositions of the form $(i, i+1)$ equal in number to that of the inversions we obtain the identity. When we multiply by $(i, i+1)$, the number of inversions increases by 1 if $k_i < k_{i+1}$, and decreases by 1 if $k_i > k_{i+1}$. Thus, by multiplying by s transpositions, the resulting permutation has at least $I(\sigma) - s$ inversions, and if this permutation has to be the identity then $0 \geq I(\sigma) - s$, i.e. $s \geq I(\sigma)$. Thus, the identity cannot be obtained with less than $I(\sigma)$ transpositions of the aforementioned form. $\diamondsuit$

Corollary 3.19. *The parity of a permutation equals that of the number of its inversions.*

The inversion table determines the permutation (M. Hall). Given the table $[b_1, b_2, \ldots, b_n]$, $b_i \leq n - i$, $i = 1, 2, \ldots, n$, the following algorithm allows one to determine the permutation admitting that table of inversions: in an n-place string, put digit i in the $(b_i + 1)$-st free place starting from the left.

Example 3.11. Consider the table [2,4,3,1,0,0] of *ex.* 3.89, and a 6-place string. Since $a_1 = 2$, insert digit 1 in the $(2+1)$-th place, i.e. in the third free place (which in this case is simply the third place):

| | | 1 | | | |

[18] They are the so called *Coxeter generators* of S^n.

Let us now insert 2. This digit has four inversions, so that it goes to the fifth free place (the last one):

		1			2

Proceeding in this way we find 541632:

5	4	1	6	3	2

Exercises

109. How many inversions does the transposition (i, j) have?

110. *i)* If σ has t inversions, how many inversions does σ^{-1} have?
ii) If $k_1 k_2 \ldots k_n$ has t inversions, how many inversions does its transpose $k_n k_{n-1} \ldots k_1$ have?
iii) Conjugation does not preserve the number of inversions.

In the next two exercises, assume that the permutations all have the same possibility.

111. What is the average number of inversions in a random permutation? [*Hint*: consider the total number of inversions of a permutation and its transpose.]

112. Let $\sigma \in S^n$ be a random permutation.
i) What is the probability that the digit 1 be contained in a cycle of length k? [*Hint*: a k-cycle containing 1 is obtained by choosing the remaining $k-1$ digits from the $n-1$ different from 1 (this can be done in $\binom{n-1}{k-1}$ ways) and then ordering these $k-1$ digits and the remaining $n-k$.]
ii) What is the probability that 1 and 2 belong to the same cycle?

113. Show that S^n imbeds in A^{n+2}.

114. Let $\sigma \in S^n$, $n \geq 3$, fixing at least two digits. Show that if $\sigma \sim \alpha$ in S^n, then $\sigma \sim \alpha$ by an element of A^n.

115. Using Serret's lemma prove that the parity of a permutation is well defined as follows:
i) the permutations of S^n divide up into two classes according to the parity of the number of their cycles (cycles of length 1 being counted);
ii) if a permutation σ is multiplied by k transpositions then the permutation obtained belongs or not to the same class of σ according as k is even or odd;
iii) conclude that a permutation σ cannot be at the same time a product of an even and of an odd number of transpositions. [*Hint*: write σ as a product of transpositions and multiply the identity by σ; the number of transpositions of σ is even if, and only if, σ belongs to the class of the identity.]

116. A permutation is even if, and only if, $n - z(\sigma)$ is even.

117. Given the table $[2, 4, 3, 1, 0, 0]$ of *Ex.* 3.10, find the corresponding permutation by devising an algorithm that sorts the digits in the order $n, n-1, \ldots, 1$.

118. For each of the inversion tables defined below devise an algorithm to find the corresponding permutation.

i) Interchanging "right" and "left" and "greater" and "less", we obtain a new table $[c_1, c_2, \ldots, c_n]$, "dual" to $[b_1, b_2, \ldots, b_n]$: c_i is the number of digits to the *right* of i that are *less* than i. The digits less than i are $i-1$ in number, so that $0 \leq c_i < i$.

ii) Instead of the inversions of a digit i, one may consider the inversions of the digit that finds itself at place i: if $x_1 x_2 \ldots x_n$ is a permutation, one obtains the table $[d_1, d_2, \ldots, d_n]$, where d_i is the number of digits to the left of x_i that are greater than x_i. Since to the left of x_i there is room for at most $i-1$ digits, one has $0 \leq d_i < i$. Dually, one has the table $[l_1, l_2, \ldots, l_n]$, where l_i is the number of digits to the right of x_i that are less than x_i. Since to the right of x_i there is room for at most $n-i$ digits, one has $l_i \leq n-i$.

119. Let (σ, α) be a pair of permutations of S^n, and let the group $\langle \sigma, \alpha \rangle$ they generate be transitive. Prove that:

i) if $z(\sigma) > 1$, then α is not the identity, and there exists a transposition τ connecting σ and disconnecting α;

ii) for σ and τ as in *i)*, the group $\langle \sigma\tau, \alpha\sigma \rangle$ is transitive; [*Hint*: prove that if $k = \alpha(h)$, then there exists $\gamma \in \langle \sigma\tau, \alpha\tau \rangle$ such that $h = \gamma(k)$.]

iii) $z(\sigma) + z(\alpha) \leq n + 1$; [*Hint*: induct on $z(\sigma)$.]

iv) $z(\sigma) + z(\alpha) + z(\alpha\sigma) \leq n + 2$; [*Hint*: induct on $z(\sigma)$ and use *i)* and *iii)*.] If the group is not transitive, but has t orbits, then $z(\sigma) + z(\alpha) + z(\alpha\sigma) \leq n + 2t$;

v) the difference between the right and left sides of *iv)* is even, $2g$ say:

$$z(\sigma) + z(\alpha) + z(\alpha\sigma) = n + 2 - 2g. \tag{3.13}$$

The non negative integer g is the *genus* of the pair (σ, α). Formula (3.13) will be referred to as the *genus formula*[19];

vi) Let $\sigma_1, \sigma_2, \ldots, \sigma_m$ generate a transitive subgroup of S^n; prove that:

$$\sum_{i=1}^{m} z(\sigma_i) + z(\sigma_1\sigma_2 \cdots \sigma_m) \leq (m-1)n + 2^{20}.$$

[*Hint*: induct on $z(\sigma_1)$. If $z(\sigma_1) =$ then, by Serret's lemma and Corollary 3.18,

$$z(\sigma_1\sigma_2 \cdots \sigma_m) \leq z(\sigma_1) + (m-1)n - \sum_{i=2}^{m} z(\sigma_i),$$

i.e. the result. Assume $z(\sigma_1) > 1$, and let k be maximum such that the group generated by $\sigma_1, \sigma_2, \ldots, \sigma_k$ is not transitive, and consider $\sigma_1\sigma_2 \cdots \sigma_k\sigma_{k+1}$. Proceed as above.]

120. Let G and G_1 be two subgroups of S^n that commute elementwise, and let Γ and Δ be the sets of orbits of G and G_1, respectively. Prove that G acts on Δ, G_1 on Γ, and that the number of orbits of these actions is the same.

---

[19] Jacques A.: Sur le genre d'une paire de substitutions. C.R. Acad. Sci. Paris **267** (1968), 625–627.

[20] Ree R.: A Theorem on Permutations. J. Comb. Theory **10** (1971), 174–175, where the resut is proved using the theory of Riemann surfaces.

121. Let $\langle \sigma, \alpha \rangle$ be transitive, and let G be a subgroup of its centralizer in S^n. G is semiregular (Theorem 3.25), so its orbits all have the same length, namely $|G|$. Prove that:

i) σ and α act on the set Γ of these orbits. Call $\widehat{\sigma}$ and $\widehat{\alpha}$ the permutations thus obtained;
ii) the group $\langle \widehat{\sigma}, \widehat{\alpha} \rangle$ is transitive on Γ. Call $(\widehat{\sigma}, \widehat{\alpha})$ the *quotient pair* of (σ, α) w.r.t. G.

122. With notation as in the previous exercise, and with g the genus of (σ, α) and γ that of $(\widehat{\sigma}, \widehat{\alpha})$ we have:

$$2g - 2 = |G|(2\gamma - 2) + \sum_{I \neq \phi \in G} \chi(\phi) \tag{3.14}$$

(the *Riemann-Hurwitz formula*[21]), where $\chi(\phi)$ is the total number of cycles of σ, α and $\alpha\sigma$ left fixed by ϕ. Prove this formula as follows:

i) $z(\widehat{\sigma})$ equals the number of orbits of G on the cycles of σ (*ex.* 120, with σ playing the role of G_1), and the latter is given by Burnside's formula $\frac{1}{|G|}\sum_{\phi \in G} \chi_\sigma(\phi)$, where χ_σ denotes the number of cycles of σ fixed by ϕ;
ii) since $\chi_\sigma(I) = z(\sigma)$, we may write $\sum_{\phi \in G} \chi_\sigma(\phi)$ as $z(\sigma) + \sum_{I \neq \phi \in G} \chi_\sigma(\phi)$. This yields:

$$z(\widehat{\sigma}) = \frac{z(\sigma)}{|G|} + \frac{1}{|G|} \sum_{I \neq \phi \in G} \chi_\sigma(\phi);$$

iii) do the same for $\widehat{\alpha}$ and $\widehat{\alpha\sigma}$;
iv) sum the equalities thus obtained. The left hand side of the sum is $z(\widehat{\sigma}) + z(\widehat{\alpha}) + z(\widehat{\alpha\sigma})$ which equals $\frac{n}{|G|} + 2 - 2\gamma$; the right hand side is

$$\frac{z(\sigma) + z(\alpha) + z(\alpha\sigma)}{|G|} + \frac{1}{|G|} \sum_{I \neq \phi \in G} \chi(\phi),$$

where the first summand equals $\frac{n+2-2g}{|G|}$. The conclusion follows.

123. With the above notation, prove that $\gamma \leq g$.

124. If G is cyclic, $G = \langle \phi \rangle$, then (3.14) becomes:

$$2g - 2 = o(\phi)(2\gamma - 2) + \sum_{i=1}^{o(\phi)-1} \chi(\phi^i). \tag{3.15}$$

125. The stabilizer S of a cycle c of σ, α or $\alpha\sigma$ in the centralizer of $\langle \sigma, \alpha \rangle$ is cyclic. [*Hint*: consider the restriction $\bar{\phi}$ to c of an element ϕ of S, and the mapping $\phi \to \bar{\phi}$.]

126. Let $I \neq \phi \in G$ and $g = 0$. Then the number of points fixed by ϕ is exactly 2[22].

[21] In the theory of Riemann surfaces this formula relates the genus g of a surface, that of the quotient surface w.r.t. a group G of automorphisms (conformal self maps), and the number of branch points for the covering $W \to W/G$.

[22] A permutation group in which every nonidentity element fixes two points, and the stabilizer of a point is cyclic, is one of the groups C_n, D_n, A^4, S^4 or S^5 (see Zassenhaus H.: The theory of groups. Chelsea Publishing, New York (1958), pp. 16–19; these groups are the finite rotation groups of the sphere). From *ex.* 125 and 126, this is the case for the centralizer of a pair (σ, α) for which $g = 0$.

In the next two exercises, by a "point" we mean a cycle of either σ, α or $\alpha\sigma$.

127. With the above notation, prove that a non trivial element $\phi \in G$ fixes at most $2g+2$ points. [*Hint*: $\chi(\phi^i) \geq \chi(\phi)$.]

128. Assume that $g \geq 2$, and that G contains an involution ϕ fixing $2g+2$ points[23]. Prove that:

i) ϕ is the only involution fixing $2g+2$ points [*Hint*: if ψ is another such involution, then the group $\langle \phi, \psi \rangle$ is dihedral, of order $2o(\phi\psi)$ (Lemma 2.1). An application of (3.14) with $G = \langle \phi\psi \rangle$ yields $o(\phi\psi) = 1$, and $\phi = \psi$.]

ii) ϕ is central. [*Hint*: an involution conjugate to ϕ also fixes $2g+2$ points, so equals ϕ.]

129. Assume that σ, α and $\alpha\sigma$ are regular permutations of orders a, b and c, respectively. Show that:

i) the genus formula (3.13) becomes

$$\frac{n}{a} + \frac{n}{b} + \frac{n}{c} = n + 2 - 2g, \text{ i.e. } n(1 - \frac{1}{a} - \frac{1}{b} - \frac{1}{c}) = 2(g-1);$$

ii) the maximum value (less than 1) of the sum $\frac{1}{a} + \frac{1}{b} + \frac{1}{c}$ is $\frac{41}{42}$, and is attained for the triple 2,3, and 7;

iii) if $g \geq 2$, the quantity $1 - \frac{1}{a} - \frac{1}{b} - \frac{1}{c}$ is positive, and has $\frac{1}{42}$ as minimum value;

iv) conclude that $n \leq 84(g-1)$. Since $|G|$ divides n we have, in particular, $|G| \leq 84(g-1)$.

130. Prove that (3.14) can be reduced to the following form:

$$2g - 2 = |G|(2\gamma - 2) + |G| \sum_{i=1}^{r} (1 - \frac{1}{n_i}) \tag{3.16}$$

where r is the number of orbits of G of cycles of σ, α and $\alpha\sigma$ fixed by nonidentity elements of G and n_i is the order of the stabilizer of a cycle in the i-th orbit.

131. By a discussion of the possible values of γ, r and n_i in (3.16), prove that for $g \geq 2$ the upper bound $|G| \leq 84(g-1)$ holds in any case (i.e. not only for regular σ, α and $\alpha\sigma$). [*Hint*: if $\gamma \geq 2$, then $2g-2 \geq 2|G|$, and $|G| \leq g-1$. If $\gamma = 1$, then $2g - 2 = |G| \sum_{i=1}^{r} (1 - \frac{1}{n_i})$; the minimum value of the sum Σ is obtained for $n_i = 2$, all i, so that $\Sigma \geq \frac{r}{2}$ and $r|G| \leq 4(g-1)$. If $\gamma = 0$, then $\frac{2g-2}{|G|} = r - 2 - \sum_{i=1}^{r} \frac{1}{n_i}$; if $r \leq 2$, then $2g-2 < 0$ and $g < 1$. Thus $r \geq 3$. If $r \geq 5$, then $|G| \leq 4(g-1)$. Similarly for $r = 4$ and $r = 5$.] (The upper bound $|G| \leq 84(g-1)$ is the same as that due to Hurwitz for the order of the automorphism group of a compact Riemann surface of genus $g \geq 2$.)

[23] In the theory of Riemann surfaces, such a ϕ is the so-called a *hyperelliptic involution*, and the $2g+2$ fixed points are the *Weierstraß points*. Various results concerning the centralizer of a pair of permutations are the same as those for the automorphisms of compact Riemann surfaces (see Chapter V of Farkas H.M., Kra I.: Riemann surfaces. GTM, Springer, Berlin-New York (1980)).

3.8 Some Simple Groups

In this section we prove the existence of simple groups other than the alternating groups.

3.8.1 The Simple Group of Order 168

Consider the group $G = GL(3,2)$ of invertible linear transformations of a 3-dimensional vector space V over $\mathbf{Z}_2$ (V contains 8 vectors, so G can also be considered as the group of automorphisms of the elementary abelian group of order 8). This group has order 168; among other things, we will show that, up to isomorphisms, it is the unique simple group of order 168. We have $|G| = (2^3-1)(2^3-2)(2^3-2^2) = 168 = 2^3 \cdot 3 \cdot 7$. Let $v, w \in V$ be nonzero, and let $\{v_1 = v, v_2, v_3\}$ and $\{w_1 = w, w_2, w_3\}$ be two basis of V. Then the function f such that $f(v_i) = w_i$, $i = 1,2,3$ extends uniquely to an invertible linear transformation of V in V, i.e. to an element x of G. Since $v^x = w$, G is transitive on the 7 nonzero vectors of V. Let $N \neq \{1\}$ be a normal subgroup of G; we will show that $N = G$. N is transitive (Theorem 3.8), so $7||N|$. The stabilizer G_v of a vector $v \neq 0$ has index 7, and therefore order 24. The action on the cosets of G_v yields a homomorphism of G in S^7, whose kernel K is contained in G_v. If $K \neq \{1\}$ and normal, its order is divisible by 7; but then $7||G_v| = 24$, impossible. Hence $K = \{1\}$, and G imbeds in S^7. A subgroup of order 7 cannot be normal in G. If it is, its image in S^7 would have a normalizer of order at least 168; but a subgroup of order 7 in S^7 is generated by a 7-cycle, and therefore its normalizer has order $7 \cdot \varphi(7) = 7 \cdot 6 = 42$ (Theorem 2.26, *ii*)), and is the holomorph of C_7. It follows that the number of 7-Sylows of G is 8, and since $7||N|$ and $N \trianglelefteq G$, the eight 7-Sylow are all contained in N. Thus, $|N|$ is divisible by 8 and by 7, and so by 56; it contains $8 \cdot (7-1) = 48$ 7-elements, and in addition the 8 elements of a 2-Sylow, and so at least 56 elements. If $|N| = 56$, the 2-Sylow is unique in N, therefore characteristic in N and so normal in G. However, a normal subgroup of G must have order divisible by 7. Hence $|N| > 56$, $|N| = 168$ and $N = G$. Incidentally, we have shown that a transitive subgroup of S^7 of order 168 is simple.

We now show that a simple group G of order 168 imbeds in A^8, and that two such groups are conjugate in A^8. Hence, up to isomorphisms, there is a unique simple group of order 168. We split the proof into various parts. We first determine the number of Sylow subgroups of G.

i) $n_7 \equiv 1 \bmod 7$ and $n_7|24$ imply $n_7 = 8$. G acts by conjugation on these eight Sylow 7-subgroups, and therefore imbeds in A^8. $\mathbf{N}_G(C_7)$ has order 21. If $C_3 \triangleleft \mathbf{N}_G(C_7)$, $\mathbf{N}_G(C_7)$ is cyclic. Then G, and hence A^8, would contain an element of order 21. But A^8 does not contain such an element. Hence $\mathbf{N}_G(C_7)$ contains seven subgroups of order 3.

ii) $n_3 = 28$. We have $n_3 = 7, 28$. If $n_3 = 7$, $\mathbf{N}_G(C_7)$ contains all the Sylow 3-subgroups, and is generated by the set of elements they contain. These subgroups being conjugate, $\mathbf{N}_G(C_7)$ would be normal, and this implies C_7 normal (Theorem 3.16), which is excluded. Thus $n_3 = 28$ and $|\mathbf{N}_G(C_3)| = 6$. If $\mathbf{N}_G(C_3)$ is cyclic, it contains two elements of order 6, and this happens for all the $\mathbf{N}_G(C_3)$ that are all conjugate. Then there would be $28 \cdot 2 = 56$ elements of order 6. But there are already $8 \cdot 6 = 48$ 7-elements and $28 \cdot 2 = 56$ 3-elements: a total of 160 elements, and there would be room for only one Sylow 2-subgroup, which is excluded. It follows $\mathbf{N}_G(C_3) \simeq S^3$. In particular, there are no elements of order 6 in G, because the subgroup generated by one of these normalizes (and in fact centralizes) the C_3 it contains.

iii) $n_2 = 21$. We have $n_2 = 7, 21$. If $n_2 = 7$, then $|\mathbf{N}_G(S_2)| = 24$. The center of S_2 is normal in $\mathbf{N}_G(S_2)$ and commutes elementwise with a C_3. The product of a central element of order 2 with an element of order 3 gives an element of order 6, which is excluded. It follows $|\mathbf{N}_G(S_2)| = 8$, i.e. the 2-Sylow is self-normalizing, and therefore $n_2 = 21$.

iv) An element of order 2 of $G \subseteq A^8$ fixes no point (i.e. no 7-Sylow: 2 should divide $|\mathbf{N}_G(C_7)| = 21$). Then in the image of G in A^8 an element of order 2 is a product of four disjoint transpositions.

An element of order 3 of A^8 is either of type (1,2,3)(4)(5)(6)(7)(8) or (1,2,3)(4,5,6)(7)(8). $\mathbf{N}_G(C_3)$ acts on the set of fixed points of C_3 (Lemma 3.5). If y is an element of order 2 of $\mathbf{N}_G(C_3)$, and $x \in C_3$ is of the first type, y acts on the five points fixed by x, and therefore one must be left fixed, which is excluded. Hence, an element of order 3 is of the second type. If C_3 is generated by $(1,2,3)(4,5,6)(7)(8)$ the three elements of order 2 of $\mathbf{N}_G(C_3)$ are $(1,4)(2,6)(3,5)(7,8)$, $(1,5)(2,4)(3,6)(7,8)$ and $(1,6)(2,5)(3,4)(7,8)$, and these are the only fixed point free involutions of A^8 that normalize C_3.

v) Let us now prove that two simple groups G and $\overline{G}$ of order 168 are conjugate in A^8. In A^8, the normalizer of a 7-Sylow has order 21. Indeed, in S^8 an element x of order 7 is a 7-cycle, and so it must fix a digit. Then the elements of the normalizer of $\langle x \rangle$ all fix the same digit, and so they belong to an S^7. In S^7 the normalizer of $\langle x \rangle$ has order 42, and contains a cycle of order 6: if $x = (1,2,3,4,5,6,7)$, the permutation $\tau = (1,4,6,5,2,7)(3)$ takes x to x^3 and therefore normalizes the subgroup $\langle x \rangle$. But τ is odd, so it cannot belong to A^8. It follows $|\mathbf{N}_{A^8}(C_7)| = 21$.

Let C_7 and $\overline{C}_7$ be two 7-Sylows of G and $\overline{G}$, respectively. These being also 7-Sylow of A^8, there exists $\sigma \in A^8$ that conjugates them: $C_7^\sigma = \overline{C}_7$, from which $\mathbf{N}_{A^8}(C_7)^\sigma = \mathbf{N}_{A^8}(\overline{C}_7)$. It follows $\mathbf{N}_{\overline{G}}(\overline{C}_7) = \mathbf{N}_{A^8}(\overline{C}_7) \cap \overline{G} \subseteq \mathbf{N}_{A^8}(\overline{C}_7)$, so that $\mathbf{N}_{\overline{G}}(\overline{C}_7) = \mathbf{N}_{A^8}(\overline{C}_7)$ since they have the same order. It follows $\mathbf{N}_{\overline{G}}(\overline{C}_7) = \mathbf{N}_G(C_7)^\sigma$. Let $o(t) = 3$, $t \in \mathbf{N}_G(C_7)$, and let I be the set of fixed-point-free involutions of A^8 that normalize $\langle t \rangle$. By the above, $|I| = 3$, and since there are three elements of order 2 of G that normalize $\langle t \rangle$ we have $I \subseteq G$. Similarly, if I' is the set of fixed-point-free involutions of A^8 that normalize $\langle t^\sigma \rangle$ we have $|I'| = 3$, and therefore $I' \subseteq \overline{G}$. But the elements of I^σ that normalize $\langle t^\sigma \rangle$ are fixed-point-free and are three in number, and so

$I^\sigma = I'$. Since $\mathbf{N}_G(C_7)$ together with an involution that normalizes $\langle t \rangle$ generate G, their conjugates by σ are contained in $\overline{G}$ and generate $\overline{G}$. It follows $\overline{G} = G^\sigma$.

vi) Let us show that a Sylow 2-subgroup is dihedral. Let H be a maximal 2-Sylow intersection. $|H| \neq 1$ (since there are already 48 7-elements, there is no room left for 147 2-elements). If $|H| = 2$, since a normal subgroup of order 2 is central, $x \in H$ of order 2 commutes with all the elements of $\mathbf{N}_G(H)$. But $|\mathbf{N}_G(H)|$ is divisible by 3 or 7, so that an element of order 3 o 7 multiplied by x would yield either an element of order 6, which is excluded, or of order 14, that does not exist in A^8 (at least nine digits are needed). Thus $|H| = 4$, and therefore $\mathbf{N}_G(H)$ contains a 2-Sylow, and hence at least three. If it contains more than three 2-Sylows, then their number is 7 or 21, so that $|\mathbf{N}_G(H)|$ is divisible by $8 \cdot 7 = 56$, $|\mathbf{N}_G(H)| \geq 56$, and G is not simple. Hence $|\mathbf{N}_G(H)| = 2^3 \cdot 3$. C_3 is not normal in $\mathbf{N}_G(H)$ because $|\mathbf{N}_G(C_3)| = 6$; therefore there are four C_3 in $\mathbf{N}_G(H)$, so the normalizer in $\mathbf{N}_G(H)$ of C_3 has order 6 and cannot be cyclic, and hence is S^3. It follows $\mathbf{N}_G(H) \simeq S^4$ (*ex.* 48), the 2-Sylow of $\mathbf{N}_G(H)$, and so of G, is dihedral. In particular, H is a Klein group.

vii) A subgroup L of G of order 24 is isomorphic to S^4. If $\mathbf{N}_L(C_3) \simeq S^3$, the result follows (*ex.* 48 again). Otherwise, since $\mathbf{N}_G(C_3) \simeq S^3$, we have $\mathbf{N}_L(C_3) = C_3$; but then $n_3 = [L : C_3] = 8 \not\equiv 1 \bmod 3$. Moreover, $\mathbf{N}_G(L) = L$.

viii) G contains two distinct conjugacy classes of subgroups isomorphic to S^4. Consider $GL(3,2)$. The stabilizers G_v of the points are all conjugate, and since $7 = [G : G_v]$, we have $|G_v| = 24$. Moreover, as a group of linear transformations of a vector space, G is transitive on the subspaces W of dimension 2. Like the points, these are 7 in number, so that $[G : G_W] = 7$ and $|G_W| = 24$. It cannot be $G_v \sim G_W$. Indeed, if $(G_v)^x = G_W$, then $G_{v^x} = G_W$. But the stabilizer of a vector u cannot coincide with the stabilizer of a subspace W. Indeed, if $u \in W$, given $u' \in W$, $u' \neq u$, and $w \notin W$, u, u' and u, w can be extended to two bases u, u', u'' and u, w, w' and the mapping taking the first three vectors to the other three in the given order extends to a linear transformation that belongs to G_u but takes an element of W outside W. Similarly if $u \notin W$ (with $w \in W$ and $u' \notin W$ and extending to two bases the pairs u, w and u, w'). The stabilizer of a vector and that of a subspace are not conjugate (but see *ex.* 132).

Let us now take a closer look at the structure of $G = GL(3,2)$. First, from the number of Sylow p-subgroups for the various p one sees at once that G has 56 elements of order 3 and 48 elements of order 7.

The elements of order 3 are all conjugate. Indeed, if x and y belong to one and the same 3-Sylow, then $y = x^{-1}$ and the two elements are conjugate in the normalizer S^3 of the 3-Sylow; if they belong to two distinct 3-Sylows, the conjugacy of the two 3-Sylows takes x either to y or to y^{-1}. If a, b, c are three generators of the space V, an element of order 3 is, for instance, $(a, b, c)(a + b, b + c, a + c)(a + b + c)$. It can be verified that this permutation of the 7 nonzero elements of V takes subspaces to subspaces (lines to lines in the representation of V as a Fano plane, i.e. it is a *collineation*).

If $o(x) = 7$, since its centralizer C coincides with that of the 7-Sylow C_7 generated by x, we have $C \subseteq \mathbf{N}_G(C_7)$. But there cannot be equality (there would be an element of order 21 (see *i*)), and therefore $C = C_7$ and x has $168/7 = 24$ conjugates. The 48 elements of order 7 split into two conjugacy classes, each with 24 elements. An element of order 7 is, for instance, the 7-cycle $x = (a, a+b, b+c, b, a+c, c, a+b+c)$, and an element of the normalizer of C_7 is $y = (a)(a+b, b+c, a+c)(b, a+b+c, c)$, which conjugates x and x^2 and together with x generates $\mathbf{N}_G(C_7)$.

An element z of order 4 belongs to a unique 2-Sylow. Indeed, the intersection of two 2-Sylows has order 1, 2 or 4; if it contains z is of order 4, and therefore is a maximal 2-Sylow intersection; by *vi*) it is a Klein group, a contradiction. Since there are 21 2-Sylows, there are 21 cyclic subgroups of order 4, all conjugate, and therefore 42 elements of order 4, also all conjugate (for the same reason as that for the elements of order 3). At this point we have 48+56+42+1=147 elements, including the identity. An element of order 4 is, for example, $z = (a, a+b+c)(a+b, b, a+c, c)(b+c)$.

The remaining 21 elements are involutions. Indeed, let $o(u) = 2$. $C = \mathbf{C}_G(u)$ has order $|C| \geq 4$ (at least the Klein group or the C_4 containing it centralizes u). If $|C| = 4$, u has $[G : C] = 84$ conjugates, which is absurd. It follows $|C| \geq 8$, and since u cannot commute with elements of order 3 (there are no elements of order 6) or 7 (the normalizer of a C_7 has order 21), we have $|C| = 8$, and hence C is a 2-Sylow. In particular, each element of order 2, together with the identity, is the center of a 2-Sylow: these elements are therefore as many as the 2-Sylows, i.e 21, and are all conjugate.

Let V be a Klein subgroup of G. Let us show that V is the intersection of two 2-Sylows. Let u and v be two elements of V. By what we have just seen, their centralizers are two 2-Sylows S_1 and S_2, and hence the intersection of the latter, containing V, has order at least 4, and therefore exactly 4, so that $S_1 \cap S_2 = V$. V being a maximal 2-Sylow intersection we have, as in *vi*), that $\mathbf{N}_G(V)$ has order 24, and is isomorphic to S^4. The conjugacy class of a Klein group thus contains $[G : S^4] = 7$ subgroups. In S^4 the three dihedral 2-Sylows intersect in a Klein group, which is therefore normal. Each dihedral group contains one more Klein group, which is normal in another S^4.

The Klein groups split into two conjugacy classes. To prove this we shall make use of the following lemma, with which we anticipate the discussion about fusion of elements or subgroups of a Sylow p-subgroup we shall deal with in Chapter 5 (see Lemma 5.6 and Definition 5.9).

Lemma 3.8 (Burnside). *Two normal subgroups of a Sylow p-subgroup S of a group G that are conjugate in G are already conjugate in the normalizer of S.*

Proof. Let H and K be the two subgroups, and let $H^x = K$, $x \in G$. Let $N = \mathbf{N}_G(K)$; then $S \subseteq N$. But K is also normal in S^x, because $K = H^x$ and

H is normal in S, so that we also have $S^x \subseteq N$. It follows that S and S^x are Sylow N, and therefore conjugate in N: $S = (S^x)^y$, $y \in N$. But then $xy \in N$, and with $g = xy$ we have $H^g = H^{xy} = (H^x)^y = K^y = K$. ◇

Let now two Klein subgroups of a Sylow 2-subgroup S of $G = GL(3,2)$ be given. Were they conjugate in G, they would already be conjugate in the normalizer N of S. However, $N = S$, and the two Klein subgroups are normal in S; therefore, they belong to two distinct conjugacy classes of G. By what we have seen above, each class contains 7 subgroups, so that we have 14 Klein subgroups that split into two conjugacy classes.

3.8.2 Projective Special Linear Groups

The simple group of order 168 belongs to an infinite class of simple groups, the *projective special linear groups.* These groups are obtained as quotients $SL(n,K)/Z$, where Z is the center of $SL(n,K)$. In the course of proving the simplicity of these groups, we will see that certain linear transformations, the so-called transvections, play the role the 3-cycles play in the proof of the simplicity of the alternating groups.

We first prove that $SL(n,K)$ is generated by these transformations, and in fact we will see that only a subset of them suffices.

Let V be a vector space over K. A linear transformation $\tau \neq 1$ of V in V is a *transvection* if it fixes a hyperplane H (i.e. a subspace of dimension $n-1$) and the quotient V/H elementwise:

$$\tau(u) = u, u \in H, \text{ and } \tau(v) - v \in H, v \in V.$$

Let H be the kernel of a linear form f, and let $u \in H$. Then

$$\tau(v) = v + f(v)u,$$

is a transvection over H. The subspace $\langle u \rangle$ of dimension 1 (the "line" $\langle u \rangle$) is the *direction* of the transvection, and u is its *center.* We shall write τ_u in case is necessary to specify the vector u. Every transvection is obtained in this way, as the following lemma shows.

Lemma 3.9. *Let $\tau \neq 1$ be a transvection over a hyperplane H and let F be a linear form with kernel H. Then there exists a vector $0 \neq u \in H$ such that $\tau(v) = v + f(v)u$, for all $v \in V$.*

Proof. Let $V = H \oplus \langle w \rangle$; w can always be chosen such that $f(w) = 1$. The vector $u = \tau(w) - w$ is the one we seek. First, $u \neq 0$ because τ is not the identity. Let $v \in V$; then $v = h + \alpha w$, $h \in H$, $\alpha \in K$, and $f(w) = f(h) + \alpha f(w) = \alpha$. It follows $\tau(v) = \tau(h + \alpha w) = h + \alpha\tau(w) = h + \alpha w + \alpha u = v + \alpha u$, as required. ◇

A matrix differing from the identity only in the (i,j) entry, $i \neq j$, where there is an element $\alpha \in K$, is called *elementary*, and is denoted $E_{i,j}(\alpha)$. Such an elementary matrix with $\alpha \neq 0$ is the matrix of a transvection. Indeed, if $e_1, e_2, \ldots, e_n$ is a basis of V, then the transformation τ with matrix $E_{i,j}(\alpha)$ fixes e_k, $k \neq j$, and sends e_j to $e_j + \alpha e_i$. If H is the hyperplane generated by the e_k, $k \neq j$, then τ fixes H elementwise, and the line $\langle e_i \rangle$ is the direction.

Multiplication of a matrix on the right (left) by a matrix $E_{i,j}(\alpha)$ amounts to adding column (row) j multiplied by α to column (row) i . The effect of this multiplication is an *elementary transformation* of the given matrix.

Lemma 3.10. *The matrices of the transvections have determinant* 1.

Proof. Let τ be a transvection over H, with matrix $T = (t_{i,j})$, and let $v_1, v_2, \ldots, v_{n-1}$ be a basis of H. Let us extend it to a basis of V by means of a vector v_n. Then, since τ fixes v_i, $i = 1, 2, \ldots, n-1$, we have $t_{i,i} = 1$ and $t_{i,j} = 0$; moreover, since $\tau(v_n) - v_n \in H$,

$$\sum_{i=1}^{n-1} t_{n,i}v_i + (t_{n,n} - 1)v_n = \sum_{i=1}^{n-1} \lambda_i v_i.$$

It follows $t_{n,i} - \lambda_i = 0, i = 1, 2, \ldots n-1$, $t_{n,n} - 1 = 0$ and $t_{n,n} = 1$. The matrix T has 1's on the main diagonal, all 0's on top of it, the λ_i on the last row up to the $(n, n-1)$ entry, and 1 in the (n,n) entry; the result follows. ♢

Therefore, the transvections are elements of $SL(n,K)$.

Lemma 3.11. *i) The conjugate of a transvection is a transvection;*

ii) all transvections are conjugate in $GL(n,K)$, and if $n \geq 3$ they are already conjugate in $SL(n,K)$.

Proof. i) Let $\tau(v) = v + f(v)u$; then $\sigma\tau\sigma^{-1}(v) = \sigma\tau(\sigma^{-1}(v)) = \sigma(\sigma^{-1}(v) + f(\sigma^{-1}(v))u = v + f(\sigma^{-1}(v))\sigma(u)$. This is the transvection over the hyperplane $\sigma(H)$, which is the kernel of the form $f\sigma^{-1}$, with vector $\sigma(u) \in \sigma(H)$.

ii) Let $\tau(v) = v + f(v)u$ and $\tau'(v) = v + g(v)u'$ be two transvections, $u \in H$, $u' \in H'$, respectively, and let $x, y \in V$ such that $f(x) = 1$ and $g(y) = 1$. Let B and B' be two bases of V obtained by adding x to a basis $\{u, u_2, \ldots, u_{n-1}\}$ of H and y to a basis $\{u', u'_2, \ldots, u'_{n-1}\}$ of H'. Then there exists a linear transformation $\sigma \in GL(n,K)$ taking u to u', u_i to u'_i and x to y, and we have $\sigma\tau\sigma^{-1} = \tau'$. Indeed, $\sigma\tau\sigma^{-1}(u') = \sigma\tau(u) = \sigma(u) = u'$, and similarly $u'_2, \ldots, u'_{n-1}$, and finally $\sigma\tau\sigma^{-1}(y) = \sigma\tau(x) = \sigma(x+u) = \sigma(x) + \sigma(u) = y + u'$. This proves that τ and τ' are conjugate in $GL(n,K)$. If $n = 2$, the two bases are $\{u, x\}$ and $\{u', y\}$, and these cannot be modified if the centers of the two transvections are to be u and u' . If $n \geq 3$, there exists $u_2 \in H$ independent of u. If A is a matrix of σ of determinant d, let η be the linear transformation sending u_2 to u_2/d and fixing all the other vectors of B; then $\sigma\eta$ has determinant 1 (in terms of matrices, it is obtained from A by dividing the elements of the second row by d). ♢

Lemma 3.12. *The elementary matrices $E_{i,j}(\alpha)$, $\alpha \in K$, generate $SL(n, K)$.*

Proof. We shall prove something more general, by showing that by means of elementary transformations of type $E_{i,j}(\alpha)$ on the rows of an invertible matrix $A \in GL(n, K)$ one obtains the diagonal matrix D=diag(1,1,...,1,d), where d is the determinant of A. If Q is the product of the $E_{i,j}(\alpha)$ corresponding to the operations performed on the rows of A, then $QA = D$, from which $A = Q^{-1}D$, and since the inverse of an $E_{i,j}(\alpha)$ is still of the same type, i.e. $E_{i,j}(-\alpha)$, the result will follow.

$A = (a_{i,j})$ being non singular, not all the elements of the first column are zero; if $a_{2,1} = 0$, let us add to the second row a row j with $a_{2,j} \neq 0$ (i.e multiply A on the left by $E_{2,j}$). Thus we may assume $a_{2,1} \neq 0$. Now multiply the second row by $a_{2,1}^{-1}(1 - a_{1,1})$ and add it to the first one, thus obtaining 1 in the $(1,1)$ entry, and subtract multiples of the first row from the other ones to obtain $a_{i,1} = 0, i \neq 1$. Proceeding similarly with the second row and column we obtain $a_{2,2} = 1$ and $a_{i,2} = 0$, $i > 2$, and subtracting a suitable multiple of the second row from the first, $a_{1,2} = 0$, and $a_{1,i} \neq 0$, $i \neq 2$. The procedure terminates when one reaches $a_{n,n}$, and at this point $QA = D$, where D=diag(1,1,...,1,$a_{n,n}$). Taking determinants, we have on one side det(QA) =det(Q)·det$(A) = 1 \cdot d$, because Q is a product of matrices $E_{i,j}(\alpha)$ that have determinant 1, and on the other det(D)=$a_{n,n}$; it follows $a_{n,n} = d$. In particular, if $A \in SL(n, K)$, then $d = 1$, D is the identity, and $A = Q$, a product of elementary transformations of type $E_{i,j}(\alpha)$. ♢

Lemma 3.13. *Let $u \in V$, $G_u = \{\sigma \in SL(n, K) \mid \sigma(u) = u\}$ be the stabilizer of vector u. Then the transvections with center u,*

$$\tau(v) = v + f(v)u,$$

as f varies in V^*, $f(u) = 0$, form an abelian normal subgroup T_u of G_u. Moreover, $SL(n, K)$ is generated by the conjugates of T_u in $SL(n, K)$.

Proof. i) T_u is abelian:

$$\begin{aligned}\tau^f\tau^g(v) &= \tau^f(v + g(v)u) = v + g(v)u + f(v + g(v)u)u \\ &= v + g(v)u + f(v) + g(v)f(u) = v + (f + g)(v)u,\end{aligned}$$

and similarly $\tau^g\tau^f(v) = v + (g+f)(v)u$. Since $f+g = g+f$, the result follows. This also shows that T_u is isomorphic to the additive group of the dual space V^* of V.

ii) $T_u \trianglelefteq G_u$. Let $\sigma \in G_u$. Then:

$$\sigma^{-1}\tau^f\sigma(v) = \sigma^{-1}(\sigma(v) + f(\sigma(v))u) = v + f(\sigma(v))\sigma^{-1}(u) = v + f(\sigma(v))u,$$

so that $\sigma^{-1}\tau^f\sigma = \tau^{f\sigma}$, which is also a transvection with center u.

iii) The conjugates of T_u generate $SL(n, K)$:

$$\langle T_u^\sigma \mid \sigma \in SL(n, K)\rangle = \langle T_{\sigma(u)} \mid \sigma \in SL(n, K)\rangle = \langle T_v \mid v \in V\rangle = SL(n, K)$$

where the second equality follows from the transitivity of $SL(n,K)$, and the third one from the fact that the transvections generate $SL(n,K)$. ◇

Lemma 3.14 (Iwasawa). *Let G be a primitive permutation group on a set Ω, and let A be an abelian normal subgroup of the stabilizer G_α of a point $\alpha \in \Omega$ such that its conjugates under the various elements of G generate G. Then every non trivial normal subgroup of G contains G', the derived subgroup of G. In particular, if $G = G'$, then G is a simple group.*

Proof. Let $N \neq \{1\}$ be a normal subgroup of G. Then $N \not\subseteq G_\alpha$, some α, and is transitive (*es.* 102). By Theorem 3.9 (or by the maximality of G_α, *ex.* 100) $G = NG_\alpha$. $A \trianglelefteq G_\alpha$ implies $NA \trianglelefteq NG_\alpha = G$, so that $A^g \subseteq NA$, all $g \in G$, and since these conjugates of A generate G, $G = NA$. It follows $G/N = NA/N \simeq A/A \cap N$, an abelian group, and therefore $N \supseteq G'$. ◇

Lemma 3.15. *If $n \geq 3$, or if $n = 2$ e $|K| > 3$, then:*

$$GL(n,K)' = SL(n,K)' = SL(n,K).$$

Proof. (Write GL for $GL(n,K)$ and SL for $SL(n,K)$). Since $GL/SL \simeq K^*$, we have $GL' \subseteq SL$, and since $SL' \subseteq GL'$ it is sufficient to show that $SL \subseteq SL'$, and for this that every transvection belongs to SL'.

i) $n \geq 3$. In this case, the transvections are conjugate in SL (Lemma 3.11, *ii)*) and therefore they all belong to the same coset of SL'. Let τ_u be a transvection with center u, $f(u) = 0$, and let $u' \neq -u$, $f(u') = 0$, and $\tau_{u'}$ a transvection with center u'. Then $\tau_u^f \tau_{u'}^f = \tau_{u+u'}^f$ is still a transvection, and therefore belongs to the same coset $\tau_{u'}$ mod SL': $SL'\tau_u\tau_{u'} = SL'\tau_{u'}$; it follows $SL'\tau_u = SL'$ and $\tau_u \in SL'$.

ii) $n = 2$. Since $SL' \trianglelefteq GL$, and since the conjugate of a transvection is still a transvection, if one of these belongs to SL' then all do. Let $\{v_1, v_2\}$ be a basis of V, and τ be the transvection $\tau(v_1) = v_1$ and $\tau(v) = v + v_1$ (center v_1, hyperplane $\langle v_1 \rangle$ and $f(v) = 1$). Since $|K| \geq 3$, there exists $d \in K$ such that $d^2 \neq 0, 1$. Let $\sigma(v_1) = dv_1$ and $\sigma(v_1) = d^{-1}v_2$; as for the commutator $[\sigma, \tau^{-1}]$ we have, on v_1, $\sigma^{-1}\tau\sigma\tau^{-1}(v_1) = \sigma^{-1}\tau\sigma(v_1) = \sigma^{-1}\tau(dv_1) = \sigma^{-1}(dv_1) = v_1$, and on v_2, $\sigma^{-1}\tau\sigma\tau^{-1}(v_2) = \sigma^{-1}\tau\sigma(v_2+v_1) = \sigma^{-1}\tau(d^{-1}v_2+dv_1)\sigma^{-1}(d^{-1}v_2 + d^{-1}v_1 - dv_1) = v_2 + d^{-2}v_1 - v_1 = v_2 + (d^{-2} - 1)v_1$. It follows that the commutator $[\sigma, \tau^{-1}]$ is the transvection $\tau_v = v + f(v)v_1$ ($f(v) = \beta(d^{-2} - 1)$, $v = \alpha v_1 + \beta v_2$), with center v_1 and hyperplane $\langle v_1 \rangle$; the result follows. ◇

We recall that if V is a vector space of dimension n, the subspaces $\langle v \rangle$ of dimension 1 (the lines) are the *points* of the *projective space* $\mathbf{P}^{n-1}(K)$. In other words, $\mathbf{P}^{n-1}(K)$ is the set of equivalence classes of non zero vectors of V w.r.t. the relation"$u\rho v$ if there exists $\lambda \neq 0$ in K such that $u = \lambda v$". If $K = \mathbf{F}_q$, each equivalence class contains $q - 1$ vectors, and since $|V| = q^n$, the space $\mathbf{P}^{n-1}(K)$ has $q^{n-1} + q^{n-2} + \cdots + q^2 + q + 1$ points. $SL(n,K)$ acts on $\mathbf{P}^{n-1}(K)$ (lines go to lines): $\langle v \rangle^\sigma = \langle v^\sigma \rangle$. However, the central elements, that are multiplications by scalars, fix all the lines, and therefore the quotient group $PSL(n, K)$, the *projective special linear group*, acts: if $a = \langle v \rangle$, $a^{Z\sigma} = \langle v \rangle^{Z\sigma} =_{def} \langle v^\sigma \rangle$ (cf. *ex.* 138).

Now let (a_1, a_2) and (b_1, b_2) be two pairs of distinct points $a_i = \langle u_i \rangle$, $b_i = \langle v_i \rangle$, $i = 1, 2$. The points being distinct, the vectors u_1 and u_2 are independent, and therefore belong to a basis $u_1, u_2, \ldots, u_n$. Similarly, we have a basis $v_1, v_2, \ldots, v_n$, and therefore there exists $\sigma \in GL(n, K)$ taking u_i to v_i. If σ has determinant $d \neq 1$, one can obtain $\sigma' \in SL(n, K)$ by proceeding as in the proof of Lemma 3.11, *ii*): consider $\sigma' = \sigma\eta$, where $\eta(u_1) = \frac{1}{d}u_1$ and $\eta(u_i) = u_i, i \geq 2$. Then σ' sends u_1 to $\frac{1}{d}v_1$ and u_i to v_i, $i \geq 2$, so that $\langle u_1 \rangle^{\sigma'} = \langle u_1^{\sigma'} \rangle = \langle (\frac{1}{d}v_1) \rangle = \langle v_1 \rangle$ In other words, $Z\sigma' \in PSL(n, K)$ takes $a_i = \langle u_i \rangle$ to $b_i = \langle v_i \rangle$, $i = 1, 2$. We have proved:

Lemma 3.16. *The group $PSL(n, K)$ is 2-transitive on the points of the projective space $\mathbf{P}^{n-1}(K)$.*

We now have all the ingredients to prove the main result of this section.

Theorem 3.32. *The group $PSL(n, K)$ is a simple group if $n > 2$, or if $n = 2$ and $|K| > 3$.*

Proof. Let us verify that PSL satisfies the hypothesis of Iwasawa's lemma. Let T_u be as in Lemma 3.13; then the conjugates of T_uZ/Z generate $SL/Z = PSL$. It follows that a non trivial normal subgroup of PSL contains PSL'. By Lemma 3.15, $SL = SL'$, and since $PSL' = (SL/Z)' = SL'Z/Z = SL/Z = PSL$, the result follows. ♢

In case the field is finite, this theorem yields a new infinite class of finite simple groups. As to the orders of the groups $SL(n, q)$ and $PSL(n.q)$ we have:

Theorem 3.33. *Let $K = F_q$. Then:*
i) $|SL(n, q)| = |GL(n, q)|/(q-1) = (q^n - 1)(q^n - q) \cdots (q^n - q^{n-1})/(q-1)$;
ii) $|PSL(n, q)| = |SL(n, q)|/(n, q-1)$.

Proof. i) The order of $GL(n, q)$ is given in *Ex.* 1.3, 2. The result follows from the fact that $SL(n, q)$ is the kernel of the homomorphism sending a linear transformation to the determinant of its matrix. The kernel of this homomorphism is the multiplicative group F_q^* of the field, of order $q - 1$.

ii) $PSL(n, q)$ is the image of $SL(n, q)$ under a homomorphism whose kernel is given by the n-th roots of unity of F_q^*, and since F_q^* is cyclic of order $q - 1$, the n-th roots form a subgroup of order $(n, q - 1)$. ♢

The groups $PSL(2, 2)$ and $PSL(2, 3)$, excluded in Theorem 3.32, are isomorphic to S^3 and A^4, respectively (cf. *ex.* 141), and therefore are not simple. In case $n = 3$ and $q = 2$, the group $GL(3, 2)$, that we have proved to be the unique simple group of order 168, is found as the group $PSL(3, 2)$

(on a 2-element field there is no distinction between $GL(n,K), SL(n,K)$ and $PSL(n,K)$) or also as $PSL(2,7)$. The alternating group A^5 is isomorphic to $PSL(2,4)$ and to $PSL(2,5)$.

Exercises

132. In the group $G = GL(3,2)$ consider the stabilizer H of the vector $u = (1,0,0)$, and the stabilizer K of the subspace $W = \{(0,0,0),(0,1,0),(0,0,1),(0,1,1)\}$. Prove that H and K are almost conjugate, i.e. they satisfy (3.10)). [*Hint:* in the matrix representation, the elements of K are the transposed of those of H. Moreover, the cardinality of $\mathbf{cl}(g) \cap H$ equals the number of fixed points of g.]

133. Prove that $GL(n,K)$ is the semidirect product of $SL(n,K)$ and the multiplicative group K^* of K.

134. Given a transvection τ, determine an $E_{i,j}(\alpha)$ conjugate to τ.

135. Prove that if the dimension of the space is 2, and -1 is not a square in K, the two transvections $E_{1,2}(1)$ and $E_{1,2}(-1)$ are not conjugate in $SL(2,K)$.

136. Let τ_u and τ_v be two transvections over the same hyperplane. Prove that $\tau_u\tau_v = \tau_{u+v}$. In this way, one obtains an isomorphism between the group of transvections over the same hyperplane and the additive group of the hyperplane.

137. Prove that if $n \geq 3$, every $E_{i,j}(\alpha)$ is a commutator.

138. Prove that the action of $PSL(n,K)$ on the points of $\mathbf{P}^{n-1}(K)$ is well defined and faithful.

The two exercises to follow allow a new proof of the simplicity of $PSL(n,K)$, $n \geq 3$[24].

139. Let $n \geq 3$, $\sigma \in SL(n,K) \setminus Z$; then there exists u such that $\sigma(u) = v \neq u$. The plane $\langle u,v\rangle$ is contained in a hyperplane H. Let τ_u be a transvection over H. Prove that for $\delta = \sigma\tau_u\sigma^{-1}\tau_u^{-1}$ one has:

i) $\delta \neq 1$; [*Hint:* use the existence of a vector $w \notin \langle u,v\rangle$.]
ii) $\delta(v) - v \in \langle u,v\rangle$, for all $v \in V$;
iii) $\delta(H) \subseteq H$, $\delta(v) = v + h$, $h \in H$ and therefore $\delta(H) = H$;
iv) if δ commutes with all the transvections over H, then it fixes H elementwise, and therefore is a transvection over H;
v) there exists a transvection τ_u over H such that $\delta' = \delta\tau_u\delta^{-1}\tau_u^{-1} \neq 1$. Thus δ' is a product of two transvections ($\delta\tau_u\delta^{-1}$ and τ_u^{-1}) over $\delta(H)$ and H; however, $\delta(H) = H$, and therefore these are two transvections over the same hyperplane, so that δ' is a transvection over H.

140. Let $n \geq 3$, $N \trianglelefteq SL(n,K)$, $N \not\subseteq Z$. Show that:

i) N contains a transvection; [*Hint:* if $\sigma \in N \setminus Z$, N either contains $\delta = \sigma(\tau_u\sigma^{-1}\tau_u^{-1})$ or $\delta' = \delta(\tau_u\delta^{-1}\tau_u^{-1})$.]
ii) N contains all the transvections (Lemma 3.11).

Conclude that $PSL(n,K)$ is a simple group.

---

[24] Cf. Lang S.: Algebra. Addison-Wesley, London (1970), p. 476.

141. Prove that $PSL(2,2) \simeq S^3$ and $PSL(2,3) \simeq A^4$. [*Hint:* $PSL(2,2)$ acts faithfully on the three points of the projective line $\mathbf{P}^1(F_2)$ (the three lines of the space $V = \{0, u, v, u+v\}$), and $PSL(2,3)$ on the four points of $\mathbf{P}^1(F_3)$ (the four lines of the space $W = \{0, u, -u, v, -v, u+v, u-v, -u+v, -u-v\}$).]

142. Prove that $PSL(2,4) \simeq PSL(2,5) \simeq A^5$.

143. Prove that the simple groups $PSL(3,4)$ and $PSL(4,2)$ have the same order 20160 but are not isomorphic. [*Hint*: prove that the center of a Sylow 2-subgroup of $PSL(3,4)$ is a Klein group by considering the upper unitriangular matrices (*Ex.* 3.4, 1); those belonging to the center are: $\begin{pmatrix} 1 & 0 & x \\ 0 & 1 & 0 \\ 0 & 0 & 1 \end{pmatrix}$, with $x \in F_4$. The center of a Sylow 2-subgroup of $PSL(4,2)$ is of order 2. (The use of matrices is justified by the fact that the Sylow p-subgroups of $PSL(n,q)$, $q = p^f$, are isomorphic to those of $SL(n,q)$.) 20160 is the smallest integer such that there exist non isomorphic simple groups of the same order.]

Remark 3.9. The following isomorphisms hold:

i) $PSL(2,2) \simeq SL(2,2) = GL(2,2) \simeq S^3$;
ii) $PSL(2,3) \simeq A^4$;
iii) $PSL(2,4) \simeq PSL(2,5) \simeq A^5$;
iv) $PSL(2,7) \simeq PSL(3,2)$;
v) $PSL(4,2) \simeq A^8$;
vi) $PSL(2,9) \simeq A^6$.

(Most of these have been proved in the text proper or in the exercises.) It can be shown[25] that these are the only isomorphisms among the groups $PSL(n,K)$ or with symmetric or alternating groups.

[25] Artin E.: Geometric Algebra. Interscience Publishers, Inc., New York (1957), pp. 170–172.

4

Generators and Relations

4.1 Generating Sets

We recall that the subgroup $\langle S\rangle$ of a group G generated by a set S of elements of G is the set of all products

$$\langle S\rangle = \{s_1 s_2 \cdots s_n,\ s_i \in S \cup S^{-1}\}, \tag{4.1}$$

$n = 0, 1, 2, \ldots$, and $S^{-1} = \{s^{-1},\ s \in S\}$. If S is empty, $\langle S\rangle = \{1\}$. In the form (4.1) the elements of $\langle S\rangle$ are *words* over the *alphabet* $S \cup S^{-1} \cup \{1\}$. Similarly, one can define the normal subgroup generate by S, also called the *normal closure* of S, denoted $\langle S\rangle^G$. The latter is also generated by the elements of S and their conjugates, and coincides with the intersection of all the normal subgroups of G containing the set S (there is at least G). If the elements of S commute, then $\langle S\rangle$ is abelian. Moreover, if S is finite, and all its elements have finite period and commute, then $\langle S\rangle$ is finite. If $N \trianglelefteq G$, and $G = \langle S\rangle$, the quotient G/N is generated by the images Ns, $s \in S$. A set S of generators (also called a *system*) is *minimal* if no proper subset of S generates $\langle S\rangle$.

As seen in Definition 1.8, a group G is said to be finitely generated (*f.g.* for short) if there exists a finite subset S such that $G = \langle S\rangle$. However, a subgroup of a f.g. group is not necessarily f.g. (see *Ex.* 4.1, 3 and 4).

Examples 4.1. 1. We know that group of integers is generated by the set $S = \{1\}$ and also by $S = \{2, 3\}$. They are both minimal systems. This shows that the cardinality of a system of generators is not determined, not even in the case of a minimal set.

2. The additive group $\mathbf{Q}$ of rational numbers. It is a group that cannot be f.g. Indeed, if $S = \{\frac{p_i}{q_i}\}$, $i = 1, 2, \ldots, n$, is a finite set of rationals, and $r \in \langle S\rangle$, $r = \sum_{i=1}^{n} m_i \frac{p_i}{q_i} = \frac{\sum_{i=1}^{n} t_i}{q_1 q_2 \cdots q_n}$ for some t_i's, after reduction to the same denominator. As r varies in $\langle S\rangle$, the elements $\sum_{i=1}^{n} t_i$ thus obtained make up a subgroup of the integers (because $\langle S\rangle$ is a subgroup), and therefore it is

Machì A.: Groups. An Introduction to Ideas and Methods of the Theory of Groups.
DOI 10.1007/978-88-470-2421-2_4, Springer-Verlag Italia 2012

cyclic, generated by a certain integer h. Then $\langle S\rangle$ is also cyclic, and is generated by $\frac{h}{q_1 q_2 \cdots q_n}$. In other words, every f.g. subgroup of the rationals is cyclic (a group with this property is called *locally cyclic*). However, $\mathbf{Q}$ is not a cyclic group: given a rational, there always exist other rationals that are not integral multiples of it.

In the group of rationals no minimal generating systems exists. In fact, every element of a generating system is superfluous. Indeed, let $\mathbf{Q} = \langle S\rangle$, $s \in S$; we show that $S \setminus \{s\}$ generates $\mathbf{Q}$. Let H be the subgroup generated by $S \setminus \{s\}$, and let $y \in H$. Then there exist two integers n and m such that $ns = my$, and therefore $ns \in H$. Now, for $n > 0$, $\frac{1}{n}s$ must be expressible in the form $\frac{1}{n}s = \sum m_i s_i + ks$ with $s_i \in S \setminus \{s\}$; but then, as a sum of two elements of H, $s = \sum n m_i s_i + nks \in H$. It follows $\mathbf{Q} = \langle S\rangle = \langle S \setminus \{s\}\rangle = H$, and s is superfluous.

Remark 4.1. In the example of $\mathbf{Q}$, or in similar cases, one could think of eliminating one by one all the elements $s_1, s_2, \ldots$ from a generating set S and arrive at the empty set; this would yield the absurd equality $\mathbf{Q} = \langle\emptyset\rangle = \{0\}$. The latter equality arises because the above operation of elimination corresponds to a proposition with an infinite number of conjunctions, which is not allowed.

From the above proof one also sees that $\mathbf{Q}$ does not have maximal subgroups. Indeed, if $M < \mathbf{Q}$ is maximal, let $x \notin M$; then $\mathbf{Q} = \langle M, x\rangle$, and as above $\mathbf{Q} = \langle M\rangle = M$.

A system of generators for $\mathbf{Q}$ is given by $S = \{\frac{1}{n}, n = 1, 2, \ldots\}$: one has $\frac{r}{s} = r\frac{1}{s}$. Another system is $\{\frac{1}{k!}\}$: here $\frac{r}{s} = (r(s-1)!)\frac{1}{s!}$, $k = 1, 2, \ldots$. The latter system allows us to represent $\mathbf{Q}$ as a union of subgroups (as is the group C_{p^∞}):

$$\langle 1\rangle \subset \langle \frac{1}{2!}\rangle \subset \langle \frac{1}{3!}\rangle \subset \cdots$$

($\langle 1\rangle$ is the group of integers). Observe that, setting $c_k = \frac{1}{k!}$, $\mathbf{Q}$ has a system of generators $c_1, c_2, \ldots$ such that $c_1 = 2c_2, c_2 = 3c_3, \ldots, c_n = (n+1)c_{n+1}, \ldots$ (see *Ex.* 1.6, 3).

In the group of integers, all proper subgroups have finite index; in $\mathbf{Q}$ we have the opposite situation: all proper subgroups have infinite index (see *Ex.* 1.7, 7). However, for $H \neq \{0\}$, the elements of $\mathbf{Q}/H$ all have finite period. Indeed, let $H + a \in \mathbf{Q}/H$, $\frac{r}{s} \in H$ and $a = \frac{p}{q}$. Then $r = s\frac{r}{s} \in H$, and therefore $pr \in H$. But $rqa = pr \in H$, and so $rq(H + a) = H$.

3. Let H be an infinite direct sum of copies of $\mathbf{Z}_2$ indexed by $\mathbf{Z}$:

$$H = \cdots \oplus \mathbf{Z}_2 \oplus \mathbf{Z}_2 \oplus \mathbf{Z}_2 \oplus \cdots.$$

H has all its elements of order 2, so is abelian, and therefore cannot be f.g. (otherwise it would be finite). Consider the automorphism of H given by $\sigma(x_i) = x_{i+1}$, where x_i is the generator of the i-th summand, $i \in \mathbf{Z}$. It is clear that the semidirect product G of $\langle\sigma\rangle$ by H is generated by one of the x_i, for instance x_0, and by σ, because s_k can then be obtained as $\sigma^k(x_0)$. Hence $G = \langle H, \sigma\rangle = \langle x_0, \sigma\rangle$[1] is generated by two elements and contains H which is not f.g. Note that we even have a descending chain of f.g. subgroups whose intersection is not f.g. Indeed, from x_0, x_1 and σ^2 one obtains all the elements of H, and therefore $\langle H, \sigma^2\rangle = \langle x_0, x_1, \sigma^2\rangle$, and in general $\langle H, \sigma^{2^k}\rangle = \langle x_0, x_1, \ldots, x_{2^k-1}, \sigma^{2^k}\rangle$. Hence we have the chain of subgroups:

$$G = \langle H, \sigma\rangle \supset \langle H, \sigma^2\rangle \supset \ldots \supset \langle H, \sigma^{2^k}\rangle \supset \ldots,$$

the intersection of all the subgroups of the chain, that are f.g., is the subgroup

$$\bigcap_{i=1}^{\infty} \langle H, \sigma^{2^k}\rangle = H,$$

which is not f.g.

4. In the affine group of the real line consider the two transformations:

$$s: \, x \to 2x, \quad t: \, x \to x+1,$$

and the subgroup $G = \langle s, t\rangle$ they generate. The elements:

$$s_k = s^k t s^{-k} \; : \; x \to x + \frac{1}{2^k}, \; k \geq 0,$$

are such that $s_k = s_{k+1}^2$, and therefore $\langle s_k\rangle \subset \langle s_{k+1}\rangle$. The subgroup of G

$$H = \bigcup_{k \geq 0} \langle s_k\rangle$$

cannot be f.g.; in particular, $H \neq G$. Observe also that $s^{-1}s_{k+1}s = s_{k+1}^2 = s_k$, and therefore $\langle s_{k+1}\rangle^s = \langle s_k\rangle$[2].

There are two cases which ensure in general that a subgroup of a f.g. group is itself f.g. This is the content of the two theorems to follow.

Theorem 4.1. *A subgroup of finite index in a f.g. group is itself f.g.*

This theorem is an immediate consequence of the following lemma.

Lemma 4.1. *Let G be a group, H a subgroup of G, T a system of representatives of the right cosets of H with $1 \in T$. Let X be a system of generators of G. Then the elements of TXT^{-1} that belong to H form a system of generators of H, i.e. $H = \langle TXT^{-1} \cap H\rangle$.*

[1] This group G is the *lamplighter* group.
[2] This gives an example of a conjugation that "shrinks" a subgroup.

Proof. Let $h \in H$ be given, $h = y_1 y_2 \cdots y_n$, $y_i \in X \cup X^{-1}$, and let $y_1 \in Ht_1$. Then $y_1 = h_1 t_1$, so that $h_1 = y_1 t_1^{-1} = 1 \cdot y_1 t_1^{-1}$. If $y_1 = x_1 \in X$, then $h_1 \in XT^{-1}$; if $y_1 = x_1^{-1}$, then $1 \cdot y_1 t_1^{-1} = (t_1 x_1 \cdot 1)^{-1}$ so that h_1 is the inverse of an element of TXT^{-1}. Similarly, $t_1 y_2 = h_2 t_2$; if $y_2 = x_2$ we have $h_2 = t_1 x_2 t_2^{-1} \in TXT^{-1}$, and if $y_2 = x_2^{-1}$, then $h_2 = (t_2 x_2 t_1^{-1})^{-1}$. It follows:

$$\begin{aligned} h &= y_1 t_1^{-1} \cdot t_1 y_2 \cdot y_3 \cdots y_n = 1 y_1 t_1^{-1} \cdot t_1 y_2 t_2^{-1} \cdot t_2 y_3 \cdots y_n = \ldots \\ &= 1 y_1 t_1^{-1} \cdot t_1 y_2 t_2^{-1} \cdots t_{n-2} y_{n-1} t_{n-1}^{-1} \cdot t_{n-1} y_n = h_1 h_2 \cdots h_{n-1} \cdot t_{n-1} y_n, \end{aligned}$$

and since $t_{n-1} y_n = t_{n-1} x_n \cdot 1 = (h_1 h_2 \cdots h_{n-1})^{-1} h \in H$, we are done. ◇

By construction, in an element $h = t_1 x_2 t_2^{-1}$, t_2 is the representative of the coset to which $t_1 x_2$ belongs. Writing $t_2 = \overline{t_1 x_2}$, h takes the form $h = t_1 x_2 (\overline{t_1 x_2})^{-1}$. If $h = t_1 x_2^{-1} t_2^{-1}$, then $h_2 t_2 x_2 = t_1$, and t_1 is the representative of the coset to which $t_2 x_2$ belongs; in this case:

$$h = t_1 x_2^{-1} t_2^{-1} = (t_2 x_2 t_1^{-1})^{-1} = (t_2 x_2 \cdot (\overline{t_2 x_2})^{-1})^{-1}.$$

In other words, the preceding lemma says that $H = \langle tx(\overline{tx})^{-1},\ t \in T, x \in X \rangle$.

Theorem 4.2. *A subgroup of a f.g. abelian group is itself f.g.*

Proof (Additive notation). By induction on the number n of generators of the group. If $n = 1$, the group is cyclic, and so is every subgroup. If $n > 1$, $x_1, x_2, \ldots, x_n$ are generators and H is a subgroup, we can write, for $h \in H$,

$$h = m_1 x_1 + m_2 x_2 \cdots + m_n x_n,\ m_i \in \mathbf{Z}. \tag{4.2}$$

Let S be the subset of $\mathbf{Z}$ consisting of the integers that appear as coefficients of x_1 in the expression of some element of H. If $h_1 = t x_1 + \cdots$ and $h_2 = r x_1 + \cdots$, then $t - r$ appears as coefficient of x_1 of the element $h_1 - h_2 \in H$. It follows that S is a subgroup of $\mathbf{Z}$, and as such is cyclic, generated by s, say. Let h' be an element of H having s as coefficient of x_1: $h' = s x_1 + \cdots$, and let h be as in (4.2). Then $m_1 = ps$ and x_1 is missing in $h - ph'$; hence $h - ph' \in K = \langle x_2, x_3, \ldots, x_n \rangle$. K being generated by $n-1$ elements, by induction every subgroup of K is f.g.; in particular, $H \cap K$ is f.g: $H \cap K = \langle y_1, y_2, \ldots, y_m \rangle$. But $h - ph' \in H \cap K$, and therefore $h \in \langle h', y_1, y_2, \ldots, y_m \rangle$; h being arbitrary in H, it follows that $H \subseteq \langle h', y_1, y_2, \ldots, y_m \rangle$. The other inclusion is obvious since h' and the y_i are elements of H. ◇

Exercises

1. Let G be an abelian group, n an integer. Show that $nG = \{na,\ a \in G\}$ and $G[n] = \{a \in G \mid na = 0\}$ are subgroups of G and that $nG \simeq G/G[n]$.

Definition 4.1. In a group G, an element x is *divisible by the integer* n if there exists $y \in G$ such that $x = ny$ ($x = y^n$, in multiplicative notation); it is *divisible* if it is divisible by every n. A group is *divisible* if every element is divisible.

In other words, G is divisible if $nG = G$ for all n (see the previous exercise). For instance, the additive groups of the rationals is divisible ($y = \frac{x}{n}$), as is the multiplicative group of the complex numbers ($y = \sqrt[n]{x}$).

2. If $(o(x), n) = 1$, then x is divisible by n.

3. Let G be an abelian divisible group. Show that:
i) G is infinite, and every proper quotient is also divisible;
ii) the torsion subgroup of G is divisible;
iii) the torsion subgroup of $\mathbf{C}^*$ consists of the n-th roots of unity, all n, and is isomorphic to the additive group $\mathbf{Q}/\mathbf{Z}$. The p-elements of this group constitute the group $\mathbf{Q}^p$ of *Ex.* 1.6, 4;
iv) if G is torsion free, then it is isomorphic to a vector space over the rationals; [*Hint*: define the scalar multiplication of $x \in G$ by a rational $\frac{m}{n}$ as my, where y is the unique element such that $x = ny$.]
v) G is divisible if, and only if, it is divisible by every prime;
vi) G is divisible if, and only if, it does not contain maximal subgroups; [*Hint*: if G is not divisible, then $pG < G$, for some prime p; consider G/pG, a direct sum of copies of $\mathbf{Z}_p$, and the sum H/pG of all summands except one.]
vii) use *i)* to show that the additive group of the rationals does not have subgroups of finite index, and in particular it does not have maximal subgroups;
viii) a direct sum of divisible groups is divisible.

4. Show that C_{p^∞} is divisible.

5. Show that C_{p^∞} is isomorphic to every proper quotient. [*Hint*: use *ex.* 1 and 3 recalling that a subgroup of C_{p^∞} consists of the the p^n-th roots of unity, all n, and hence of the elements z such that $z^{p^n} = 1$, some n.]

6. Using the generators $X = \{1/k!,\ k = 1, 2, \ldots\}$ of $\mathbf{Q}$ give a new proof that $\mathbf{Q}$ is locally cyclic (see *Ex.* 4.1, 2).

7. Let $p_1, p_2, \ldots, p_n$ be distinct primes. Show that the products

$$\overline{p}_i = p_1 p_2 \cdots p_{i-1} p_{i+1} \cdots p_n,\ i = 1, 2, \ldots, n,$$

form a minimal generating system of $\mathbf{Z}$.

8. If $\langle X \rangle = G$ and α is an automorphism of G, then $\langle X^\alpha \rangle = G$.

9. A f.g. group is countable.

10. A group is *locally finite* if every f.g. subgroup is finite. Show (Schmidt) that if $N \trianglelefteq G$ and G/N are locally finite, so is G. [*Hint*: let $\{x_i\}$ be finite; the images $x_i N$ generate a finite subgroup of G/N.]

11. Show that:
i) the rationals of the form m/p^r, p prime, m, r integers, form a subgroup $\mathbf{Q}_p$ of $\mathbf{Q}$ which is p-divisible but not divisible; [*Hint*: let n be coprime to m and p.]
ii) $\mathbf{Q}_p/\mathbf{Z} \simeq C_{p^\infty}$, and therefore is divisible (see *Ex.* 1.6, 4);
iii) $\mathbf{Q}_p \neq \mathbf{Q}_q$, $p \neq q$;
iv) $\mathbf{Q}_p \bigcap \mathbf{Q}_q = \mathbf{Z}$.

12. Show that the subgroup H of *Ex.* 4.1, 4, is normal and abelian (and in fact locally cyclic), and that G/H is cyclic.

4.2 The Frattini Subgroup

An element x of a group is a *nongenerator* if whenever x together with a set S generate G, than S alone generates G:

$$\langle S, x\rangle = G \Rightarrow \langle S\rangle = G;$$

in other words, a nongenerator is an element that can be removed from any generating set of the group.

The identity element is a nongenerator: if $\langle S, 1\rangle = G$, then $\langle S\rangle$ being a subgroup, contains 1, and therefore $\langle S\rangle = G$ (if $S = \{1\}$, $S\backslash\{1\} = \emptyset$, and we know that $\langle\emptyset\rangle = \{1\}$). If x is a nongenerator, and if x^{-1} together with S generate G, then since $x = (x^{-1})^{-1} \in \langle S, x^{-1}\rangle$, we have $G = \langle S, x^{-1}\rangle = \langle S, x\rangle = \langle S\rangle$. The more interesting fact is that if x and y are two nongenerators, their product xy also is a nongenerator. Indeed, let $G = \langle S, xy\rangle$; since $\langle S, xy\rangle \subseteq \langle S, x, y\rangle$, we have $G \subseteq \langle S, x, y\rangle$ and hence $G = \langle S, x, y\rangle$. But y is a nongenerator, so that $\langle S, x\rangle = G$, and x being a nongenerator, $\langle S\rangle = G$, as required. Thus, the set of nongenerators is a subgroup.

Definition 4.2. The subgroup of nongenerators of a group G is the *Frattini subgroup* of G, and is denoted $\mathbf{\Phi}$(G).

This subgroup is normal, and in fact characteristic. Indeed, let α be an automorphism of G, $x \in \mathbf{\Phi}$ and $\langle S, x^{\alpha}\rangle = G$. Then:

$$\langle S^{\alpha^{-1}}, x\rangle = G^{\alpha^{-1}} = G \Rightarrow \langle S^{\alpha^{-1}}\rangle = G \Rightarrow \langle S\rangle = G^{\alpha} = G.$$

It may happen that the Frattini subgroup coincides with the whole group. This is the case, for example, of the additive group of rationals. As we have seen, every element of a generating set of this group is superfluous (*Ex.* 4.1, 2: if $\mathbf{Q} = \langle S\rangle$ then $H = \langle S \setminus \{s\}\rangle = \mathbf{Q}$). We already know that $\mathbf{Q}$ does not have maximal subgroups; in general, the fact that the Frattini subgroup is or not a proper subgroup is related to the existence of maximal subgroups. If $M < G$ is maximal, and $x \notin M$, then x cannot be a nongenerator: indeed, $\langle M, x\rangle = G$ but $\langle M\rangle = M \neq G$. For every element x not belonging to a maximal subgroup, there is at least one generating system of G from which x cannot be removed. It is natural to suspect that the nongenerators belong to every maximal subgroups; this is indeed the case. For the proof we need Zorn's lemma, in the following form:

let $\mathcal{F}$ be a non empty family of subsets of a set ordered by inclusion, and such that every well ordered subfamily (a chain) $\{H_\alpha\}_{\alpha\in I}$ of elements of $\mathcal{F}$ has an upper bound H belonging to $\mathcal{F}$, i.e. there exists $H \in \mathcal{F}$ such that $H_\alpha \subseteq H$, for all $\alpha \in I$. Then there exists in $\mathcal{F}$ a maximal element, i.e. a subset $M \in \mathcal{F}$ which is not properly contained in any other subset of $\mathcal{F}$.

Lemma 4.2. *If $H < G$ and $x \notin H$, there exists a subgroup M of G containing H and maximal with respect to the property of not containing x.*

Proof. The family $\mathcal{F}$ of subgroups of G containing H and not x is not empty (there is at least H). If $\{H_\alpha\}$ is a chain of elements of $\mathcal{F}$ then the union $U = \bigcup_\alpha H_\alpha$ also belongs to $\mathcal{F}$, since any H_α contains H, and if $x \in U$ then $x \in H_\alpha$, some α, contradicting $H_\alpha \in \mathcal{F}$. The family $\mathcal{F}$ satisfies the hypothesis of Zorn's lemma, and therefore there exists in $\mathcal{F}$ a maximal element M. ◇

This lemma allows us to prove that a f.g. group always admits maximal subgroups. More precisely:

Theorem 4.3. *Let G be a f.g. group, H a proper subgroup of G. Then there exists a maximal subgroup of G containing H.*

Proof. Let $s_1, s_2, \ldots, s_n$ generate G, and let x_1 be the first of the s_i's not belonging to H. Let $M_1 \supseteq H$ be maximal with respect to the property of not containing s_1. If $\langle M_1, s_1\rangle = G$, M_1 is the required subgroup, since every subgroup of G properly containing M_1 contains s_1, and therefore coincides with G. If $\langle M_1, s_1\rangle < G$, let s_2 be the first of the s_i not contained in $\langle M_1, s_1\rangle$, and let $M_2 \supseteq \langle M_1, s_1\rangle$ be maximal with respect to the property of not containing s_2. By continuing in this way we reach a subgroup M_i such that $M_i \supseteq \langle M_{i-1}, s_{i-1}\rangle \supseteq \ldots \supseteq H$ with $\langle M_i, s_i\rangle = G$. M_i is the required maximal subgroup. ◇

But there are non f.g. groups for which this theorem holds, as the following example shows.

Example 4.2. Let H be the group of *Ex.* 4.1, 3, K a subgroup of H, x an element not in K, M a subgroup containing K and maximal w.r.t. the property of excluding x (Lemma 4.2). We show that M is maximal in H. Let y be an arbitrary element outside M; then x belongs to $\langle M, y\rangle$, otherwise this subgroup, which properly contains K would exclude x, contrary to the maximality of M. Hence x is of the form $yh_1yh_2\cdots yh_n$, $h_i \in M$. H being abelian, we may collect the h_i and the y: $x = y^k h$, some $h \in M$, and y being of order 2, we have $k = 0$ or $k = 1$. But $k = 0$ is impossible, otherwise $x \in M$, so $x = yh$, i.e. $y = xh$, and y being arbitrary, we have $\langle M, x\rangle = H$. By the maximality of M, a subgroup containing M must contain x and therefore is the whole group H.

We observed above that if there exists a maximal subgroup in a group then there exist elements that are not nongenerators. Conversely,

Theorem 4.4. *i) Let $\mathbf{\Phi}(G) < G$; then G contains a maximal subgroup;*
ii) let $G \neq \{1\}$, and let $\mathbf{\Phi}(G)$ be finitely generated. Then G contains a maximal subgroup.

Proof. i) Let $x \in G$, $x \notin \mathbf{\Phi}$. Then there exists a set S such that:

$$\langle S, x\rangle = G,\ \langle S\rangle \neq G. \tag{4.3}$$

In particular, $x \notin \langle S \rangle$. Let $M \supseteq \langle S \rangle$ be maximal with respect to the property of excluding x. Then M is maximal *tout court* (i.e. maximal in the family of all subgroups of G). Indeed, if $M < L$, then $S \subset L$ and therefore L must also contain x. Thus L contains S and x, and therefore also $\langle S, x \rangle = G$, so that $L = G$.

ii) If not, $\mathbf{\Phi}(G) = G$, and therefore if $\mathbf{\Phi}(G) = \langle x_1, x_2, \ldots, x_n \rangle$ we have $G = \mathbf{\Phi}(G) = \langle x_1, x_2, \ldots, x_n \rangle = \langle x_1, x_2, \ldots, x_{n-1} \rangle = \ldots = \langle x_1 \rangle = \langle \emptyset \rangle = \{1\}$ (see Remark 4.1). ◇

With the same proof of Theorem 4.4:

Theorem 4.5. *If $\mathbf{\Phi}(G) < G$, then $\mathbf{\Phi}(G)$ is the intersection of all maximal subgroups of G.*

Proof. If $\mathbf{\Phi}(G) < G$ then by Theorem 4.4, *i)*, there exists at least one maximal subgroup M. We already know that a nongenerator cannot lie outside a maximal subgroup, so $\mathbf{\Phi}(G) \subseteq \bigcap M$. Conversely, let $x \in \bigcap M$ and $x \notin \mathbf{\Phi}(G)$. We are in the situation (4.3), and proceeding as in Theorem 4.4, *i)*, we find a maximal subgroup L that does not contain x. ◇

Example 4.3. *The Heisenberg group $\mathcal{H}$ over* $\mathbf{Z}$. It is the group of 3×3 matrices with integer coefficients:

$$\begin{pmatrix} 1 & a & c \\ 0 & 1 & b \\ 0 & 0 & 1 \end{pmatrix}, \quad a, b, c \in \mathbf{Z},$$

with the usual matrix product. $\mathcal{H}$ can also be viewed as the set of integer triples (a, b, c) with the product:

$$(a_1, b_1, c_1)(a_2, b_2, c_2) = (a_1 + a_2, b_1 + b_2, c_1 + c_2 + a_1 b_2).$$

By letting the triple (a, b, c) correspond to the above matrix, this product corresponds to the matrix product. The unit element is $(0, 0, 0)$, and the inverse of (a, b, c) is $(-a, -b, -c + ab)$. Moreover, for $n \geq 2$,

$$(a, b, c)^n = (na, nb, nc + \binom{n}{2} ab)$$

so that $(a, b, c) = (0, b, 0)(a, 0, 0)(0, 0, c) = (0, 1, 0)^b (1, 0, 0)^a (0, 0, 1)^c$, and since $(0, 0, 1) = [(1, 0, 0), (0, 1, 0)]$, $\mathcal{H}$ is generated by the two elements $(1, 0, 0)$ and $(0, 1, 0)$. It is torsion free. The commutator of two elements is $[(a_1, b_1, c_1), (a_2, b_2, c_2)] = (0, 0, a_1 b_2 - a_2 b_1)$, and therefore is of type $(0, 0, c)$. Since $(a, 0, 0)^n = (na, 0, 0)$, $(0, 0, c)^n = (0, 0, nc)$, the elements $(a, 0, c)$, with $a, c \in \mathbf{Z}$, form a subgroup isomorphic to the direct product $\mathbf{Z} \times \mathbf{Z}$. If (a, b, c) commutes with $(1, 0, 0)$, then $b = 0$, as is easily verified, and if it commutes with $(0, 1, 0)$ then $a = 0$. An element commuting with both of them, and therefore with all the elements of the group, is of the form $(0,0,c)$, and since such an element actually commutes with all the elements of $\mathcal{H}$, it follows that the center of $\mathcal{H}$ is given by:

$$\mathbf{Z}(\mathcal{H}) = \{(0,0,c),\ c \in \mathbf{Z}\}.$$

Thus, this center is cyclic, generated by $(0,0,1)$. However, the commutator of two elements x, y is $[x,y] = (0,0,c)$ and therefore $\mathcal{H}' \subseteq \mathbf{Z}(\mathcal{H})$. On the other hand, $[(1,0,0),(0,1,0)] = (0,0,1)$, which generates $\mathbf{Z}(\mathcal{H})$: it follows $\mathcal{H}' = \mathbf{Z}(\mathcal{H})$. Moreover,

$$\mathcal{H}/Z(\mathcal{H}) \simeq \mathbf{Z} \times \mathbf{Z},$$

and this quotient is generated by the images of the two generators of $\mathcal{H}$ (the powers of these two elements only have $(0,0,0)$ in common with $Z(\mathcal{H})$).

Consider now the subgroup:

$$H_p = \{(hp,b,c),\ p \text{ prime},\ h,b,c \in \mathbf{Z}\};$$

we show that it is a maximal subgroup. Indeed, let $H_p < L$, and let $(a,b,c) \in L \backslash H_p$. Then p does not divide a, and therefore there exist two integers m and n such that $na + mp = 1$. Let $g = (a,b,c)^n = (na, nb, c')$, $h = (mp, -nb, n^2ab - c')$; then $h \in H_p$ and $gh = (1,0,0) \in L$. But $(0,1,0) \in L$, so that $L = \mathcal{H}$. This holds for all primes p. Let $H = \bigcap H_p = \{(0,b,c),\ b,c \in \mathbf{Z}\}$, where the intersection is over all primes. Similarly, $K_p = \{(a,kp,c),\ p \text{ prime},\ k,b,c \in \mathbf{Z}\}$ is maximal. Let $K = \bigcap K_p = \{(a,0,c),\ a,c \in \mathbf{Z}\}$, then $H \cap K = \{(0,0,c),\ c \in \mathbf{Z}\} = \mathbf{Z}(\mathcal{H})$, so that $\mathbf{\Phi}(\mathcal{H}) \subseteq \mathbf{Z}(\mathcal{H})$. If $\mathbf{Z}(\mathcal{H}) \not\subseteq \mathbf{\Phi}(\mathcal{H})$, let M be maximal and $\mathbf{Z}(\mathcal{H}) \not\subseteq M$. Then $M\mathbf{Z}(\mathcal{H}) = \mathcal{H}$ and therefore $M \triangleleft \mathcal{H}$; it follows $|\mathcal{H}/M| = p$, a prime, and therefore $\mathcal{H}' \subseteq M$. But $\mathcal{H}' = \mathbf{Z}(\mathcal{H})$, so that $\mathbf{Z}(\mathcal{H}) \subseteq M$, a contradiction. In conclusion, $\mathbf{\Phi}(\mathcal{H}) = \mathbf{Z}(\mathcal{H}) = \mathcal{H}'$.

Exercises

13. Let G be a f.g. group having a unique maximal subgroup. Show that G is a cyclic p-group (in particular G is finite).

14. The group C_{p^∞} equals its Frattini subgroup.

15. The Frattini subgroup of the integers is trivial.

16. If G is a finite cyclic group, $G = \langle x \rangle$, then the Frattini subgroup G is generated by $x^{p_1 p_2 \cdots p_n}$, where the p_i are the prime divisors of the order of G.

17. If $H < G$ and $\mathbf{\Phi}(G)$ is f.g., then $H\mathbf{\Phi}(G) < G$.

18. Let $\mathcal{H}_p$ be the Heisenberg group defined as in the text but with coefficients in $\mathbf{Z}_p$. Show that for $p > 2$, $\mathcal{H}_p$ is a nonabelian group of order p^3 in which all the elements have order p, and that $\mathcal{H}_2$ is the dihedral group D_4.

19. Let G be a group, S a subset of G, H a subgroup of G, and let $S \cap H = K$. Show that there exists a maximal subgroup M of G containing H and such that $M \cap H = K$.

20. A f.g. group contains a maximal normal subgroup (possibly trivial).

21. A divisible abelian group cannot be f.g.

4.3 Finitely Generated Abelian Groups

For the finitely generated abelian groups there is a structure theorem that we now prove. The main tool is the following lemma.

Lemma 4.3. *Let $u_1, u_2, \ldots, u_n$ be generators of the abelian group G, and let*

$$v = a_1u_1 + a_2u_2 + \cdots + a_nu_n$$

be an element of G such that the gcd of the a_i is 1. Then there exists a generating system of G, also with n elements, one of which is v.

Proof. If $n = 1$, then $v = a_1u_1$, and the conditions on the gcd implies $a_1 = \pm 1$, and therefore $G = \langle v \rangle$. If $n = 2$, there exist two integers e and f such that $ea_1 + fa_2 = 1$; then the two elements

$$v = a_1u_1 + a_2u_2,$$
$$v' = -fu_1 + eu_2,$$

generate G because the matrix of this system has determinant 1 so that u_1 and u_2 can be expressed as functions of v and v':

$$u_1 = ev - a_2v',$$
$$u_2 = fv + a_1v',$$

Assume now $n > 2$. Setting

$$v' = a_1'u_1 + a_2'u_2 + \cdots + a_{n-1}'u_{n-1},$$

where $a_i' = a_i/d$, $i = 1, 2, \ldots, n-1$, and d is the gcd of the first $n-1$ of the a_i, by induction on n we have that in the subgroup H generated by $u_1, u_2, \ldots, u_{n-1}$ there exist $n-2$ elements $v_2, v_3, \ldots, v_{n-1}$ such that $H = \langle v', v_2, v_3, \ldots, v_{n-1} \rangle$. It follows

$$G = \langle u_1, u_2, \ldots, u_n \rangle = \langle H, u_n \rangle = \langle v', v_2, v_3, \ldots, v_{n-1}, u_n \rangle.$$

Now, $v = dv' + a_nu_n$ and $(d, a_n) = 1$. As in the case $n = 2$, $\langle v', u_n \rangle = \langle v, v_n \rangle$, for a certain v_n. Hence $G = \langle v, v_2, \ldots, v_n \rangle$. ◇

Remarks 4.2. **1.** This lemma shows, in a different language, that every n-tuple of integers $(a_1, a_2, \ldots, a_n)$ with gcd equal to ± 1 (a *unimodular row of integers*) is a row of an $n \times n$ integer matrix invertible over the integers. That the row be unimodular is necessary since the cofactor expansion along the n-tuple shows that the determinant of the matrix is a linear combination of the a_i's.

2. In the polynomial ring $K[x_1, x_2, \ldots, x_n]$ over a field K an n-tuple $(f_1, f_2, \ldots, f_n)$ of polynomials that generate the whole ring (*a unimodular row of polynomials*) is a row of a polynomial matrix with nonzero constant determinant. (This was proved in 1976 by Quillen and Suslin, independently, thereby answering in the affirmative a question of Serre as to whether finitely generated projective modules over the ring $K[x_1, x_2, \ldots, x_n]$ are free.)

3. It may very well happen that the element v of the lemma is zero. For instance, for $G = \mathbf{Z}$, $G = \langle u_1, u_2 \rangle$ with $u_1 = 2$, $u_2 = 3$, and $v = a_1u_1 + a_2u_2$, with $a_1 = 3$ and $a_2 = -2$. Here $e = f = 1$ and $v' = -f \cdot u_1 + e \cdot u_2 = -1 \cdot 2 + 1 \cdot 3 = 1$. Then the lemma yields $G = \mathbf{Z} = \langle v, v' \rangle = \langle 0, 1 \rangle = \langle 1 \rangle$. In other words, if $v = 0$ the given system of generators is not minimal.

The proof of the lemma contains an algorithm that allows the construction of a system of n generators containing v. Let us see an example of how this algorithm runs.

Example 4.4. Let $G = \langle u_1, u_2, u_3 \rangle$, and let $v = 4u_1 + 6u_2 + 5u_3$. Let us seek a generating system of G with three elements, one of which is v. Consider the subgroup $H = \langle u_1, u_2 \rangle$. The gcd of the coefficients of u_1 and u_2 in v is $d = 2$; then let

$$v' = \frac{4}{2}u_1 + \frac{6}{2}u_2 = 2u_1 + 3u_2.$$

Since (2,3)=1, v' is part of a generating system with two vectors; the other one is obtained using Bézout's relation $2e_1 + 3f_1 = 1$; with $e_1 = -1, f_1 = 1$ we have:

$$v_1' = -fu_1 + e_1u_2 = -u_1 - u_2.$$

It follows $H = \langle u_1, u_2 \rangle = \langle v', v_1' \rangle$. Now, $v = dv' + 5u_3 = 2v' + 5u_3$, and since gcd(2,5)=1, by Bézout we have $2e_2 + 5f_2 = 1$; with $e_2 = -2, f_2 = 1$ let

$$v'' = -f_2v_1' + e_2u_3 = -v' - 2u_3 = -2u_1 - 3u_2 - 2u_3.$$

Now $G = \langle u_1, u_2, u_3 \rangle = \langle H, u_3 \rangle = \langle v', v_1', u_3 \rangle$, and since $\langle v', u_3 \rangle = \langle v, v'' \rangle$, we have $G = \langle v, v_1', v'' \rangle$, as required. Summing up:

$$\begin{aligned} v &= 4u_1 + 6u_2 + 5u_3, \\ v_1' &= -u_1 - u_2 + 0u_3, \\ v'' &= -2u_1 - 3u_2 - 2u_3. \end{aligned}$$

The matrix of this system $M = \begin{pmatrix} 4 & 6 & 5 \\ -1 & -1 & 0 \\ -2 & -3 & -2 \end{pmatrix}$ has determinant 1 and inverse $M^{-1} = \begin{pmatrix} 2 & -3 & 5 \\ -2 & 2 & -5 \\ 1 & 0 & 2 \end{pmatrix}$; by multiplying M^{-1} on the right by the column vector $[v, v_1', v'']^t$ we have $M^{-1}[v, v_1', v'']^t = [u_1, u_2, u_3]^t$ (t denotes transposition).

Theorem 4.6. *Let G be a f.g. abelian group and let n be the smallest integer such that G is generated by n elements. Then G is a direct sum of n cyclic groups.*

Proof. For every generating system S with n elements, let k_S be the minimum among the orders of the elements of S. Let k be the minimum of the k_S as S varies, and let $u_1, u_2, \ldots, u_n$ be a system of generators in which one of the u_i, u_n say, has order k (k infinite is not excluded). If $n = 1$, G is cyclic and there is nothing to prove. Let $n \geq 2$; the subgroup $H = \langle u_1, u_2, \ldots, u_{n-1} \rangle$ cannot be generated by less than $n-1$ elements, otherwise $G = \langle H, u_n \rangle$ would be generated by less than n elements. By induction on n, H is a direct sum of cyclic groups; it is therefore sufficient to prove that $H \cap \langle u_n \rangle = \{0\}$. Indeed, if this is not the case, then $a_1u_1 + a_2u_2 + \cdots + a_{n-1}u_{n-1} - a_nu_n = 0$, for certain integers a_i, and where we may assume $a_n > 0$ and $a_n < o(u_n)$ (otherwise we divide by $o(u_n)$ and obtain a relation with the remainder of the division, which is less than $o(u_n)$, in place of a_n). If $d = (a_1, a_2, \ldots, a_n)$, the element $v = a'_1u_1 + a'_2u_2 + \cdots + a'_{n-1}u_{n-1} - a'_nu_n$, with $a'_i = a_i/d$, is such that $(a'_1, a'_2, \ldots, a'_n)$=1. By Lemma 4.3 there exist $v_2, \ldots, v_n$ such $G = \langle v, v_2, \ldots, v_n \rangle$. But $dv = 0$, so that $o(v)$ is finite; if $o(u_n)$ is infinite, this contradicts the choice of the system $\{u_i\}$, and the same happens if $o(u_n)$ is finite because $o(v) \leq d \leq a_n < o(u_n)$. ♢

Note that the fact that n be minimum is essential. For instance, for **Z** we have $n = 1$; but **Z** is also generated by 2 and 3, that form a minimal but not minimum system, and in fact **Z** is not the direct sum of $\langle 2 \rangle$ and $\langle 3 \rangle$.

In an abelian group the elements of finite period form a subgroup, the *torsion subgroup*, denoted T. If G is f.g., so is T (Theorem 4.2), and therefore finite. In any case, the quotient G/T is torsion free, because if $r(T + x) = T$ then $rx \in T$, and therefore $s(rx) = 0$ for some s, $(sr)x = s(rx) = 0$, i.e. $x \in T$. Now G/T is f.g., and by Theorem 4.6 we have

$$G/T = \langle T + x_1 \rangle \oplus \langle T + x_2 \rangle \oplus \cdots \oplus \langle T + x_m \rangle,$$

with m minimum. The subgroup H of G generated by the x_i

$$H = \langle x_1, x_2, \ldots, x_m \rangle$$

is torsion free because if $\sum h_ix_i \in T$, then $\sum h_ix_i + T = \sum h_i(x_i + T) = T$, and the sum being direct, $h_i(x_i + T) = T$ all i, and G/T would be a torsion group. It follows $T \cap H = \{0\}$ and $G = H \oplus T$. By applying Theorem 4.6 to H, and H being torsion-free, the summands must be infinite cyclic. Hence we have m copies of **Z**:

$$H = \mathbf{Z} \oplus \mathbf{Z} \oplus \cdots \oplus \mathbf{Z}.$$

As for T, we have if $|T| = p_1^{n_1}p_2^{n_2} \cdots p_t^{n_t}$,

$$T = S_1 \oplus S_2 \oplus \cdots \oplus S_t,$$

where the S_i are Sylow p_i-subgroups of order $p_i^{n_i}$, and by Theorem 4.6 each

S_i is a direct sum of cyclic groups, and so of cyclic p_i-groups:

$$S_i = C_{p_i^{h_1}} \oplus C_{p_i^{h_2}} \oplus \cdots \oplus C_{p_i^{h_r}},$$

where $h_1 + h_2 + \cdots + h_r = n_i$. In conclusion:

Corollary 4.1 (Fundamental theorem of f.g. abelian groups). *A f.g. abelian group is isomorphic to a direct sum of cyclic group in which each summand is either isomorphic to* $\mathbf{Z}$ *or is a* p*-group.*

A decomposition as that of Corollary 4.1 cannot be refined: neither $\mathbf{Z}$ nor a cyclic p-group can be decomposed into a direct sum of non trivial subgroups. By Theorem 4.6, the latter are the only f.g. abelian groups for which this happens, and therefore the only ones that are irreducible with respect to direct sum. Corollary 4.1 also provides a means to determine the number of abelian groups of order n. Indeed, let P be an abelian p-group of order p^m. We know that P is a direct product of cyclic groups (we switch to multiplicative notation) $P = C_{p^{h_1}} \times C_{p^{h_2}} \times \cdots \times C_{p^{h_t}}$, where $h_1 + h_2 + \cdots + h_t = m$; the p-group P is said to be *of type* $(h_1, h_2, \ldots, h_t)$ and the orders $p^{h_1}, p^{h_2}, \ldots, p^{h_t}$ are its *elementary divisors*. The integers h_i appearing in the decomposition make up a partition of m. Conversely, given a partition of m like the above, we have an abelian p-group as a direct product of cyclic groups of order p^{h_i}, $i = 1, 2, \ldots, t$. It follows that the number of abelian p-groups of order p^m equals the number $\pi(m)$ of partitions of m. In particular, this number does not depend on the prime p, but only on the exponent m. Thus, for example, there are five non isomorphic abelian groups of order $16 = 2^4$ because there are five partitions of the number 4: to the partition 4=4 there corresponds the cyclic group C_{16}, to the 4=1+1+2 the group $C_2 \times C_2 \times C_4$, etc. For the same reason, there are five abelian groups of order $3^4, 19^4$, etc. An abelian group G of order $n = p_1^{m_1} p_2^{m_2} \cdots p_r^{m_r}$ is the direct product of its Sylow subgroups. As we have seen, for each Sylow subgroup of order $p_i^{m_i}$, expressed as a direct product of cyclic groups, we have a partition of m_i, and viceversa. This holds for every Sylow subgroup. Hence:

Corollary 4.2. *The number of abelian groups of order* $n = p_1^{m_1} p_2^{m_2} \cdots p_r^{m_r}$ *is equal to the product* $\pi(m_1)\pi(m_2)\cdots\pi(m_r)$*, where* $\pi(m_i)$ *is the number of partitions of* m_i*. In particular, the number of abelian groups of order* n *does not depend on* n *but only on the exponents of the prime numbers appearing in its prime decomposition.*

For example, there are four abelian groups of order 36: since $36 = 2^2 \cdot 3^2$, we have $\pi(2)\pi(2) = 2 \cdot 2 = 4$. Corresponding to the two partitions of 2 we have the 2-groups C_4 (2=2) and $C_2 \times C_2$ (2=1+1), and the 3-groups $C_3 \times C_3$ and C_9. The direct products of a 2- and a 3- group give the four groups.

The number of summands isomorphic to $\mathbf{Z}$ given by Corollary 4.1 is the (torsion-free) *rank* of the group G. This number is an invariant of the group, that is, there cannot be another decomposition of G with a different number of summands isomorphic to $\mathbf{Z}$, as the following theorem shows.

Theorem 4.7. *If a group is at the same time a direct sum of m copies and n copies of $\mathbf{Z}$, then $m = n$.*

Proof. Let K be the subgroup of the multiples px of the elements x of the given group G, for a certain prime p. Then G/K is generated by $K + x_1, K + x_2, \ldots, K + x_m$, where x_i generates the i-th copy of $\mathbf{Z}$ in the sum of m copies, and is a finite p-group in which all nonzero elements have order p. The sum of the $\langle K + x_i \rangle$ is direct. Indeed, if $\sum a_i(K + x_i) = K$, then $\sum a_i x_i \in K$ and therefore $\sum a_i x_i = px$, some x. Let $x = \sum b_i x_i$; it follows $\sum(a_i - pb_i)x_i = 0$, and the sum of the copies of $\mathbf{Z}$ being direct, one has $(a_i - pb_i)x_i = 0$, from which, since $o(x_i) = \infty$, $a_i - pb_i = 0$, $a_i = pb_i$, $a_i x_i = pb_i x_i \in K$, and all the summands of $\sum a_i(K + x_i)$ are equal to K. It follows that H/K is a group of order p^m. Similarly, H/K has order p^n, so that $m = n$. ♢

G/K is a vector space over $\mathbf{Z}_p$, and m is its dimension, which therefore is well determined.

Let us now give a new proof of the decomposition of Theorem 4.6 in the case of a finite group.

Lemma 4.4. *Let G be a group, $x, y \in G$, $y^m = x^t$ and $m|o(y)$. Then $o(y) = mo(x)/(o(x), t)$, and if $o(y)|o(x)$ then $m|(o(x), t)$.*

Proof. (If $m = 1$, this is Theorem 1.9, *vii*).) We have:

$$o(y^m) = \frac{o(y)}{(o(y), m)} = \frac{o(y)}{m},\ o(x^t) = \frac{o(x)}{(o(x), t)},$$

and the result. Writing $\frac{(o(x),t)}{m} = \frac{o(x)}{o(y)}$ we have that if $o(y)|o(x)$, then $m|(o(x), t)$. ♢

With the same proof of Theorem 2.4 we have:

Lemma 4.5. *Let G be abelian, x an element of G of maximal order, y any element of G. Then $o(y)$ divides $o(x)$.*

The crucial result is given by the following lemma.

Lemma 4.6. *Let G be a finite abelian group, x an element of maximal order, $H = \langle x \rangle$, and let Hy be a coset of order m. Then there exists in Hy an element of order m.*

Proof. Let $(Hy)^m = H$, that is $y^m \in H$ and $y^m = x^t$, some t. Since the order of a coset divides the order of every element of the coset, we have $m|o(y)$. Then, by the two previous lemmas, $m|(o(x), t)$. Now consider $z = yx^{-((o(x),t)/m)(t/(o(x),t))} = yx^{-t/m}$. The element z belongs to the coset Hy, and therefore $m|o(z)$, and is such that $z^m = y^m x^{-t} = 1$, so that $o(z)|m$. Hence $o(z) = m$, and z is the element we seek. ♢

The previous lemma is false for nonabelian groups. In the quaternion group, the coset $\{j, -j, k, -k\}$ of the quotient with respect to $\langle i \rangle$ has order 2 but does not contain elements of order 2.

Theorem 4.8 (Fundamental theorem of finite abelian groups). *Let G be a finite abelian group. Then*

$$G = G_1 \times G_2 \times \cdots \times G_t,$$

where the G_i are cyclic of order e_i. Moreover:

i) e_{i+1} *divides* e_i*;*

ii) *the product of the* e_i *equals the order of* G*;*

iii) *the* e_i *are uniquely determined by properties i) and ii).*

Proof. By induction on $|G|$, the theorem being trivially true for the identity group. Let $|G| > 1$, g_1 an element of G of maximal order e_1, and $G_1 = \langle g_1 \rangle$. By induction, the result holds for G/G_1:

$$G/G_1 = \langle G_1 g_2 \rangle \times \langle G_1 g_3 \rangle \times \ldots \times \langle G_1 g_t \rangle,$$

where the groups $\langle G_1 g_i \rangle$ are of order e_i, with $e_{i+1}|e_i$, $i = 2, 3, \ldots, t-1$, and where, by Lemma 4.6, we can take g_i of order equal to the order of the coset $G_1 g_i$ to which it belongs, i.e. $o(g_i) = e_i$. Let H be the product of the subgroups $\langle g_i \rangle$:

$$H = \langle g_2 \rangle \langle g_3 \rangle \cdots \langle g_t \rangle,$$

and let us show that this product is direct. We have to show that the order of H equals the product of the orders of the subgroups $\langle g_i \rangle$, i.e. $|H| = e_2 e_3 \cdots e_t$. However, $|H| \leq |\langle g_2 \rangle||\langle g_3 \rangle| \cdots |\langle g_t \rangle| = e_2 e_3 \cdots e_t$, and in the canonical homomorphism $\varphi : G \rightarrow G/G_1$ the image of H contains $G_1 g_2, G_1 g_3, \ldots, G_1 g_t$, and therefore equals the whole group G/G_1 which is generated by these cosets: $\varphi(H) = HG_1/G_1 = G/G_1$, so that $|\varphi(H)| \geq |G/G_1|$ and a fortiori $|H| \geq |G/G_1| = e_2 e_3 \cdots e_t$. Moreover, $HG_1 = G$ and $|H| = |G/G_1| = |HG_1/G_1| = |H/H \cap G_1| = |H|/|H \cap G_1|$, and hence $H \cap G_1 = \{1\}$. It follows

$$G = G_1 H = G_1 \times H = G_1 \times \langle g_2 \rangle \times \langle g_3 \rangle \times \cdots \times \langle g_t \rangle,$$

and with $G_i = \langle g_i \rangle$ the result follows. As for the uniqueness of the e_i, let $G = G_1 \times G_2 \times \cdots \times G_t = H_1 \times H_2 \times \cdots \times H_s$, with $|H_j| = f_j$, and the f_j satisfying *i)* and *ii)*. As $f_j | f_1$ for all j, h_1 is an element of maximal order of G, and so is g_1. Then, $e_1 = f_1$. Let k be the first index for which $e_k \neq f_k$, and suppose $e_k > f_k$. The set A of the f_k-th powers of the elements of G is a subgroup. If $x \in G$, then $x = g_1^{r_1} g_2^{r_2} \ldots g_t^{r_t}$, some r_i, and therefore $x^{f_k} = (g_1^{f_k})^{r_1} (g_2^{f_k})^{r_2} \cdots (g_t^{f_k})^{r_t}$. It follows that x^{f_k} belongs to the product of the subgroups generated by $g_i^{f_k}$, which is a direct product. Conversely, this product is contained in A, and therefore $A \simeq \langle g_1^{f_k} \rangle \times \langle g_2^{f_k} \rangle \times \cdots \times \langle g_k^{f_k} \rangle \times \cdots$, and the assumption $e_k > f_k$ implies that all the displayed factors are different from $\{1\}$. Similarly, $A \simeq \langle h_1^{f_k}\rangle \times \langle h_2^{f_k}\rangle \times \cdots \times \langle h_{k-1}^{f_k}\rangle$, because $h_j^{f_k} = 1$ if $j \geq k$, since f_k is a multiple of the order of h_j. Now, if $j < k$, then $f_k | f_j$; it follows $o(h_j^{f_k}) = \frac{f_j}{(f_j,f_k)} = \frac{f_j}{f_k}$. From the second decomposition of A we then have $|A| = \frac{f_1}{f_k} \cdot \frac{f_2}{f_k} \cdots \frac{f_{k-1}}{f_k} = \frac{e_1 e_2 \cdots e_{k-1}}{f_k^{k-1}}$, recalling that $f_j = e_j$ if $j < k$. Similarly, from the first decomposition of A it follows $|A| = \frac{e_1 e_2 \cdots e_{k-1}}{f_k^{k-1}} \cdot \frac{e_k}{(e_k,f_k)} \cdots$ But $\frac{e_k}{(e_k,f_k)} > 1$, and recalling that $e_k > f_k$, comparison of the two expressions for A gives the contradiction $|A| > |A|$. ♢

Like Corollary 4.1, this theorem provides a way to determine the number of abelian groups of order a given integer n. Indeed:

Corollary 4.3. *The number of abelian groups of order n equals the number of ways of decomposing n into a product $e_1 e_2 \cdots e_t$ of integers e_i such that e_{i+1} divides e_i.*

For example, the four abelian groups of order 36 seen above may be recovered from the decompositions $36 = 36$ (which gives C_{36}), $36 = 18 \cdot 2$ ($C_{18} \times C_2$), $36 = 12 \cdot 3$ ($C_{12} \times C_3$) and $36 = 6 \cdot 6$ ($C_6 \times C_6$).

4.4 Free abelian groups

Definition 4.3. An abelian group A is *free* if it is a direct sum of infinite cyclic groups:

$$A = \sum_{x \in X}^{\oplus} \langle x \rangle, \ \langle x \rangle \simeq \mathbf{Z},$$

for a subset X of A.

Every element g of A can be written in a unique way as a linear combination with integer coefficients of a finite number of the x, $g = \sum n_x x$. The elements of X are *free generators* and constitute a *free basis* of A, and the group is said to be *free on X* of *rank* the cardinality $|X|$ of X. The basic property of free generators is given in the following theorem.

Theorem 4.9. *Let A be a free abelian group on a set X, and let G be any abelian group. Let f be a function $f : X \to G$. Then f admits a unique extension to a homomorphism $\phi : A \to G$:*

$$\begin{array}{ccc} & & A \\ & \overset{\iota}{\nearrow} & \downarrow \phi \\ X & \xrightarrow[f]{} & G \end{array}$$

where $\phi \circ \iota = f$ and ι is the inclusion[3]. *Conversely, if A is an abelian group, X a subset of A such that any function f from X to an abelian group G admits a unique extension to a homomorphism $\phi : A \to G$, then A is free on X.*

Proof. If such a ϕ exists, it is unique because if $a \in A$, then $a = \sum h_i x_i$, so

$$\phi(a) = \phi(\sum h_i x_i) = \sum h_i \phi(x_i) = \sum h_i f(x_i), \tag{4.4}$$

and ϕ has the form (4.4). If ϕ is defined as in (4.4), then it is well defined because the h_i are uniquely determined by a, and is a homomorphism (see *ex.* 28).

As for the converse, we first prove that X generates A. Let $G = \langle X \rangle$, j the inclusion of G in A and $f : X \to G$ any function. The composition $j\phi$ yields a homomorphism $A \to A$ extending f: $j\phi\iota(x) = jf(x) = f(x)$. But the identity $id : A \to A$ also extends f:

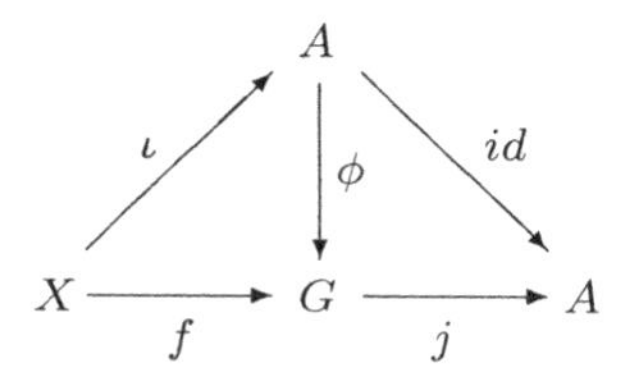

and uniqueness then implies $j\phi = id$. If $G < A$, then $\phi(A) \subseteq G < A$, and we have the contradiction $A = id(A) = j\phi(A) < j(A) = A$. Thus $G = \langle X \rangle = A$.

We now prove that A is free on X. Take as G a direct sum of copies of the integers indexed by the elements of $|X|$, $G = \sum_{x \in X}^{\oplus} \langle a_x \rangle$; the function f which sends $x \in X$ to a generator a_x extends to a homomorphism $\phi : A \to G$, which is onto: an element $\sum_{x \in X} n_x a_x$ equals $\sum_{x \in X} n_x f(x) = \sum_{x \in X} n_x \phi\iota(x) = \phi(\sum_{x \in X} n_x \iota(x))$. Moreover, the kernel of ϕ is zero because if $\phi(a) = 0$, then with $a = \sum_{x \in X} n_x \iota(x)$ we have

$$\phi(\sum_{x \in X} n_x \iota(x)) = \sum_{x \in X} n_x \phi\iota(x) = \sum_{x \in X} n_x f(x) = \sum_{x \in X} n_x a_x = 0,$$

and the sum being direct all the coefficients n_x are zero. Thus $a = 0$, $G \simeq A$, and A is free on X. ◇

If instead of a direct sum one considers a countable direct product $\prod_k \langle a_k \rangle$, $k = 1, 2, \ldots$ of copies of $\mathbf{Z}$ (functions from the natural numbers to $\mathbf{Z}$, or infinite sequences of integers with the componentwise sum) one has a torsion-free abelian group of infinite rank which is not free (Baer–Specker group)[4].

---

[3] This is the so called "universal property" of free abelian groups. It will be used to define freeness for nonabelian groups (Section 4.6).

[4] Kaplanski, Theorem 21.

Theorem 4.10. *The rank of a finitely generated free abelian group is well defined, i.e., if the group is the direct sum of $|X|$ and at the same time of $|Y|$ copies of the integers, then $|X| = |Y|$.*

Proof. The proof relies on the invariance of the dimension of a vector space. If A is the group, the quotient A/pA is a vector space over $\mathbf{Z}_p$, and the proof that the cosets $x + pA$, $x \in X$, are linearly independent and generate A/pA is as in Theorem 4.7. Hence $|X|$ is the dimension of A/pA, and the same holds for $|Y|$. ◇

The fact that a free abelian group has a basis over the integers implies that the are differences with vector spaces. For example, for $A = \mathbf{Z}$, the generating set {2,3} does not contain a basis (the only bases are {1} and {-1}). Moreover, an independent subset like {2} cannot be extended to a basis.

Theorem 4.11. *Every abelian group is a quotient of a free abelian group.*

Proof. Let G be an abelian group, A the direct sum of $|G|$ copies of $\mathbf{Z}$:

$$A = \sum_{g \in G}^{\oplus} \langle x_g \rangle,\ \langle x_g \rangle \simeq \mathbf{Z}.$$

Then A is free on $X = \{x_g,\ g \in G\}$ and by Theorem 4.9 the mapping $f : x_g \to g$ extends to a homomorphism $\phi : A \to G$ which is surjective because already f is surjective. Therefore $G \simeq A/ker(\phi)$. (It is clear that it is not necessary to take all the elements of G; a generating set suffices). ◇

Let G be an abelian group generated by a set S, and let A be a free abelian group on a set X such that $|X| = |S|$. The homomorphism $\phi : A \to G$ induced as in the above theorem by a bijection $X \to S$, $x_i \to s_i$, is surjective; the elements of the kernel are the integral linear combinations $\sum_{i=1}^{n} n_i x_i$ such that $\sum_{i=1}^{n} n_i s_i = 0$. A generating set $\{r_i\}$ of the kernel is a set of *relators* and their evaluation at the s_i, $\sum_{i=1}^{n} n_i s_i = 0$, a set of *relations* among the s_i. In the group G, every relation among the generators $\{s_i\}$ is a consequence of the relations $\phi(r_i)$, i.e. it is an element of the subgroup generated by the $\phi(r_i)$ (the logical relation of consequence $r \dashv R$ translates into the set theoretic relation of inclusion). Conversely, knowing that G is generated by a set S subject to a number of relations $\sum_{i=1}^{n} a_i s_i = 0$, consider the subgroup of the free abelian group A on X, $|X| = |S|$, generated by the elements $\sum_{i=1}^{n} a_i x_i$; then the quotient of A with respect to this subgroup is isomorphic to G.

If an abelian group G is described as a quotient of a free abelian group on a set X with respect to a subgroup generated by a set R as above, then we say that the pair X, R is a *presentation* of G, and we write $G = \langle X \mid R \rangle$. The presentation is *finite* if both X and R are finite, and in this case the group is said to be *finitely presented.* Sometimes, by *abus de langage*, we shall denote by R also the subgroup generated by the set R.

We now consider the problem of determining the structure of a finitely presented abelian group G. By saying "to determine the structure" of G we mean "to establish which cyclic groups appear in the decomposition given by Theorem 4.6". This is what we shall now see. First a theorem.

Theorem 4.12. *A subgroup H of a f.g free abelian group A of rank n is free of rank at most n.*

Proof. By induction on n. If $n = 1$, $A \simeq \mathbf{Z}$ and the result is obvious. Assume the result for $n-1$ and let $\{x_1, x_2, \ldots, x_n\}$ be a basis of A. Consider the subgroups $A_1 = \langle x_1, x_2, \ldots, x_{n-1}\rangle$ and $\langle x_n\rangle$. Then $A = A_1 \oplus \langle x_n\rangle$, A_1 is free by induction and so is $A_1 \cap H$, on elements $y_1, \ldots, y_m$, $m \leq n-1$, say. Moreover, $H/(A_1 \cap H) \simeq (A_1+H)/A_1 \subseteq A/A_1 \simeq \langle x_n\rangle \simeq \mathbf{Z}$, so that $H/(A_1 \cap H)$ is either $\{0\}$ or isomorphic to $\mathbf{Z}$. In the first case $H = A_1 \cap H$ and we are done. Otherwise, let $H/(A_1 \cap H)$ be generated by a coset $h + A_1 \cap H$, and let $h = a + px_n$, $a \in A_1$ and $p \neq 0$ an integer. We show that H is free on $\{y_1, \ldots, y_m, h\}$, by proving that the only element equal to zero in these elements is zero. It is clear that these elements generate H. But they are linearly independent, because if $\sum_{i=1}^{s} t_i y_i + qh = 0$, then $qpx_n = q(h-a) = -\sum_{i=1}^{s} t_i y_i - qh \in A_1$, whence $qpx_n = 0$ because $A_1 \cap \langle x_n\rangle = 0$. Since $p \neq 0$ we must have $q = 0$ and $\sum_{i=1}^{s} t_i y_i = 0$, but as the y_i are free generators of $A_1 \cap H$, $t_i = 0$, all i. The rank of H is $m+1 \leq n-1+1 = n$, as required[5]. ♢

An *elementary matrix* is a matrix obtained by applying the following *elementary transformations* to the columns (rows) of the identity matrix:

- interchange two columns (rows) i and j;
- multiply an element of column (row) i by a number $k \neq 0$;
- add to column (row) j column (row) i multiplied by k.

For example, the matrix $E_{i,j}(k)$, $i \neq j$, obtained by the third transformation. differs from the identity matrix in that has k at the (i, j) entry. Multiplication on the right (left) of a matrix M by $E_{i,j}(k)$ amounts to adding to column j (row i) column i (row j) multiplied by k.

Over a field, if V is a vector space with basis $u_1, u_2, \ldots, u_n$, and W is subspace of V with basis $w_1, w_2, \ldots, w_m$, $m \leq n$, then it is possible to change the basis of V to $v_1, v_2, \ldots, v_n$ in such a way that W is generated by $v_1, v_2, \ldots, v_m$. If the w_k are expressed as $w_k = \sum_{i=1}^{n} \alpha_{k,i} u_i$, $k = 1, \ldots, m$, then the required basis of V is obtained by reducing the matrix of the $\alpha_{k,i}$, by means of elementary operations to a matrix having all entries zero except for the first m

[5] This result also holds in case A is a free abelian group of infinite rank. Using the axiom of choice, one can prove that a subgroup of A is also free, of rank at most equal to that of A (see Rotman, Theorem 10.18).

entries 1 along the main diagonal:

$$D = \begin{pmatrix} 1\,0\,0 & \ldots 0\,0 \ldots 0 \\ 0\,1\,0 & \ldots 0\,0 \ldots 0 \\ \vdots\,\vdots\,\ddots & \vdots\,\vdots\,\vdots \quad \vdots\; 0 \\ 0\,0\,0 \quad 1 & 0\,0 \ldots 0 \end{pmatrix},$$

and $[v_1, v_2, \ldots, v_m]^t = D[v_1, v_2, \ldots, v_n]^t$. This is not possible in general for a free abelian group, which is a module over the integers, because of the non-invertibility of the integer coefficients. However, it is possible to find such a basis with the elements multiplied by a coefficient, leading to a matrix D with integers possibly different from 1 on the diagonal.

Theorem 4.13. *It is always possible by means of elementary operations to bring an $m \times n$ matrix $M = (a_{i,j})$ with integer coefficients to the form:*

$$D = \begin{pmatrix} e_1\; 0\;\; 0 & \ldots 0\,0 \ldots 0 \\ 0\;\; e_2\; 0 & \ldots 0\,0 \ldots 0 \\ \vdots\;\; \vdots\;\; \ddots & \vdots\,\vdots\,\vdots \quad \vdots\; 0 \\ 0\;\; 0\;\; 0 \quad e_m & 0\,0 \ldots 0 \end{pmatrix}$$

with $e_i \mid e_{i+1}, i = 1, 2, \ldots, m-1$.

Proof. Let d be the smallest element of M (in modulus) and by suitable permutations of rows and columns, if necessary, let us move it to the upper left position (1,1). By applying an operation of type $2'$, if necessary, we may assume that d is positive. We now set to zero the other entries of the first row as follows. If these entries are all multiples of d, $a_{1,j} = k_j d$, say, then subtract from column j the first column multiplied by $k_j, j = 2, 3, \ldots, n$ (an operation of type $3'$), and we are done. If, for some j, $a_{1,j}$ is not a multiple of d, division by d yields a quotient q and a remainder $0 \leq r < d$. Subtract from the j-th column the first multiplied by q, and then interchange this column with the first: we have a matrix with the positive integer $r < d$ in the upper left position. Proceeding in this way we arrive at a matrix having $\gcd(a_{1,j}, d)$ as upper left entry (the last nonzero remainder of Euclidean division), and a multiple of it in position $(1, j)$. As above, set to zero the entry $(1, j)$, and all the entries of the first row different from the first one. Note that in this process the (1,1)-entry is an integer that decreases at each step; let r' be the (1,1)-entry at the end of the process.

Now set to zero the entries of the first column except the first one. If these are all multiples of r', by subtracting from the various rows the first one multiplied by the corresponding multiple we are done. Otherwise, if $a_{j,1}$ is not a multiple of r', by dividing and subtracting as above and by interchanging the two rows we have a new matrix whose first row, in general, does not have all zeros from the second entry on any longer, but whose first entry is an integer less than r'. By repeating the above seen annulling process of the elements of the first row we obtain a matrix whose (1,1)-entry is an integer $r'' < r'$. Now we start again with the entries of the first column; the process terminates when these are all multiples of the first one, and this will eventually happen because the entry (1,1) becomes smaller and smaller. Let M' be the matrix obtained at the end of the process, and let a' be its (1,1)-entry. If some entry b of the matrix B of dimension $(m-1) \times (n-1)$ obtained by crossing out the first row and column of M' is not a multiple of a', add to the first row of M' the row where b lies, and by performing once more the previous process we obtain $\gcd(a', b)$ as (1,1)-entry, and zero in place of b. Similarly, we set to zero all the other elements of the first row, and the (1,1)-entry is now an integer that divides all the entries of the row of b. By repeating the process, we end up with a matrix M'' whose entries of the first row and column are all zero, except the (1,1)-entry. This entry, say e_1, divides all the entries of the matrix B' obtained from M'' by crossing out the first row and column. By operating in the same way on B', we end up with a matrix with an (1,1)-entry e_2 that divides the remaining elements. Since e_2 is obtained by taking greatest common divisors, and since e_1 divides all the elements of B', e_1 divides e_2. The other e_i are obtained similarly. $\diamondsuit$

Matrix D of Theorem 4.13 is the *Smith normal form* of matrix M. The products of the elementary matrices corresponding to the transformations operated on the columns (rows) of matrix M are two matrices Q and P^{-1} such that $D = QMP^{-1}$ (M and D are *equivalent*).

The e_i of Theorem 4.13 are invariant. This can be proved by considering the effect of the elementary operations on the matrix M. Indeed, these operations do not change the value d_h of the gcd of the minors of order h of M, $h = 1, 2, \ldots, m$. This is obvious for the operations of type $1'$ and $2'$. As for an operation of type $3'$, if a minor does not contain entries of the columns (rows) i and j then the operation does not change its value. If it contains entries of both columns, then it is well known that a determinant does not change if a column (row) multiplied by a number k is added to another one. If it contains elements of column i but not of column j, the value of the new minor is the sum of the original one and of the same minor multiplied by k (as it can be seen by expanding according to the entries of column i). Thus we see that d_h divides all the minors of order h of the new matrix, and therefore also their gcd. At the end of the process, d_h divides the gcd of the minors of order h of matrix D. Since $e_i | e_{i+1}$, a moment's reflection shows that this gcd is the product $e_1 e_2 \cdots e_h$. However, the elementary operations are invertible, and going back from D to M the same argument shows that $e_1 e_2 \cdots e_h$ divides d_h. Therefore, we have equality. Note that, in particular, e_1 is the gcd of the elements of M.

Theorem 4.14. *Let A be a free abelian group of rank n, and H a subgroup of rank m. Then it is always possible to find a basis $\{v_1, v_2, \ldots, v_n\}$ of A and m positive integers $e_1, e_2, \ldots, e_m$, where e_i divides e_{i+1}*[6]*, $i = 1, 2, \ldots, m-1$, and such that $e_1v_1, e_2v_2, \ldots, e_mv_m$ is a basis for H.*

Proof. Let $\{u_i\}$, $i = 1, \ldots, n$ be a basis for A and $\{v_i\}$, $i = 1, \ldots, m$ one for H. Then $y_k = \sum_{i=1}^n a_{k,i}x_i$, $k = 1, \ldots, m$. If M is the matrix of the $a_{k,i}$, then reducing M to the Smith normal form D we have the result. ◊

The following example will illustrate the previous theorem. Let A be the free abelian group with basis u_1, u_2, u_3, and let the subgroup H be freely generated by $w_1 = 2u_1 - u_2 + 3u_3$ and $w_2 = 3u_1 + 5u_2 - 2u_3$. We have:

$$\begin{bmatrix} w_1 \\ w_2 \end{bmatrix} = \begin{pmatrix} 2 & -1 & 3 \\ 3 & 5 & -2 \end{pmatrix} \begin{bmatrix} u_1 \\ u_2 \\ u_3 \end{bmatrix}.$$

Interchange the first and second column of $M = \begin{pmatrix} 2 & -1 & 3 \\ 3 & 5 & -2 \end{pmatrix}$:

$$\begin{pmatrix} -1 & 2 & 3 \\ 5 & 3 & -2 \end{pmatrix},$$

in the new basis u_2, u_1, u_3. Multiply the first column by -1:

$$\begin{pmatrix} 1 & 2 & 3 \\ -5 & 3 & -2 \end{pmatrix},$$

in the basis $-u_2, u_1, u_3$, and subtract from the second column the first multiplied by 2:

$$\begin{pmatrix} 1 & 0 & 3 \\ -5 & 13 & -2 \end{pmatrix},$$

and now the basis is $-u_2 + 2u_1, u_1, u_3$. Subtract from the third column the first multiplied by 3:

$$\begin{pmatrix} 1 & 0 & 0 \\ -5 & 13 & 13 \end{pmatrix},$$

with basis $-u_2 + 2u_1 + 3u_3, u_1, u_3$, i.e., w_1, u_1, u_3. At this point,

$$\begin{bmatrix} w_1 \\ w_2 \end{bmatrix} = \begin{pmatrix} 1 & 0 & 0 \\ -5 & 13 & 13 \end{pmatrix} \begin{bmatrix} w_1 \\ u_1 \\ u_3 \end{bmatrix}.$$

Up to now, the basis $\{w_1, w_2\}$ of H has not been modified. Let us add to the second row the first multiplied by 5:

$$\begin{pmatrix} 1 & 0 & 0 \\ 0 & 13 & 13 \end{pmatrix}.$$

[6] The e_i are the same as those of Theorem 4.8 taken in the reverse order.

The generators of A are now those of the previous step, but those of H are w_1 and $w_2 + 5w_1$. Finally, we subtract the second column from the third:

$$\begin{pmatrix} 1 & 0 & 0 \\ 0 & 13 & 0 \end{pmatrix};$$

now the generators of A become $v_1 = -u_2 + 2u_1 + 3u_3$, $v_2 = u_1 + u_3$, $v_3 = u_3$, and those of H have not changed. The bases of Theorem 4.14 are $v_1 = 2u_1 - u_2 + 3u_3$, $v_2 = u_1 + u_3$, $v_3 = u_3$ for A, and $1 \cdot v_1$, $13 \cdot v_2$ for H. Note that $w_2 + 5w_1 = 3u_1 + 5u_2 - 2u_3 + 10u_1 - 5u_2 + 15u_3 = 13u_1 + 13u_3 = 13v_2$.

The equality we started with has become:

$$\begin{bmatrix} 1 \cdot v_1 \\ 13 \cdot v_2 \end{bmatrix} = \begin{pmatrix} 1 & 0 & 0 \\ 0 & 13 & 0 \end{pmatrix} \begin{bmatrix} v_1 \\ v_2 \\ v_3 \end{bmatrix}.$$

Consider now a finitely generated abelian group G presented as a quotient A/R, where A is free on a set of n elements and R is freely generated by a set of m relators. Changing bases in A and R as in Theorem 4.14 we have $A = \langle v_1 \rangle \oplus \langle v_2 \rangle \oplus \cdots \oplus \langle v_n \rangle$ and $R = \langle e_1 v_1 \rangle \oplus \langle e_2 v_2 \rangle \oplus \cdots \oplus \langle e_m v_m \rangle$. But A is the direct sum of the v_i which are isomorphic to $\mathbf{Z}$, so we see (Theorem 2.30) that G is the direct product of the groups $\mathbf{Z}/\langle e_i \rangle$, $i = 1, 2, \ldots n$. Hence:

Theorem 4.15 (Basis theorem for f.g. abelian groups). *If G is a f.g. abelian group, then there exist positive integers m, n and integers $e_i \geq 2$, with $e_i | e_{i+1}$, $i = 1, 2, \ldots, m-1$, such that*

$$G = \mathbf{Z}_{e_1} \oplus \mathbf{Z}_{e_2} \oplus \cdots \oplus \mathbf{Z}_{e_m} \oplus \mathbf{Z} \oplus \mathbf{Z} \oplus \cdots \oplus \mathbf{Z}$$

with $n - m$ copies of $\mathbf{Z}$.

The integers e_i are the *invariant factors* of the group, the number of copies of $\mathbf{Z}$ is the *rank* of the group.

In the above example, with $H = R$ the group A/R is

$$\{0\} \oplus \mathbf{Z}_{13} \oplus \mathbf{Z} \simeq \mathbf{Z}_{13} \oplus \mathbf{Z}$$

(note that $\{0\}$ appears as $\mathbf{Z}/\langle 1 \rangle$ and $\mathbf{Z}$ as $\mathbf{Z}/\{0\}$).

Theorem 4.16. *If H is a subgroup of finite index in the free abelian group A of rank n, then H is also of rank n.*

Proof. If the rank m of H il less than n, in the quotient A/H there are $n - m$ summands isomorphic to $\mathbf{Z}$, and A/H cannot be finite. ◇

Theorem 4.15 no longer holds if A is a free abelian group of infinite rank as the following example shows. Let A be a countable direct sum $\sum_k \langle a_k \rangle$, $k = 1, 2, \ldots$, of copies of $\mathbf{Z}$, and consider the homomorphism ϕ of A in the additive group of the rationals $\mathbf{Q}$ induced by $a_k \to \frac{1}{k!}$. The $b_k = \frac{1}{k!}$ are a system of generators of $\mathbf{Q}$ (*Ex.* 4.1, 2) and satisfy the relations $2c_2 - c_1 = 0, 3c_3 - c_2 = 0, \ldots, kc_k - c_{k-1} = 0, \ldots$. Then the kernel R of ϕ is generated by the elements $2a_2 - a_1, 3a_3 - a_2, \ldots, ka_k - a_{k-1}, \ldots$, and $A/R \simeq \mathbf{Q}$. However, $\mathbf{Q}$ is not a direct sum, and therefore the v_i and e_i of Theorem 4.15 do not exist.

4.5 Projective and Injective Abelian Groups

An abelian group P is *projective* if the following holds: if G is an abelian group, $H \leq G$, and ϕ a homomorphism $P \to G/H$, then there exists a homomorphism $\psi : P \to G$ such that $\gamma\psi = \phi$, where γ is the canonical homomorphism $G \longrightarrow G/H$:

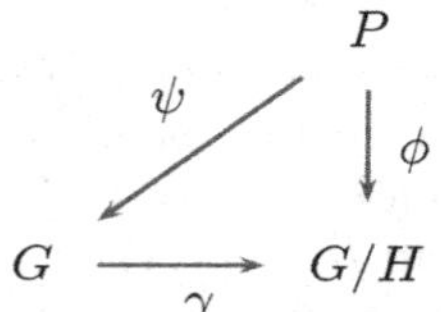

More briefly: P is projective if every homomorphism of P into a quotient G/H lifts to a homomorphism of P into G.

Theorem 4.17. *An abelian group is projective if, and only if, is free.*

Proof. Let P be a projective abelian group, quotient of a free abelian group A. The identity $id : P \to P$ extends to a homomorphism $\phi : P \to A$ such that $\gamma\phi = id$, where γ is the canonical homomorphism $A \to P$. ϕ is injective: if $\phi(x) = \phi(y)$, then $\gamma\phi(x) = \gamma\phi(y)$, i.e. $id(x) = id(y)$ and $x = y$. This shows that P imbeds in A, and therefore, as a subgroup of an abelian free group, is free[7]. Conversely, consider the above diagram. Let A be free, X a basis of A, $\phi(x)$ the image of an element x of X and g_x an element of G such that $\gamma(g_x) = \phi(x)$. The mapping $x \to g_x$ lifts to a homomorphism ψ such that $\gamma\psi = \phi$. Thus A is projective. ◇

Corollary 4.4. *Let $G/H = A$ with A free abelian. Then $G \simeq H \oplus A$. In other words, a free abelian group is a direct summand of every abelian group of which it is a quotient.*

Proof. Consider the above diagram with ϕ the identity. By the theorem, there exists ψ such that $\gamma\psi = id$. ψ is injective because the identity is injective, and therefore $Im(\psi) \simeq A$. If $g \in Im(\psi) \cap H$, then $g = \psi(x)$ and $0 = \gamma(g) = \gamma\psi(x) = id(x) = x$. The result follows[8]. ◇

[7] For this implication, assume that the subgroup theorem for free abelian groups, that we have proved for f.g. groups, holds in general; see footnote 5.

[8] In connection with this result it is interesting to point out the following *Whitehead's problem*: is it true that if all surjective homomorphisms of an abelian group

Given two presentations for an abelian group G:

$$G \simeq A/K, \;\; G \simeq A_1/K_1 \tag{4.5}$$

with A and A_1 free abelian, which relation is there between them? The following lemma gives an answer.

Lemma 4.7 (Schanuel). *Let the two presentations* (4.5) *of the abelian group G be given. Then:*

$$K \oplus A_1 \simeq K_1 \oplus A.$$

Proof. Let $f : A \to A/K$ and $f_1 : A_1 \to A_1/K_1$ be the canonical homomorphisms. Since $f : A \longrightarrow A/K \simeq G \simeq A_1/K_1$, we have a homomorphism $A \to A_1/K_1$ of A (that we keep calling f). By Theorem 4.17, A is projective, so that there exists a homomorphism $\gamma : A \to A_1$ such that $f_1\gamma = f$. Now, given $a_1 \in A_1$ and its image $f_1(a_1)$, there exists $a \in A$ such that $f_1(a_1) = f(a)$ (f is surjective). But $f = f_1\gamma$, and therefore $f_1(a_1) = f_1\gamma(a)$, i.e. $a_1 - \gamma(a) \in ker(f_1) = K_1$. Every element $a_1 \in A_1$ is thus of the form $a_1 = \gamma(a) + k_1, \;\; a \in A, \; k_1 \in K_1$. Then consider the mapping $\psi : A \oplus K_1 \longrightarrow A_1$ given by $(a, k_1) \to \gamma(a) + k_1$, that we have just seen to be surjective, and that, A_1 being abelian, is a homomorphism. Let us find out what its kernel is. If $\gamma(a) + k_1 = 0$, then $f_1\gamma(a) + f_1(k_1) = 0$, and therefore $f_1\gamma(a) = 0 = f(a)$ and $a \in ker(f) = K$. It follows $ker(\psi) = \{(k, k_1) \mid \gamma(k) = -k_1\}$. The mapping $K \to ker(\psi)$ given by $k \to (k, k_1)$ is certainly injective. But it is also surjective; given $k \in K$, there exists $k_1 \in K_1$ such that $\gamma(k) = -k_1$. Indeed, $f_1\gamma(k) = f(k) = 0$ so that $\gamma(k) \in K_1$, and $\gamma(k) = -k_1$ for $k_1 \in K_1$. We have proved that $ker(\psi) \simeq K$ and $A_1 \simeq A \oplus K_1/K$. A_1 being free, by Corollary 4.4 (and its proof) we have $K \oplus A_1 \simeq K_1 \oplus A$. ♢

If in the diagram of Theorem 4.17 we replace "quotient" with "subgroup" and invert all arrows, we obtain the diagram:

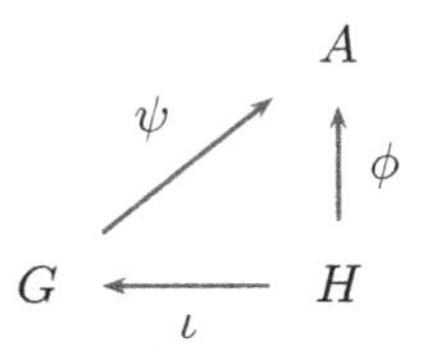

which furnishes the notion dual to projective: an abelian group A is said to be *injective* if every homomorphism φ from a subgroup H of an abelian group G into A extends to a homomorphism of G into A. The rational numbers are an example of an injective group, as a consequence of the following theorem.

G to A with kernel $\mathbf{Z}$ are such that $G \simeq A \oplus \mathbf{Z}$, then A is free? This problem has been proved to be undecidable.

Theorem 4.18 (Baer). *An abelian group is injective if, and only if, is divisible.*

Proof. Let A be injective, $x \in A$ and n an integer. We have to show that n divides x, i.e that there exists $y \in A$ such that $x = ny$. The mapping $\varphi : n\mathbf{Z} \to A$, given by $nk \to xk$, is a homomorphism, and therefore extends to a homomorphism ψ of $\mathbf{Z}$ in A. Let $y = \psi(1)$ and let ι be the inclusion $n\mathbf{Z} \to \mathbf{Z}$; we have: $x = \varphi(n) = \psi\iota(n) = \psi(n) = \psi(1+1+\cdots+1) = n\psi(1) = ny$.

For the converse we use Zorn's lemma. Let A be divisible, H, G and φ as above, and let $\mathcal{S}$ be the set of pairs (S, ψ), where $H \leq S \leq G$ and ψ extends φ to S. $\mathcal{S}$ is not empty because it contains at least (H, φ). Define $(S, \psi) \leq (S', \psi')$ if $S \subseteq S'$ and ψ' extends ψ to S'. Let $\widetilde{S} = \bigcup S_\alpha$, and let $\widetilde{\psi}$ be defined on $\widetilde{S}$ as follows: if $s \in \widetilde{S}$, $s \in S_\alpha$ for some α; then $\widetilde{\psi}(s) = \psi_\alpha(s)$. The pair $(\widetilde{S}, \widetilde{\psi})$ is an upper bound for $\widetilde{S}$; by Zorn's lemma there exists a maximal pair (M, ψ) in $\widetilde{S}$. We show that $M = G$. If $x \in G$, $x \notin M$, then $M < M'$, and we shall reach a contradiction by showing that ψ can be extended to M'.

i) Let $M \cap \langle x \rangle = \{0\}$. Then $M' = M \oplus \langle x \rangle$ and $\psi' : m + kb \to \varphi(m)$ is a homomorphism extending φ.

ii) $M \cap \langle x \rangle \neq \{0\}$, and let k be the least positive integer such that $kx \in M$. If $y \in M'$, $y = m + tx$, $m \in M$, $0 \leq t < k$ (and this expression is unique: if $m + tx = m' + t'x$, $t < t'$, then $(t' - t)x = m - m' \in M$ with $t' - t < k$). Now, $\psi(kx) \in A$; by the divisibility of A, given k there exists $a \in A$ such that $ka = h(kx)$; if we define $\psi'(y) = \psi(m) + ta$, we immediately verify that ψ' extends ψ to M'. ◇

Dually to Corollary4.4 we have:

Corollary 4.5. *An injective abelian group A is a direct summand of every group G in which it is contained as a subgroup.*

Proof.

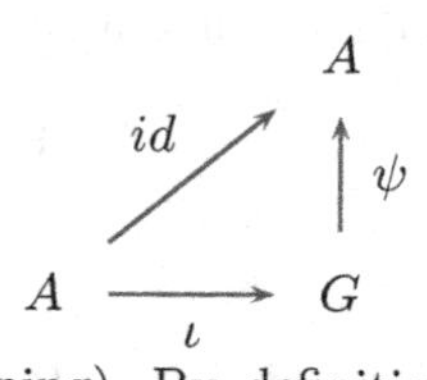

(id denotes the identity mapping). By definition, there exists ψ such that $\psi\iota = id$, and therefore $\psi(a) = a$, for all $a \in A$ and $Im(\psi) \simeq A$. Now $ker(\psi) \cap Im(\psi) = \{0\}$, and ψ is surjective because the identity is surjective. Hence $G = ker(\psi) \oplus Im(\psi) \simeq ker(\psi) \oplus A$. ◇

4.6 Characters of Abelian Groups

Let A be an abelian group, $\mathbf{C}^*$ the multiplicative group of the complex numbers. A *character* of A is a homomorphism χ of A in C^*. These homomorphisms form a multiplicative group (*ex.* 32, *i*)), the *character group* of A, denoted $\widehat{A}$.

Theorem 4.19 (Duality theorem). *If A is finite, $\widehat{A} \simeq A$.*

Proof. If $A = \langle a \rangle$ is cyclic and $a^n = 1$, let $\chi \in \widehat{A}$; then $\chi(a)^n = 1$, and therefore $\chi(a)^n$ is an n-th root of unity, and since these are n in number and χ is determined by the value it has on a, we have n possible characters. If $\chi_k(a) = w^k$, where w is a primitive root, the mapping $A \to \widehat{A}$ given by $a^k \to \chi_k$ is the sought isomorphism. In the general case, A is a direct product of cyclic groups $A_i = \langle a_i \rangle$, $|A_i| = n_i$, $i = 1, 2, \dots t$. Fix $a \in A$; a admits a unique expression $a = a_1^{k_1} a_2^{k_2} \cdots a_t^{k_t}$. If $b = a_1^{h_1} a_2^{h_2} \cdots a_t^{h_t}$ and w_i is a primitive n_ith root of unity, define:

$$\chi_a : b \to w_1^{k_1 h_1} w_2^{k_2 h_2} \cdots w_t^{k_t h_t}.$$

It is clear that $\chi_{aa'} = \chi_a \chi_{a'}$. If $a \neq a'$, then $k_i \neq k'_i$ for at least an i, and therefore $\chi_a(a_i) = w_i^{k_i} \neq w_i^{k'_i} = \chi_{a'}(a_i)$. As a varies, the χ_a are therefore all distinct, and the mapping $A \to \widehat{A}$, given by $a \to \chi_a$, is an isomorphism. Note that $\chi_a(a') = \chi_{a'}(a)$. ◇

Similarly, $\widehat{A} \simeq \widehat{\widehat{A}}$. Thus, $\widehat{A} \simeq A$ and $A \simeq \widehat{\widehat{A}}$; but while the first isomorphism is not "natural" (meaning that, in order to define it, it is necessary to choose a set of generators of the A_i, and therefore it depends on this choice), the second is natural, in the sense that it can be defined directly on the elements: if $\widehat{A} = \{\chi_a, a \in A\}$ and $a' \in A$, then $a' \to \widehat{\chi}_{a'} : \chi_a \to \chi_a(a')$ (for all $a \in A$, $\widehat{\chi}_{a'}$ sends χ_a in the value that χ_a takes on a'). The situation is similar to that of the isomorphisms between a vector space and its dual and double dual.

If $H \leq A$, then every character χ of A/H extends to one of A by composing it with the projection $p : A \to A/H$:

$$\begin{array}{ccc} A & \xrightarrow{p} & A/H \\ & \searrow^{\chi p} & \downarrow \chi \\ & & C^* \end{array}$$

The mapping $\widehat{A/H} \to \widehat{A}$ given by $\chi \to \chi p$ is injective: if $\chi p = \chi_1 p$ then $\chi p(a) = \chi_1 p(a)$ for all $a \in A$. But every coset of H is of the form $p(a)$ for some $a \in A$, and therefore χ and χ_1 coincide on all the cosets of H; hence they are the same character, $\chi = \chi_1$. If A is finite, by the theorem $\widehat{A/H} \simeq A/H$ and $\widehat{A} \simeq A$. We have proved:

Theorem 4.20. *If A is a finite abelian group, and $H \leq A$, then A contains a subgroup isomorphic to A/H.*

We recall that this result no longer holds if the group is not abelian (see *Ex.* 2.1, 3).

The group C^* being divisible, we have the following result.

Lemma 4.8. *Let $B \leq A$, $\chi_1 \in \widehat{B}$. Then χ_1 extends to a character $\chi \in \widehat{A}$, and the mapping $\chi \to \chi_1$ is a homomorphism $\widehat{A} \to \widehat{B}$ whose kernel is:*

$$B^{\perp} = \{\chi \in \widehat{A} \mid \chi(b) = 1, \forall b \in B\}.$$

Proof. Every homomorphism of B into C^* lifts to one of A in C^*. The stated properties follow. ◇

Exercises

22. Determine the groups:

$$G = \langle a, b \mid 2a + 3b = 0 \rangle \text{ and } G = \langle a, b \mid 2a + 3b = 0,\ 5a - 2b = 0 \rangle.$$

23. Give an example that shows that if $H \leq G$, where G is finite abelian, it is not true that there exist cyclic subgroups H_i and G_i such that $H_i \leq G_i$ and G and H are isomorphic to the direct products of the G_i and the H_i, respectively. [*Hint*: consider $G = C_2 \times C_8$.]

24. Using the fundamental theorem of finite abelian groups, show that if H is a subgroup of a finite abelian group G then G contains a subgroup isomorphic to the quotient G/H (see Theorem 4.20).

25. Show that the multiplicative group of the positive rationals $\mathbf{Q}^+$ is free (cf. Chapter 1, *ex.* 22).

26. Let A be free abelian of rank n. Define the *height* $h(x)$ of an element $0 \neq x \in A$ to be the least positive integer appearing as a coefficient of x as the basis ranges over all bases. Show that $h(x)$ exists, and that the gcd of the coefficients of x in the various bases is always the same, and is equal to $h(x)$.

27. Let A be as in the preceding exercise, $H \leq A$ and let $h(H)$ be the least height among all of the elements of H. Show that $h(H)$ divides the height $h(x)$ of every element $x \in H$.

28. In the definition of the universal property of a free abelian group consider, instead of the inclusion $\iota : X \to A$, any mapping $\sigma : X \to A$. Show that the universal property implies that σ is necessarily injective, and that the image $\sigma(X)$ of X generates A. Moreover, A is free on $\sigma(X)$.

29. Show that if an abelian group is generated by n elements, a subgroup is generated by n or fewer elements.

30. Every abelian group embeds in a divisible group. [*Hint*: write $G = A/R$, A free, $A = \sum_{\lambda\in\Lambda}^{\oplus} \mathbf{Z}_\lambda$ and imbed each copy $\mathbf{Z}_\lambda$ in a copy $\mathbf{Q}_\lambda$ of the rationals $\mathbf{Q}$. Then $G = A/R = (\sum_{\lambda\in\Lambda}^{\oplus} \mathbf{Z}_\lambda)/R \subseteq (\sum_{\lambda\in\Lambda}^{\oplus} \mathbf{Q}_\lambda)/R$, and the latter group is divisible.]

31. *i*) If $A = \mathbf{Z} \oplus \mathbf{Z} \oplus \cdots \oplus \mathbf{Z}$ (n copies of $\mathbf{Z}$), then $\mathbf{Aut}(A)$ is isomorphic to the group $GL(n, \mathbf{Z})$.

ii) If $A = \mathbf{Z}_m \oplus \mathbf{Z}_m \oplus \cdots \oplus \mathbf{Z}_m$ (n copies of $\mathbf{Z}_m$), then $\mathbf{Aut}(A)$ is isomorphic to the group $GL(n, \mathbf{Z}_m)$.

32. If A is an abelian group, then:

i) the characters of A form a multiplicative group;

ii) the multiplicative group of the homomorphisms of $\mathbf{Z}$ into A is a group isomorphic to A;

iii) (Artin) Distinct characters are linearly independent over $\mathbf{C}$, i.e, if $\sum_{i=1}^{m} c_i\chi_i(a) = 0$, all $a \in A$, with $c_i \in \mathbf{C}$, then $c_i = 0$, all i. [*Hint*: let $\sum_{i=1}^{m} c_i\chi_i = 0$, m minimum, $a \in A$ such that $\chi_1(a) \neq \chi_2(a)$, and consider $\sum_{i=1}^{m} c_i\chi_i(xa) = 0$, $x \in A$. By dividing the first relation by $\chi(a)$, and subtracting the two relations one obtains a relation which is shorter than the chosen one.]

33. Let G be an abelian group of order p^m, $m > 1$. Show that G has an automorphism of order p.

34. If m is the exponent of a finite abelian group G, then the order of every element of G divides m.

35. (Fedorov) Show that the group of integers is the only infinite group in which every non trivial proper subgroup has finite index. [*Hint*: if $x \neq 1$, $[G : \langle x\rangle]$ is finite, so G is f.g.; the centralizer of every generator has finite index, hence the center, which is the intersection of these centralizers, has finite index. Apply Schur's theorem (Theorem 2.38) and Corollary 4.1.]

4.7 Free Groups

In the preceding chapter we have seen free abelian groups and abelian groups given by generators and relations. In this chapter we extend these concepts to nonabelian groups. The universal property we have seen in the case of a free abelian group, which embodies the basic property of a free set of generators, will be taken as definition.

Definition 4.4. Let F be a group and X a subset of F. Then F is *free* on X, if, for every group G, every function $f : X \to G$ has a unique extension to a homomorphism $\phi : F \to G$:

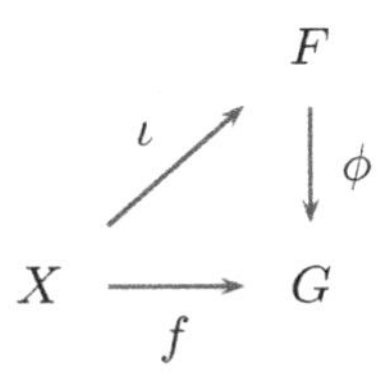

(ι is the inclusion.) The set X is also called a *basis* of F, and its cardinality $|X|$ the *rank* of F.

The next theorem shows that if free groups exist at all, then they are determined by the cardinality of a generating set.

Theorem 4.21. *Let F_1 and F_2 be free groups on X_1 and X_2, respectively. Then F_1 and F_2 are isomorphic if, and only if, X_1 and X_2 have the same cardinality.*

Proof. Let $f : X_1 \to X_2$ be a bijection. Then f determines a mapping, that we keep calling f, $f : X_1 \to F_2$. F_1 being free, f extends to a homomorphism $\phi : F_1 \to F_2$. Similarly, the inverse f^* of f determines $\phi^* : F_2 \to F_1$. By composing ϕ with ϕ^* we have a homomorphism $\phi^*\phi : F_1 \to F_1$, whose value on X_1 is $\phi^*\phi(x) = \phi^*(f(x)) = f^*f(x) = id(x)$, where id is the identity on X_1, because f^* is the inverse of f. It follows that $\phi^*\phi$ extends the inclusion $j : X_1 \to F_1$, and since the identity $I_{F_1} : F_1 \to F_1$ also extends j, and this extension is unique because F_1 is free, we have $\phi^*\phi = I_{F_1}$. Similarly, $\phi\phi^* = I_{F_2}$, and therefore ϕ is an isomorphism: $F_1 \simeq F_2$.

Conversely, let F be free on X. By definition, the homomorphisms of F into a group G are in one-to-one correspondence with the mappings $X \to G$. With $G = \mathbf{Z}_2$, the 2-element group, the latter are $2^{|X|}$ in number, and therefore there are exactly $2^{|X|}$ homomorphisms of F in $\mathbf{Z}_2$. Since this number is invariant under isomorphisms, $2^{|X|}$, and therefore $|X|$, is determined by the isomorphism class of F. ♢

Having proved uniqueness, let us now prove existence.

Let X^{-1} be a set in one to one correspondence with X and disjoint from it and let $A = X \cup X^{-1}$; we call the set A the *alphabet*, and *letters* its elements. A *word* is a finite sequence of elements of A. We write such a sequence as:

$$w = x_1^{\epsilon_1} x_2^{\epsilon_2} \dots x_n^{\epsilon_n},\ n \geq 0, \tag{4.6}$$

where $x_i \in X, \epsilon_i = 0, \pm 1$ $(x^0 = 1), i = 1, 2, \dots, n-1, \epsilon_n = \pm 1$; the integer n is the *length* $l(w)$ of w. If the sequence is empty, the word is the *empty word*; we denote it by 1, and set $l(1) = 0$. A word (4.6) is reduced if $w = 1$ or if $\epsilon \neq 0$ for all ϵ and x and x^{-1} are never adjacent. Let F be the set of reduced words.

A product may be defined between two words by simply juxtaposing them; but this product cannot be defined in the set of reduced words, because the product of two reduced word need not be reduced. One may define the product as the word obtained after reduction, i.e. after cancellation of adjacent x and x^{-1}. With respect to this new product the reduced words form a group (the free group on X we seek). However, verification of the associative law is rather tedious, so we resort to the *van der Waerden trick*. The free group on X will be constructed as the set of permutations of the set F of reduced words.

Consider, for each $x \in X$ and $x^{-1} \in X^{-1}$, the functions $\alpha_x, \alpha_{x^{-1}}$ defined as follows: let $w = yu \in F$, then[9]

$$\alpha_x.w = \begin{cases} xw, \text{ if } y \neq x^{-1}, \\ u, \quad \text{if } y = x^{-1}, \end{cases} \quad \alpha_{x^{-1}}.w = \begin{cases} x^{-1}w, \text{ if } y \neq x, \\ u, \qquad \text{if } y = x. \end{cases}$$

Now consider $\alpha_x\alpha_{x^{-1}}.w$, $w = yzu$. We have:

i) if $y \neq x^{-1}$, then $\alpha_{x^{-1}}\alpha_x.w = \alpha_{x^{-1}}.xyzu = yzu = w$

ii) if $y = x^{-1}$, then $\alpha_{x^{-1}}\alpha_x.w = \alpha_{x^{-1}}.zu = yzu = w$.

In *ii)*, $z \neq x$ because w is reduced. It follows $\alpha_x\alpha_{x^{-1}} = I$. Similarly, $\alpha_{x^{-1}}\alpha_x = I$ so that α_x is bijective and therefore a permutation of F, with inverse $\alpha_{x^{-1}}$. In the group S^F of all permutations of F consider the subgroup generated by the α_x:

$$\mathcal{F} = \{\alpha_x, \ x \in X\}.$$

Let $g \in \mathcal{F}$, $g \neq 1$; g admits the factorization:

$$g = \alpha_{x_1}^{\epsilon_1} \circ \alpha_{x_2}^{\epsilon_2} \circ \ldots \circ \alpha_{x_n}^{\epsilon_n},$$

with α_x^{ϵ} and $\alpha_x^{-\epsilon}$ never adjacent. This factorization is unique; indeed, by applying g to the empty word 1 we have $g.1 = x_1^{\epsilon_1}x_2^{\epsilon_2}\ldots x_n^{\epsilon_n}$, which being a reduced word has a unique spelling. We now show that $\mathcal{F}$ is a free group on the set $[X] = \{\alpha_x, x \in X\}$. Let G be a group, f a function $[X] \to G$. The function $\phi : \mathcal{F} \to G$ defined as follows:

$$\phi(g) = \phi(\alpha_{x_1}^{\epsilon_1} \circ \alpha_{x_2}^{\epsilon_2} \circ \ldots \circ \alpha_{x_n}^{\epsilon_n}) =_{def} f(\alpha_{x_1}^{\epsilon_1})f(\alpha_{x_2}^{\epsilon_2})\ldots f(\alpha_{x_n}^{\epsilon_n}),$$

is well defined (because the spelling of g is unique) and extends f. We show that it is a homomorphism. Let $g, h \in \mathcal{F}$; we have, when the juxtaposition $gh = g \circ h$ is reduced,

$$\phi(g \circ h) = \phi(g)\phi(h). \tag{4.7}$$

If gh is not reduced, $g = u \circ v$ and $h = v^{-1} \circ s$, say, then $\phi(g) = \phi(u)\phi(v)$ and $\phi(h) = \phi(v^{-1})\phi(s)$, because g and h are reduced. Hence in the group G:

$$\phi(g)\phi(h) = \phi(u)\phi(v)\phi(v^{-1})\phi(s) = \phi(u)\phi(s). \tag{4.8}$$

But by definition $\phi(v^{-1}) = \phi(v)^{-1}$ and therefore, by (4.7) and (4.8),

$$\phi(g \circ h) = \phi(g)\phi(h),$$

and ϕ is a homomorphism. If ψ is another homomorphism extending f, then ϕ and ψ coincide on $[X]$, and this set being a generating set for $\mathcal{F}$ we have

[9] Following van der Waerden we write $\alpha_x.w$ for $\alpha_x(w)$.

$\psi = \phi$. Thus, the group $\mathcal{F}$ is a free group on $[X]$. Note that, since the mapping $\mathcal{F} \to F$ defined by:

$$\alpha_{x_1^{\epsilon_1}} \circ \alpha_{x_2^{\epsilon_2}} \circ \cdots \circ \alpha_{x_n^{\epsilon_n}} \to x_1^{\epsilon_1} x_2^{\epsilon_2} \ldots x_n^{\epsilon_n}$$

is a bijection, we have $\mathcal{F} \simeq F$ so that the group F of reduced words is free on $|X|$. In conclusion:

Theorem 4.22. *For every set X there exists a group free on X.*

The unicity of the extension in the definition of a free group implies that the set X generates F (same proof as for the abelian case; see Theorem 4.9). This can also be seen from the above theorem; indeed, the bijection $\mathcal{F} \to F$ takes $[X]$ onto X, and since $[X]$ generates $\mathcal{F}$, X generates F.

A reduced word is *cyclically reduced* if it starts with a letter x_i^ϵ and ends with a letter different from $x_i^{-\epsilon}$. If w is reduced, but not cyclically reduced, then

$$w = y_r^{-\eta_r} y_{r-1}^{-\eta_{r-1}} \ldots y_1^{-\eta_1} (x_1^{\epsilon_1} x_2^{\epsilon_2} \ldots x_m^{\epsilon_m}) y_1^{\eta_1} y_2^{\eta_2} \ldots y_r^{\eta_r},$$

$x_i, y_j \in X$, where $u = x_1^{\epsilon_1} x_2^{\epsilon_2} \ldots x_m^{\epsilon_m}$ is cyclically reduced. Thus, a reduced but not cyclically reduced word is conjugate to a cyclically reduced word: $w = v^{-1}uv$, where $v = y_1^{\eta_1} y_2^{\eta_2} \ldots y_r^{\eta_r}$.

The basis X of F is such that if a word $x_1^{\epsilon_1} x_2^{\epsilon_2} \ldots x_m^{\epsilon_m}$, $x_i \in X$, equals 1, then $m = 0$ and the word is empty. This is equivalent to saying that the expression of a word in terms of the x^ϵ is unique. There are other subsets besides X that have this property. For instance, let $X = \{x, y\}$, and let $X' = \{x, y^x\}$. X' generates F, since the subgroup $\langle X' \rangle$ contains x and y and therefore X. Moreover, it can be seen that a reduced word in x and y^x is conjugate to a reduced word in x and y; for example,

$$x^{-1} y^x y^x x x (y^x)^{-1} = x^{-1}(x^{-1} y y x x y^{-1}) x.$$

Hence, the only reduced word in X' which equals 1 is the empty word. Note that the word $x^{-1}yx$ has length 3 in the basis X and length 1 in X'.

If a subset Y of F generates F and is such that the only word equal to 1 in the elements of Y is the empty word, then Y is a *free basis*, or a set of *free generators*, or that F is *freely generated* by Y.

Lemma 4.9 (Ping pong lemma). *Let a group G act on a set Ω. Let a and b be two elements of G, $o(a) > 2$, and suppose that Ω contains two disjoint non-empty subsets Γ and Δ such that each non trivial element of $A = \langle a \rangle$ maps Γ into Δ, and each non trivial element of $B = \langle b \rangle$ maps Δ into Γ. Then the subgroup of G generated by a and b is free on the set $\{a, b\}$.*

Proof. Let $a_1, a_2 \in A$ be two non identity distinct elements of A. Since $\Gamma a_1 a_2^{-1} \cap \Gamma$ is empty, Γa_1 and Γa_2 are disjoint non-empty subsets of Δ; it follows that $\Gamma a_1 \subset \Delta$, a proper subset. Let $w \in G$, and assume that w begins with an element of A and ends with an element of B, $w = a_1 b_1 a_2 b_2 \ldots a_n b_n$, $n \neq 1, 1 \neq a_i, 1 \neq b_i$ all i. Then $\Gamma a_1 \subset \Delta, \Gamma a_1 b_1 \subset \Delta b_1 \subset \Gamma \subset \ldots \subset \Gamma w \subset \Gamma$. Thus $w \neq 1$. If w begins with an element of A and ends with an element of A, then $\Gamma w \subset \Delta$, and again $w \neq 1$. The remaining cases are treated similarly. No word in the elements a and b is the identity, so G is free on $\{a, b\}$ (cf. *ex.* 39).[10] ◊

Example 4.5. Consider the two matrices $A = \begin{pmatrix} 1 & 2 \\ 0 & 1 \end{pmatrix}$ and $B = \begin{pmatrix} 1 & 0 \\ 2 & 1 \end{pmatrix}$ of $SL(2, \mathbf{Z})$ acting on the plane $\mathbf{R}^2$. Let Δ be the subset of $\mathbf{R}^2$ of the points (x, y) with $|x| > |y|$, and let Γ be that of the points with $|x| < |y|$. Now, the elements of the subgroup $\langle A \rangle$ map Γ to Δ, and those of $\langle B \rangle$ map Δ to Γ. By the lemma, the subgroup $\langle A, B \rangle$ is free on the set $\{A, B\}$.

Exercises

36. Using the universal property of free groups prove that:
i) a free group is projective (as in the abelian case);
ii) the group $\mathbf{Z}$ is free on the set $X = \{1\}$;
iii) a free abelian group is not a free group, except when it is isomorphic to $\{1\}$ (empty basis) or to $\mathbf{Z}$ (rank 1);
iv) no finite group can be free.

37. If G is a group containing a normal subgroup N such that G/N is free, then G contains a free subgroup F such that $G = NF$ and $N \cap F = \{1\}$. [*Hint*: Corollary 4.4.]

38. Prove that the free group of rank 2 contains exactly three subgroups of index 2. (The number $N_{n,r}$ of subgroups of index n in the free group of rank r is given recursively by $N_{1,r} = 1$, $N_{n,r} = N(n!)^{r-1} - \sum_{i=1}^{n-i}(n-i)!^{r-1} N_{i,r}$. For instance, the free group of rank 2 contains 13 subgroups of index 3.[11])

39. Let X be a subset of a group G such that each element of G has a unique spelling $x_1^{k_1} x_2^{k_2} \cdots x_t^{k_t}$, with $x_i \in X$, $t \geq 0$ and $x_i \neq x_{i+1}$. Then G is free on X.

40. A free group is torsion free. [*Hint*: let $w^n = 1, n > 0$; distinguish the two cases w cyclically reduced or not.]

41. Prove, as in *Ex.* 4.5, that for $k \geq 2$ the matrices $A = \begin{pmatrix} 1 & k \\ 0 & 1 \end{pmatrix}$ e $B = \begin{pmatrix} 1 & 0 \\ k & 1 \end{pmatrix}$ of $SL(2, \mathbf{Z})$ generate a free group. For $k = 1$ consider $A^{-1}BA^{-1}$.[12]

---

[10] If $o(a) = o(b) = 2$ the subgroup generated by a and b is dihedral; see next section, *Ex.* 4.6, 2 and 3.

[11] M. Hall, p. 105–106; P. de la Harpe, p. 23.

[12] P. de la Harpe, p. 26.

4.8 Relations

As in the abelian case we have:

Theorem 4.23. *Every group is a quotient of a free group.*

Proof. Let G be a group. Consider the set $X = \{x_g,\ g \in G\}$ and the free group F on X. The bijection $j : X \to G$ extends to a homomorphism $\phi : F \to G$ which is surjective because already j is surjective. Thus $G \simeq F/K$, where $K = ker(\phi)$. More precisely, a group generated by a set X is a quotient of a free group on a set having the same cardinality as X: if F is free on a set Y, $|Y| = |X|$, the mapping $\phi : F \to G$ which associates a reduced word in Y with the corresponding word in X of G:

$$y_1^{\epsilon_1} y_2^{\epsilon_2} \cdots y_n^{\epsilon_n} \to x_1^{\epsilon_1} x_2^{\epsilon_2} \cdots x_n^{\epsilon_n}$$

is the homomorphism extending $Y \to X \to G$, and is surjective (because X generates G, so every element of G is of the said form). ◇

As an application of the above theorem we prove that *if F is the free group of rank n, then F/F' is a free abelian group of rank n.* Indeed, let A be the free abelian group of finite rank n. Then A is a quotient of the free group F of rank n, $A \simeq F/K$. F/K being abelian, the derived group F' of F is contained in K, and therefore:

$$F/K \simeq (F/F')/(K/F').$$

But F/F' is an abelian group with n generators (and no less, otherwise A would be generated by less than n elements), and therefore $F/F' \simeq A$.

Let R be a set of elements of the free group F, and let R^F be the normal closure of R in F). If $G = F/R^F$, then we call the pair X, R a *presentation* of G; we write $G = \langle X|R\rangle$. An element of R^F is a *relator*, and one of R a *defining relator*; an equations $r = 1$ for $r \in R^F$ is a *relation*, and one of R a *defining relation*. If X and R are finite, the presentation is finite and G is *finitely presented*. The group G thus obtained is the "largest" group generated by X and satisfying the relations R, in the sense of Theorem 4.24 below. For the free group we write $F = \langle X|\emptyset\rangle$ or $F = \langle X \mid \rangle$.

The generators of a free group satisfy no relations because a reduced word has a unique spelling, and this is true in particular for the empty word 1: if $w = 1$, then all factors of w are already equal to 1. Moreover, a free group being torsion free (*ex.* 40) it cannot be $w^n = 1$ unless w is already empty.

Theorem 4.24. *Let $G = \langle X|R\rangle$ and $G_1 = \langle X|R_1\rangle$ where $R \subseteq R_1$, then G_1 is a quotient of G. (In other words, by adding relations to a presentation of a group one obtains homomorphic images of the group.)*

Proof. In the free group F on X one has $R \leq R_1$, and therefore, $(G/R^G)/(R_1^G/R_G) \simeq G/R_1^G = G_1$ is a quotient of G. ◇

Taking $G = F$ in this theorem one obtains Theorem 4.23.

Theorem 4.24 may be stated in the following form, called the *substitution test*.

Theorem 4.25. *Let $G = \langle X|R\rangle$ and H be two groups, and let f be a mapping $X \to H$. Then f extends to a homomorphism $\phi : G \to H$ if, and only if, for all $x \in X$ and $r \in R$, by substituting $f(x)$ in r one obtains the identity of H. ϕ is unique because G is generated by X; moreover, if H is generated by $\phi(x)$, $x \in X$, then ϕ is surjective.*

Examples 4.6. 1. Let $G = \langle X|R\rangle$, where $X = \{a, b\}$ and R is given by the three defining relations:

$$a^3 = 1,\ b^2 = 1,\ abab = 1. \tag{4.9}$$

In the group $F = \langle x, y\rangle$ these three relations correspond to the elements $r_1 = xxx$, $r_2 = yy$, and $r_3 = xyxy$. Writing R for the normal closure in F of the set of these three elements, then $G = F/R$. G consists of the cosets Rw, $w \in F$, and it is easily seen that these cosets are equal to the following ones:

$$R,\ xR,\ x^2R,\ yR,\ yxR,\ yx^2R.$$

For example, $xyR = yx^2R$ since $xy = yx^2r$, where r is the element of R obtained as $r = r_1^{-1}(xr_2^{-1}x^{-1})r_3$. Thus $|G| \leq 6$. However, we cannot conclude that G has six elements because there could be hidden consequences of (4.9) that make some of the cosets equal. However, by Theorem 4.24, every group H with two generators satisfying (4.9) is a quotient of G. Such a group is $H = S^3$, generated by the two permutations $\sigma = (1, 2, 3)$ and $\tau = (1, 2)$, for which one actually has $\sigma^3 = 1$, $\tau^2 = 1$ and $(\sigma\tau)^2 = 1$. Then from $|G| \geq |S^3| = 6$ it follows $|G| = 6$, and S^3 being a quotient di G, $G \simeq S^3$. In the terms of Theorem 4.25, we have applied the substitution test with $H = S^3$ and $f : a \to \sigma,\ b \to \tau$.

2. The preceding example extends to all dihedral groups. Let:

$$G = \langle a, b \mid a^n = 1,\ b^2 = 1,\ (ab)^2 = 1\rangle,$$

and let us show that this is the group D_n. From $abab = 1$ we have that a is conjugate to its inverse $b^{-1}ab = a^{-1}$. It follows that $b^{-1}a^kb = a^{-k}$, for all integers k, and that

$$ba^k = a^{-k}b. \tag{4.10}$$

The generic element of G is of the form $a^{h_1}ba^{h_2}\cdots a^{h_m}b^{\epsilon}$, $0 \leq h_i \leq n-1$ (here $-h = n - h$) and $\epsilon = 0, 1$. By (4.10), a factor ba^h can be replaced with $a^{-h}b$. In other words, the given relations allow one to bring all the a's on one side (on the left, say), and all the b's on the other, and therefore to express the elements of the group generated by a and b in the form:

$$a^hb^{\epsilon},\ h = 0, 1, \ldots, n-1,\ \epsilon = 0, 1. \tag{4.11}$$

It follows $|G| \leq 2n$. The dihedral group D_n is generated by the rotation r of $2\pi/n$ and by a flip s around an axis, so that $r^n = 1$ and $s^2 = 1$, and moreover $(rs)^2 = 1$. It follows $|G| = 2n$ and $G \simeq D_n$. Finally, note that the relations allow one to determine the product of two elements of the form (4.11): if $x = a^h b^\epsilon, y = a^k b^\eta$ then $xy = a^{h+k}b^\eta$, if $\epsilon = 0$, and $xy = a^h \cdot a^{-k}b \cdot b^\eta = a^{h-k}b^{\eta+1}$ ($\eta + 1 \bmod 2$) if $\epsilon = 1$.

3. *Infinite dihedral group* (see *Ex.* 2.10, 3). Let

$$G = \langle a, b \mid b^2 = 1,\ (ab)^2 = 1\rangle$$

(in the preceding example remove the relation $a^n = 1$). Also in this case, the relation $(ab)^2 = 1$ implies that the elements of G can be written in the form $a^h b^\epsilon$. A group satisfying the given relations is the group D_∞ of matrices $\begin{pmatrix} \epsilon & k \\ 0 & 1 \end{pmatrix}$, $\epsilon = \pm 1$, $k \in \mathbf{Z}$, with $a \to A=\begin{pmatrix} 1 & 1 \\ 0 & 1 \end{pmatrix}$, $b \to B = \begin{pmatrix} -1 & 0 \\ & 0\,1 \end{pmatrix}$. Consider the homomorphism $\phi : G \to D_\infty$ induced by $a \to A$ and $b \to B$. A is of infinite order, and therefore so is a, hence the powers a^h, $h \in \mathbf{Z}$, are all distinct. The $a^h b^\epsilon$ are also distinct because if $a^h b^\epsilon = a^k b^\eta$, then $a^{h-k} = b^{\epsilon-\eta}$, that in D_∞ becomes $A^{h-k} = B^{\epsilon-\eta}$, which is only possible if both sides equal the identity matrix; thus $h = k$ and $\epsilon = \eta$. This also proves that $ker(\phi) = \{1\}$ (if $A^h B^\epsilon = I$ then $h = \epsilon = 0$), so that G is isomorphic to D_∞.

Setting $c = ab$, G also has the presentation $\langle b, c \,|\, b^2 = 1,\ c^2 = 1\rangle$; the element $bc = a^{-1}$ has infinite order.

Ex. 2 and 3 above show that *any two involutions generate a dihedral group.* This is finite if the product of the two involutions is of finite order, and is the infinite dihedral group in the other case.

4. *The quaternion group.* Let

$$G = \langle a, b \mid ab = b^{-1}a,\ ba = a^{-1}b\rangle.$$

We have:

i) $a^2 = b^2$: $a^{-2}b^2 = a^{-1}(a^{-1}b)b = a^{-1}(ba)b = (a^{-1}b)(ab) = (ba)(b^{-1}a) = (a^{-1}b)(b^{-1}a) = a^{-1}a = 1$;

ii) $a^4 = 1$ and $b^4 = 1$: $a^4 = a(a^2)a = a(b^2)a = (ab)(ba) = (b^{-1}a)(a^{-1}b) = b^{-1}b = 1$, and similarly $b^4 = 1$;

iii) every element of G has the form $a^h b^k$, $h = 0, 1, 2, 3,\ k = 0, 1$. Indeed, the generic element is of type $a^{h_1}b^{k_1}a^{h_2}\cdots a^{h_m}b^{k_m}$, $h_i, k_i \in \mathbf{Z}$, and we have:

$$ba = a^{-1}b,$$

$$b^{-1}a = ab,$$

$$ba^{-1} = ba^3 = b(a^2)a = b(b^2)a = b^3a = b^{-1}a = ab,$$

$$b^{-1}a^{-1} = b^{-1}a^3 = b^{-1}(a^2)a = b^{-1}b^2a = ba = a^{-1}b;$$

iv) moreover, $ab^2 = aa^2 = a^3$, $ab^3 = a(b^2)b = a(a^2)b = a^3b$; hence an element of G is equal to one of $1, a, a^2, a^3, b, ab, a^2b, a^3b$, so $|G| \leq 8$. The mapping $G \to Q$, where $Q = \langle i, j \rangle$ is the quaternion group, given by $a \to i$, $b \to j$, extends to a homomorphism; it follows $G \simeq Q$.

5. Let $G = \langle a, b, c \mid a^3 = b^3 = c^4 = 1, ac = ca^{-1}, aba^{-1} = bcb^{-1} \rangle$. We show that G is the identity group $\{1\}$. We have $1 = ab^3a^{-1} = bc^3b^{-1}$, so $c^3 = 1$, and with $c^4 = 1$ we have $c = 1$. The relation $aba^{-1} = bcb^{-1}$ yields $aba^{-1} = bb^{-1} = 1$, and therefore $b = 1$, and from $ac = ca^{-1}$ we have $a = a^{-1}$ i.e. $a^2 = 1$, which together with $a^3 = 1$ implies $a = 1$.

The *deficiency* of a finite presentation $\langle X|R \rangle$ is defined as the difference $|X| - |R|$. The deficiency of a finitely presented group is the maximum of the deficiencies of all its finite presentations[13]. It is bounded above by the torsion-free rank of the commutator quotient of the group[14].

Theorem 4.26. *A group of positive deficiency is infinite.*

Proof. If $|X| = n$ and $|R| = m$, let F be the free group with basis $X = \{x_1, x_2, \ldots, x_n\}$ and K the normal closure of the defining relators $\{r_1, r_2, \ldots, r_m\}$. Let $\alpha_{i,j}$ be the sum of the exponents of x_i appearing in r_j, and consider the m equations:

$$\sum_{i=1}^{n} \alpha_{i,j} y_i = 0, \; j = 1, 2, \ldots, m.$$

Since $n > m$, there exists a nonzero solution $\beta_1, \beta_2, \ldots, \beta_n$; in general, the β_i are rational, but we may assume that they are integer. To prove that G is infinite we construct a homomorphism from G to a nonzero subgroup of $\mathbf{Z}$. Let $f : X \to \mathbf{Z}$ be given by $f(x_i) = \beta_i$; then f extends to a homomorphism $\phi : F \to \mathbf{Z}$. The choice of the $\alpha_{i,j}$ and β_i ensures that $\phi(r_j) = 0$, $j = 1, 2, \ldots, m$, and so also $\phi(k) = 0$, all $k \in K$. Thus we have a homomorphism $\psi : F/K \to \mathbf{Z}$ given by $\psi(Kg) = \phi(g)$, all $g \in F$. The fact that the β_i are not all zero shows that the image of F and of F/K is a nonzero subgroup of $\mathbf{Z}$; hence F/K, and therefore G, is infinite. ◇

The question now arises as to whether a group with negative deficiency is necessarily finite. The answer is no, as the following example shows.

Example 4.7. Let $G = \langle x, y \mid x^3 = y^3 = (xy)^3 = 1 \rangle$, and consider the two permutations of the set of integers:

$$\sigma = \cdots (0, 1, 2)(3, 4, 5) \cdots, \; \tau = \cdots (1, 2, 3)(4, 5, 6) \cdots.$$

[13] A brief discussion about the definition of deficiency can be found in Macdonald, pp. 163–165.

[14] Chapter 7, inequality (7.29).

We have $\sigma\tau = \cdots(-5,-3,-1)(-2,0,2)(1,3,5)(4,6,8)\cdots$, so $(\sigma\tau)^3 = 1$. Therefore, there is a homomorphism of G onto the group $\langle\sigma,\tau\rangle$, and the latter group is infinite because it contains the element

$$\sigma\tau^2 = (\ldots,-12,-9,-6,-3,0,3,6,9,12,\ldots)\ldots$$

(the other digits are left fixed) which has infinite order.

Exercises

42. Let $G = \langle a, b \mid a^2 = b^2\rangle$. Show that:
i) adding the relation $b^2 = (ab)^2$ the quaternion group is obtained;
ii) adding $b^2 = 1$ the group D_∞ is obtained.

43. For $n = 1, 2, 3, 4$, determine the groups:

$$G = \langle x_1, x_2, \ldots, x_n \mid x_1x_2 = x_3, x_2x_3 = x_4, \ldots, x_{n-1}x_n = x_1, x_nx_1 = x_2\rangle.$$

44. Determine the group $G = \langle a, b \mid a = (ab)^3, b = (ab)^4\rangle$.

45. Show that $G = \langle x, y \mid x^{-1}yx = y^2, y^{-1}xy = x^2\rangle$ is the identity group.

46. (Ph. Hall). Show that if $N \trianglelefteq G$ and G/N are finitely presented, so is G.

4.8.1 Relations and simple Groups

In this section we consider the notion of a simple group in the framework of groups given by generators and relations. There are interesting analogies with some notions of logic and topology.

Theorem 4.27. *Let $G = \langle X|R\rangle$ a group given by generators and relations. Then G is simple if, and only if, the group obtained by adding a word $w \neq 1$ in G to the relators R is the identity:*

$$G_1 = \langle X|R, w\rangle = \{1\}.$$

Proof. Let G be simple, $w \neq 1$, and w^G the normal subgroup generated by w (i.e. the smallest normal subgroup of G containing w). Since $w \neq 1$, we have $w^G \neq \{1\}$, and therefore by the simplicity of G, $w^G = G$. But $G_1 = \langle X|R, w\rangle = G/w^G$ (Theorem 4.24) and therefore $G_1 = \{1\}$. Conversely, let $G \neq \{1\}$, G not simple, $\{1\} < N \triangleleft G$ and $1 \neq w \in N$. From $w^G \subseteq N$ it follows $(G/w^G)/(N/w^G) \simeq G/N \neq \{1\}$, and a fortiori $G/w^G \neq \{1\}$. But $G/w^G = \langle X|R, w\rangle$, which by assumption equals $\{1\}$, and we have a contradiction. ◇

Theorem 4.28. *Let $G \neq \{1\}$, $G = \langle X|R\rangle$, and $G_1 = \langle X|R \setminus r\rangle$ (remove a relator). Then if G is simple, G_1 is not.*

Proof. From $G = G_1/r^G$ it follows, since $G \neq \{1\}$, that $r^G \neq G_1$, and $r \neq 1$ since r is not a relator of G_1. Then r^G is a proper normal subgroup of G_1. ♢

The two theorems above have analogies in logic and topology.

A logical theory T is said to be *complete* if, given a formula ψ, either ψ or its negation $\neg\psi$ can be deduced in T. The theory is *consistent* (or *non-contradictory*) if ψ and $\neg\psi$ cannot both be deduced in T.

Remark 4.3. A theory which is complete and consistent is such that "a lot" can be deduced (either ψ or $\neg\psi$) but "not too much" (not both ψ and $\neg\psi$).

Theorem 4.29. *If a formula not derivable from the axioms is added to the complete and consistent theory T, the resulting theory T' is no longer consistent.*

Proof. Since ψ is not derivable in T and T is complete, $\neg\psi$ is derivable in T. Then, in T', ψ and $\neg\psi$ are both derivable (ψ is derivable because is an axiom). ♢

Theorem 4.30. *Let T be complete and consistent. Then if an axiom ψ is removed, the resulting theory T' is no longer complete.*

Proof. (We assume that the axioms of T are independent.) ψ is not derivable in $T' = T \setminus \psi$ because of the independence of the axioms, nor is $\neg\psi$ otherwise T, which contains ψ and T', would not be consistent. ♢

In topology we have the following results[15].

Lemma 4.10. *Let Y be a compact topological space and Z a Hausdorff space. If $f : Y \to Z$ is continuous we have:*
i) *f is closed;*
ii) *if f is bijective, then it is a homeomorphism.*

Proof. *i)* Let $A \subseteq Y$ be closed; it is compact, and consequently so is $f(A)$, and Z being Hausdorff, $f(A)$ is closed;

ii) f continuous, closed and bijective, and therefore a homeomorphism. ♢

Theorem 4.31. *Let $(X, \mathcal{T})$ be a topological space, where $\mathcal{T}$ is compact Hausdorff. Then:*
i) *if $\mathcal{T} \subset \mathcal{T}_+$, then $(X, \mathcal{T}_+)$ is no longer compact;*
ii) *if $\mathcal{T}_- \subset \mathcal{T}$, then $(X, \mathcal{T}_-)$ is no longer Hausdorff.*

Proof. *i)* If $(X, \mathcal{T}_+)$ is compact, then the identity $x \to x$ yields a function $(X, \mathcal{T}_+) \to (X, \mathcal{T})$ which is continuous because the inverse image of an open set U of $\mathcal{T}$ is U itself, which is also one of the open sets of $\mathcal{T}_+$ because $\mathcal{T}_+$ contains $\mathcal{T}$. Since it is bijective, by the lemma is a homeomorphism. It follows $\mathcal{T} = \mathcal{T}_+$, against the hypothesis.

[15] We assume known the basic notions of topology.

ii) If $(X, \mathcal{T}_-)$ is Hausdorff, consider the mapping $(X, \mathcal{T}) \to (X, \mathcal{T}_-)$ induced by the identity on X. As in *i)*, this mapping is continuous and therefore a homeomorphism. ◇

Remarks 4.4. **1.** In a compact Hausdorff space there are "many" open sets (so that all pairs of points can be separated) but "not too many" (so that a finite covering can be extracted from every open covering). Compare with Remark 4.3.

2. It can be said that if "an open set is removed" from a compact Hausdorff topology, the space is no longer Hausdorff, and if "an open set is added", the space is no longer compact. The situation corresponds to that of simple groups with "open set" in place of "relation" (Theorems 4.28 and 4.27), and to that of logical theories with "open set" in place of "axiom" (Theorems 4.29 and 4.30).

Summing up, a correspondence among groups, logical theories and topological spaces may be established. These structures are defined, respectively, by

Groups: generators and sets of relations.
Logical theories: terms and sets of axioms.
Topological spaces: points and sets of points (the open sets).

We have the following analogies:

Simple groups – complete theories – Hausdorff spaces.
Group $\neq \{1\}$ – consistent theory – compact spaces.

More precisely, a set R of relations for a group G such that $G \neq \{1\}$ (G is simple) corresponds to a system of axioms A for a theory T such that T is consistent (T is complete), and to a family of open sets $\mathcal{T}$ (topology) for a topological space X such that X is compact (X is Hausdorff):

Group G	*Logical theory T*	*Topological space X*
$R : G \neq 1$	$A : T$ consistent	$\mathcal{T} : X$ compact
$R : G$ simple	$A : T$ complete	$\mathcal{T} : X$ Hausdorff

Finally, observe that the identity group $\{1\}$ plays the role of a group defined by "contradictory" relations.

4.9 Subgroups of Free Groups

In this section we prove the Nielsen-Schreier theorem, according to which a subgroup of a free group is again free.

A set S of reduced words of a free group $F = \langle X \rangle$ is a *Schreier system* if together with an element it contains all its initial segments:

$$a_1 a_2 \cdots a_t \in S \Rightarrow a_1 a_2 \cdots a_i \in S,$$

$i \leq t$, $a_i \in X \cup X^{-1}$. In particular, S contains 1, the empty word.

Lemma 4.11. *Let F be free on X, $H \leq F$. Then there exists a set of representatives of the right cosets of H which is a Schreier system.*

Proof. We will construct inductively a choice function of representatives as follows. Define the *length* $l(Hu)$ of a coset Hu as the shortest length of its elements.

0. If $l(Hu) = 0$, then $1 \in Hu$ and therefore $Hu = H$; we choose 1 as representative of H.

1. If $l(Hu) = 1$, then there exists an element of $X \cup X^{-1}$ belonging to Hu, and we choose any of these as representative of Hu.

2. If $l(Hu) = 2$, there exist a_1, a_2 of $X \cup X^{-1}$ such that $Hu = Ha_1a_2$. As $l(Ha_1) = l(a_1) = 1$, let $\overline{a_1}$ be the representative chosen at the preceding step for this coset; we have $\overline{a_1} = ha_1$, some $h \in H$, so that $\overline{a_1}a_2 = ha_1a_2 \in Hu$. Then choose $\overline{a_1a_2} = \overline{a_1}a_2$ as representative of Hu.

3. Assume the representatives of the cosets of length n have been chosen, and let $l(Hu) = n+1$. Then there exist $a_1, a_2, \ldots a_n, a_{n+1} \in X \cup X^{-1}$ such that $Hu = a_1a_2 \cdots a_na_{n+1}$. Let s be the representative chosen for the coset $Ha_1a_2 \cdots a_n$; then $s = ha_1a_2 \cdots a_n$, some $h \in H$, and $sa_{n+1} = ha_1a_2 \cdots a_na_{n+1} \in Hu$. Then choose $\overline{a_1a_2 \cdots a_na_{n+1}} = sa_{n+1}$ as representative of Hu.

By construction, the system thus obtained is a Schreier system. Moreover, such a system is minimal, i.e. it is a Schreier system in which the length of the words does not exceed that of any word they represent. ◇

The system of the lemma is also called a *Schreier transversal.*

Examples 4.8. 1. Let $F = \langle a, b \rangle$, R the normal closure of the set $\{a^2, b^2, a^{-1}b^{-1}ab\}$. F/R is a Klein group; thus R has four cosets and taking as representatives $\{1, a, b, ab\}$ we have a Schreier system. The set $\{1, a, b, ab^{-1}\}$ is also a Schreier system, whereas $\{1, a, b, a^{-1}b^{-1}\}$ is not because the initial segment a^{-1} of $a^{-1}b^{-1}$ is not a representative. The system $\{1, a, ab, aba\}$ is Schreier but not minimal because the coset represented by aba contains the word b, a word of shorter length.

2. For R the normal closure of $\{a^n, b^2, abab\}$ (F/R is the dihedral group D_n) we have the Schreier system $1, a, a^2, \ldots, a^{n-1}, b, ab, a^2b, \ldots, a^{n-1}b$.

In the lemmas to follow, S will be a Schreier transversal of the right cosets of a subgroup H of a free group $F = \langle X \rangle$.

Lemma 4.12. *i) If $s \in S$, $x \in X \cup X^{-1}$, and $u = sx(\overline{sx})^{-1}$, then either $u = 1$ or else u is reduced;*

ii) if $u \neq 1$ the spelling of u is unique.

Proof. i) s and $\overline{sx}$ are reduced, so that if there is a cancellation in u it can only take place between s and x or between x and $(\overline{sx})^{-1}$. In the former case, s ends with x^{-1}, i.e. $s = a_1a_2 \cdots a_tx^{-1}$, and hence, S being Schreier,

$sx = a_1a_2\cdots a_t \in S$; then $\overline{sx} = sx$ and $u = 1$. In the latter, $(\overline{sx})^{-1}$ begins with x^{-1} and therefore $\overline{sx}$ ends with x, that is $\overline{sx} = a_1a_2\cdots a_tx$. It follows $Hsx = Ha_1a_2\cdots a_tx$, $Hs = Ha_1a_2\cdots a_t$, so that $\overline{a_1a_2\cdots a_t} = s$, $\overline{sx} = sx$ and $u = 1$.

ii) Let $sx(\overline{sx})^{-1} = s_1x_1(\overline{s_1x_1})^{-1}$. If s and s_1 have the same length, the two words being reduced by *i*), we have $s = s_1$ and $x = x_1$. If $l(s) < l(s_1)$, sx in an initial segment of s_1, and therefore $sx \in S$. Then $sx = \overline{sx}$ and $u = 1$. ◇

Observe that setting $s_1 = \overline{sx}$ we have $(sx(\overline{sx})^{-1})^{-1} = s_1x(\overline{s_1x^{-1}})^{-1}$.

Lemma 4.13. *Let $v_1 = sx^\epsilon(\overline{sx^\epsilon})^{-1}$, $v_2 = ty^\delta(\overline{ty^\delta})^{-1}$, with $s,t \in S$, $x,y \in X$, $\epsilon,\delta = 1,-1$, $v_1 \neq 1$, $v_2 \neq 1$, $v_2 \neq v_1^{-1}$. Then in the product v_1v_2 neither x^ϵ nor y^δ can be cancelled.*

Proof. Assume the contrary, and suppose that y^δ is cancelled first. Then $t = a_1a_2\cdots a_h$ and $(\overline{sx^\epsilon})^{-1} = a_k^{-1}\cdots y^{-\delta}a_h^{-1}\cdots a_1^{-1}$. Thus ty^δ is an initial segment of $\overline{sx^\epsilon}$, and therefore $ty^\delta = \overline{ty^\delta}$, from which $v_2 = 1$, contrary to $v_2 \neq 1$. If x^ϵ is cancelled first, then $\overline{sx^\epsilon}\cdot x^{-\epsilon} = s$ (the coset of $\overline{sx^\epsilon}$ is Hsx^ϵ and therefore that of $\overline{sx^\epsilon}\cdot x^{-\epsilon}$ is $Hsx^\epsilon \cdot x^{-\epsilon} = Hs$, and hence its representative is s). $\overline{sx^\epsilon} = sx^\epsilon$ and $v_1 = 1$, contrary to $v_1 \neq 1$. If x^ϵ and ty^δ are simultaneously cancelled, then $t = \overline{sx^\epsilon}$, $y^\delta = x^{-\epsilon}$ and $v_2 = v_1^{-1}$. ◇

Corollary 4.6. *A product $v_1v_2\cdots v_m$, $v_i \neq 1$, $v_{i+1} \neq v_i^{-1}$, with the v_i of the form of the preceding lemma, can never be the identity.*

Proof. By the lemma, the cancellations between v_i and v_{i+1} cannot concern the x^ϵ and y^δ of v_i and v_{i+1}. It follows that when the v_i are expressed in terms of the generators and their inverses, these elements remain, and the product can never be the empty word. ◇

If $H \leq F = \langle X\rangle$, taking as transversal of the cosets of H a Schreier system S, we know (Lemma 4.1) that the elements $sx(\overline{sx})^{-1}$, with $s \in S$ and $x \in X$, form a system of generators for H.

Theorem 4.32 (Nielsen–Schreier). *Every subgroup of a free group is free.*

Proof. Let $H \leq F$; the elements $u = sx(\overline{sx})^{-1} \neq 1$ generate H, and will be free generators of H if no reduced product u or u^{-1} equals the identity (i.e. it reduces to the empty word when expressed in terms of the generators). By Corollary 4.6 we have the result. ◇

Theorem 4.33. *Let F be free of rank r and let H be a subgroup of finite index j. Then H is free of rank $1 + (r-1)j$.*

Proof. The elements $u = sx(\overline{sx})^{-1}$ are in this case rj in number; in order to obtain the rank of H we have to remove the u equal to 1. Let $S_0 = S \setminus \{1\}$ and consider the following mapping τ of S_0 into the set of the u equal to 1:

if $s \in S_0$ and s ends with x^{-1}, then $\tau : s \to sx(\overline{sx})^{-1}$, and if s ends with x, then $s = s_1x$ and $\tau : s \to s_1x(\overline{s_1x})^{-1}$. τ is injective; let us show that is also surjective. If $u = s'x(\overline{s'x})^{-1} = 1$, x must cancel either with an element of s' or with one of $(\overline{s'x})^{-1}$. In the former case, $s' = s_1x^{-1}$ so that u is the image of s'. In the latter, $(\overline{s'x})^{-1}$ begins with x^{-1}, and therefore $\overline{s'x}$ ends with x (and in fact $\overline{s'x} = s'x$); then u is the image of $\overline{s'x}$. The number of $u = 1$ is $|S_0| = j - 1$, and therefore the rank of H is $rj - (j-1) = 1 + (r-1)j$, as required. Note in particular that if F is of rank $r > 1$, then every subgroup of finite index has rank greater than r. ◊

Examples 4.9. 1. The normal closure R of the subgroup generated by a^2, b^2, and $[a^{-1}b^{-1}ab]$ in *Ex.* 4.8, 1 has index 4, and since F ha rank 2, the rank of R is $1 + (2-1)4 = 5$. We determine five free generators of R. Let us take $S = \{1, a, b, ab\}$ as Schreier system; then, with $X = \{a, b\}$, we have:

$$\begin{aligned}
1a \cdot (\overline{1a})^{-1} &= a \cdot a^{-1} = 1,\\
1b \cdot (\overline{1b})^{-1} &= b \cdot b^{-1} = 1,\\
aa \cdot (\overline{aa})^{-1} &= a^2 \cdot (\overline{a^2})^{-1} = a^2 \cdot 1^{-1} = a^2,\\
ab \cdot (\overline{ab})^{-1} &= ab \cdot (ab)^{-1} = 1,\\
ba \cdot (\overline{ba})^{-1} &= ba \cdot (ab)^{-1} = bab^{-1}a^{-1},\\
bb \cdot (\overline{bb})^{-1} &= b^2 \cdot (\overline{b^2})^{-1} = b^2 \cdot 1^{-1} = b^2,\\
aba \cdot (\overline{aba})^{-1} &= aba \cdot (\bar{b})^{-1} = abab^{-1},\\
abb \cdot (\overline{abb})^{-1} &= ab^2 \cdot (\overline{ab^2})^{-1} = ab^2(\bar{a})^{-1} = ab^2a^{-1}.
\end{aligned}$$

Therefore, the normal closure R has the five free generators:

$$a^2,\ b^2,\ bab^{-1}a^{-1},\ abab^{-1},\ ab^2a^{-1}.$$

2. Let $|G| = n$, $G = \langle X \rangle$ with $X = G$. If $G = F/K$ with F free on X, by Theorem 4.33 K has rank $\rho = (n-1)n + 1$, and $G = \langle X | R \rangle$, with $\rho(R) = \rho$.

Theorem 4.34. *If $F = \langle x, y \rangle$ is the free group of rank 2, the derived subgroup F' has infinite rank.*

Proof. F/F' is free abelian of rank 2, generated by $F'x$ and $F'y$. Thus, every element of F/F' has a unique expression as $F'x^m \cdot F'y^n = F'x^my^n$, $m, n \in \mathbf{Z}$. Every coset of F' then contains a unique element of the form x^my^n (first the x and then the y); these elements form a Schreier system S. An element of the form y^nx does not belong to S, $\overline{y^nx} \neq y^nx$. It follows $u = y^nx(\overline{y^nx})^{-1} \neq 1$, for all $n > 0$, so that the basis of F', given by $sx(\overline{sx})^{-1}$, contains infinite elements. ◊

Corollary 4.7. *The free group of rank 2 contains free subgroups of any finite or countable rank.*

Proof. The derived subgroup has infinite rank and contains free subgroups of any finite or countable rank. ◊

Exercises

47. Using the Nielsen–Schreier theorem prove that two commuting elements u and v of a free group are powers of a common element[16]. [*Hint*: consider $\langle u, v\rangle$ and apply *ex.* 36 *iii*).] More generally, if two powers of u and v commute, then u and v are powers of a common element.

48. In a free group, the relation "$x\rho y$ if x and y commute" is an equivalence relation in the set of nonidentity elements of the group.

49. In a free group F the centralizer of every element is cyclic. In particular, if F is of rank greater than 1, then the center of F is the identity.

50. A nonidentity element of a free group cannot be conjugate to its inverse.

4.10 The Word Problem

The *word problem*[17] may be stated as follows: given a f.g. group G, does there exist an algorithm to establish for an arbitrary word w in the generators of G whether or not $w = 1$? This problem in equivalent to the *equality problem*: given two words $w_1, w_2 \in G$, does there exist an algorithm to determine whether or not $w_1 = w_2$? (Clearly, equality holds if, and only if, $w_1 w_2^{-1} = 1$.)

We need a few notions of recursion theory. A subset S of the natural numbers is said to be *recursive* if there exists an algorithm to determine of an arbitrary element whether or not it belongs to S. It is *recursively enumerable* if there exists an algorithm which enumerates the elements of S in some order, i.e. if a list can be made of its elements. Every recursive set S is recursively enumerable, and S is recursive if both S and its complement are recursively enumerable. Thus, if S is recursive, a given natural number will eventually appear either in the list for S, or in that for its complement. By a diagonal type argument it can be shown that there exist sets of natural numbers that are recursively enumerable but not recursive. Now consider a set X indexed by a set of natural numbers. The above notions can be carried over to the set of words over X.

The words over a finite alphabet $X \cup X^{-1}$ can be listed as follows: first the word 1, of length 0; those of length 1 are the letters of the alphabet, that can be listed, for instance, as

$$x_1, x_1^{-1}, x_2, x_2^{-1}, \ldots, x_n, x_n^{-1}.$$

[16] For a direct proof see Magnus-Karrass-Solitar, p. 42 *ex.* 6, or Lyndon-Schutzenberger, Michigan Math J. **9** (1962), p. 289.

[17] Posed by Max Dehn (1911).

Then all the words can be listed using the lexicographic order; for instance, those of length 2:

$$x_1x_1, x_1x_1^{-1}, x_1x_2, \ldots, x_n^{-1}x_n^{-1},$$

and so on. In a finitely presented group, using the list $w_1, w_2, \ldots$ we can make a list of the words equal to 1 in G as the following theorem shows.

Theorem 4.35. *If G is finitely presented, $G = \langle X|R\rangle$, and Ω is the set of all words over X, then the set*

$$W = \{w \in \Omega \mid w = 1 \text{ in } G\}$$

is recursively enumerable.

Proof. The words on X can be listed in the lexicographic order, and so can the relators of R. Then the elements of Ω are of the form:

$$w_1^{-1}r_1^{\epsilon_1}w_1w_2^{-1}r_2^{\epsilon_2}w_2 \ldots w_i^{-1}r_i^{\epsilon_i}w_i,$$

for all $i = 1, 2, \ldots$, and sequences $w_1, w_2, \ldots, w_i$ of words on X, $r_1, r_2, \ldots$ of relators of R, and $\epsilon_i = \pm 1$. Thus Ω is recursively enumerable. ◇

Now G being finitely presented, W is recursively enumerable, and the word problem for G is solvable if $\{w \in \Omega \mid w \neq 1\}$ is also recursively enumerable. However, in general the word problem is not solvable[18].

Remark 4.5. A f.g. group is *recursively presented* if the set of relations is recursively enumerable. The word problem for these groups can be defined as in the case of finitely presented groups, and is also a recursively enumerable problem. An important theorem of G. Higman states that a finitely generated group can be embedded in a finitely presented group if, and only if, it has a recursively enumerable set of defining relators.

It is clear that for a free group the word problem is solvable. Given a word, the required algorithm is simply the reduction process: a word in the free generators x_i is 1 if after reduction the word is empty; otherwise, $w \neq 1$.

Theorem 4.36 (Kuznetsov). *For a finitely presented simple group $G = \langle X|R\rangle$ the word problem is solvable.*

Proof. Let $w \in G$. If $G = \{1\}$ there is nothing to prove. Let $G \neq \{1\}$, $x \neq 1$ a fixed element of G, and $G_w = \langle X \mid R, w\rangle$. If $w = 1$ in G, then obviously $G \simeq G_w$. If $w \neq 1$, then G being simple by Theorem 4.27 we have $G_w = \{1\}$. It follows that $x = 1$ in G_w if and only if $w \neq 1$ in G. The group G_w is also finitely presented. Then, given the word w of G, we can simultaneously form two lists:

- words equal to 1 in G;
- words equal to 1 in G_w.

[18] E.L. Post, 1945; P.S. Novikov, 1955.

Hence:

- if $w = 1$ in G, w appears on the first list;
- if $w \neq 1$ in G, x appears on the second list.

Just wait and see which of the two events occurs. $\diamondsuit$

4.11 Residual Properties

Let G be a group, ρ a relation among elements and subsets defined on G and its homomorphic images, and let $\mathcal{P}$ be an abstract group property[19]. Then G has *residually* property $\mathcal{P}$ *with respect to* ρ, or that G is *residually* $\mathcal{P}$, if for any pair of elements x, y that *are not* in the relation ρ there exists a surjective homomorphism ϕ of G in a group K having property $\mathcal{P}$ and such that $\phi(x) \neq \phi(y)$ (both ϕ and the group depend on the pair x, y).

Examples 4.10. **1.** Let ρ be the relation of equality, and let $\mathcal{P}$ be an abstract group property. Then G is residually $\mathcal{P}$ with respect to ρ if for any pair of distinct elements x and y there exists a homomorphism of G in a group K having property $\mathcal{P}$ and such that the images are also distinct. Since $x \neq y$ is equivalent to $xy^{-1} \neq 1$, this is equivalent to saying that an element of G different from 1 remains different from 1 in a group having property $\mathcal{P}$. Thus G is residually $\mathcal{P}$ with respect to ρ if for all $g \neq 1$, there exists $N \trianglelefteq G$, depending on x, such that $g \notin N$ and G/N has property $\mathcal{P}$. (The fact that $g \neq 1$ in the group G is witnessed in some quotient of G which has property $\mathcal{P}$.)

2. The previous example is a special case of the membership relation between elements and sets "$x\rho A$ if $x \in A$"; the equality relation $x = y$ is obtained with $A = \{y\}$. Thus, if $\mathcal{P}$ is an abstract group property then G is residually $\mathcal{P}$ with respect to this relation if for all $x \notin A$ there exists $N \trianglelefteq G$ depending on x and A such that $Nx \notin \{Ny,\, y \in A\}$. This means that $xy^{-1} \notin A$ for all $y \in A$, and therefore $\phi(x) \neq \phi(y)$, where ϕ is the canonical homomorphism $G \to G/N$, G/N has property $\mathcal{P}$, and we are in the previous case. The set A can be a subgroup, a f.g. subgroup, etc.

3. A further example is the conjugacy relation. In this case, G is residually $\mathcal{P}$ if, for all pairs of non conjugate elements $x \not\sim y$, there exists a homomorphism ϕ, depending on x and y, of G onto a group having property $\mathcal{P}$ and such that $\phi(x) \not\sim \phi(y)$.

4. A specially important residual property is *residual finiteness* (if the relation ρ is not specified, the equality relation is meant). For example, the integers are residually finite. Indeed, given $n \in\mathbf{Z}$, $n \neq 0$, there exists a finite group in which the image of n is nonzero: just take the quotient $\mathbf{Z}/\langle m\rangle$, with $m \nmid n$. This property is shared by all free groups (Theorem 4.37 below).

[19] An *abstract group property* is a property which is preserved under isomorphism.

5. If $\mathcal{P}$ is a property inherited by subgroups, then being residually $\mathcal{P}$ is inherited by subgroups. Indeed, let $1 \neq x \in H$, and $x \notin N \lhd G$; then $x \notin H \cap N$, $H \cap N \lhd H$, $H/(H \cap N) \simeq HN/N \leq G/N$. But G/N has $\mathcal{P}$, and therefore also HN/N and $H/(H \cap N) \simeq HN/N$.

Let K be the intersection of all normal subgroups N of G such that G/N has property $\mathcal{P}$. If G is residually $\mathcal{P}$, given $1 \neq x \in G$ there exists N_x such that $x \notin N_x$ and G/N_x has $\mathcal{P}$; it follows $K \subseteq \bigcap_{1 \neq x \in G} N_x = \{1\}$. Conversely, if $K = \{1\}$, an element $x \neq 1$ cannot belong to all N, and hence $x \notin N$ for at least one N. In other words, *G is residually $\mathcal{P}$ if, and only if, the intersection of the normal subgroups N such that G/N has property $\mathcal{P}$ is the the identity.*

Theorem 4.37. *A free group is residually finite.*

Proof.[20] We must show that if $1 \neq x \in F$ there exists a normal subgroup N_x depending on x such that $x \notin N_x$ and the quotient F/N_x is finite. Let $X = \{x_\lambda,\ \lambda \in \Lambda\}$ be a basis of F and let $x = x_{\lambda_1}^{\epsilon_1} x_{\lambda_2}^{\epsilon_2} \cdots x_{\lambda_n}^{\epsilon_n}$ be the reduced form of x, with $\epsilon_i = \pm 1$. Let us then define a function f from X to the symmetric group S^{n+1}, where n is the number of the x_λ appearing in x, as follows: $f(x_\lambda) = 1$, if x_λ does not appear in x, and $f(x_{\lambda_i}^{\epsilon_i})$ is a permutation σ_i sending i to $i+1$ if $\epsilon_i = 1$, and $i+1$ to i if $\epsilon_i = -1$. If an index λ_i equals λ_{i+1}, then $\sigma_i = \sigma_{i+1}$, and if $\sigma_i(i) = i+1$ and $\sigma_{i+1}(i+1) = i$ then in the expression of x there would appear $x_{\lambda_i}^{\epsilon_i} = x_{\lambda_i}^{-\epsilon_i}$, which is excluded because the expression is reduced. Thus, up to this point, the permutation σ_i is well defined; the images of the other elements under σ_i may be prescribed in an arbitrary way (obviously the resulting function must be a permutation) and such that the product of the σ_i is different from 1. Then f extends to a homomorphism $F \to S^{n+1}$, whose kernel K does not contain x; with $N_x = K$ we have the result. ◊

Corollary 4.8. *The intersection of all subgroups of finite index in a free group is the identity.*

Proof. For any element $x \neq 1$ there exists a subgroup N_x of finite index not containing x. ◊

An important feature of a residual property is that it allows the solution of the decision problem corresponding to a relation ρ; that is, it is possible to establish whether or not two elements are in the relation ρ. If ρ is the equality relation the following theorem holds.

Theorem 4.38. *For a finitely presented and residually finite group G the word problem is solvable.*

---
[20] We follow Macdonald, p. 167.

Proof. Let F be free of rank n, R finitely generated and $G = F/R$; we have to decide whether $w \in R$ or not. Then we begin two procedures: with the first one, we enumerate the elements of R; with the second, we make a list of the multiplication tables of the finite quotients of F/R. Then necessarily either w appears in the first list, and then $w \in R$, or if $w \notin R$ then $wR \neq R$, and therefore, F/R being residually finite, there exists $H/R \trianglelefteq F/R$ such that $(F/R)/(H/R)$ is finite and $wR \notin H/R$, and hence wR will appear in the list of finite quotients of F/R. Just wait and see which occurs. ♢

Definition 4.5. A group is *hopfian* if it is not isomorphic to a proper quotient. In other words, a group is hopfian if every surjective homomorphism of the group in itself is injective (and therefore is an automorphism).

Theorem 4.39 (Malcev). *A f.g. and residually finite group is hopfian.*

Proof. Let G be the group, and let $\phi : G \to G$ be a surjective homomorphism. For a fixed n let $H_1, H_2, \ldots, H_k$ be the subgroups of index n of G (Theorem 3.5, *i*)). Now, H_i thought of as belonging to $\phi(G)$, is the image of a subgroup L_i: $H_i = L_i/K$, where $K = ker(\phi)$, and

$$n = [G : H_i] = [\phi(G) : \phi(L_i)] = [G/K : L_i/K] = [G : L_i].$$

If $L_i = L_j$, then $H_i = H_j$, and therefore the L_i are all distinct, and having index n the set of the L_i coincides with that of the H_i. It follows that the kernel K is contained in all the H_i. Since n is any integer, the above argument applies to all subgroups of finite index, and hence $K \subseteq \bigcap_{[G:H]<\infty} H$. But this intersection is $\{1\}$; thus $K = \{1\}$ and ϕ is injective. ♢

Theorem 4.40 (Baumslag). *The automorphism group of a f.g. and residually finite group G is residually finite.*

Proof. If $1 \neq \alpha \in \mathbf{Aut}(G)$, there exists $x \in G$ such that $\alpha(x) \neq x$, that is $\alpha(x)x^{-1} \neq 1$. Let $N \lhd G$ of finite index with $\alpha(x)x^{-1} \notin N$; by Theorem 3.5, *ii*), N contains a subgroup K which is characteristic in G and of finite index. Then $\mathbf{Aut}(G/K)$ is finite and every automorphism β of G induces an automorphism $\overline{\beta}$ of G/K: $\overline{\beta}(Kx) = K\beta(x)$. The mapping $\phi : \mathbf{Aut}(G) \to \mathbf{Aut}(G/K)$ given by $\beta \to \overline{\beta}$ is a homomorphism and $\overline{\alpha}$ is not the identity of $\mathbf{Aut}(G/K)$. Indeed, for the coset Kx we have $\overline{\alpha}(Kx) = K\alpha(x) \neq Kx$, since $\alpha(x)x^{-1} \notin K$. ♢

Corollary 4.9. *The automorphism group of a free group of finite rank is residually finite, and if it is f.g. is hopfian.*

Given a family $\{G_\lambda\}$ of groups, a *subdirect product* G of G_λ is a subgroup of the cartesian product $\prod_\lambda G_\lambda$ such that for each $g_\lambda \in G_\lambda$ there is at least one $g \in G$ having g_λ as its λ-th component. Now if N_λ, $\lambda \in \Lambda$, is a family of normal subgroups of a group G we have a homomorphism of G in the cartesian product of the quotient groups $G_\lambda = G/G_\lambda$ obtained by sending $x \in G$

to the function which associates with λ the element xN_λ of G_λ. The kernel is $\bigcap_\lambda N_\lambda$, so that if this intersection is $\{1\}$ the group G is a subdirect product of the G_λ. It follows:

Theorem 4.41. *A group G having property $\mathcal{P}$ residually is a subdirect product of groups having property $\mathcal{P}$.*

Proof. Consider the family G_λ of normal subgroups such that G/G_λ has property $\mathcal{P}$. ♢

Corollary 4.10. *A free group is a subdirect product of finite groups.*

Remarks 4.6. 1. Finitely generated linear groups over a field are residually finite (Malcev).

2. The simplest example of a finitely presented non-hopfian group is the Baumslag-Solitar group. It is the group generated by two elements a, b with a single defining relation $a^{-1}b^2a = b^3$.[21]

Exercises

51. Subgroups, direct products and direct sums of residually finite groups are residually finite.

52. *i)* A f.g. abelian group is residually finite, and therefore hopfian;
ii) the automorphism group of a f.g. abelian group is residually finite.

53. A nontrivial divisible group cannot be residually finite.

54. If the group is not f.g., Theorem 4.39 is no longer true. [*Hint*: consider the free group F of infinite rank on $x_1, x_2, \ldots$; if $N = \langle x_1 \rangle^F$, then $F/N \simeq F$.]

55. Prove that the group $GL(n, \mathbf{Z})$ is residually finite. [*Hint*: let $A = (a_{i,j})$, and let m be an integer such that $|a_{i,j}| < m$.]

56. Use *ex.* 51 and 55, and *Ex.* 4.5 to prove that a free group of finite or countable rank is residually finite.

57. (Nielsen) If F is free of finite rank n, and $a_1, a_2, \ldots, a_n$ generate F, then the a_i are a free basis. [*Hint*: Theorem 4.39.]

Definition 4.6. A group is *co-hopfian* if it is not isomorphic to a proper subgroup. In other words, a group is co-hopfian if every injective self homomorphism is surjective, and therefore an automorphism. (It is the notion dual to hopfian.)

58. Show that:
i) a finite group is both hopfian and co-hopfian;
ii) the additive group of rational numbers is both hopfian and co-hopfian;
iii) the multiplicative group Q^* of the rationals is not co-hopfian; [*Hint*: consider the mapping $r \to r^3$.]

---

[21] Cf. Lyndon-Schupp, Theorem 4.3.

iv) the additive group of the reals is not co-hopfian;
v) the group of integers is hopfian but not co-hopfian, and this holds for every f.g. infinite abelian group;
vi) the group C_{p^∞} is co-hopfian but not hopfian;
vii) a f.g. free group is hopfian (Theorems 4.37 and 4.39) but not co-hopfian.

59. A group acting faithfully on a rooted tree is residually finite.

60. Determine a subdirect product of the group $C_2 \times C_4$.

5

Nilpotent Groups and Solvable Groups

There are two important properties of groups that are stronger than commutativity: they are solvability and nilpotence. Solvable[1] groups are obtained by forming successive extensions of abelian groups; nilpotent groups lie midway between abelian and solvable groups.

5.1 Central Series and Nilpotent Groups

In Chapter 2 we defined the commutator of two elements and the subgroup generated by all the commutators. We now extend these notions.

Definition 5.1. If x, y and z are three elements of a group G, define $[x, y, z] = [[x, y], z]$, and inductively

$$[x_1, x_2, \ldots, x_n] = [[x_1, x_2, \ldots, x_{n-1}], x_n].$$

(Hence $[[x_1, x_2, \ldots, x_{n-1}], x_n] = [\cdots[[x_1, x_2], x_3] \ldots, x_n])$. If H, K are subgroups of G, define $[H, K] = \langle [h, k],\ h \in H,\ k \in K \rangle$, and inductively $[H_1, H_2, \ldots, H_n] = [[H_1, H_2, \ldots, H_{n-1}], H_n]$.

We have $[H, K] = [K, H]$ and $[H, K] \trianglelefteq \langle H, K \rangle$. If $H, K \trianglelefteq G$, then $[H, K] \trianglelefteq G$ and $[H, K] \subseteq H \cap K$. If φ is a homomorphism of G, then $[H, K]^\varphi = [H^\varphi, K^\varphi]$. Obviously, $[H_1, H_2, \ldots, H_n] \supseteq \langle [h_1, h_2, \ldots, h_n], h_i \in H_i \rangle$, but in general the other inclusion does not hold, as the following example shows.

Example 5.1. In A^5, let $H_1 = \{I, (1,2)(3,4)\}$, $H_2 = \{I, (1,3)(2,5)\}$, $H_3 = \{I, (1,3)(2,4)\}$. With $h_1 = (1,2)(3,4)$, $h_2 = (1,3)(2,5)$ and $h_3 = (1,3)(2,4)$, we have $[h_1, h_2, h_3] = (1,4,5,2,3)$, whereas $[h_1, h_2, h_3] = 1$ if one of the three elements is 1. It follows $\langle [h_1, h_2, h_3],\ h_i \in H_i \rangle = \langle (1,4,5,2,3) \rangle$. Now $[H_1, H_2]$ contains $[h_1, h_2]^2 = (1,2,4,5,3)$ and therefore $[H_1, H_2, H_3]$ contains $[[h_1, h_2]^2, h_3] = (1,3,5)$.

[1] British English speakers often say "soluble" instead of "solvable".

Machì A.: Groups. An Introduction to Ideas and Methods of the Theory of Groups.
DOI 10.1007/978-88-470-2421-2_5, Springer-Verlag Italia 2012

The next lemma follows from the definitions.

Lemma 5.1. *i) If* $H, K \leq G$ *then* $K \subseteq \mathbf{N}_G(H)$ *if and only if* $[K, H] \subseteq H$. *In particular,* $H \trianglelefteq G$ *if and only if* $[G, H] \subseteq H$.
ii) Let $K, H \leq G$ *with* $K \subseteq H$. *Then the following are equivalent:*
a) $K \trianglelefteq G$ *and* $H/K \subseteq \mathbf{Z}(G/K)$*; and*
b) $[H, G] \subseteq K$.

Definition 5.2. A series of subgroups of a group G:

$$\{1\} = H_0 \subseteq \ldots \subseteq H_1 \subseteq \ldots \subseteq H_i \subseteq H_{i+1} \subseteq \ldots \tag{5.1}$$

is said to be *central* if, for all i,

$$H_{i+1}/H_i \subseteq \mathbf{Z}(G/H_i), \tag{5.2}$$

i.e. xH_i, for $x \in H_{i+1}$, commutes with all the elements of G/H_i. This means that

$$x \in H_{i+1} \Rightarrow [x, g] \in H_i, \tag{5.3}$$

for all $g \in G$, i.e.

$$[H_{i+1}, G] \subseteq H_i. \tag{5.4}$$

In other words, the action of G on G/H_i given by $(xH_i)^g = x^g H_i$ is trivial on H_{i+1}/H_i: $(xH_i)^g = xH_i$, i.e. $x^{-1}x^g \in H_i$. Conversely, (5.4) implies (5.2). The H_i are normal in G.

We now consider two special kinds of central series. The first one is an ascending series. If in (5.2) we take the entire center $\mathbf{Z}(G/H_i)$ of G/H_i, then setting $H_i = Z_i$ we have:

$$\{1\} = Z_0 \subseteq Z_1 = \mathbf{Z}(G) \subseteq Z_2 \subseteq \ldots \subseteq Z_i \subseteq Z_{i+1} \subseteq \ldots$$

where $Z_{i+1}/Z_i = \mathbf{Z}(G/Z_i)$. Hence Z_{i+1} consists of all the elements $x \in G$ for which (5.3) holds:

$$Z_{i+1} = \{x \in G \mid [x, g] \in Z_i, \ \forall g \in G\}.$$

Obviously, if the series (5.1) is central, we have $H_i \subseteq Z_i$.

The second series is descending. Starting from the group G and using (5.4), define the subgroups $\Gamma_i = \Gamma_i(G)$ as follows:

$$\Gamma_1 = G, \ \Gamma_2 = [\Gamma_1, G], \ldots, \Gamma_{i+1} = [\Gamma_i, G], \ldots$$

(Note that $\Gamma_2 = [\Gamma_1, \Gamma_1] = [G, G] = G'$, the derived group of G.) We have:

$$G = \Gamma_1 \supseteq \Gamma_2 \supseteq \ldots \supseteq \Gamma_i \supseteq \Gamma_{i+1} \supseteq \ldots$$

If $x \in \Gamma_i$, then $[x, g] \in \Gamma_{i+1}$ for all $g \in G$, so that

$$[x\Gamma_{i+1}, g\Gamma_{i+1}] = [x, g]\Gamma_{i+1} = \Gamma_{i+1},$$

i.e. $x\Gamma_{i+1}$ commutes with all the elements of G/Γ_{i+1}. In other words,

$$\Gamma_i/\Gamma_{i+1} \subseteq \mathbf{Z}(G/\Gamma_{i+1}),$$

which proves that the series of the Γ_i is central.

By definition of Z_i, if $H_n = G$ then $Z_n = G$. The next theorem shows that if $Z_n = G$ then $\Gamma_{n+1} = \{1\}$.

Theorem 5.1. *Let (5.1) be a central series of a group G with $H_n = G$. Then:*

$$\Gamma_{n-i+1} \subseteq H_i \subseteq Z_i,\ i = 0, 1, \ldots, n. \tag{5.5}$$

Proof. The second inclusion has already been seen. As to the first, we prove by induction on $j = n - i$ that $\Gamma_{j+1} \subseteq H_{n-j}$; this inclusion holds for $j = 0$. Assume it is true for j; then $\Gamma_{j+2} = [\Gamma_{j+1}, G] \subseteq [H_{n-j}, G] \subseteq H_{n-(j+1)}$, as required. ◇

Corollary 5.1. *Let*

$$\{1\} = Z_0 \subset Z_1 \subset \ldots \subset Z_{c-1} \subset Z_c = G,$$

and

$$G = \Gamma_1 \supset \Gamma_2 \supset \ldots \supset \Gamma_r \supset \Gamma_{r+1} = \{1\}.$$

Then $r = c$.

Proof. Let $n = c$ in (5.5). If $i = 0$, then $\Gamma_{c+1} \subseteq Z_0 = \{1\}$, so $c \geq r$. Now let us take for the series of the H_i that of the Γ_i in ascending order:

$$H_0 = \Gamma_{r+1},\ H_1 = \Gamma_r,\ \ldots, H_i = \Gamma_{r-i+1}, \ldots$$

for $i = r$ we have $H_r = \Gamma_1 = G$ and $H_r \subseteq Z_r$. It follows $Z_r = G$ and $r \geq c$. ◇

In words, this corollary states that in a nilpotent group the series of the Z_i and of the Γ_i reach G and $\{1\}$, respectively, in the same number $c + 1$ of steps.

Definition 5.3. In a nilpotent group G the integer c of Corollary 5.1 is the *nilpotence class* of G. The series of the Z_i is the *upper central series* (the ascending central series that reaches G in the least number of steps); the series of the Γ_i is the *lower central series* (the descending central series that reaches $\{1\}$ in the least number of steps). It is clear that the terms of both series are characteristic.

The identity group is nilpotent of class $c = 0$. Abelian groups are nilpotent of class $c \leq 1$ (and conversely). If $c \leq 2$ then $\{1\} \subset Z \subset G$, so $G/Z = \mathbf{Z}(G/Z)$, G/Z is abelian and $G' \subseteq Z$; conversely, the latter inclusion implies $c \leq 2$. This is the case, for instance, of the group D_4 or of the quaternion group. The integer c is a measure of how far the group is from being abelian. Note that if G is of class c, G/Z is of class $c-1$.

It is clear that if G is nilpotent of class $c \leq m$, then $[x_1, x_2, \ldots, x_{m+1}] = 1$ for all $x_1, x_2, \ldots, x_{m+1}$ in G, and conversely.

Nilpotence is a property inherited by subgroups and homomorphic images.

Theorem 5.2. *Subgroups and homomorphic images of a class c nilpotent group are nilpotent of class at most c.*

Proof. i) If $H \leq G$, then $\Gamma_{i+1}(H) = [\Gamma_i(H), H] \subseteq [\Gamma_i(G), G] \subseteq \Gamma_{i+1}(G)$, so $\Gamma_{i+1}(G) = \{1\}$, implies $\Gamma_{i+1}(H) = \{1\}$.

ii) If φ is a homomorphism of G, and $H, K \leq G$, then $[h,k]^\varphi = [h^\varphi, k^\varphi]$, thus $\Gamma_{i+1}(G^\varphi) = [\Gamma_i^\varphi, G^\varphi] = \Gamma_{i+1}(G)^\varphi$. Hence, if $\Gamma_{i+1}(G) = \{1\}$, its image equals $\{1\}$ as well. ◇

Example 5.2. The inclusions (5.5) are strict in general. Let $G = D_4 \times C_2$ (cf. *Ex.* 2.11, 3). The center of G is the product of the centers of the factors, so it is a Klein group V, and G/V has order 4 and so is abelian. G is nilpotent with upper central series $\{1\} \subset V \subset G$. The derived group of G coincides with that of D_4; hence $\Gamma_2 = C_2$, and this C_2 being contained in the center of G it follows $\Gamma_3 = [\Gamma_2, G] = \{1\}$. The lower central series is therefore $G \supset \Gamma_2 \supset \{1\}$. Hence Γ_2 is properly contained in Z_1.

An ideal J of an associative ring R is nilpotent if a power J^n is $\{0\}$. By definition,

$$J^n = \{\sum x_1 x_2 \cdots x_n,\ x_i \in J\},$$

so J is nilpotent if any product $x_1 x_2 \cdots x_n$ of n elements is zero. We now show that in a ring with unity a nilpotent ideal gives rise to a nilpotent group (a subgroup of the invertible elements of the ring).

Theorem 5.3. *Let J be a nilpotent ideal of a ring R with unity* 1, $J^n = \{0\}$. *Then $G = 1 + J = \{1 + x,\ x \in J\}$ is a nilpotent group.*

We prove that G satisfies the three properties of a group.

i) closure: $(1+x)(1+y) = 1 + (x + y + xy)$, and $xy \in J$ if $x, y \in J$;
ii) identity: 1;
iii) inverse: $(1+x)(1 - x + x^2 - \cdots \pm x^{n-1}) = 1$.

Moreover, G is nilpotent. Let $H_k = 1 + J^k$. The ideal J^k is nilpotent because $(J^k)^n = J^{nk} = \{0\}$, so by 1 the H_k are subgroups. Moreover, $J^k = J^{k-1} \cdot J \subseteq J^{k-1}$, thus $H_k \subseteq H_{k-1}$. Finally, the series

$$\{1\} = H_n \subseteq H_{n-1} \subseteq \cdots \subseteq H_1 = G$$

is central. Let us show that

$$a \in H_k \Rightarrow [a,g] \in H_{k+1}, \ \forall g \in G,$$

i.e. $[a,g] \in 1 + J^{k+1}$ or $[a,g] - 1 \in J^{k+1}$. Now,

$$[a,g] - 1 = a^{-1}g^{-1}ag - 1 = a^{-1}g^{-1}(ag - ga),$$

and with $a = 1 + x$, $x \in J^k$, and $g = 1 + y$, $y \in J$, we have

$$\begin{aligned} ag - ga &= (1+x)(1+y) - (1+y)(1+x) \\ &= 1 + y + x + xy - 1 - x - y - yx \\ &= xy - yx \in J^k \cdot J - J \cdot J^k \subseteq J^{k+1}, \end{aligned}$$

so $[a,g] - 1 \in J^{k+1}$. ◇

Examples 5.3. 1. In the ring of $n \times n$ upper triangular matrices, those having zero diagonal form a nilpotent ideal J with $J^n = 0$. If I is the identity matrix, $G = I + J$ is the group of upper unitriangular matrices. Over a finite field, G is a nilpotent group of order $q^{n(n-1)/2}$ (see *Ex.* 3.4, 1).

2. If V is a vector space of dimension n over a field K and

$$V = V_n \supset V_{n-1} \supset \cdots \supset V_1 \supset \{0\},$$

where $V = \langle v_1, v_2, \ldots, v_n \rangle$ and $V_{n-i} = \langle v_{i+1}, \ldots, v_n \rangle$, the linear transformations ϕ such that $V_i\phi \subseteq V_{i-1}$ form an ideal J which is nilpotent because

$$v\phi_1\phi_2 \cdots \phi_n = (v\phi_1)\phi_2 \cdots \phi_n \in V_{n-1}\phi_2 \cdots \phi_n \subseteq \ldots \subseteq \{0\}$$

for all $v \in V$, so $\phi_1\phi_2 \cdots \phi_n = 0$ and $J^n = \{0\}$. The matrices of the linear transformations ϕ are the matrices of the ideal J of the previous example. The group G of Theorem 5.3 stabilizes the series of subspaces V_i (we recall that this means that if $v \in V_i$ and $g \in G$, then $v^g \in V_i$ and $v^g - v \in V_{i-1}$, i.e. G fixes the subspaces and acts trivially on the quotient spaces).

Let us consider a few properties of nilpotent groups.

Theorem 5.4. *If G is nilpotent and $H < G$, then $H < \mathbf{N}_G(H)$ ("normalizers grow").*

Proof. Let $\{\Gamma_k\}$ be the lower central series, and let i be such that $\Gamma_i \not\subseteq H$ and $\Gamma_{i+1} \subseteq H$. Then $[\Gamma_i, H] \subseteq [\Gamma_i, G] = \Gamma_{i+1} \subseteq H$, so Γ_i normalizes H and is not contained in H. ◇

Corollary 5.2. *A maximal subgroup of a nilpotent group is normal, and therefore is of prime index.*

Corollary 5.3. *In a nilpotent group, the derived group is contained in the Frattini subgroup.*

Proof. A maximal subgroup M is normal, hence G/M is of prime order and $G' \subseteq M$. This holds for all maximal subgroups, so $G' \subseteq \cap M = \mathbf{\Phi}(G)$. ♢

Theorem 5.5. *Let G be nilpotent, $N \neq \{1\}$ a normal subgroup. Then the intersection $N \cap \mathbf{Z}(G) \neq \{1\}$. In particular, every minimal normal subgroup of G is contained in the center and has prime order.*

Proof. Let $\{\Gamma_k\}$ be the lower central series of G and let i be such that $\Gamma_i \cap N \neq \{1\}$ and $\Gamma_{i+1} \cap N = \{1\}$. Then $[\Gamma_i \cap N, G] \subseteq [\Gamma_i, G] \subseteq \Gamma_{i+1}$ because $\{\Gamma_k\}$ is central, and $[\Gamma_i \cap N, G] \subseteq N$ because N is normal. It follows $[\Gamma_i \cap N, G] \subseteq \Gamma_{i+1} \cap N = \{1\}$, and therefore $N \cap \mathbf{Z}(G) \neq \{1\}$ because it contains $\Gamma_i \cap N \neq \{1\}$. If N is minimal, then $N \subseteq \mathbf{Z}(G)$: the subgroups of N are normal in G, so N has no proper subgroups and $|N| = p$. ♢

These properties have already been encountered in the case of finite p-groups. The reason is in the following theorem.

Theorem 5.6. *A finite p-group of order p^n is nilpotent, and is of class at most $n-1$.*

Proof. Since the center of a p-group is nontrivial, the upper central series stops when $G/Z_i = \bar{1}$, i.e. $Z_i = G$. Hence the group is nilpotent. If it is of class c, then:

$$|Z_1/Z_0| \cdot |Z_2/Z_1| \cdots |Z_c/Z_{c-1}| = |Z_c|/|Z_0| = |G| = p^n.$$

All the quotients have order at least p because they are all nontrivial, but not all of them have order p. Indeed, $Z_c/Z_{c-1} = G/Z_{c-1}$ cannot be cyclic; if it were, the quotient of G/Z_{c-2} by the center Z_{c-1}/Z_{c-2} would be cyclic because it is isomorphic to G/Z_{c-1}, and G/Z_{c-2} would be abelian (Chapter 2, *ex.* 23) and would coincide with its center Z_{c-1}/Z_{c-2}. Then $G = Z_{c-1}$, against G being of class c. There being c quotients, the product of their orders is greater than p^c, so $p^n > p^c$, $c < n$ and $c \leq n-1$. ♢

A p-group of order p^n always admits a central series of length $n+1$ (*ex.* 2), and therefore with cyclic quotients of order p. For example, in D_4 we have the series $\{1\} = Z_0 \subset Z_1 \subset V \subset D_4$, with V in between Z_1 and $Z_2 = D_4$.

An infinite p-group is not necessarily nilpotent (*ex.* 5). The next theorem shows that in the finite case nilpotent groups are a generalization of p-groups.

Theorem 5.7. *In a finite nilpotent group a Sylow p-subgroup is normal, and therefore is the unique Sylow p-subgroup. It follows that the group is the direct product of its Sylow p-subgroups, for the various primes p dividing the order of the group.*

Proof. If S is a Sylow p-subgroup, then by Theorem 3.16 $\mathbf{N}_G(\mathbf{N}_G(S)) = \mathbf{N}_G(S)$. But by Theorem 5.4, if $\mathbf{N}_G(S) \neq G$ then $\mathbf{N}_G(\mathbf{N}_G(S)) > \mathbf{N}_G(S)$. Hence $\mathbf{N}_G(S) = G$, and S is normal. The product of these p-Sylows for the various p is then a subgroup that coincides with G because its order is divisible by the whole power of p dividing $|G|$, for all p. An element x_i of a p_i-Sylow S_i cannot equal a product of elements x_j of p_j-Sylow S_j, $j \neq i$, because since the p_i-Sylows are pairwise commuting (they are normal and of trivial intersection) this product has order the product of the orders (the x_j have coprime orders), and therefore is coprime to $o(x_i)$. Hence a Sylow p-subgroup has trivial intersection with the product of the others, and therefore the product of the Sylow p-subgroups is direct. ♢

Theorem 5.8. *The Frattini subgroup* $\mathbf{\Phi} = \mathbf{\Phi}(G)$ *of a finite group* G *is nilpotent.*

Proof. Let P be a p-Sylow of $\mathbf{\Phi}$, and let us show that P is normal in $\mathbf{\Phi}$ (in fact $P \trianglelefteq G$). Since $\mathbf{\Phi} \trianglelefteq G$, by the Frattini argument (Theorem 3.15) $G = \mathbf{\Phi}\mathbf{N}_G(P)$. A fortiori, $G = \langle \mathbf{\Phi}, \mathbf{N}_G(P)\rangle$, so by the property of $\mathbf{\Phi}$, $G = \langle \mathbf{N}_G(P)\rangle = \mathbf{N}_G(P)$, i.e. $P \trianglelefteq G$. It follows that the group is the direct product of its Sylows and so (*ex.* 3) is nilpotent. ♢

It is not true in general that if N is a nilpotent normal subgroup of a group G such that G/N is nilpotent then G is nilpotent[2]; the group S^3 is a counterexample. However, if N is the center or (in the finite case) the Frattini subgroup then G is nilpotent (see also *ex.* 20, *iii*)).

Theorem 5.9. *i*) *If* $G/\mathbf{Z}(G)$ *is nilpotent, then so is* G*;*
ii) *if* G *is finite and* $G/\mathbf{\Phi}(G)$ *is nilpotent, then so is* G.

Proof. i) Let $Z = \mathbf{Z}(G)$. From $\Gamma_{i+1}(G/Z) = \Gamma_i Z/Z$ and $\Gamma_{n+1}(G/Z) = Z$ it follows $\Gamma_n Z/Z = Z$, from which $\Gamma_n Z \subseteq Z$, $\Gamma_n \subseteq Z$ and $\Gamma_{n+1} = \{1\}$.

In the finite case, the proof can go as follows. Let $Z = \mathbf{Z}(G)$, with G/Z nilpotent, and let SZ/Z be a p-Sylow of G/Z, some p. Since $SZ/Z \trianglelefteq G/Z$ we have $SZ \trianglelefteq G$. Then, by the Frattini argument, $G = Z\mathbf{N}_G(S)$, and since $Z \subseteq \mathbf{N}_G(S)$, we have $G = \mathbf{N}_G(S)$ and $S \trianglelefteq G$. But this holds for all p, so G is nilpotent.

ii) Let $\mathbf{\Phi} = \mathbf{\Phi}(G)$, and let $G/\mathbf{\Phi}(G)$ be nilpotent. As above, $G = \mathbf{\Phi}\mathbf{N}_G(S)$. Hence $G = \langle \mathbf{\Phi}, \mathbf{N}_G(S)\rangle = \langle \mathbf{N}_G(S)\rangle = \mathbf{N}_G(S)$, and $S \trianglelefteq G$. ♢

Lemma 5.2. *i*) Let $x, y, z \in G$; then:

$$[x, y^{-1}, z]^y [y, z^{-1}, x]^z [z, x^{-1}, y]^x = 1;$$

(Hall-Witt identity).

ii) (Hall's three subgroups lemma) Let $H, K, L \leq G$ and $N \trianglelefteq G$. If $[H, K, L] \subseteq N$ and $[K, L, H] \subseteq N$, then $[L, H, K] \subseteq N$.

iii) $[xy, z] = [x, z]^y [y, z] = [x, z][[x, z], y][y, z]$,
$[x, yz] = [x, z][x, y]^z = [x, z][x, y][[x, y], z]$.

[2] But see *ex.* 20, *iii*).

Proof. *i*) We have

$$[x, y^{-1}, z] = [x^{-1}yxy^{-1}, z] = yx^{-1}y^{-1}x \cdot z^{-1} \cdot x^{-1}yxy^{-1} \cdot z,$$

and therefore $[x, y^{-1}, z]^y = x^{-1}y^{-1}xz^{-1}x^{-1} \cdot yxy^{-1}zy = a \cdot b$. From *i*) we have, similarly,

$$[y, z^{-1}, x]^z = b^{-1} \cdot c, [z, x^{-1}, y]^x = c^{-1}a^{-1}.$$

ii) For $x \in H$, $y \in K$ and $z \in L$ we have, by assumption,

$$[x, y^{-1}, z]^y \in N, \quad [y, z^{-1}, x]^z \in N.$$

Then, by *i*), is also $[z, x^{-1}, y]^x \in N$ and therefore $[z, x^{-1}, y] \in N$ for all $z \in L, x \in H, y \in K$, and so also $[L, H, K] \subseteq N$.

iii) follows by calculation. ◇

The linear transformations of a vector space that stabilize the series of subspaces given in *Ex.* 5.3, 2 form a nilpotent group. The next theorem shows that in the fact of stabilizing a series of subgroups lies the true reason for nilpotence. We prove the theorem in the case of an invariant series of subgroups, but by a result of Hall it also holds without the hypothesis of normality on the subgroups[3] (in this case, however, the upper bound for the nilpotence class is far worse: if the series has $n+1$ terms, one obtains a nilpotent group of class at most $\binom{n}{2}$).

Theorem 5.10. *Let $G = G_0 \supseteq G_1 \supseteq \ldots \supseteq G_n = \{1\}$ be an invariant series*[4] *of G having $n+1$ terms, and let A be the group of automorphisms of G that stabilize the series, that is,*

$$G_i^\alpha = G_i, \quad x^{-1}x^\alpha \in G_{i+1}, \; x \in G_i \tag{5.6}$$

for $\alpha \in A$ and $i = 0, 1, \ldots, n-1$. Then A is nilpotent of class less than n.

Proof. Consider the subgroups A_j of A defined as follows:

$$A_j = \{\alpha \in A \mid x^{-1}x^\alpha \in G_{i+j}, \text{ if } \; x \in G_i\}.$$

Therefore A_j consists of the elements of A acting trivially on the quotients G_i/G_{i+j}, $i = 0, 1, \ldots, n-j$. Clearly,

$$A = A_1 \supseteq A_2 \supseteq \ldots \supseteq A_n = \{1\}. \tag{5.7}$$

Let us prove that (5.7) is a central series, i.e.

$$[A_j, A] \subseteq A_{j+1}, \tag{5.8}$$

[3] M. Kargapolov, Iou. Merzliakov, Theorem 16.3.2.

[4] Definition 2.19.

for all j. Consider the semidirect product $\overline{G}$ of G by A. Then an element of the form $x^{-1}x^{\alpha}$ is a commutator of $\overline{G}$, and the second condition of (5.6) means that $[G_i, A] \subseteq G_{i+1}$. Then (5.8) is equivalent to

$$[[A_j, A], G_i] \subseteq G_{i+j+1}, \tag{5.9}$$

because, by definition, A_{j+1} consists of all elements α such that $[G_i, \alpha] \subseteq G_{i+j+1}$. But

$$[[A, G_i], A_j] \subseteq [A_j, G_{i+1}] \subseteq G_{i+j+1},$$

and

$$[[A_j, G_i], A] \subseteq [G_{i+j}, A] \subseteq G_{i+j+1},$$

and being $G_{i+j+1} \trianglelefteq G$, by Lemma 5.2, *ii*), we have (5.9). Since (5.7) is of length at most n the result follows. ◇

Lemma 5.3. *If H, K and L are normal in a group G, then $[HK, L] = [H, L]$ $[K, L]$ and $[H, KL] = [H, K][H, L]$.*

Proof. For $x \in H$, $y \in K$ and $z \in L$ we have $[xy, z] = [x, z]^y[y, z] = [x^y, z^y][y, z]$ (Lemma 5.2, *iii*). H and L being normal, $[x^y, z^y] \in [H, L]$, from which $[HK, L] \subseteq [H, L][K, L]$. The other inclusion is obtained by observing that $[H, L], [K, L] \subseteq [HK, L]$. Similarly for the second equality. ◇

Theorem 5.11 (Fitting). *The product of two normal nilpotent subgroups H and K of a group G, of class c_1 and c_2, respectively, is a normal nilpotent subgroup of class at most $c_1 + c_2$.*

Proof. We may assume $G = HK$. We have

$$\Gamma_n(G) = [HK, HK, \ldots, HK].$$

Repeated application of Lemma 5.3 yields $\Gamma_n(G)$ as a product of 2^n terms each of which has the form $A = [A_1, A_2, \ldots, A_n]$, where the A_i are equal to either H or K. H and K are normal in G, and $\Gamma_i(H)$ is characteristic in H and therefore normal in G, from which $[\Gamma_i(H), K] \subseteq H$; similarly for K. It follows that if in A there are l of the A_i that are equal to H, then $A \subseteq \Gamma_l(H)$; similarly $A \subseteq \Gamma_{n-l}(K)$, from which $A \subseteq \Gamma_l(H) \cap \Gamma_{n-l}(K)$. With $n = c_1 + c_2 + 1$, we have either $l \geq c_1 + 1$ or $n - l \geq c_2 + 1$. In any case $A = \{1\}$. ◇

Let us prove directly that under the hypothesis of the previous theorem the center of G is nontrivial. If $H \cap K = \{1\}$, then H and K commute elementwise. So $\{1\} \neq \mathbf{Z}(H)$ centralizes K, and since it also centralizes H, it centralizes G. Hence $\mathbf{Z}(G)$ contains $\mathbf{Z}(H)$ that is different from $\{1\}$. Thus assume $H \cap K \neq \{1\}$. $H \cap K$ is a nontrivial normal subgroup of H, so it meets the center of H non trivially (Theorem 5.5): $\mathbf{Z}(H) \cap H \cap K \neq \{1\}$. It follows, $\mathbf{Z}(H) \cap K \neq \{1\}$, and since $\mathbf{Z}(H) \trianglelefteq G$ (it is characteristic in $H \trianglelefteq G$), we have that $\mathbf{Z}(H) \cap K$ is normal in G, and in particular in K. Hence $\mathbf{Z}(K) \cap \mathbf{Z}(H) \cap K \neq \{1\}$, so

$\mathbf{Z}(K)\cap\mathbf{Z}(H) \neq \{1\}$. The elements of this intersection centralize H and K,and therefore G. It follows that $\mathbf{Z}(G) \neq \{1\}$.

In the finite case it is possible to deduce from this proof that G is nilpotent. Indeed, set $Z = \mathbf{Z}(G)$; then G/Z is the product of the normal nilpotent subgroups HZ/Z and KZ/Z. But $|G/Z| < |G|$, so by induction G/Z is nilpotent, and so also is G (Theorem 5.9, *i*)).

The next theorem characterizes in various ways the nilpotent finite groups.

Theorem 5.12. *The following properties of a finite group G are equivalent:*

i) *G is nilpotent;*

ii) *if $H < G$ then $H < \mathbf{N}_G(H)$;*

iii) *maximal subgroups are normal;*

iv) *$G' \subseteq \mathbf{\Phi}(G)$;*

v) *Sylow p-subgroups are normal;*

vi) *G is the direct product of its Sylow subgroups;*

vii) *G is the product of normal p-subgroups, for various p.*

Proof. The implications *i*) $\Rightarrow$ *ii*) $\Rightarrow$ *iii*) $\Rightarrow$ *iv*) are those of Theorem 5.4 and its Corollaries 5.14 and 5.15;

iv) $\Rightarrow$ *v*) (Wielandt) $SG' \trianglelefteq G$, so by the Frattini argument it follows $G = SG'\mathbf{N}_G(S) = \langle S, G', \mathbf{N}_G(S)\rangle = \langle G', \mathbf{N}_G(S)\rangle = \mathbf{N}_G(S)$, where the last equality follows from the assumption $G' \subseteq \mathbf{\Phi}(G)$;

v) $\Rightarrow$ *vi*) Theorem 5.7;

vi) $\Rightarrow$ *vii*) obvious;

vii) $\Rightarrow$ *i*) Theorem 5.6 and Theorem 5.11. ◇

Definition 5.4. The maximal nilpotent normal subgroup of a group G is called the *Fitting subgroup* of G. If G is finite such a subgroup always exists, and is the product of all normal nilpotent subgroups of G. It is denoted $\mathbf{F}(G)$. (It is not excluded that $\mathbf{F}(G) = \{1\}$, as is the case of a simple group.)

In the finite case, the Fitting subgroup has the following characterization (recall that $O_p(G)$ is the maximal normal p-subgroup of G[5]).

Theorem 5.13. *Let G be a finite group, and let p_i, $i = 1, 2, \ldots, t$, be the prime divisors of the order of G. Then:*

$$\mathbf{F}(G) = O_{p_1} \times O_{p_2} \times \cdots \times O_{p_t}.$$

Proof. Since $O_p \subseteq \mathbf{F}(G)$ for all p we have one inclusion. As to the other, observe that $\mathbf{F}(G)$ being nilpotent, it has a normal Sylow p-subgroup, which as such is contained in all the Sylow p-subgroups of G, and so in O_p. But $\mathbf{F}(G)$ is the product of its Sylow p-subgroups, and the inclusion follows. ◇

[5] See *ex.* 31 of Chapter 3.

Theorem 5.14. *Let G be a finite group, and let $\mathbf{F} = \mathbf{F}(G), \mathbf{\Phi} = \mathbf{\Phi}(G), \mathbf{Z} = \mathbf{Z}(G)$. Then:*

i) $\mathbf{\Phi} \subseteq \mathbf{F}$, $\mathbf{Z} \subseteq \mathbf{F}$;

ii) $\mathbf{F}/\mathbf{\Phi} = \mathbf{F}(G/\mathbf{\Phi})$;

iii) $\mathbf{F}/\mathbf{Z} = \mathbf{F}(G/\mathbf{Z})$.

Proof. *i)* is obvious. As for *ii)* we have, for one thing, $\mathbf{F}/\mathbf{\Phi} \subseteq \mathbf{F}(G/\mathbf{\Phi})$ because $\mathbf{F}/\mathbf{\Phi}$ is the image of the nilpotent normal subgroup $\mathbf{F}$ under the canonical homomorphism. As to the other inclusion, observe that if $H/\mathbf{\Phi}$ is the Fitting subgroup of $G/\mathbf{\Phi}$, and P is a Sylow p-subgroup of H, then $P\mathbf{\Phi}/\mathbf{\Phi}$ is Sylow in $H/\mathbf{\Phi}$, and therefore characteristic, by the nilpotence of $H/\mathbf{\Phi}$; hence it is normal in $G/\mathbf{\Phi}$. It follows $P\mathbf{\Phi} \trianglelefteq G$. Now apply the Frattini argument as in Theorem 5.8. *iii)* is proved similarly. ♢

Theorem 5.15. *The Fitting subgroup of a finite group G centralizes all the minimal normal subgroups of G.*

Proof. Let N be minimal normal in G, and let H be normal and nilpotent. If $N \cap H = \{1\}$, then N and H commute elementwise, so H centralizes N. If $N \cap H \neq \{1\}$, then $N \subseteq H$ by the minimality of N. H being nilpotent, $N \cap \mathbf{Z}(H) \neq \{1\}$. But $\mathbf{Z}(H)$ characteristic in H and $H \trianglelefteq G$ imply $\mathbf{Z}(H) \trianglelefteq G$, and always because of the minimality of N, $N \subseteq \mathbf{Z}(H)$; hence H centralizes N in this case too. ♢

Corollary 5.4. *Let H/K be a chief factor of a group G. Then $\mathbf{F}(G/K) \subseteq \mathbf{C}_{G/K}(H/K)$.*

An element gK *centralizes the quotient* H/K if $[h, g] \in K$ for all $h \in H$.

Definition 5.5. If $K \subseteq H$ are two normal subgroups of a group G, the *centralizer in G of the factor H/K* is the subgroup

$$\mathbf{C}_G(H/K) = \{g \in G \mid [h, g] \in K,\ \forall h \in H\}.$$

Hence, by definition, $\mathbf{C}_{G/K}(H/K) = \mathbf{C}_G(H/K)/K$; the quotient H/K being normal in G/K its centralizer $\mathbf{C}_{G/K}(H/K)$ is also normal. The inclusion of Corollary 5.4 may therefore be written as $\mathbf{F}(G/K) \subseteq \mathbf{C}_G(H/K)$. $\mathbf{F}K/K$ is normal in G/K and nilpotent, so is contained in $\mathbf{F}(G/K)$, and therefore $\mathbf{F}(G) \subseteq \mathbf{C}_G(H/K)$. Summing up:

Corollary 5.5. *The Fitting subgroup of a finite group G centralizes all minimal normal subgroups of a quotient of G. In particular, it centralizes every chief factor of G.*

Chief series and nilpotent normal subgroups are related by the following theorem.

Theorem 5.16. *Let $\{G_i\}$ be a chief series of a finite group G, $i = 1, 2, \ldots, n$. Then:*

$$\mathbf{F}(G) = \bigcap_{i=0}^{n-1} \mathbf{C}_G(G_i/G_{i+1}).$$

Proof. Set $L = \bigcap_{i=0}^{n-1} \mathbf{C}_G(G_i/G_{i+1})$. By Corollary 5.5 we have the inclusion $\mathbf{F}(G) \subseteq L$. As to the other inclusion, let us prove that L is normal and nilpotent. Obviously, it is normal. Let us consider the series $\{L_i\}, i = 0, 1, \ldots, n-1$, where $L_i = L \cap G_i$, and let us prove that it is a central series of L. By definition of $\mathbf{C}_G(G_i/G_{i+1})$, we have $[G_i, \mathbf{C}_G(G_i/G_{i+1})] \subseteq G_{i+1}$, and since $L_i \subseteq G_i$, it follows:

$$[L_i, L] \subseteq [G_i, \mathbf{C}_G(G_i/G_{i+1})] \subseteq G_{i+1},$$

which, together with $[L_i, L] \subseteq L$, yields the inclusion $[L_i, L] \subseteq L_{i+1}$, showing that the series of the L_i is central. Hence, L is a nilpotent normal subgroup of G, so $L \subseteq \mathbf{F}(G)$. ◇

We now consider a few properties of nilpotent groups in the general case.

Theorem 5.17. *A f.g. periodic nilpotent group G is finite.*

Proof. By induction on the nilpotence class c of the group. If $c = 1$, then G is abelian, f.g. and periodic, and therefore finite. Let $c>1$. G/Z_{c-1} is f.g., periodic and abelian, and therefore finite; Z_{c-1} being of finite index is f.g., and being periodic and of class $c - 1$ by induction is finite, and so is G. ◇

Theorem 5.18. *A f.g. nilpotent group with finite center is finite.*

Proof. Fix $g \in G$ and consider the mapping from Z_2 to the center of G given by $x \to [x, g]$. This mapping is a homomorphism:

$$[xy, g] = [x, g]^y[y, g] = [x, g][y, g],$$

because $[x, y] \in Z$ and therefore $[x, g]^y = [x, g]$. If $G = \langle g_1, g_2, \ldots, g_n \rangle$, there are n homomorphisms $Z_2 \to Z$ given by $x \to [x, g_i]$, with kernels $Z_2 \cap \mathbf{C}_G(g_i)$, each of which has finite index (the images are contained in the center, which is finite). Their intersection is $Z_2 \bigcap(\bigcap_{i=1}^n \mathbf{C}_G(g_i)) = Z_2 \cap Z = Z$, so Z has finite index in Z_2, and therefore Z_2 is finite. Now $\mathbf{Z}(G/Z) = Z_2/Z$, which is finite, and G/Z has class $c - 1$ and is f.g. By induction on c, G/Z is finite and so also is G (if $c = 1$, then $G = Z$, and there is nothing to prove). ◇

Let now G be a group generated by a set X. The group $G/G' = \Gamma_1/\Gamma_2$ is generated by the images $x\Gamma_2$, $x \in X$, and since Γ_2 is generated by the commutators $[x, y]$, $x, y \in G$, the quotient group Γ_2/Γ_3 is generated by the images of these commutators. Let us show that the commutators between elements of X suffice, i.e. that

$$\Gamma_2 = \langle [x_i, x_j], \Gamma_3;\ x_i, x_j \in X \rangle. \tag{5.10}$$

Indeed,

$$[x,y]=[x_ig,x_jh],\ x_i,x_j\in X,\ g,h\in \Gamma_2,$$

which, by the first equality of Lemma 5.2, *iii*), and setting $x=x_i, y=g, z=x_jh$, equals

$$[x_i,x_jh][[x_i,x_jh],g][g,x_jh],$$

where the last two terms belong to Γ_3. As to the first term we have, by the second equality of Lemma 5.2, *iii*),

$$[x_i,x_jh]=[x,h][x_i,x_j][[x_i,x_j],h],$$

where the first and third term belong to Γ_3. It follows $[x,y]\in\langle[x_i,x_j],\Gamma_3\rangle$ and (5.10). In general, let us prove by induction that

$$\Gamma_i=\langle[x_1,x_2,\dots,x_i],\ x_k\in X,\ \Gamma_{i+1}\rangle, \tag{5.11}$$

(recall that $[x_1,x_2,\dots,x_i]$ means $[[\dots[x_1,x_2],x_3],\dots,x_i])$, which we call a *commutator of weight i*. Assume (5.11) holds. Now

$$\Gamma_{i+1}=\langle[x,y],\ x\in\Gamma_i, y\in G\rangle,$$

and by (5.11) $x=t_1t_2\cdots t_ks$, where the t_k are commutators of weight i in the generators in X and $s\in\Gamma_{i+1}$. It follows

$$[x,y]=[t_1t_2\cdots t_ks,y]=[t_1t_2\cdots t_k,y][t_1t_2\cdots t_k,y,s][s,y],$$

with the last two terms in Γ_{i+2}. If we now write y in terms of the generators in X we have:

$$[t_1t_2\cdots t_k,x_1x_2\cdots x_m]=$$

$$[t_1t_2\cdots t_k,x_m][t_1t_2\cdots t_k,x_1x_2\cdots x_{m-1}][t_1t_2\cdots t_k,x_1x_2\cdots x_{m-1},x_m],$$

where the last term of the right hand side is in Γ_{i+2}. As to the first term observe that

$$\begin{aligned}[t_1t_2\cdots t_k,x_m]&=[t_1,x_m][t_1,x_m,t_2\cdots t_k][t_2\cdots t_k,x_m],\\ &=[t_1,x_m][t_2\cdots t_k,x_m]\ \mathrm{mod}\ \Gamma_{i+2},\end{aligned}$$

and similarly for the second. Proceeding in this way we obtain:

$$\Gamma_{i+1}=\langle[x_1,x_2,\dots,x_i,x_{i+1}],\ x_k\in X,\ \Gamma_{i+2}\rangle,$$

and the following result:

Theorem 5.19. *If a group is generated by a set X, then the i-th term of the lower central series Γ_i is generated by the commutators of weight i in the elements of X and by the elements of Γ_{i+1}.*

Corollary 5.6. *If a group is f.g., the quotients of the lower central series are also f.g.*

Corollary 5.7. *A f.g. nilpotent group admits a central series with cyclic quotients.*

Proof. The quotients of the lower central series are f.g. and abelian, and as such are direct product of cyclic group. For $i = 1, 2, \ldots, c$, let:

$$\Gamma_i/\Gamma_{i+1} = A_1/\Gamma_{i+1} \times A_2/\Gamma_{i+1} \times \cdots \times A_n/\Gamma_{i+1}$$

with the A_k/Γ_{i+1} cyclic. Then, setting $G_k/\Gamma_{i+1} = A_k/\Gamma_{i+1} \times \cdots \times A_n/\Gamma_{i+1}$, we have (Theorem 2.30, *ii*)):

$$\begin{aligned} G_k/G_{k+1} &= (A_k/\Gamma_{i+1} \times \cdots \times A_n/\Gamma_{i+1})/(A_{k+1}/\Gamma_{i+1} \times \cdots \times A_n/\Gamma_{i+1}) \\ &\simeq A_k/\Gamma_{i+1}, \end{aligned}$$

a cyclic group. Hence the quotients of the series $\Gamma_k = G_1 \supset G_2 \supset \ldots \supset G_n = A_n \supset \Gamma_{i+1}$ are cyclic. ◊

By virtue of the last corollary, we have that *a f.g. nilpotent group is built up by finitely many successive extensions of cyclic groups*; we call *cyclic* an extensions H of a group K with H/K cyclic. We have the following theorem:

Theorem 5.20. *A cyclic extension of a f.g. and residually finite group G is residually finite.*

Proof. Let $G = \langle a, H \rangle$, with $H \trianglelefteq G$ residually finite, and let $1 \neq g \in G$. If $g \notin H$, G/H being cyclic, and therefore residually finite, there exists $L/H \lhd G/H$ such that $(G/H)/(L/H)$ is finite and $gH \notin L/H$; then $g \notin L$, with G/L finite. Therefore we may assume $g \in H$. Let $g \notin K \lhd H$ and H/K finite. H being f.g., there exists a characteristic subgroup of H, still denoted K, of finite index in H (Theorem 3.5, *ii*)) not containing g (there are only a finite number of subgroups of a given finite index (Theorem 3.5, *i*))). Moreover, K characteristic in H and $H \lhd G$ imply $K \lhd G$. If G/H is finite, G/K is also finite, and $g \notin K$. Thus we assume $G/H = \langle aH \rangle$ infinite, so that $\langle aH \rangle = \langle a \rangle H$ with $\langle a \rangle \cap H = \{1\}$. By the N/C theorem, since H/K is normal in G/K and finite, the centralizer $\mathbf{C}_{G/K}(H/K)$ has finite index in G/K. It follows that a power of aK centralizes H/K, $(aK)^m \in \mathbf{C}_{G/K}(H/K)$, $m \neq 0$, and this $(aK)^m$ has infinite order because if $a^{mt} \in K$, then $a^{mt} \in H$, whereas $\langle aH \rangle$ is infinite cyclic. It follows $gK \notin \langle a^m K \rangle$ (gK belongs to the finite group H/K). Now $a^m K$ commutes with all the elements hK, and obviously with aK, and therefore with all the elements of G/K; in particular $\langle a^m K \rangle \lhd G/K$. The quotient $(G/K)/(\langle a^m K \rangle/K)$ is finite, and the image of g in this group is not the identity. ◊

Definition 5.6. A group is *polycyclic* if it admits a normal series with cyclic quotients.

Theorem 5.21. *A polycyclic group is f.g., and so are its subgroup.*

Proof. Let $G = G_0 \supset G_1 \supset \ldots \supset G_i \supset \ldots \supset G_n = \{1\}$ be a normal series of G with cyclic quotients. We have $G/G_1 = \langle xG_1 \rangle$ for a certain $x \in G$, and therefore $G = \langle x, G_1 \rangle$. By induction on n, G_1 is f.g., and therefore G is. Similarly, all the other G_i of the series are f.g. If $H \leq G$, then $HG_1/G_1 \simeq H/(H \cap G_1)$, and therefore $H/(H \cap G_1)$ is cyclic because is isomorphic to a subgroup of the cyclic group G/G_1. Then $H = \langle x, H \cap G_1 \rangle$, for some x. But by induction $H \cap G_1$ is f.g. because is a subgroup of G_1 which is f.g. and polycyclic, with a normal series with cyclic quotients of length $n-1$. It follows that H is f.g. ♢

As in the abelian case we have (cf. Theorem 4.2):

Corollary 5.8. *In a f.g. nilpotent group every subgroup is f.g.*

Corollary 5.9. *A f.g. nilpotent group is residually finite, and therefore hopfian.*

Proof. By Theorem 5.20 a polycyclic group is residually finite, and since by Corollary 5.7 a f.g. nilpotent group is polycyclic, by Theorem 4.39 we have the result. ♢

Corollary 5.10. *For a f.g. nilpotent group, and more generally for polycyclic groups, the word problem is solvable.*

Proof. Recall that if $N \trianglelefteq G$ and G/N are finitely presented, then so also is G (Chapter 4, *ex.* 46). Since a polycyclic group is obtained by a sequence of extensions of finitely presented groups by means of cyclic groups, a polycyclic group is finitely presented; it being residually finite by Corollary 5.9, the result follows from Theorem 4.38. ♢

Lemma 5.4. *If G is nilpotent of class $c \geq 2$ and $x \in G$, then $H = \langle x, G' \rangle$ is of class at most $c-1$.*

Proof. Let us show that $H = Z_{c-1}(H)$. We have $G' \subseteq Z_{c-1}(G) \cap H \subseteq Z_{c-1}(H)$, so $H/Z_{c-1}(H) \simeq (H/G')/(Z_{c-1}(H)/G')$ is a cyclic group, being a quotient of the cyclic group H/G'. But

$$H/Z_{c-1}(H) \simeq (H/Z_{c-2}(H))/((Z_{c-1}(H)/(Z_{c-2}(H)),$$

and therefore $H/Z_{c-2}(H)$ is abelian (the quotient w.r.t. the center is cyclic). Hence, the quotient appearing on the right hand side of the previous expression is the identity, so that $H = Z_{c-1}(H)$.

(Note that this lemma holds not only for G' but also for every normal subgroup of G contained in Z_{c-1}.) ♢

Theorem 5.22. *In a torsion-free nilpotent group we have:*

i) *an element $x \neq 1$ cannot be conjugate to its inverse;*

ii) *if $x^n = y^n$ then $x = y$;*

iii) *if $x^h y^k = y^k x^h$ then $xy = yx$.*

Proof. *i*) Let us show that if $x \sim x^{-1}$, either $x = 1$, or else the order of x is power of 2. By induction on the nilpotence class c. If $c = 1$, then G is abelian, and $x \sim x^{-1}$ means $x = x^{-1}$, i.e. $x^2 = 1$. Let $c > 1$. If x belongs to the center Z we still have $x^2 = 1$; if $x \notin Z$, then $xZ \sim x^{-1}Z$ and by induction $(xZ)^{2^k} = Z$, i.e. $x^{2^k} \in Z$. But since $x^{2^k} \sim x^{-2^k}$, we have $x^{2^k} = x^{-2^k}$ and $x^{2^{k+1}} = 1$.

ii) By Lemma 5.4, the subgroup $H = \langle x, G' \rangle$ is nilpotent of class at most $c - 1$. Now, $x^{-1} \cdot y^{-1}xy \in G'$, and therefore $x \cdot x^{-1} \cdot y^{-1}xy = y^{-1}xy \in H$, so $(y^{-1}xy)^n = x^n$ since $(y^{-1}xy)^n = y^{-1}x^n y = y^{-1}y^n y = y^n = x^n$. By induction on c we have $y^{-1}xy = x$, i.e. x and y commute; from $x^n = y^n$ then follows $(xy^{-1})^n = 1$, and torsion-freeness implies $xy^{-1} = 1$ and $x = y$.

iii) We have $y^{-k}x^h y^k = x^h$ i.e. $(y^{-k}xy^k)^h = x^h$. By *ii*), it follows $y^{-k}xy^k = x$, from which $x^{-1}y^k x = y^k$, i.e. $(x^{-1}yx)^k = y^k$, and again by *ii*) $x^{-1}yx = y$, i.e. $xy = yx$. ◇

Remark 5.1. The properties of Theorem 5.22 are shared by free groups.

Theorem 5.23. *In a torsion-free nilpotent group G the quotients of the upper central series are torsion-free.*

Proof. Consider Z_2/Z_1, and let $(xZ_1)^h = Z_1$ for some $x \in Z_2$. Then $x^h \in Z_1$ and therefore $x^h y = yx^h$, all $y \in G$. By *iii*) of Theorem 5.22, $xy = yx$, all $y \in G$, i.e. $x \in Z_1$. Similarly, if $x \in Z_3$ and $x^h \in Z_2$ then, by definition, $[x, y] \in Z_2$ and $[x^h, y] \in Z_1$, all $y \in G$, i.e. $x^{-h}y^{-1}x^h y = x^{-h}(y^{-1}xy)^h \in Z_1$. But then x^{-h} and $(y^{-1}xy)^h$ commute, so x^{-1} and $y^{-1}xy$ also commute. Hence $x^{-h}(y^{-1}xy)^h = [x, y]^h = (x^{-1}y^{-1}xy)^h \in Z_1$, and being $[x, y] \in Z_2$, by the previous case we have $[x, y] \in Z$ and $x \in Z_2$. This holds in general: if Z_i/Z_{i-1} is torsion-free Z_{i+1}/Z_i is also torsion-free. We must show that if $x \in Z_{i+1}$, i.e. $[x, y] \in Z_i$, all $y \in G$, and $x^h \in Z_i$, i.e. $[x^h, y] \in Z_{i-1}$, then $x \in Z_i$, i.e. $[x, y] \in Z_{i-1}$. The quotient Z_i/Z_{i-1} being torsion-free, it suffices to show that $[x, y]^h \in Z_{i-1}$, i.e.:

$$x \in Z_{i+1},\ [x^h, y] \in Z_{i-1} \Rightarrow [x^h, y] \in Z_{i-1}.$$

Since $x^h \in Z_i$, we have $y^{-1}x^{-h}y = (y^{-1}x^{-1}y)^h \in Z_i$ and also $(y^{-1}x^{-1}y)^h x^h \in Z_i$, and since Z_i/Z_{i-1} is the center of G/Z_{i-1}, the two cosets $(y^{-1}x^{-1}y)^h Z_{i-1}$ and $x^h Z_{i-1}$ commute. But Z_i/Z_{i-1} is torsion-free, so $y^{-1}x^{-1}yZ_{i-1}$ and xZ_{i-1} also commute; it follows:

$$(y^{-1}x^{-1}y)^h x^h Z_{i-1} = (y^{-1}x^{-1}yx)^h Z_{i-1} = [y, x]^h Z_{i-1}.$$

But $y^{-1}x^{-h}yx^h = [y, x^h] = [x^h, y]^{-1} \in Z_{i-1}$, and so also $[y, x]^h \in Z_{i-1}$ and $[x, y]^h \in Z_{i-1}$. ◇

Theorem 5.23 is not true in general for the quotients of the lower central series Γ_i/Γ_{i+1}, as the following example shows.

Example 5.4. The Heisenberg group $\mathcal{H}$ (*Ex.* 4.3) is torsion free, and is nilpotent of class 2 because the factor group w.r.t. the center is abelian. Consider in $\mathcal{H}$ the subgroup of the triples having in the first position a multiple of a fixed integer n:

$$G = \Gamma_1 = \{(kn, b, c),\ k, b, c \in \mathbf{Z}\}.$$

For the commutator of two elements of G we have:

$$[(kn, b, c), (k'n, b', c')] = (0, 0, (kb' - k'b)n),$$

and on the other hand $(0, 0, n) = [(2n, 1, c), (n, 1, c')]$. It follows:

$$\Gamma_2 = G' = \{(0, 0, tn),\ t \in \mathbf{Z}\}.$$

The element $(0, 0, 1)$ of G is such that $(0, 0, 1)^n = (0, 0, n) \in \Gamma_2$, and therefore in Γ_1/Γ_2 the coset to which $(0, 0, 1)$ belongs has finite period.

We close this section with one more application of Lemma 5.4, which shows a property of nilpotent groups shared by abelian groups.

Theorem 5.24. *The elements of finite period of a nilpotent group G form a subgroup.*

Proof. By induction on the nilpotence class c of the group G. If c=1, the group is abelian. Let $c>1$, and let x and y be two elements of finite period. We must show that xy also has finite period. Let $H = \langle x, G'\rangle$; by Lemma 5.4, H is of class less than c, so by induction its elements of finite period form a subgroup $t(H)$. This is characteristic in H, which is normal in G (it contains G') and therefore is normal in G. Similarly, if $K = \langle y, G'\rangle$, then $t(K) \trianglelefteq G$. Let $m = o(x)$; then:

$$\begin{aligned}(xy)^m &= xyxy\cdots xy = x \cdot xy^x \cdots xy = \ldots = x^m y^{x^{m-1}} y^{x^{m-2}} \cdots y^x y \\ &= x^m y' = y',\end{aligned}$$

with $y' \in t(K)$. If $o(y') = n$ then $(xy)^{mn} = 1$, as required. ◇

Exercises

1. The subgroups Z_i and Γ_i are characteristic in any group.

2. A group of order p^n admits a central series of length $n+1$ in which all the quotients have order p. [*Hint*: with H central of order p, the group G/H admits by induction a central series of length n.]

3. The direct product of a finite number of nilpotent groups is nilpotent.

4. *i*) Prove by induction on m, that for p prime, $(1+p)^{p^{m-1}} \equiv 1 \bmod p^m$; [*Hint*: for $0 < i < p$, $\binom{p}{i}$ is divisible by p.]

ii) prove that in the cyclic group $\langle a\rangle$ of order p^m the mapping $b: a \to a^{1+p}$ is an automorphism of order p^{m-1};

iii) prove that the semidirect product of $\langle a\rangle$ by $\langle b\rangle$ of *ii*) is a nilpotent p-group of class n. [*Hint*: prove that $G \supset \langle a^p\rangle \supset \langle a^{p^2}\rangle \supset \ldots \supset \langle a^{p^{m-1}}\rangle \supset \{1\}$ is a central series of G.] Hence, for each n, we have a nilpotent group of class n, and in fact a p-group.

5. Let p be a prime, and let $G = P_1 \times P_2 \times \cdots \times P_n \times \cdots$ be an infinite direct product of p-groups P_n of class n, $n = 1, 2, \ldots$. Prove that:

i) G is not nilpotent; (Hence an infinite p-group is not necessarily nilpotent; moreover, *ex.* 3 is not true for an infinite number of groups.)
ii) the union of a chain of normal nilpotent subgroups is normal but not necessarily nilpotent;
iii) G has no maximal normal nilpotent subgroups.

6. If G is nilpotent, and $H < G$, then $HG' < G$. [*Hint*: let $HZ_{i+1} = G$ and $HZ_i < G$; then $HZ_i \trianglelefteq HZ_{i+1}$ with abelian quotient, so that $G' \subseteq HZ_i < G$.]

7. If $x \in Z_2$, then x commutes with all its conjugates.

8. In a finite p-group of order p^n and class $n-1$ we have $Z_i \subseteq \Gamma_{n-i}$.

9. In a finite nilpotent group, a normal subgroup of order p^m, p a prime, is contained in the m-th term of the upper central series. [*Hint*: Theorem 5.5.]

10. A finite group is nilpotent if and only if any two elements of coprime order commute. [*Hint*: the centralizer of a p-Sylow contains all the q-Sylows, $q \neq p$.]

11. A dihedral group D_n is nilpotent if and only if n is a power of 2.

12. If G is a nonabelian nilpotent group, and A is a maximal abelian normal subgroup, then $\mathbf{C}_G(A) = A$.

13. $G/\mathbf{F}(G)$ nilpotent does not imply G nilpotent.

14. Let G be a finite nilpotent subgroup (not cyclic). Then:
i) there exist in G two subgroups of the same order that are not conjugate;
ii) G cannot be generated by conjugate elements.

15. Let M be a nilpotent maximal subgroup of a finite group G. Prove that either M is of order coprime to its index, or for some p a p-Sylow of M is normal in G.

16. Let H be a maximal nilpotent subgroup of a finite group G. Prove that $\mathbf{N}_G(\mathbf{N}_G(H)) = \mathbf{N}_G(H)$.

17. Let G be a finite group, $N \trianglelefteq G$, and H minimal among the subgroups whose product with N is G. Prove that:
i) $N \cap H \subseteq \mathbf{\Phi}(H)$;
ii) if G/N is nilpotent (cyclic), then H is nilpotent (cyclic).

18. Let $\mathcal{C}$ be a class of finite groups such that:
i) if $G/\mathbf{\Phi}(G) \in \mathcal{C}$ then $G \in \mathcal{C}$;
ii) if $G \in \mathcal{C}$ and $N \trianglelefteq G$ then $G/N \in \mathcal{C}$.

Prove that if $N \trianglelefteq G$ and $G/N \in \mathcal{C}$, then there exists $H \leq G$ with $H \in \mathcal{C}$ such that $G = NH$. [*Hint*: let H be minimal such that $G = NH$; prove that $N \cap H \subseteq \mathbf{\Phi}(H)$.] (Examples of classes with the given properties are the class of nilpotent groups and that of cyclic groups.)

19. Let A, B, C be three subgroups of a group G such that $A \subseteq C \subseteq AB$. Prove that $C = AB \cap C = A(B \cap C)$ (*Dedekind's identity*). [*Hint*: consider $c = ab$.]

20. *i*) Let $N \trianglelefteq G$. Prove that $\mathbf{\Phi}(N) \subseteq \mathbf{\Phi}(G)$. [*Hint*: use Dedekind's identity.]
ii) Give an example to show that that in *i*) the normality assumption is necessary. [*Hint*: let G be the semidirect product of $C_5 = \langle a \rangle$ by $C_4 = \langle b \rangle$, where $b^{-1}ab = a^2$.]
iii) If $N \trianglelefteq G$ and G/N' are nilpotent, then G is nilpotent. [*Hint*: use *i*).]

21. A finite group such that for each prime p there exists a composition series a term of which is a p-Sylow is nilpotent.

22. Let G be a finite nonabelian group with non trivial center and such that every proper quotient is abelian. Prove that G is a p-group.

23. Let $H \leq G$, and let A be the group of automorphisms of G such that $x^{-1}x^{\alpha} \in H$ for all $x \in G$ and $h^{\alpha} = h$ for all $h \in H$ (this is the case $n = 2$ of Theorem 5.10 with G_1 not necessarily normal). Prove that A is abelian.

24. The following are equivalent:
i) $[[a, b], b] = 1$, for all $a, b \in G$;
ii) two conjugate elements commute.

25. If H is a subnormal subgroup of a group G (Definition 2.16) and N a minimal normal subgroup of G, then N normalizes H. [*Hint*: let $H \triangleleft H_1 \triangleleft H_2 \triangleleft \cdots \triangleleft H_n \triangleleft G$; by induction on n. If $n = 1$, $H \trianglelefteq G$; if $n > 1$ and $N \cap H_n = \{1\}$, $N \subseteq \mathbf{C}_G(H_n) \subseteq \mathbf{C}_G(H) \subseteq \mathbf{N}_G(H)$. If $N \subseteq H_n$, take $N_1 \triangleleft N$ minimal normal in H_n and consider the conjugates N_1^g, $g \in G$.]

26. A finite group is nilpotent if and only if it contains a normal subgroup for each divisor of the order.

27. Prove that in a finite group G there always exists a nilpotent subgroup K such that $G = \langle K^g,\ g \in G \rangle$.

28. *i*) Prove that a group is residually nilpotent if and only if $\bigcap_{i \geq 0} \Gamma_i = \{1\}$;
ii) prove that the infinite dihedral group (*Ex.* 2.10, 3) is residually nilpotent. On the other hand, this group admits nonnilpotent images.

5.2 p-Nilpotent Groups

Theorem 5.25 (Burnside's basis theorem). *Let P be a finite p-group, and let $\mathbf{\Phi} = \mathbf{\Phi}(P)$. Then:*

i) *$P/\mathbf{\Phi}$ is elementary abelian, and therefore a vector space over $\mathbf{F}_p$;*

ii) *the minimal systems of generators of P all have the same cardinality d, where $p^d = |P/\mathbf{\Phi}|$;*

iii) *an element x of P, not belonging to $\mathbf{\Phi}$, is part of a minimal system of generators of P.*

Proof. *i*) The quotient $P/\mathbf{\Phi}$ is abelian because $P' \subseteq \mathbf{\Phi}$. Moreover, if M is maximal then P/M has order p and therefore $(Mx)^p = M$, i.e. $x^p \in M$, for all M. Hence $x^p \in \mathbf{\Phi}$, so all the elements of $P/\mathbf{\Phi}$ have order p. (This could also be seen by observing that $P/\mathbf{\Phi} = P/\cap M_i \leq P/M_1 \times P/M_2 \times \cdots \times P/M_k = Z_1 \times Z_2 \times \cdots \times Z_k$, where the M_i are all the maximal subgroups of P.)

ii) By *i*), $P/\mathbf{\Phi}$ is elementary abelian, so is generated by d elements, and not less. Hence P cannot be generated by less than d elements either, otherwise this would also be true for $P/\mathbf{\Phi}$.

iii) If $x \notin \mathbf{\Phi}$, then $\mathbf{\Phi}x$ is a nonzero vector of $P/\mathbf{\Phi}$, so is part of a basis of this vector space: $P/\mathbf{\Phi} = \langle \mathbf{\Phi}x_1, \mathbf{\Phi}x_2, \ldots, \mathbf{\Phi}x_d \rangle$, where $x_1 = x$. It follows $P = \langle \mathbf{\Phi}, x_1, x_2, \ldots, x_d \rangle = \langle x_1, x_2, \ldots, x_d \rangle$. ♢

In general, $P/\mathbf{\Phi}$ is distinct from the product of the P/M_i (think of the Klein group). Moreover, if the group is not a p-group the theorem is no longer true. The group $C_6 = \langle a \rangle$ admits, besides $\{a\}$, of cardinality 1, the minimal generating system $\{a^2, a^3\}$, of cardinality 2.

Lemma 5.5. *Let $x_1, x_2, \ldots, x_m$ be a generating system of a group G, and let $a_1, a_2, \ldots, a_m$ be elements of $\mathbf{\Phi} = \mathbf{\Phi}(G)$. Then $a_1x_1, a_2x_2, \ldots, a_mx_m$ is also a generating system of G.*

Proof. From $G/\mathbf{\Phi} = \langle \mathbf{\Phi}x_1, \mathbf{\Phi}x_2, \ldots, \mathbf{\Phi}x_m \rangle$, and $\mathbf{\Phi}x_i = \mathbf{\Phi}a_ix_i$ it follows

$$G/\mathbf{\Phi} = \langle \mathbf{\Phi}a_1x_1, \mathbf{\Phi}a_2x_2, \ldots, \mathbf{\Phi}a_mx_m \rangle,$$

so $G = \langle \mathbf{\Phi}, a_1x_1, a_2x_2, \ldots, a_mx_m \rangle = \langle a_1x_1, a_2x_2, \ldots, a_mx_m \rangle$. ♢

Theorem 5.26. *Let σ be an automorphism of a finite group G inducing the identity on $G/\mathbf{\Phi}$, where $\mathbf{\Phi} = \mathbf{\Phi}(G)$ (i.e. $x^{-1}x^\sigma \in \mathbf{\Phi}$ for all $x \in G$). Then the order of σ divides $|\mathbf{\Phi}|^d$, where d is the cardinality of a generating system of the group G.*

Proof. Let $G = \langle x_1, x_2, \ldots, x_d \rangle$. By the previous lemma, the components of the ordered d-tuples $(a_1x_1, a_2x_2, \ldots, a_dx_d)$, $a_i \in \mathbf{\Phi}$, also generate G. As the a_i vary in all possible ways in $\mathbf{\Phi}$ we have a set Ω of $|\mathbf{\Phi}|^d$ d-tuples. Now the group $\langle \sigma \rangle$ acts on this set according to $(a_1x_1, a_2x_2, \ldots, a_dx_d)^\sigma = ((a_1x_1)^\sigma, (a_2x_2)^\sigma, \ldots, (a_dx_d)^\sigma)$; this is actually an action on Ω because by assumption $x_i^\sigma x_i^{-1} \in \mathbf{\Phi}$, and therefore $x_i^\sigma = a_i'x_i$, with $a_i' \in \mathbf{\Phi}$, $(a_ix_i)^\sigma = a_i^\sigma x_i^\sigma = a_i^\sigma a_i' x_i = a_i'' x_i$, where $a_i'' \in \mathbf{\Phi}$. If an element $\langle \sigma \rangle$ fixes a d-tuple, then it fixes all its components, and so all the elements of a generating system, and therefore is the identity. Thus the stabilizer of an element of Ω is the identity, i.e. $\langle \sigma \rangle$ is semiregular on Ω, so its orbits all have the same cardinality $o(\sigma)$; hence $|\Omega|$ is a multiple of $o(\sigma)$. ♢

Corollary 5.11. *If an automorphism σ of a finite p-group P induces the identity on $P/\mathbf{\Phi}(P)$, then $o(\sigma)$ is a power of p.*

The latter corollary applies in particular when the p-group is a Sylow p-subgroup S of a finite group G. If $g \in \mathbf{N}_G(S)$, then g induces by conjugation an automorphism σ of S whose order divides $o(g)$. σ is also an automorphism of $S/\mathbf{\Phi}(S)$, and if it is trivial its order is a power of p. If g is a p'-element, then $p \nmid o(\sigma)$. The only possibility is $o(\sigma) = 1$, and since g acts by conjugation this means that g centralizes S. Let us see a few examples of this situation.

We recall that an element g of finite order of a group is a product $g = x_1 x_2 \cdots x_m$ where the x_i are commuting p_i-elements, $p_i \neq p_j$ if $i \neq j$,

Theorem 5.27. *Let S be a Sylow p-subgroup of a finite group G, and let $S \cap G' \subseteq \mathbf{\Phi}(S)$. Then $\mathbf{N}_G(S) = S \cdot \mathbf{C}_G(S)$.*

Proof. Let $g \in \mathbf{N}_G(S)$, and let $g = xy$ where x is a p-element and y a p'-element. Let σ be the automorphism of S induced by conjugation by y. For all $z \in S$ we have

$$z^{-1}z^{\sigma} = z^{-1}z^{y} = z^{-1}y^{-1}zy = [z, y] \in S \cap G'.$$

Then by assumption $z^{-1}z^{\sigma} \in \mathbf{\Phi}(S)$, and therefore the automorphism σ induced by y on S is the identity. In other words, y centralizes S and so $g = xy$ with $x \in S$ and $y \in \mathbf{C}_G(S)$. ◊

Definition 5.7. Let S be a Sylow p-subgroup of a group G. A subgroup K of G such that $S \cap K = \{1\}$ and $G = SK$ is called a *p-complement*. If G has a normal p-complement then G is *p-nilpotent*.

A normal p-complement is unique. Indeed, it contains all the Sylow q-subgroups with $q \neq p$, and therefore it consists of the elements of G of order coprime to p.

If a Sylow p-subgroup S is abelian, and $S \cap G' \subseteq \mathbf{\Phi}(S)$, the previous theorem implies $\mathbf{N}_G(S) = \mathbf{C}_G(S)$. As we shall see in Theorem 5.30, this condition implies the existence of a normal p-complement K. This subgroup K will be determined as the kernel of a homomorphism called the transfer, that we now define.

Let H be a subgroup of finite index n of a group G, and let $x_1, x_2, \ldots, x_n$ be representatives of the right cosets of H. If $g \in G$, then $x_i g \in Hx_j$ for some j, and the mapping $x_i \to x_j$ thus obtained is a permutation $\sigma = \sigma_g$ of the x_i:

$$x_i g = h_i x_{\sigma(i)}, \tag{5.12}$$

where $h_i \in H$. In this equality, not only the $x_{\sigma(i)}$ but also the h_i are uniquely determined by g and x_i.

Theorem 5.28. *Let H be abelian. The mapping $V : G \to H$ given by $g \to \prod_{i=1}^{n} h_i$, where the h_i are determined by g as in (5.12), is a homomorphism.*

Proof. Let $g, g' \in G$, and let x_i be as above and $x_i g' = h'_i x_{\tau(i)}$. If $x_i g = h_i x_{\sigma(i)}$, then

$$x_i g g' = (x_i g) g' = h_i x_{\sigma(i)} g' = h_i h'_{\sigma(i)} x_{\tau\sigma(i)}$$

so that

$$V(gg') = \prod_{i=1}^{n} h_i h'_{\sigma(i)} = \prod_{i=1}^{n} h_i \cdot \prod_{i=1}^{n} h'_i = V(g)V(g'),$$

where the second equality follows from the fact that σ being a permutation, the set of the $h'_{\sigma(i)}$ coincides, as i varies, with the set of the h'_i. ◇

Definition 5.8. The homomorphism $V : G \to H$ of Theorem 5.28 is called the *transfer*[6] of G in H.

Remark 5.2. The transfer does not depend on the choice of the representatives x_i of the right cosets of H in G (cf. *ex.* 30).

Now let $x_i g = h_i x_{\sigma(i)}$, and let $(i_1, i_2, \ldots, i_{r-1}, i_r)$ be a cycle of σ. Then $x_{i_k} g = h_{i_k} x_{i_{k+1}}$, $k = 1, 2, \ldots, r$ (indices mod r), and the contribution of these h_{i_k} to the product $\prod h_i$ is $h_{i_1} h_{i_2} \cdots h_{i_{r-1}} h_{i_r}$, that is,

$$x_{i_1} g x_{i_2}^{-1} \cdot x_{i_2} g x_{i_3}^{-1} \cdots x_{i_r} g x_{i_1}^{-1} = x_{i_1} g^r x_{i_1}^{-1}.$$

If σ has t cycles, each of length r_i, let us choose for each cycle an element x_i such that

$$V(g) = \prod_{i=1}^{t} x_i g^{r_i} x_i^{-1}, \tag{5.13}$$

where $\sum_{i=1}^{t} r_i = n = [G : H]$. Note that r_i is the least integer such that $x_i g^{r_i} x_i^{-1} \in H$.

Let us now see two applications of the transfer. The first one is Schur's theorem, that we have already proved in a different way (Theorem 2.38); the second is a theorem of Burnside (Theorem 5.30).

Theorem 5.29 (Schur). *If the center of a group G is of finite index, then the derived group G' is finite.*

Proof. G' is finitely generate (by a set of representatives of the cosets of the center). Now, setting $Z = \mathbf{Z}(G)$, we have $G'/(G' \cap Z) \simeq G'Z/Z \leq G/Z$ and therefore $G'/G' \cap Z$ has finite index in G', and as such is finitely generated. It will be enough to show that it is torsion because, being abelian, it will be finite. Consider the transfer of G in Z. If $g \in G' \cap Z$, then $x_i g^{r_i} x_i^{-1} = g^{r_i}$ and (5.13) becomes

$$V(g) = \prod_{i=1}^{t} g^{r_i} = g^{\sum r_i} = g^n,$$

[6] V is the initial of *Verlagerung*, the German term for transfer.

where $n = [G : Z]$. The image of V being abelian, we have $G' \subseteq ker(V)$, and $g^n = 1$ since $g \in G'$. ♢

Lemma 5.6 (Burnside). [7] *Two elements of the center of a Sylow p-subgroup S of a group G that are conjugate in G are conjugate in the normalizer of S.*

Proof. Let $x, y \in \mathbf{Z}(S)$, and let $y = x^g$. Then y belongs to the center of S and S^g, so these two subgroups are Sylow in the centralizer $\mathbf{C}_G(y)$ of y and therefore $S = (S^g)^h$, $h \in \mathbf{C}_G(y)$. Then $gh \in \mathbf{N}_G(S)$, and $x^{gh} = (x^g)^h = y^h = y$, as required. ♢

Theorem 5.30 (Burnside). *Let G be a finite group, S a Sylow p-subgroup of G, and let* $\mathbf{N}_G(S) = \mathbf{C}_G(S)$ *(i.e. S is contained in the center of its own normalizer). Then G has a normal p-complement.*

Proof. S is abelian, since $S \subseteq \mathbf{N}_G(S) = \mathbf{C}_G(S)$ and therefore is contained in its centralizer. The subgroup K we look for will be the kernel of the transfer V of G in S. Consider (5.13) with $1 \neq g \in S$. The elements g^{r_i} and $x_i g^{r_i} x_i^{-1}$ are elements of S that are conjugate in G, and since S coincides with its center, by Lemma 5.6 they are conjugate in the normalizer of S: $y g^{r_i} y^{-1} = x_i g^{r_i} x_i^{-1}$, with $y \in \mathbf{N}_G(S)$. But being $\mathbf{N}_G(S) = \mathbf{C}_G(S)$, y commutes with g^{r_i}, and therefore $g^{r_i} = x_i g^{r_i} x_i^{-1}$. Then (5.13) becomes:

$$V(g) = \prod_{i=1}^{t} x_i g^{r_i} x_i^{-1} = \prod_{i=1}^{t} g^{r_i} = g^{\sum_{i=1}^{t} r_i} = g^n.$$

However g is a p-element, and $n = [G : S]$ being coprime to p we have $V(g) \neq 1$; thus $ker(V) \cap S = \{1\}$. Let $K = ker(V)$; then $V(S) = SK/K \simeq S/S \cap K \simeq S$, and $V(S) = S$. *A fortiori*, $V(G) = S$ so $G/K \simeq S$. It follows $G = SK$ with $S \cap K = \{1\}$, as required. ♢

Let us now consider a few examples of application of Theorem 5.30.

Theorem 5.31. *i) Let G be a finite group, and let p be the smallest prime divisor of the order of G. Assume that a Sylow p-subgroup S is cyclic. Then G has a normal p-complement.*

ii) Let S be a cyclic Sylow p-subgroup of a group G. Then either $S \cap G' = S$ *or* $S \cap G' = \{1\}$.

Proof. i) Let us show that $N = \mathbf{N}_G(S)$ and $C = \mathbf{C}_G(S)$ coincide. We know that $|N/C| \leq |\mathbf{Aut}(S)|$. If $|S| = p^n$, $|\mathbf{Aut}(S)| = \varphi(p^n) = p^{n-1}(p-1)$. S being abelian, $S \subseteq C$, so p^n divides C. Hence $|N/C|$ is coprime to p, and therefore divides $p-1$. The prime p being the smallest divisor of $|G|$ it follows $|N/C| = 1$ and $N = C$.

[7] Cf. Lemma 3.8.

ii) If $S \not\subseteq G'$, $S \cap G'$ is a proper subgroup of S and therefore is contained in a maximal subgroup of S. Being a cyclic p-group S has a unique maximal subgroup, which therefore coincides with the Frattini subgroup. Hence $S \cap G' \subseteq \mathbf{\Phi}(S)$. By Theorem 5.27, $\mathbf{N}_G(S) = S \cdot \mathbf{C}_G(S)$, and being $S \subseteq \mathbf{C}_G(S)$ because S is abelian, $\mathbf{N}_G(S) = \mathbf{C}_G(S)$. By Burnside's theorem, G has a normal p-complement: $G = SK$, $K \trianglelefteq G$, $S \cap K = \{1\}$. Since $G/K \simeq S$ is cyclic, we have $G' \subseteq K$, and $S \cap G' = \{1\}$. $\diamondsuit$

$S \cap G'$ is the *focal* subgroup of S in G. The quotient $S/(S \cap G')$ is isomorphic to the largest abelian p-factor group of G (cf. *ex.* 40).

Theorem 5.32. *If G is a finite simple group, then either* 12 *divides* $|G|$, *or* p^3 *divides* $|G|$, *where p is the smallest prime divisor of* $|G|$.

Proof. Let p be the smallest prime divisor of $|G|$ and let $|S| = p$ or p^2. S is abelian. If it is cyclic, by the previous theorem G is not simple; if it is elementary, then $|N/C| \leq |\mathbf{Aut}(S)| = (p^2-1)(p^2-p) = p(p-1)^2(p+1)$. If $p > 2$, no prime greater than p divides the latter integer, so no prime divides $|N/C|$; it follows $|N/C| = 1$ and $N = C$. If $p = 2$, S is a Klein group, and $\mathbf{Aut}(S)$ is isomorphic to S^3. If $N/C \neq \{1\}$, since $V \subseteq C$ we have that 4 divides $|C|$ and therefore $|N/C| = 3$. Hence $|G|$ is divisible by 4 and by 3, and so by 12. $\diamondsuit$

Example 5.5. Let us prove that A^5 is the only simple group of order $p^2qr, p < q < r$. By the previous theorem, the order of the group must be divisible by 12, hence $p = 2$, $q = 3$ and the order is $12r$. We have already seen that the unique prime number r compatible with simplicity is 5 (Chapter 3, *ex.* 43). Hence the order of the group is 60, and $G \simeq A^5$ (*Ex.* 3.5, 3).

Exercises

29. *i*) If a group is p-nilpotent for all primes p, then it is nilpotent;
ii) if a group is p-nilpotent, then so also are subgroups and factor groups.

30. Prove that the transfer does not depend on the choice of a transversal.

31. Let G be p-nilpotent, and let N be a minimal normal subgroup of G with $p||N|$. Prove that N is a p-group contained in the center of G.

32. Let G be p-nilpotent, and let $G = SK$. Prove that $\mathbf{Z}(S)K \trianglelefteq G$.

33. If $S, S_1 \in Syl_p(G)$, then $\mathbf{Z}(S) \trianglelefteq S_1$ implies $\mathbf{Z}(S) = \mathbf{Z}(S_1)$.

34. G is said to be *p-normal* if given two Sylow p-subgroups S and S_1, then $\mathbf{Z}(S) \subseteq S_1$ implies $\mathbf{Z}(S) = \mathbf{Z}(S_1)$. Prove that:
i) if the Sylow p-subgroups are abelian, then G is p-normal;
ii) if any two Sylow p-subgroups have trivial intersection, then G is p-normal;
iii) if G is p-nilpotent, then G is p-normal;
iv) S^4 is not 2-normal.

35. Let G be a finite group such that the normalizer of every abelian subgroup coincides with its centralizer. Prove that G is abelian. [*Hint*: prove that the Sylows are abelian by considering an abelian maximal subgroup of a Sylow p-subgroup and then prove that G is p-nilpotent by applying Burnside's theorem.]

36. Let S be a Sylow p-subgroup of order p of a group G, and let $x \neq 1$ be an element of S. Prove that:

i) under the action of G on its Sylow p-subgroups by conjugation, the permutation induced by x consists of a fixed point and of cycles of length p;

ii) an element of G that normalizes but does not centralize S cannot fix two elements belonging to the same orbit of S, and therefore the number of Sylow p-subgroups it fixes is at most equal to the number of orbits of S.

37. Use *i*) of the previous exercise to prove that groups of order 264, 420 or 760 cannot be simple.

38. Using *ex.* 36 and (3.9) prove that a group of order 1008 or 2016 cannot be simple.

39. If p^2, p a prime, divides $|G|$, then G has an automorphism of order p.

40. Let $S \in Syl_p(G)$. Prove that:

i) if $H \trianglelefteq G$ is such that G/H is an abelian p-group, then $S \cap G' \subseteq H$ and G/H is isomorphic to a homomorphic image of $S/(S \cap H)$;

ii) there exists a subgroup $H \trianglelefteq G$ such that G/H is isomorphic to $S/(S \cap H)$ [*Hint*: consider the inverse image H of $O^p(G/G')$ (cf. Chapter 3, *ex.* 32); also $S \cap G' = S \cap H$.]

5.3 Fusion

If two elements of a finite group are conjugate, then two suitable powers of them are p-elements, for some p, and are also conjugate. Since any two Sylow p-subgroups are conjugate, a conjugation of these two powers reduces to that of two elements belonging to the same Sylow p-subgroup. In this section, we shall be considering the converse problem: when are two elements of a Sylow p-subgroup S that are not conjugate in S conjugate in G?

Definition 5.9. Two elements or subgroups of a Sylow p-subgroup S of a group G are said to be *fused* in G if they are conjugate in G but not in S. (The two conjugacy classes of S to which the two elements or subgroups belong fuse into one conjugacy class of G.)

We have already seen two results about fusion: one is Lemma 5.6 of Burnside, concerning the conjugation of two elements of the center of a Sylow p-subgroup. The other one, again by Burnside, is Lemma 3.8, concerning conjugation in the group of two normal subgroups of a Sylow p-subgroups. (This result extends immediately to the case of two normal sets, that is, sets that coincide with the set of conjugates of their elements.) If the normality assumption is dropped, two subsets of S, although not necessarily conjugate in the normalizer of S any more, are conjugate by an element that is a product of elements belonging to normalizers of subgroups of S. This is the content of a theorem of Alperin that we shall prove in a moment.

Lemma 5.7. *Let S be a Sylow p-subgroup of a group G, and let T be a subgroup of S. Then there exists a subgroup U of S which is conjugate to T in $G$and such that $\mathbf{N}_S(U)$ is Sylow in $\mathbf{N}_G(U)$.*

Proof. T is contained in a Sylow p-subgroup P of its normalizer $\mathbf{N}_G(T)$, and there exists $g \in G$ such that $P^g \subseteq S$. The subgroup we seek is $U = T^g$. Indeed, P^g is Sylow in $\mathbf{N}_G(T)^g = \mathbf{N}_G(T^g) = \mathbf{N}_G(U)$, is contained in S, and therefore is contained in $\mathbf{N}_S(U)$, but since it contains a Sylow p-subgroup of $\mathbf{N}_G(U)$ it coincides with it. ◇

Definition 5.10. Let S be a Sylow p-subgroup of a group G, $\mathcal{F}$ a family of subgroups of S. Let A and B be two nonempty subsets of S and let g be an element of S. Then A is said to be *$\mathcal{F}$-conjugate to B via g* if there exist subgroups $T_1, T_2, \ldots, T_n$ of the family $\mathcal{F}$ and elements $g_1, g_2, \ldots, g_n$ of G such that:

i) $g_i \in \mathbf{N}_G(T_i)$, $i = 1, 2, \ldots, n$;
ii) $\langle A \rangle \subseteq T_1$, $\langle A \rangle^{g_1 g_2 \cdots g_i} \subseteq T_{i+1}$, $i = 1, 2, \ldots, n-1$;
iii) $A^g = B$, where $g = g_1 g_2 \cdots g_n$.

The family $\mathcal{F}$ is called a *conjugation family* (for S in G) if, given any two subsets A and B of S that are conjugate in G by an element g, then A is $\mathcal{F}$-conjugate to B by g.

Lemma 5.7 suggests how to determine a conjugation family.

Theorem 5.33 (Alperin[8]). *Let S be a Sylow p-subgroup of a group G, and let $\mathcal{F}$ the family of subgroups T of S such that $\mathbf{N}_S(T)$ is Sylow in $\mathbf{N}_G(T)$. Then $\mathcal{F}$ is a conjugation family for S in G.*

Proof. Let $A, B \subseteq S$, $A^g = B$; we prove that A is $\mathcal{F}$-conjugate to B via g. By induction on the index of $\langle A \rangle$ in S. Setting $T = \langle A \rangle$ and $V = \langle B \rangle$, we have $T^g = V$. Let $[S : T] = 1$; then $S = T$, $S^g = T^g = V \subseteq S$, and therefore $S^g = S$, i.e. $g \in \mathbf{N}_G(S)$. We obtain $i), ii)$ and $iii)$ of Definition 5.10 by putting $n = 1$, $T_1 = S$ and $g_1 = g$, and by observing that S certainly belongs to $\mathcal{F}$. Now let $[S : T] > 1$, i.e. $T < S$. Then $T < \mathbf{N}_S(T)$ and $V < \mathbf{N}_S(V)$. Let $U \in \mathcal{F}$ be conjugate to T^g (Lemma 5.7); then there exists $h_1 \in G$ such that $T^{gh_1} = V^{h_1} = U$, from which $\mathbf{N}_S(T)^{gh_1} \subseteq \mathbf{N}_G(T)^{gh_1} = \mathbf{N}_G(T^{gh_1}) = \mathbf{N}_G(U)$. Since $\mathbf{N}_S(U)$ is Sylow in $\mathbf{N}_G(U)$, there exists $h_2 \in \mathbf{N}_G(U)$ such that $\mathbf{N}_S(T)^{gh_1h_2} \subseteq \mathbf{N}_S(U)$. Setting $h = h_1h_2$, we have $T^{gh} = U$, $\mathbf{N}_S(T)^{gh} \subseteq \mathbf{N}_S(U)$. Similarly, there exists $k \in G$ such that $V^k = U$, $\mathbf{N}_S(T)^k \subseteq \mathbf{N}_S(U)$. The index of $\mathbf{N}_S(T)$ in S being less than that of T in S, by induction $\mathbf{N}_S(T)$

[8] Alperin J.L.: Sylow Intersections and Fusion. J. Algebra **6**, 222–241 (1967), and Glauberman G.: Global and Local Properties of Finite Groups. In: Powell M.B., Higman G. (eds.), Finite Simple groups. Academic Press, London and New York (1971).

is $\mathcal{F}$-conjugate to $\mathbf{N}_S(T)^{gh}$ via gh, and $\mathbf{N}_S(V)^k$ is $\mathcal{F}$-conjugate to $\mathbf{N}_S(V)$ via k^{-1}. Then (*ex.* 46) A is $\mathcal{F}$-conjugate to B^h via gh and B^k is $\mathcal{F}$-conjugate to B via k^{-1}. Moreover, since $h^{-1}k \in \mathbf{N}_G(U)$ and $U \in \mathcal{F}$, B^h is $\mathcal{F}$-conjugate to B^k via $h^{-1}k$. From the sequence of $\mathcal{F}$-conjugations:

$$A \xrightarrow{gh} B^h \xrightarrow{h^{-1}k} B^k \xrightarrow{k^{-1}} B$$

and by *ex.* 47, A is $\mathcal{F}$-conjugate to B via $(gh)(h^{-1}k)k^{-1} = g$. ◇

This theorem may be expressed by saying that *conjugation has a local character.* Normalizers of nonidentity p-subgroups are called p-*local* subgroups. Thus, Alperin's theorem may be expressed by saying that two elements or subsets of a Sylow p-subgroup S of a group G are conjugate in G if and only if they are locally conjugate in G.

Exercises

41. Two element of the centralizer of a p-Sylow S that are conjugate in G are conjugate in the normalizer of S.

42. Let S be Sylow in G, and let $H \leq G$ be such that $H^g \subseteq S$ implies $H^g = H$ (H is said to be *weakly closed* in S w.r.t. G). Prove that two elements of the centralizer of H that are conjugate in G are conjugate in the normalizer of H. (Note that since $H = S$ is certainly weakly closed in S, this exercise generalizes the case of two elements of the center of a Sylow (Lemma 5.6) already generalized in the previous exercise.

43. Let the p-Sylows of a group G have pairwise trivial intersection. Prove that two elements x and y of a p-Sylow S that are conjugate in G are conjugate in the normalizer of S. In fact, any element that conjugates them belongs to $\mathbf{N}_G(S)$.

44. Let G be p-nilpotent. Then:

i) two p-Sylows are conjugate via an element of the centralizer of their intersection;

ii) if S is a p-Sylow, two elements of S that are conjugate in G are conjugate in S[9];

iii) if P is a p-subgroup of G, then $\mathbf{N}_G(P)/\mathbf{C}_G(P)$ is a p-group.

45. If $(n, \varphi(n)) = 1$, then all group of order n are cyclic. Conversely, if all groups of order n are cyclic, then $(n, \varphi(n)) = 1$.

46. Let $\mathcal{F}$ be a family of subgroups of a p-Sylow S of a group G, and let A and B be two nonempty subsets of S with A $\mathcal{F}$-conjugate to B via g. If C is a nonempty subset of A, prove that C is $\mathcal{F}$-conjugate to C^g via g.

47. With S and $\mathcal{F}$ as in the previous exercise, let $A_1, A_2, \ldots, A_{m+1}$ be subsets of S, and let $h_1, h_2, \ldots, h_m$ be elements of G such that A_i is $\mathcal{F}$-conjugate to A_{i+1} via h_i, $i = 1, 2, \ldots, m$. Prove that A_i is $\mathcal{F}$-conjugate to A_{m+1} via the product $h_1h_2 \cdots h_m$.

---

[9] This fact characterizes p-nilpotent groups (Huppert, p. 432).

48. Consider the simple group G of order 168. In the dihedral Sylow 2-subgroup S let x and y be two involutions that generate S. These two involutions are not conjugate in S. They are conjugate in G (all involutions are conjugate in G), but not in $N_G(S)$ (because the latter equals S). Hence there is no subgroup H of S containing x and y such that the two elements are conjugate in the normalizer of H. Prove that if z be the central involution of S, then x and z are conjugate in the normalizer of the Klein group they generate, and the same happens for z and y. The product of the two conjugating elements conjugates x and y.

5.4 Fixed-Point-Free Automorphisms and Frobenius Groups

Definition 5.11. An automorphism σ of a group is *fixed-point-free* (f.p.f.) if it fixes only the identity of the group.

Theorem 5.34. *Let σ be a f.p.f. automorphism of a finite group G. Then:*

i) *the mapping $G \to G$ given by $x \to x^{-1}x^\sigma$ (or $x \to x^\sigma x^{-1}$) is one-to-one (but not an automorphism, in general);*

ii) *if N is a normal σ-invariant subgroup G, σ acts f.p.f. on G/N;*

iii) *for each prime p dividing $|G|$ there exists one and only one Sylow p-subgroup of G fixed by σ.*

Proof. *i)* If $x^{-1}x^\sigma = y^{-1}y^\sigma$, then $(yx^{-1})^\sigma = yx^{-1}$ and therefore $yx^{-1} = 1$ and $y = x$. The mapping is injective, and since the group is finite also surjective.

ii) If $(Nx)^\sigma = Nx\ x^{-1}x^\sigma \in N$. As to *i)*, there exists $h \in N$ such that $x^{-1}x^\sigma = h^{-1}h^\sigma$; as above $x = h$, so that $Nx = N$.

iii) If S is a p-Sylow, $S^\sigma = S^x$ for some $x \in G$, and by *i)* we have $x = y^{-1}y^\sigma$. Hence $S^\sigma = S^x = (y^{-1})^\sigma ySy^{-1}y^\sigma$, from which $y^\sigma S^\sigma (y^{-1})^\sigma = ySy^{-1}$, or $(ySy^{-1})^\sigma = ySy^{-1}$, and the latter is a p-Sylow fixed by σ. If σ fixes S and S^x, then $S^x = (S^x)^\sigma = (S^\sigma)^{x^\sigma} = S^{x^\sigma}$, from which $x^\sigma x^{-1} \in \mathbf{N}_G(S)$. Since $\mathbf{N}_G(S)$ is σ-invariant if S is, there exists $y \in \mathbf{N}_G(S)$ such that $x^\sigma x^{-1} = y^\sigma y^{-1}$. But then $x = y$ and $x \in \mathbf{N}_G(S)$ so that $S^x = S$. ◇

In Theorem 5.31, *i)*, we have seen that if a Sylow p-subgroup of a group G is cyclic, where p is the smallest prime dividing the order of G, then G has a normal p-complement. The next theorem shows that the minimality assumption on the prime p may be replaced with G admits a f.p.f. automorphism. First a lemma.

Lemma 5.8. *Let σ be a f.p.f. automorphism of a group G, and let H be a normal cyclic and σ-invariant subgroup. Then $H \subseteq \mathbf{Z}(G)$.*

Proof. H being cyclic, $\mathbf{Aut}(H)$ is abelian, hence the restriction of σ to H (still denoted by σ) commutes with all the elements of $\mathbf{Aut}(H)$, and in particular with those induced by conjugation by the elements of G: $h^{\sigma\gamma_g} = h^{\gamma_g\sigma}$, for all $h \in H$. It follows $g^{-1}h^\sigma g = (g^\sigma)^{-1}h^\sigma g^\sigma$, that is $g^\sigma g^{-1}$ commutes with all the h^σ, and so with all $h \in H$: $g^\sigma g^{-1} \in \mathbf{C}_G(H)$. $\mathbf{C}_G(H)$ being σ-invariant, there exists $x \in \mathbf{C}_G(H)$ such that $g^\sigma g^{-1} = x^\sigma x^{-1}$, so $g = x$. Then every $g \in G$ centralizes H, and the result follows. $\diamondsuit$

Theorem 5.35. *Let G be a group admitting a f.p.f. automorphism σ and let a Sylow p-subgroup be G cyclic. Then G ha a normal p-complement.*

Proof. By Theorem 5.34, *iii*), there exists a σ-invariant Sylow p-subgroup whose normalizer N is also σ-invariant. By the lemma above $S \subseteq \mathbf{Z}(N)$, and by Burnside's theorem (Theorem 5.30), G has a normal p-complement. $\diamondsuit$

The notion of a f.p.f. automorphism presents itself naturally when dealing with a special class of groups, the Frobenius groups.

Definition 5.12. A finite transitive permutation group on a set Ω in which only the identity fixes more than one point, but the subgroup fixing a point is non trivial, is called a *Frobenius group*.

Example 5.6. The dihedral group D_n, n odd, viewed as a permutation group of the vertices of a regular n-gon, is a Frobenius group. It is transitive, and a vertex is fixed only by the symmetry w.r.t. the axis passing through that vertex. The remaining symmetries are rotations, so fix no vertex. If n is even, the symmetry w.r.t. the axis passing through two opposite vertices fixes these two vertices, so this group is not a Frobenius group. See also *Ex.* 5.7.

Theorem 5.36. *Let G be a Frobenius group. Then:*

i) *the elements of G fixing no point are regular permutations, and they are $n-1$ in number;*

ii) *a permutation fixing a point is regular on the the remaining ones, so the subgroup $H = G_\alpha$ fixing a point is semiregular on $n-1$ points. Hence its order divides $n-1$;*

iii) $N_G(H) = H$.

Proof. i) The cycles of a permutation σ fixing no point have length at least 2, and if there are cycles of length h and k, with $h > k$, then the k-th power of this permutation fixes at least $k \geq 2$ elements (those of the k-cycle), and this power is not the identity because the k-th power of the h-cycle is not the identity. Now, and since $G_\alpha \cap G_\beta = \{1\}$ the elements fixing no point are those of the set $G \setminus (\bigcup_{\alpha\in\Omega} G_\alpha)$. G being transitive, $[G : G_\alpha] = n$, so $|\bigcup_{\alpha\in\Omega} G_\alpha)| = (|G_\alpha|-1)n$, and the result follows:

ii) follows from *i*);

iii) the elements of H^g fix α^g, so if $H^g = H$ then $\alpha^g = \alpha$ that is $g \in H$. $\diamondsuit$

Theorem 5.37. *Let G be a finite group containing a subgroup H disjoint from its conjugate and self normalizing:*

i) $H \cap H^x = \{1\}$, $x \notin H$;

ii) $\mathbf{N}_G(H) = H$.

Then G is faithfully represented as a Frobenius group by right multiplication on the cosets of H.

Proof. In the action of G on the cosets of H the stabilizer of a coset Hx is the subgroup H^x, and therefore is not the identity. If $g \in G$ fixes two cosets Hx and Hy, then $g \in H^x \cap H^y$, and by *i)* either $g = 1$ or $H^x = H^y$, i.e. $xy^{-1} \in \mathbf{N}_G(H) = H$ and $Hx = Hy$. ◇

The mapping $\varphi : Hg \to \alpha^g$ establishes an equivalence between the action of the group G on the cosets of H, and that of G on the set Ω of Definition 5.12 with $H = G_\alpha$. (φ is well defined because H fixes α.) In other words for a finite group G the following conditions:

i) G acts transitively on a set and in this action only the identity fixes more than one point, but the subgroup fixing a point is nontrivial;

ii) G is a group containing a subgroup H disjoint from its conjugate and self normalizing,

are equivalent. Hence we may define a Frobenius group as a group satisfying either condition.

The elements fixing no point form, together with the identity, a subgroup K. This is the content of a theorem of Frobenius, that we shall prove in the next chapter (Theorem 6.21). The subgroup K is normal (two conjugate elements fix the same number of points (zero, in this case).

Definition 5.13. The subgroup H of a Frobenius group is a *Frobenius complement.* The subgroup K is the *Frobenius kernel.*

Theorem 5.38. *The Frobenius kernel of a Frobenius group G is regular.*

Proof. K being semiregular (Theorem 5.36, *i)*), it suffices to show that it is transitive. By Theorem 5.36, *i)*, $|K| = n$. The index of the stabilizer of a point in K being the identity, the orbit of a point contains $|K| = n$ points, so K is transitive. ◇

Theorem 5.39. *Let G be a Frobenius group with complement H and kernel K. Then:*

i) $G = HK$, *with* $H \cap K = \{1\}$ *and* $K \trianglelefteq G$, *so that G is the semidirect product of K by H;*

ii) $|H|$ *divides* $|K| - 1$.

Proof. i) Follows from the above quoted Frobenius theorem and from the fact that the transitivity of K implies $G = HK$.

ii) This is point *ii)* of Theorem 5.36, recalling that $|K| = n$. ◇

Theorem 5.40. *In a Frobenius group,*

i) *the conjugacy class in G of a non identity element $h \in H$ coincides with the conjugacy class of h in H;*

ii) *the centralizer of an element $1 \neq h \in H$ is contained in H, so $\mathbf{C}_G(h) = \mathbf{C}_H(h)$;*

iii) *an element not in K is conjugate to an element of H. Hence a set of representatives of the conjugacy classes of H is also a set of representatives of the conjugacy classes of the elements not in K;*

iv) *the centralizer of an element $1 \neq k \in K$ of K is contained in K, so $\mathbf{C}_G(k) = \mathbf{C}_K(k)$.*

Proof. i) Let $h = g^{-1}h_1g$, for $h, h_1 \in H$ and $g \in G$. Then $h \in H \cap H^g$, so h fixes both α and α^g, forcing $\alpha = \alpha^g$ and $g \in H$;

ii) the equality of the centralizers follows by taking $h' = h$ in *i)*,

iii) an element not in K belongs to the stabilizer of a point which is conjugate to H,

iv) if $g \in \mathbf{C}_G(k)$ then, by *ii)*, g fixes no element of a conjugate of H. But the conjugates of H are the stabilizers of points, so $g \in G_\alpha$ for some α, and again by *ii)* it cannot commute with an element of K. ◇

By *iii)* of this theorem the nonidentity elements of H induce by conjugation f.p.f. automorphisms of K. The next theorem shows that this fact characterizes Frobenius groups among semidirect products.

Theorem 5.41. *Let $G = HK$ with $K \trianglelefteq G$ be a semidirect product, and suppose that the action of H on K is a f.p.f. action. Then G is a Frobenius group with kernel K and complement H.*

Proof. Let us show that H is disjoint from its conjugate and is self-normalizing. The result will follow from Theorem 5.37. If $g \in G$, then $g = hk$, so $H^g = H^k$. Let $H \cap H^k \neq \{1\}$; then $h_1 = k^{-1}h_2k$, and multiplying by h_2^{-1} we obtain $h_1h_2^{-1} = k^{-1}h_2kh_2^{-1} = k^{-1}(h_2kh_2^{-1}) \in K$, by the normality of K. Hence $h_1h_2^{-1} \in H \cap K = \{1\}$ and $h_1 = h_2$. This means that k commutes with h_1, and the fact that the action is f.p.f. implies $k = 1$ and $H^k = H$. Hence, if $g \notin H$ then $H \cap H^g = \{1\}$, and moreover $H^g \neq H$ if $g \notin H$, i.e. no element outside H normalizes H. ◇

Lemma 5.9. *If G is a finite group admitting a f.p.f. automorphism σ of order 2. Then G is abelian and of odd order.*

Proof. If $x \in G$, then $x = y^{-1}y^\sigma$ for some $y \in G$. It follows

$$x^\sigma = y^{-\sigma}y^{\sigma^2} = y^{-\sigma}y = (y^{-1}y^\sigma)^{-1} = x^{-1}.$$

Then σ is the automorphism $x \to x^{-1}$, and a group admitting such an automorphism is abelian. Moreover, σ being f.p.f., if $x \neq 1$ we have $x \neq x^{-1}$, therefore there are no elements of order 2. Hence G is of odd order. ◇

Theorem 5.42. *Let G be a Frobenius group with complement H of even order. Then the kernel K is abelian.*

Proof. An element of order 2 of H induces a f.p.f. automorphism of K of order 2. Apply the previous lemma. ◇

Examples 5.7. 1. (Cf. *Ex.* 2.10, 2) Let K be a finite field. The group $\mathcal{A}$ of affine transformations over K is defined, as in the case of the field of real numbers, as the semidirect product of the subgroup $\mathcal{A}_0$ of the homotheties and the normal subgroup of the translations $\mathcal{T}$. With ψ_a a homothety and φ_b a translation, let $\psi_a^{-1}\varphi_b\psi_a = \varphi_b$. Then $x + a^{-1}b = x + b$, so either $b = 0$, and φ_b is the identity, or $a = 1$, and ψ_a is the identity. Hence $\mathcal{A}_0$ acts fixed-point-freely on the subgroup of translations, so $\mathcal{A}$ is a Frobenius group with kernel $\mathcal{T}$ and complement $\mathcal{A}_0$. The subgroup $\mathcal{A}_0$ is the stabilizer of 0; the other complements are the stabilizers of the nonzero elements of K. If $s \neq 0$, then s is fixed by the transformations $\phi_{a,s-as}$ as a varies in K, and these form a subgroup A_s which is conjugate to $\mathcal{A}_0$ via the translation φ_s. If an element $\phi \in A_s$ fixes t, then $\phi(t) = at + s - as = t$, that is $(a-1)s = (a-1)t$. If $a \neq 1$, then $s = t$, and if $a = 1$, then $\phi(x) = x$, for all x, and therefore $\phi = 1$. If an affinity $\phi_{a,b}$ is not a translation, then $a \neq 1$ and $\phi_{a,b}$ fixes $\frac{b}{1-a}$. Hence the translations are the only affinities fixing no point. The group is transitive because already $\mathcal{T}$ is transitive. $\mathcal{A}_0$ being isomorphic to the multiplicative group of K, if $|K| = q$ we have $|\mathcal{A}_0| = q - 1$, and $\mathcal{T}$ being isomorphic to the additive group of K we have $|\mathcal{T}| = q$ and $|\mathcal{A}| = (q-1)q$.

2. A^4 is a Frobenius group with kernel a Klein group and complement a cyclic group. It can be seen as the group of affine transformations over the 4-element field $K = \{0, 1, x, x+1\}$, with the product modulo the polynomial $x^2 + x + 1$.

3. The cyclic group $C_5 = \langle a \rangle$ admits the automorphism $\sigma = (a, a^2, a^4, a^3)$ of order 4, and the semidirect product $C_5\langle\sigma\rangle$ is isomorphic to the group of affine transformations over the field $\mathbf{Z}_5$. Indeed, the additive group of this field is generated by 1, and by identifying 1 with the translation $\tau_1 : x \to x + 1$, and conjugating it by the homothety ψ_2, we have $\psi_2\tau_1\psi_2^{-1} : x \to x + 2$. In this way we have the automorphism of $\mathbf{Z}_5(+)$ given by the cyclic permutation $(1, 2, 4, 3)$, and by letting correspond the latter to σ and i (or τ_i) to a^i we have the required isomorphism.

4. Let G be the semidirect product of $C_7 = \langle b \rangle$ and $C_3 = \langle a \rangle$, with the action of C_3 on C_7 given by $a^{-1}ba = b^2$. The permutation induced by a on C_7 is $(1)(b, b^2, b^4)(b^3, b^6, b^5)$. Only 1 is fixed by a, and therefore also by a^{-1}; hence G is Frobenius. It has order 21, and is a subgroup of the simple group of order 168 (the normalizer of a 7-Sylow). Another group containing it is the group of affinities over the field $\mathbf{Z}_7$, of order 42, and is the semidirect product of C_6 and $\mathbf{Z}_7(+)$. The group G is the semidirect product of the C_3 of C_6 with $\mathbf{Z}_7(+)$.

Exercises

49. If σ is f.p.f. and x^σ is conjugate to x, then $x = 1$, that is, σ permutes the conjugacy classes of the group and fixes only the class of the identity. [*Hint*: Theorem 5.34, *i*).]

50. If σ is f.p.f $o(\sigma) = n$, then $xx^\sigma x^{\sigma^2} x^{\sigma^3} \cdots x^{\sigma^{n-1}} = 1$.

51. If σ is f.p.f and $o(\sigma) = 3$, then x and x^σ commute. (It can be proved that the group is nilpotent.)

52. If σ is f.p.f and I is the group of inner automorphisms of G, then the coset σI of I in $\mathbf{Aut}(G)$ consists of f.p.f. automorphisms, and these are all conjugate.

53. A p-subgroup P σ-invariant of G is contained in the unique p-Sylow S σ-invariant of G. [*Hint*: consider a subgroup H maximal w.r.t. the property of containing P and being σ-invariant, and prove that $H = S$.]

54. Two commuting f.p.f. automorphisms fix the same Sylow p-subgroup (cf. Theorem 5.34, *iii*)). In particular, if H is an abelian group of f.p.f. automorphisms of a group G, the elements of H all fix the same Sylow p-subgroup of G.

55. Let $\alpha \in \mathbf{Aut}(G)$ (not necessarily f.p.f.) fixing only one p-Sylow S for all primes p. Then $\mathbf{C}_G(\alpha) = \{x \in G \mid x^\alpha = x\}$ is nilpotent. [*Hint*: prove that $\mathbf{C}_G(\alpha) \subseteq \mathbf{N}_G(S)$.]

56. If G is infinite, Lemma 5.9 is no longer true. [*Hint*: consider D_∞ (*Ex.* 4.6, 3).]

57. Write down the 12 affinities of *Ex.* 5.7, 2.

58. Let G be a Frobenius group of kernel K and complement H. Prove that if $N \lhd G$, and $N \cap H = \{1\}$, then $N \subseteq K$.

5.5 Solvable Groups

A central series of a group G is an invariant series in which the quotient H/K is contained in the center of G/K, and therefore is abelian. Hence we may generalize the notion of a central series to that of an invariant series with abelian quotients. The next generalization, that of a normal series with abelian quotients is fictitious because it is not obtained a larger class of groups. This is what we shall see in a moment; it is related to the existence of the derived series, that we now define.

Definition 5.14. The series of subgroups:

$$G = G^{(0)} \supseteq G' = [G,G] \supseteq G'' = [G',G'] \supseteq \ldots \supseteq G^{(i)} \supseteq \ldots \tag{5.14}$$

is the *derived series* of G. The subgroup $G^{(i+1)}$ is the derived group of $G^{(i)}$, so the quotients $G^{(i)}/G^{(i+1)}$ are abelian.

The subgroups $G^{(i)}$ are characteristic in G: if α is an automorphism of $G = G^{(0)}$ then $(G^{(0)})^\alpha = G^{(0)}$; by induction, $(G^{(i)})^\alpha = G^{(i)}$, and therefore $(G^{(i+1)})^\alpha = [G^{(i)}, G^{(i)}]^\alpha = [(G^{(i)})^\alpha, (G^{(i)})^\alpha] = [G^{(i)}, G^{(i)}] = G^{(i+1)}$.

Theorem 5.43. *Let $G = H_0 \supseteq H_1 \supseteq H_2 \supseteq \ldots \supseteq H^{(m-1)} \supseteq H^{(m)} = \{1\}$, be a normal series with abelian quotients. Then $G^{(i)} \subseteq H^{(i)}$, $i = 1, 2, \ldots, m$. In other words, if a normal series with abelian quotients reaches the identity in m steps, the derived series reaches the identity in at most m steps.*

Proof. By induction on m. We have $G = G^{(0)} \subseteq H^{(0)} = G$; assume $G^{(i)} \subseteq H^{(i)}$. The quotient H_i/H_{i+1} being abelian, we have $H_i' \subseteq H_{i+1}$, and in particular $(G^{(i)})' = G^{(i+1)} \subseteq H_{i+1}$. ◇

Theorem 5.44. *In a group consider:*

i) *an invariant series with abelian quotients;*
ii) *a normal series with abelian quotients;*
iii) *the derived series.*

Then if one of these series reaches the identity, then so do the other two.

Definition 5.15. A group G is *solvable* if, for some n, $G^{(n)} = \{1\}$ in (5.14). The smallest such n is the *derived length* of G.

Examples 5.8. **1.** An abelian group is solvable, with derived length 1. A nilpotent group is solvable: $G' = \Gamma_2$, and by induction

$$G^{(i)} = [G^{(i-1)}, G^{(i-1)}] \subseteq [\Gamma_i, G] = \Gamma_{i+1}.$$

Polycyclic groups are solvable.

2. S^3 is solvable, with derived series $S^3 \supset C_3 \supset \{1\}$, and so is S^4, with derived series $S^4 \supset A^4 \supset V \supset \{1\}$. The simplicity of A^n, $n \geq 5$, implies that S^n is not solvable: the terms of the derived series coincide with A^n from $G' = A^n$ on.

3. A solvable simple group has prime order.

Theorem 5.45. *i) Subgroups and quotient groups of a solvable group G are solvable, with derived length at most that of G;*

ii) if $N \trianglelefteq G$ and G/N are solvable, so is G.

Proof. i) If $H \leq G$ then $H^{(i)} \subseteq G^{(i)}$, so if $G^{(n)} = \{1\}$ then $H^{(n)} = \{1\}$. If $N \trianglelefteq G$ we have $(G/N)^{(i)} = G^{(i)}N/N$, as follows easily by induction.

ii) If $(G/N)^{(m)} = N$, then $G^{(m)}N/N = N$ and therefore $G^{(m)} \subseteq N$. If $N^{(r)} = \{1\}$, then $G^{(m+r)} = (G^{(m)})^{(r)} \subseteq N^{(r)} = \{1\}$. ◇

Remark 5.3. As we know, *ii)* of the above theorem is not true if "solvable" is replaced by "nilpotent".

Let now N be a minimal normal subgroup of a solvable group G. N being solvable, the derived group N' is a proper subgroup of N, and as a characteristic subgroup is normal in G. By minimality, $N' = \{1\}$, so N is abelian. If N is finite, let p be a prime dividing its order. The elements of order p of N form a subgroup, which is characteristic and nontrivial and therefore coincides with N. Hence N is an elementary abelian p-group. We have proved:

Theorem 5.46. *Let N be a minimal normal subgroup of a solvable group G. Then:*

i) *N is abelian;*

ii) *if N is finite, N is an elementary abelian p-group.*

Hence, in a finite solvable group there always exists a non trivial normal p-subgroup H, for some prime p; in particular, the intersection of the Sylow p-subgroups for that prime p is nontrivial because it contains H. Moreover, $H \subseteq \mathbf{F}(G)$, so $\mathbf{F}(G)$ is nontrivial. Another nontrivial subgroup contained in $\mathbf{F}(G)$ is the second to the last term of a composition series.

Remark 5.4. In a nilpotent group, finite or infinite, a minimal normal subgroup is a $\mathbf{Z}_p$. In in a finite solvable group a minimal normal subgroup is a direct sum of copies of $\mathbf{Z}_p$.

Corollary 5.12. *The chief factors of a finite solvable group are elementary abelian.*

Corollary 5.13. *In a finite solvable group G a minimal normal subgroup is contained in the center of $\mathbf{F}(G)$. Moreover, a normal subgroup of G meets $\mathbf{F}(G)$ non trivially.*

Proof. By Theorem 5.15, $\mathbf{F}(G)$ centralizes the minimal normal subgroups of G. If N is such a subgroup, N is a p-group, and therefore nilpotent. It follows $N \subseteq \mathbf{F}(G)$ and therefore $N \subseteq \mathbf{Z}(\mathbf{F}(G))$. If $H \trianglelefteq G$, then H contains a minimal normal subgroup N; but $N \subseteq \mathbf{F}(G)$, so $H \cap \mathbf{F}(G) \neq \{1\}$. ♢

Remark 5.5. It follows from the above that in finite solvable groups the Fitting subgroup plays the role the center plays in nilpotent groups.

If a solvable group has a composition series, the chief factors are solvable and simple, so are of prime order (in particular the group is finite). Conversely, a composition series with factor groups of prime order is a normal series with abelian factor groups reaching the identity; a group with such a series is solvable. Taking into account Corollary 5.12, we have the following characterization of finite solvable groups.

Theorem 5.47. *Let G be a finite group. Then the following are equivalent:*

i) *G is solvable;*

ii) *the chief factors of G are elementary abelian;*

iii) *the composition factors are of prime order.*

From *iii)* it follows that a finite solvable group always contains a normal subgroup of prime index (the second term of a composition series).

Corollary 5.14. *A solvable group admits a composition series if and only if is finite.*

The fact that in a solvable group the Fitting subgroup is nontrivial may be used to construct an invariant series of a solvable group whose length may be considered as a measure of how far the group is from being nilpotent. Define inductively:

$$F_0 = \{1\},\ F_1 = \mathbf{F}(G),\ F_{i+1}/F_i = \mathbf{F}(G/F_i).$$

Hence F_{i+1} is the inverse image of the Fitting subgroup $\mathbf{F}(G/F_i)$ in the canonical homomorphism $G \to G/F_i$. We have a series $\{1\} = F_0 \subset F_1 \subset F_2 \subset \ldots$, in which if G is solvable the inclusions are strict (Corollary 5.13), and therefore it stops only when it reaches G.

Definition 5.16. Let G be a finite solvable group. The series

$$\{1\} = F_0 \subset F_1 \subset F_2 \subset \ldots \subset F_{n-1} \subset F_n = G$$

is the *Fitting series* of G, and the integer n the *Fitting length* or *Fitting height* of G.

Examples 5.9. **1.** $F_0 = G$ if and only if $G = \{1\}$; $F_1 = G$ if and only if G is nilpotent. Moreover, if $F_n = G$ for a certain n, then G is solvable. Indeed, F_1 is solvable because is nilpotent; assuming by induction that F_i is solvable, we have that F_{i+1}/F_i is nilpotent, hence solvable, from which it follows F_{i+1} solvable (Theorem 5.45, *ii*)).

2. The Fitting series of S^4 is $\{1\} \subset V \subset A^4 \subset S^4$.

If $K \subseteq H$ are two normal subgroups of a group G, the action of G on H/K induces a homomorphism $G \to \mathbf{Aut}(H/K)$ with kernel $\mathbf{C}_G(H/K)$ (cf. Definition 5.5). Denoting the image by $\mathbf{Aut}_G(H/K)$ we have:

$$\mathbf{Aut}_G(H/K) \simeq G/\mathbf{C}_G(H/K).$$

In Theorem 5.16 we saw that the Fitting subgroup is the intersection of the centralizers in G of the quotients of a chief series $\{H_i\}$, $i = 1, 2, \ldots, l$, of G. It follows:

$$G/\mathbf{F}(G) \subseteq \prod_{i=1}^{l} \mathbf{Aut}_G(H_i/H_{i+1}).$$

If G is solvable, the quotients H_i/H_{i+1} are elementary abelian, hence they are vector spaces over a field with a prime number of elements (the primes may vary). Then $\mathbf{Aut}_G(H_i/H_{i+1})$ is a linear group. Moreover, since there are no normal subgroups of G in between H_i and H_{i+1}, these spaces have no G-invariant subspaces: G is *irreducible.* Therefore, if G is a solvable finite group, the structure of $G/\mathbf{F}(G)$ is determined by that of the irreducible solvable subgroups of the finite dimensional linear groups over finite prime fields.

Let us now consider a few properties of the Fitting subgroup of a finite solvable group.

Theorem 5.48. *Let G be a finite solvable group, $F = \mathbf{F}(G)$ its Fitting subgroup. Then:*

i) $\mathbf{C}_G(F) = \mathbf{Z}(F)$.

Let F be cyclic; then:

ii) *G' is cyclic;*

iii) *if $p \nmid |F|$, a Sylow p-subgroup is abelian;*

iv) *if F is a p-group, then p is the largest prime divisor of $|G|$.*

Proof. *i)* Let $H = F\mathbf{C}_G(F)$. Then $\mathbf{F}(H) = F$, $\mathbf{C}_H(F) = \mathbf{C}_G(F)$, and therefore if $H < G$, by induction $\mathbf{C}_H(F) = \mathbf{Z}(F(H))$ and the claim follows. Then let $G = F\mathbf{C}_G(F)$; it follows:.

$$\{1\} \neq \mathbf{Z}(F) \subseteq F \cap \mathbf{C}_G(F) \subseteq \mathbf{Z}(G) = Z,$$

and

$$\mathbf{C}_G(F)/Z \subseteq \mathbf{C}_{G/Z}(F/Z) = \mathbf{Z}(\mathbf{F}(G/Z)) = F/Z,$$

where the first equality follows by induction. Then $\mathbf{C}_G(F) \subseteq F$ and the claim follows.

ii) F being abelian, $F \subseteq \mathbf{C}_G(F) = \mathbf{Z}(F) \subseteq F$, so $\mathbf{C}_G(F) = F$. Since $G/F = \mathbf{N}_G(F)/\mathbf{C}_G(F)$, which is isomorphic to a subgroup of $\mathbf{Aut}(F)$, which is abelian because F is cyclic, we have $G' \subseteq F$, and so G' is cyclic.

iii) G/F being abelian, SF/F is also abelian, where $S \in Syl_p(G)$, and $SF/F \simeq S/S \cap F \simeq S$ implies S abelian ($S \cap F = \{1\}$ because p does not divide $|F|$).

iv) If $|F| = p^n$, $|\mathbf{Aut}(F)| = \varphi(p^n) = p^n - p^{n-1} = p^{n-1}(p-1)$. Since G/F is isomorphic to a subgroup of $\mathbf{Aut}(F)$, if $q \neq p$ is a prime divisor of $|G|$, then q divides $|G/F|$ and so also $|\mathbf{Aut}(F)| = p^{n-1}(p-1)$. Hence q divides $p-1$, so $q < p$. ◇

The theorem we now prove may be viewed as a generalization of Sylow's theorem in the case of solvable groups. If $|G| = p^n m$, where $p \nmid m$, Sylow's theorem ensures the existence of a subgroup of order p^n. However, if we consider a different decomposition of the order of G into relatively prime integers, $|G| = ab$, with $(a,b) = 1$, nothing can be said, in general, about the existence of a subgroup of order a. For instance, the group A_5 has order $60 = 15 \cdot 4$, $(15,4) = 1$, but A_5, has no subgroups of order 15. However, if the group is solvable, then such a subgroup always exists, and there also suitable generalization of the other parts of Sylow theorem.

Theorem 5.49 (Ph. Hall). *Let G be a finite solvable group of order $|G| = ab$, where $(a,b) = 1$. Then:*

i) *there exists in G a subgroup of order a;*

ii) *any two subgroups of order a are conjugate;*

iii) *if A' is a subgroup whose order divides a, then A' is contained in a subgroup of order a.*

Proof. By induction on the order of G. Let N be a minimal normal subgroup of G, hence a p-group: $|N| = p^k$ for some prime p. We shall prove *i)* and *ii)* simultaneously by distinguishing two cases, according as $p|a$ or $p|b$.

1. $p|a$. Consider G/N. Then:

$$|G/N| = \frac{a}{p^k} \cdot b.$$

By induction, there exists $A/N \leq G/N$ of order $|A/N| = a/p^k$; hence $|A| = a$, and we have *i)*. As for *ii)*, the order of A being coprime to the index and $p||A|$, the highest power of p dividing $|G|$ also divides $|A|$. In other words, A contains a p-Sylow S of G. If A_1 is another subgroup of order a, then for the same reason it contains a p-Sylow of G, S_1 say. N being normal, N is contained in all the p-Sylows, and in particular in S and S_1, and so also in A and A_1. Then A/N and A_1/N are two subgroups of G/N of order a/p^k, and therefore they are conjugate in G/N. It follows that A and A_1 are conjugate in G, and this proves *ii)*.

2. $p|b$. In this case,

$$|G/N| = a \cdot \frac{b}{p^k},$$

and always by induction there exists in G/N a subgroup L/N of order a. Then $|L| = a \cdot p^k$. If $L < G$, again by induction there exists in L (and so in G) a subgroup of order a. Now let $L = G$; we have:

$$|G| = a \cdot p^k,$$

so N is Sylow in G. Consider now a minimal normal subgroup K/N of G/N. Then $|K/N| = q^t$, where q is a prime different from p because N being a p-Sylow, $p \nmid |G/N|$. Hence $|K| = q^t p^k$ and $K = QN$, where Q is a Sylow q-subgroup of K. Now $K \trianglelefteq G$, so the Frattini argument implies

$$G = K \cdot \mathbf{N}_G(Q) = NQ \cdot \mathbf{N}_G(Q) = N \cdot \mathbf{N}_G(Q).$$

Hence,

$$a \cdot p^k = |G| = \frac{|N| \cdot |\mathbf{N}_G(Q)|}{|N \cap \mathbf{N}_G(Q)|},$$

from which $a = |\mathbf{N}_G(Q)|/|N \cap \mathbf{N}_G(Q)|$, and therefore a divides $|\mathbf{N}_G(Q)|$. If $\mathbf{N}_G(Q) < G$, the existence of a subgroup of order a in $\mathbf{N}_G(Q)$, and so in G, is obtained by induction. If $\mathbf{N}_G(Q) = G$, then G contains a normal q-subgroup whose order divides a, and we proceed as in case 1, with p replaced by q. The existence in case 2 is proved.

As to conjugation, let A and A_1 be of order a; then AN/N and A_1N/N are also of order a ($A \cap N = \{1\}$ because $p \nmid a$) and therefore are conjugate in G/N. It follows that AN and A_1N are conjugate in G: $AN = (A_1N)^g = A_1^g N$, and so $A, A_1^g \subseteq AN$. If $AN < G$, since $|AN| = a \cdot p^k$, by induction A and A_1^g

are conjugate in AN, $A = A_1^{gx}$, $x \in AN$, and so in G. Then let $AN = G$. N is Sylow and normal, and we may assume that it is the unique minimal normal subgroup (if there is another one, it cannot be a p-group for the same p of $|N|$, otherwise it would be contained in N, because N is the unique p-Sylow, contradicting the minimality of N; hence it is a q-group, $q \neq p$, and therefore its order divides a and we are in case 1).

Let K/N be minimal normal in G/N. Then K/N is a q-group, with $q \neq p$ (N is Sylow, so $p \nmid |K/N|$), and therefore $K = NQ$, where Q is a q-Sylow of K. Since $K \trianglelefteq G$, the Frattini argument yields $G = N \cdot \mathbf{N}_G(Q)$. We will show that $\mathbf{N}_G(Q)$ has order a, and that any subgroup of order a is conjugate to it. Set $\mathbf{N}_G(Q) = H$, and consider $N \cap H$; this subgroup centralizes Q ($N \cap H$ which is normal in H because $N \trianglelefteq G$, and $Q \trianglelefteq H$, have trivial intersection and are normal in H) and also N (because N is abelian). Hence, $N \cap H \subseteq \mathbf{Z}(K)$, and $\mathbf{Z}(K) \trianglelefteq G$ because $\mathbf{Z}(K)$ is characteristic in $K \trianglelefteq G$. If $N \cap H \neq \{1\}$, then $\mathbf{Z}(K) \neq \{1\}$, so the latter subgroup contains a minimal normal subgroup, that is N, which is the unique such. Hence N centralizes K and therefore Q, so $G = N \cdot H = H$, $Q \trianglelefteq G$, $N \trianglelefteq Q$, which is absurd. Hence $N \cap H = \{1\}$, and $G = N \cdot H$ implies $|G| = |N| \cdot |H|$, so $|H| = a$.

Let A be of order a, and consider AK. We have $AK = G = AH$, $|AK/K| = |G/K| = ap^k/|Q|p^k = a/|Q|$, and from $|AK/K| = |A/(A \cap K)| = a/|Q|$ it follows $|A \cap K| = |Q|$. Therefore $Q_1 = A \cap K$ is Sylow in K and hence is conjugate to Q in K. Moreover, from $K \trianglelefteq G$ it follows $A \cap K \trianglelefteq A$, so $A \subseteq \mathbf{N}_G(Q_1) = H_1$. But $Q_1 \sim Q$ implies $H_1 \sim H$, and therefore $|H_1| = |H| = a$. Then $A = H_1$ and $A \sim H$.

We now pass to the proof of *iii*). As above, we distinguish two cases.

1. $p^k | a$. Here $|A'N|$ divides a; otherwise there is a prime q that divides $|A'N|$ and b, and so also either $|A'|$ and b, or $|N|$ and b, both cases being excluded. It follows that $|A'N/N|$ divides a/p^k; by induction, $A'N/N \subseteq A/N$, where $|A/N| = a/p^k$. Then $A'N \subseteq A$, so $A' \subseteq A$, where $|A| = a$.

2. $p^k | b$. In this case, $A' \cap N = \{1\}$ so that $|A'N/N| = |A'|$, and therefore, by induction, $A'N/N \subseteq B/N$, with $|B/N| = a$. Hence $A'N \subseteq B$, and $|B| = a \cdot p^k$. If $B < G$, by induction A' is contained in a subgroup of B, and therefore of G, of order a. So let $B = G$. The order of G is now $|G| = a \cdot p^k$, and by *i*) there exists in G a subgroup A of order a. It follows $G = AN$, and a fortiori $G = A \cdot NA'$. Hence:

$$a \cdot p^k = |G| = \frac{|A| \cdot |NA'|}{|A \cap NA'|},$$

from which $|A \cap NA'| = |A'|$, because $|NA'| = p^k|A'|$ since $p \nmid a$. The two subgroups $A \cap NA'$ and A' have order $|A'|$, they are contained in NA', and their order being coprime to the index in NA' by *ii*) they are conjugate there: $A' = (NA' \cap A)^g$, with $g \in NA'$. Then $A' = NA' \cap A^g$, from which $A' \subseteq A^g$, and A^g is the required subgroup. ♢

If the group is not solvable, there are counterexamples to all three parts of the theorem. We have seen that the group A^5, of order $60 = 4 \cdot 15$, has no subgroups of order 15. Also, $60 = 12 \cdot 5$, and A^5 contains subgroups of order 6 (for instance the group S^3 generated by $(1,2,3)$ and (1,2)(4,5)) not contained in a subgroup of order 12 (these are all of type A^4). Hence *i*) and *iii*) do not hold. As for *ii*), in the simple group $GL(3,2)$ of order $168 = 24 \cdot 7$, there are subgroups of order 24 that are not conjugate (cf. 3.8.1, *viii*)).

Definition 5.17. A subgroup of order coprime to its index is called a *Hall subgroup.*

If H is a Hall subgroup of a group G of index a power of a prime p (and therefore necessarily the highest power of p dividing $|G|$), and S is a Sylow p-subgroup, then $S \cap H = \{1\}$ and $G = SH$. Hence H is a p-complement of G (cf. Definition 5.7). With this terminology, *i*) of Hall's theorem may be expressed by saying that *a solvable group has a p-complement for all p*[10]. Unlike normal p-complements, in a solvable group a p-complement is not in general unique (think of the 2-complements of S^3).

Definition 5.18. A *Sylow basis* for a finite group is a family of pairwise commuting Sylow p-subgroups, one for each prime dividing the order of the group.

Theorem 5.50. *In a finite solvable group there always exists a Sylow basis.*

Proof. Let $|G| = \prod_{i=1}^{t} p_i^{k_i}$, and let K_i be p_i-complements, one for each p_i. Then the K_i are of relatively prime indices, so the index of the intersection of any number of them is the product of the indices (Corollary 2.1). Consider $\bigcap_{i \neq j} K_i$; the index of this subgroup is the product of all the $p_i^{k_i}$, except $p_j^{k_j}$, so its order is $p_j^{k_j}$, and therefore is a Sylow p_j-subgroup. The Sylow obtained in this way form a Sylow basis. Indeed, let $S_j = \bigcap_{i \neq j} K_i$ and $S_h = \bigcap_{i \neq h} K_i$; we show that $S_j S_h = S_h S_j$. First, $S_j = \bigcap_{i \neq j} K_i \subseteq \bigcap_{i \neq j,h} K_i$, and the same for S_h. Then $S_j, S_h \subseteq \bigcap_{i \neq j,h} K_i$; this intersection is a subgroup of order $p_j^{k_j} p_h^{k_h}$, and since it contains the set $S_j S_h$, that has the same number of elements, they are equal. Then $S_j S_h$ is a subgroup, and therefore $S_j S_h = S_h S_j$. ◇

By the theorem above, a finite solvable group is the product of pairwise commuting nilpotent groups.

Theorem 5.49 shows how in a solvable group the Hall subgroups have properties similar to those of the Sylow subgroups. In the case of a Sylow S of a group G we know that $\mathbf{N}_G(\mathbf{N}_G(S)) = \mathbf{N}_G(S)$, and we have proved this result as an application of the Frattini argument. A suitable generalization of this argument will yield the same result for the Hall subgroups of a solvable group.

---

[10] The converse also holds: if a group has a p-complement for all p, then it is solvable.

Theorem 5.51 (Generalized Frattini Argument). *Let K_1 and K_2 be subgroups of a normal subgroup H of a finite group G with the property that if they are conjugate in G, then they are already conjugate in H. Then:*

$$G = H \cdot \mathbf{N}_G(K_1).$$

Proof. Let $g \in G$. Then $K_1^g = K_2$ is contained in H, so there exists $h \in H$ such that $K_1 = K_1^{gh}$. It follows $gh \in \mathbf{N}_G(K_1)$ and the result. ◊

If $H \trianglelefteq G$ and K_1 is Hall in H, then $K_2 = K_1^g$ is contained in H and is Hall in it. By *ii)* of the Hall theorem applied to H, K_1 and K_2 are conjugate in H, and the assumptions of Theorem 5.51 are met.

Corollary 5.15. *Let H be a Hall subgroup of a solvable group G. Then:*

$$\mathbf{N}_G(\mathbf{N}_G(H)) = \mathbf{N}_G(H).$$

Proof. Setting $K_1 = H$, $H = \mathbf{N}_G(H)$ and $G = \mathbf{N}_G(\mathbf{N}_G(H))$ in Theorem 5.51 gives the result. ◊

Definition 5.19. A subgroup of a group is a *Carter subgroup* if it is nilpotent and self-normalizing.

A Carter subgroup H is a maximal nilpotent subgroup. Indeed, if $H < K$ and K is nilpotent, then $H < \mathbf{N}_K(H)$, against H being self-normalizing.

Theorem 5.52 (Carter). *A finite solvable group always contains Carter subgroups, and any two such subgroups are conjugate.*

Proof. We first prove existence. Let N be a minimal normal subgroup of G, $|N| = p^k$. By induction, G/N contains a Carter subgroup K/N. By nilpotence, $K/N = S/N \times Q'/N$, where S/N is a Sylow p-subgroup of K/N and Q'/N is Hall. Then $K = SQ'$ and $Q' = QN$, where Q is Hall. From $Q'/N \triangleleft K/N$ it follows $Q' = QN \triangleleft K$, and Q being Hall in QN the generalized Frattini argument yields $K = QN\mathbf{N}_K(Q) = N\mathbf{N}_K(Q)$. The subgroup $H = \mathbf{N}_K(Q)$ is a Carter subgroup. Indeed:

1. H is nilpotent.

We have $K = SQ' = SNQ = SQ$, so $H = H \cap K = Q \cdot (H \cap S)$ (Dedekind's identity, *ex.* 19 with $A = Q, B = S$ and $C = H$). Moreover, $S \triangleleft K$ because $S/N \triangleleft K/N$, and therefore $H \cap S \triangleleft H$. Q is nilpotent because Q'/N is nilpotent: indeed, for this quotient we have $Q'/N = QN/N \simeq Q/(Q \cap N) \simeq Q$, because $Q \cap N = \{1\}$. Since $Q \cap (H \cap S) = \{1\}$, the two subgroups commute elementwise $H = Q \times (H \cap S)$, so H is the direct product of two nilpotent subgroups, and as such is nilpotent.

2. $\mathbf{N}_G(H) = H$.

Let $g \in G$ be such that $H^g = H$. From $K = NH$ and $N \triangleleft G$, we have $K^g = N^g H^g = NH = K$. Then $(K/N)^{gN} = K/N$, and since K/N is Carter in G/N it is self-normalizing in G/N. Hence $gN \in K/N$, and $g \in K$. But in K the subgroup Q is Hall, and therefore its normalizer does not grow: $\mathbf{N}_K(H) = H$, and $g \in H$.

Let us now prove that two Carter subgroups are conjugate. Assume by induction that this holds for a solvable group of order less that $|G|$, and under this inductive hypothesis let us show that if H is Carter in G and $N \trianglelefteq G$, then the image HN/N of H is Carter in G/N. Nilpotence is obvious. Let $(HN/N)^{gN} = HN/N$. Then $(HN)^g = HN$, i.e. $H^gN = HN$. If $g \notin HN$, then $HN < G$, and the two Carter subgroups H^g and H of HN are conjugate in HN by induction: $(H^g)^y = H$, $y \in HN$. It follows $gy \in \mathbf{N}_G(H) = H$, $g \in Hy^{-1} \subseteq H \cdot HN = HN$. Hence we have the contradiction that from $g \notin HN$ follows $g \in HN$. Therefore $g \in HN$, $gN \in HN/N$, and HN/N is self-normalizing.

Always under the above inductive hypothesis, let H_1 and H_2 be two Carter subgroups of G. Then H_1N/N and H_2N/N are Carter in G/N, and therefore by induction they are conjugate in G/N. Then H_1N and H_2N are conjugate in G: $(H_1N)^g = H_2N$, $H_1^gN = H_2N$, and if this subgroup is not the whole of G, by induction H_1^g and H_2 are conjugate in it, and therefore in G. Hence we may assume $G = H_1N = H_2N$, and N abelian (by taking N minimal normal). Then $H_1 \cap N \trianglelefteq N$ (because N is abelian), $H_1 \cap N \trianglelefteq H_1$ (because $N \trianglelefteq G$, and therefore $H_1 \cap N \trianglelefteq G$). By the minimality of N, $H_1 \cap N = N$ or $H_1 \cap N = \{1\}$. If $H_1 \cap N = N$, then $N \subseteq H_1$, so $G = H_1$ and there is nothing to prove. Then let $H_1 \cap N = \{1\}$, and similarly $H_2 \cap N = \{1\}$.

Furthermore, H_1 and H_2 are maximal subgroups. Indeed, if $H_1 \subseteq M \subseteq G$, then $M \cap N \trianglelefteq M$ (because N is normal) and $M \cap N \trianglelefteq N$ (because N is abelian), and therefore $M \cap N \trianglelefteq G$ because $G = H_1N = MN$. By the minimality of N, either $M \cap N = \{1\}$ or $M \cap N = N$. In the former case, $G/N = MN/N \simeq M$ and $G/N = H_1N/N \simeq H_1$ (if $M \cap N = \{1\}$, then a fortiori $H_1 \cap N = \{1\}$), and $M = H_1$; in the latter, $N \subseteq M$ and $M = G$.

Finally, let Q_1 and Q_2 be two p-complements of H_1 and H_2, respectively. Then they are two p-complements of G and therefore by Hall's theorem they are conjugate: $Q_1 = g^{-1}Q_2g$. Let us prove that g conjugates H_2 and H_1. Indeed, let $H_1 \neq g^{-1}H_2g$; then $Q_1 \subseteq H_1$, $Q_1 \subseteq g^{-1}H_2g$, and Q_1 is Hall in these two subgroups, and therefore normal by nilpotence. Hence it is normal in the subgroup they generate, which is G because H_1 is maximal. The factor group G/Q_1 is a p-group (Q_1 is a p-complement) and properly contains H_1/Q_1, which is Carter in G/Q_1, and hence self-normalizing; but this is impossible in a p-group. This contradiction shows that H_1 and H_2 are conjugate. ♢

In nonsolvable groups Carter subgroups may not exist. In the simple group A_5 the only nilpotent subgroups are the Sylow subgroups and their subgroups. But $\mathbf{N}_G(C_2)$ is a Klein group, $\mathbf{N}_G(V) = A_4$, $\mathbf{N}_G(C_3) = S_3$ and $\mathbf{N}_G(C_5) = D_5$.

Remark 5.6. The notion of a nilpotent self-normalizing subgroup corresponds to that of a Cartan subalgebra of a Lie algebra. We give a brief explanation of this relationship. An *algebra* is a ring that is also a vector space. More precisely, let K be a field and let A be a ring with a vector space structure on the additive group of A such that $\alpha(ab) = (a\alpha)b = a(\alpha b)$, $\alpha \in K$, $a, b \in A$. This algebra is a *Lie algebra* $\mathcal{L}$ if in addition:

i) $ab + ba = 0$;
ii) $(ab)c + (bc)a + (ca)b = 0$

(equality *ii)* is the *Jacobi identity*). Then the product ab is denoted $[a, b]$ (*Lie product*). A *subalgebra* $\mathcal{B}$ of a Lie algebra is a subspace which is closed under the Lie product, i.e. if $a, b \in \mathcal{B}$, then $[a, b] \in \mathcal{B}$. An *ideal* of a Lie algebra is a subspace I such that if $x \in I$, then $[a, x] \in I$ for each $a \in \mathcal{L}$ (since, by *i)*, $[a, b] = -[b, a]$, there is no distinction between right and left ideals). Starting from an associative algebra one obtains a Lie algebra by setting $[a, b] = ab - ba$. This is the case, for instance, of the matrix algebra over a field. Subalgebras are defined in the obvious way. Let $\mathcal{B}$ be a subalgebra of $\mathcal{L}$. Denote $\mathcal{B}' = [\mathcal{B}, \mathcal{B}]$ the subalgebra generated by all products $[b_1, b_2]$ as b_1 and b_2 vary in $\mathcal{B}$, that is the smallest subalgebra of $\mathcal{L}$ containing all these products, and define inductively $\mathcal{B}^{(k)} = [\mathcal{B}^{(k-1)}, \mathcal{B}]$ ($\mathcal{B}^{(0)} = \mathcal{B}$). We have $\mathcal{B}^{(k)} \supseteq \mathcal{B}^{(k+1)}$, and if $\mathcal{B}^{(n)} = \{0\}$ for some n the subalgebra is *nilpotent* (*abelian*, if $n = 1$). The *normalizer* $\mathbf{N}(\mathcal{B})$ of the subalgebra $\mathcal{B}$ is the set $\mathbf{N}(\mathcal{B}) = \{a \in \mathcal{L} \mid [a, b] \in \mathcal{B},\ \forall b \in \mathcal{B}\}$, which is a subalgebra (as follows from Jacobi identity). It is the largest subalgebra of $\mathcal{L}$ containing $\mathcal{B}$ as an ideal (as in the case of groups, where the normalizer is the largest subgroup in which a subgroup is normal). A nilpotent selfnormalizing subalgebra $\mathcal{B}$, $\mathbf{N}(\mathcal{B}) = \mathcal{B}$, is called a *Cartan subalgebra*. An example of a Cartan subalgebra in the matrix algebra (with the Lie product $[A, B] = AB - BA$, where AB is the usual matrix product) is the subalgebra $\mathcal{B}$ of diagonal matrices. It is a subalgebra, as is easily seen. Moreover, $[A, B] = 0$, because two diagonal matrices commute, so $\mathcal{B}' = 0$, and $\mathcal{B}$ is nilpotent (and in fact abelian). Let us show that it is selfnormalizing. Let $[A, B] \in \mathcal{B}$, for all $B \in \mathcal{B}$, that is, let $[A, B]$ be a diagonal matrix if B is. Then $AB - BA$ is diagonal; however, this matrix has all zeroes on the main diagonal, and since it is a diagonal matrix, it is the zero matrix. It follows $AB = BA$, and since a matrix commuting with all the diagonal matrices is itself diagonal, we have $A \in \mathcal{B}$.

We close this chapter with a sufficient condition for the solvability of a finite group due to I. N. Herstein. First a lemma.

Lemma 5.10. *In* Theorem 5.39 *assume further that H is abelian. Then:*

i) *the restriction to H of the transfer $V : G \to H$ is the identity on H, that is $V(h) = h$, for all $h \in H$;*

ii) *$G = HK$, $K = ker(V)$ and by i), $H \cap K = \{1\}$.*

Proof. i) We know that $V(h) = \prod_{i=1}^{t} x_i h^{r_i} x_i^{-1}$, where $x_i h^{r_i} x_i^{-1} \in H$. If $x_i \neq \{1\}$, then $h^{r_i} \in H \cap H^{x_i^{-1}} = \{1\}$. In the product, the only contribution is that of $x_1 = 1$, i.e. $V(h) = h^{r_1}$. By (5.13) with $x_1 = 1$, the cycle of σ to which 1 belongs contains only 1; indeed, $1 \cdot h = h_1 x_{\sigma(1)}$, from which $x_{\sigma(1)} = h_1^{-1} h \in H$, so $x_{\sigma(1)} = 1$, $\sigma(1) = 1$ and $r_1 = 1$. It follows $V(h) = h$. *ii)* By *i)*, V is surjective (H is already surjective), so $H \simeq G/K$, $K = ker(V)$, from which $|G| = |H| \cdot |K| = |HK|$ because $H \cap K = \{1\}$ (since by *i*) $V(h) = 1$ implies $h = 1$). ◊

If H is not abelian, this lemma yields $V(G) = H/H'$, and the restriction of V to H is the canonical homomorphism $h \to hH'$.

If in a Frobenius group the stabilizer of a points is abelian, then by the lemma above $G_\alpha = H$ has $|K|$ conjugates, where K is the kernel of the transfer of G in H; therefore, there are $(|H| - 1)|K| = |G| - |K|$ elements that move points. Hence, those moving no points are $|K|$ in number, and are the elements of the subgroup K.

Theorem 5.53 (Herstein). *Let G be a finite group admitting an abelian maximal subgroup H. Then G is solvable.*

Proof.[11] If $\mathbf{N}_G(H) > H$, then $\mathbf{N}_G(H) = G$ by the maximality of H; therefore $H \trianglelefteq G$ and G/H is of prime order (always because of the maximality of H). Hence H and G/H are both solvable, and so also is G. Then let $\mathbf{N}_G(H) = H$, and let $B = H \cap H^x \neq \{1\}$ for $x \notin H$. Then B is normal in both H and H^x (these are abelian groups), so $\mathbf{N}_G(B)$ properly contains H; hence $B \trianglelefteq G$. G/B contains the abelian maximal subgroup H/B; by induction on $|G|$, G/B is solvable, and B being solvable, so also is G. Hence we may assume $H \cap H^x = \{1\}$ for $x \notin H$. Let K be the kernel of the transfer of G in H, and for a given p let S be the Sylow p-subgroup of K fixed by conjugation by the elements of H (*es.* 54; $1 \neq h \in H$ acts f.p.f. on K). Then HS is a subgroup properly containing H, so $HS = G$, and since H normalizes S we have $S \trianglelefteq G$ (from $HK = G$ we have $S = K$). S being nilpotent, and therefore solvable, and $G/S \simeq H$ being abelian, G is solvable. ◊

Remark 5.7. If the hypothesis "H abelian" is replaced by "H nilpotent", then Theorem 5.53 is not true in general. A deep result of J.G. Thompson allows one to establish solvability if the group admits a maximal nilpotent subgroup of odd order.

Theorem 5.54 (Schmidt-Iwasawa). *A finite group in which every proper subgroup is nilpotent is solvable.*

Proof. Let G be a minimal counterexample. *i*) G is simple. Deny, and let $\{1\} < N \triangleleft G$. N is nilpotent; if $H/N < G/N$, H is nilpotent and so also is H/N; by induction, G/N is solvable, and N being nilpotent, G is solvable, against the choice of G. *ii*) If L and M are two distinct maximal subgroups of G, then $L \cap M = \{1\}$. Deny, and among all pairs of maximal subgroups let L and M be such that $L \cap M = I$ is of maximal order. We have $I < L$ and $I < M$ (because $L \neq M$), and by nilpotence $I < \mathbf{N}_L(I)$ and $I < \mathbf{N}_M(I)$. Moreover $\mathbf{N}_G(I) < G$, because G is simple, and therefore there exists a maximal subgroup H that contains $\mathbf{N}_G(I)$; it follows $L \cap H \supseteq L \cap \mathbf{N}_G(I) = \mathbf{N}_L(I) > I$,

[11] The original proof makes use of the Frobenius theorem (Theorem 6.21).

from which $|L \cap H| > |I| = |L \cap M|$. The choice of L and M implies $H = L$; similarly, $H = M$, and therefore $L = M$, whereas $L \neq M$. *iii)* A group G satisfying *i)* and *ii)* does not exist. Deny; then every element of G belongs to a maximal subgroup, and by *ii)* to only one. By *i)*, none of these is normal, hence by maximality they are self-normalizing. Then the group is a union of subgroups pairwise with trivial intersection and self-normalizing. But such a group cannot exist (Chapter 2, *ex.* 39).

This theorem applies in particular to the study of nonnilpotent groups in which every subgroup is nilpotent. Such a group is called a *minimal nonnilpotent* group.

Corollary 5.16. *The order of a minimal nonnilpotent finite group G is divisible by exactly two primes.*

Proof. By the theorem above G is solvable, hence it admits a normal subgroup N of prime index p. If Q is a Sylow q-subgroup $q \neq p$, then $Q \subset N$ and is Sylow in N, therefore characteristic in N and so normal in G. If $PQ < G$ for all Q, then these subgroups PQ are nilpotent, so P is normalized by all the Sylow q-subgroups, and so is normal in G. We already know that the Sylow q-subgroups are normal, so G would be nilpotent, against the assumption. It follows $PQ = G$, for some Q. ♢

Exercises

59. *i)* The nonsimple groups of *ex.* 38, 41, 43 e 44 of Chapter 3 (except those of order 180 and 900) are solvable.

ii) the dihedral groups D_n are solvable, for all n.

60. The inverse of Lagrange's theorem does not hold in general for solvable groups.

61. *i)* The product of two solvable groups, with one of which normal, is solvable (cf. Theorem 5.11);

ii) the direct product of a finite number of solvable groups is solvable.

62. If a group G admits an automorphism sending an element to itself or to its inverse, then G is solvable. [*Hint*: if H is the subgroup of the elements fixed by the automorphism, then H is normal and abelian, and G/H is also abelian.]

63. Let G be a finite solvable group and let σ be an automorphism fixing every element of the Fitting subgroup F of G. Prove that $(o(\sigma), |G|) \neq 1$. [*Hint*: given $g \in G$, let $g^\sigma = gx$; letting g act on F by conjugation, prove that x centralizes F, hence $x \in F$ and therefore is fixed by σ. Moreover, $g^{\sigma^k} = gx^k$, for all k, and $o(x)|o(\sigma)$.]

64. Prove the equivalence of the following propositions:

i) a group of odd order is solvable[12];

ii) a nonabelian simple group has even order.

[12] This proposition is the content of a celebrated theorem of W. Feit and J.G. Thompson.

65. Let G be solvable, $H \trianglelefteq G$. Then $HG' < G$ (cf. *ex.* 6).

66. Let G be a solvable finite group in which every Sylow subgroup coincides with its normalizer. Prove that G is a p-group[13]. [*Hint*: let H be a normal subgroup of index a prime p. Then H contains a q-Sylow Q; apply the Frattini argument.]

67. Let G be a solvable primitive permutation group of degree n. Prove that n is a power of a prime.

68. A maximal subgroup in a finite solvable group has prime power index (cf. Corollary 5.2).

69. Let H and K be two maximal subgroups of relatively prime order of a finite solvable group. Prove that their indices are also relatively prime. The converse is not true.

70. Let G be a finite solvable group in which every maximal subgroup has order coprime to the index. Prove that for some p a p-Sylow is normal.

71. Let G be a finite solvable group. Prove that if $x_1x_2\cdots x_n = 1$ and the x_i are of coprime order, then $x_i = 1$ for all i.[14]

72. Let G be a finite solvable group, and let $S_1, S_2, \ldots, S_t$ be Sylow subgroups, one for each prime dividing $|G|$. Prove that the product of the S_i, in any order, is the whole group G (cf. *Ex.* 3.4, 12).

73. Let G be a group. If two consecutive factor groups of the derived series G^{i-1}/G^i and G^i/G^{i+1}, $i > 1$ are cyclic, then $G^i = G^{i+1}$. [*Hint*: prove that setting $H = G^{i-2}/G^{i+1}$, the factor group $H'/\mathbf{Z}(H')$ is cyclic, and H' abelian.]

74. If G has all its Sylow cyclic, then:
i) G is solvable;
ii) G' and G/G' are cyclic (such a group is called *metacyclic*).

75. A maximal solvable subgroup of a finite group is self-normalizing.

76. (Galois) A transitive subgroup G of S^p, p a prime, is solvable if and only if it contains a normal subgroup of order p. In other words, G is solvable if and only if its action is affine (Chapter 3, *ex.* 67).

Remark 5.8. In Galois theory, the result of the previous exercise has the following meaning: an irreducible equation of prime degree is solvable by radicals if and only if its roots may be expressed as rational function of any two of them. Let us see why[15]. If the Galois group is solvable, then being a subgroup of S^p, by the above is an affine group, so if an element fixes two points it is the identity. Adjoining two roots α_i and α_j to the field, the Galois group reduces to a subgroup that fixes these two roots, and so is the identity. Hence $K(\alpha_1, \alpha_2, \ldots, \alpha_p) = K(\alpha_i, \alpha_j)$, so every root is a rational function of α_i and α_j. Conversely, if the roots are all rational functions of two of them, α_i and α_j, say, then $K(\alpha_1, \alpha_2, \ldots, \alpha_p) = K(\alpha_i, \alpha_j)$; this equality

[13] The result also holds without the hypothesis of solvability (Glauberman).
[14] This fact characterizes finite solvable groups (J.G. Thompson).
[15] We recall that an equation is solvable by radicals if and only if its Galois group is a solvable group (Remark 2 of 3.68).

means that if an element of the Galois group fixes two roots it fixes all of them, and therefore is the identity. Hence the group is an affine group, and therefore solvable.

77. Prove that, under the hypothesis of Theorem 5.53, $G''' = \{1\}$.

78. Let $p_1, p_2, \ldots, p_n$ be the composition factors of a solvable group G (possibly with repetitions). Then G is nilpotent if and only if for every permutation $p_{i_1}, p_{i_2}, \ldots, p_{i_n}$ of the p_i, G admits a composition series whose composition factors appear in the order $p_{i_1}, p_{i_2}, \ldots, p_{i_n}$.

79. A finitely generated periodic solvable group is finite (cf. Theorem 5.17).

80. Let G be finite, $G = ABA$, where A is an abelian subgroup and B is of prime order. Prove that G is solvable. [*Hint*: use Theorem 5.53.]

81. Prove that:
i) a subgroup A of a group G is a maximal abelian subgroup if and only if it equals its centralizer;
ii) if the Fitting subgroup F of G is abelian, then it is the only maximal abelian normal subgroup;
iii) if F is abelian, and G is solvable, then F is a maximal abelian subgroup.

82. Let G be a finite group in which every maximal subgroup has prime index. Prove that if q is the largest prime dividing $|G|$, then a Sylow q-subgroup is normal, and G is solvable. [*Hint*: if Q is not normal, let M be maximal, $M \supseteq \mathbf{N}_G(Q)$; consider the number of Sylow q-subgroups of G and M.]

83. *i)* Let G be a simple group, $G = HK$, where H and K are two proper subgroups with H abelian. Prove that $H \cap K = \{1\}$.

ii) Let $|G| = p^a q^b$ and let the Sylows be abelian. Prove that the group is solvable. [*Hint*: it is sufficient to prove that G is not simple (use *i*)) and induct.] We shall see in the next chapter that the result also holds without the assumption of the Sylows being abelian (Theorem 6.20).

84. A free group of rank greater than 1 is not solvable.

85. The group G of *Ex.* 4.1, 4 is solvable (Chapter 4, *ex.* 12), but not polycyclic (Theorem 5.21).

Definition 5.20. A group is *supersolvable* if it has an invariant series with cyclic quotients. (In the definition of a polycyclic group replace "normal" with "invariant".)

86. Subgroups and factor groups of supersolvable groups are supersolvable.

87. (Wendt) The derived group G' of a supersolvable group is nilpotent. [*Hint*: let H/K be a quotient of the series of G' obtained by intersecting G' with a series of G with cyclic quotients. Two elements of G' induce on H/K commuting automorphism, so their commutator induces the identity, and the series of G' is central.]

88. Let G be a supersolvable finite group, and let $|G| = p_1^{h_1} p_2^{h_2} \cdots p_t^{h_t}$ where $p_1 > p_2 > \ldots > p_t$, and for each p_i let S_i be a Sylow p_i-subgroup. Prove that G contains a set of normal subgroups

$$S_1, S_1S_2, \ldots, S_1S_2 \cdots S_t, \tag{5.15}$$

where the S_i are Sylow p_i-subgroups, as follows:

i) G has a normal subgroup N of prime order;

ii) by induction, G/N admits a normal Sylow p-subgroup SN/N where p is the largest prime divisor of $|G/N|$;

iii) if $p = p_1$, then $S = S_1$ is normal in G;

iv) if $p \neq p_1$, then S is the unique p_1-Sylow of SN, so $S = S_1$ is characteristic in SN and therefore normal in G;

v) by induction, G/S contains the normal subgroups $S_2, S_2S_3, \ldots, S_2S_3 \cdots S_t$.

(A set of normal subgroups as in (5.15) is called a *Sylow tower.*)

89. *i)* The symmetric group S^4 is solvable but not supersolvable;

ii) the derived group of S^4 is not nilpotent;

iii) S^4 has a normal subgroup N such that both N and S^4/N are supersolvable.

90. *i)* A maximal subgroup of a supersolvable group has prime index. [*Hint*: use *i)* of *ex.* 88.]

ii) Let G be a finite group in which the maximal subgroups are of prime index. Prove that, if q is the largest prime dividing $|G|$, a q-Sylow is normal and G is solvable. [*Hint*: if Q is not normal, let M a maximal subgroup containing $\mathbf{N}_G(Q)$; consider the number of q-Sylow of G and of M.] (By a theorem of Huppert, a finite group in which the maximal subgroups are of prime index is supersolvable.)

6

Representations

6.1 Definitions and examples

In Chapter 3 we have seen that the action of a group on a set gives rise to a homomorphism of the group into a symmetric group. This homomorphism allows the representation of the elements of the group by means of permutations; the properties of the latter may then be exploited to study the structure of the group. Instead of permutations, one may turn to linear transformations of a vector space, and consider homomorphisms of the group into a general linear group; in this way many group-theoretic problems may be reduced to problems in linear algebra.

Definition 6.1. Let V be a finite dimensional vector space of dimension n over a field K. A (*linear*) *representation* of a group G with *representation space* V is a homomorphism of G into the group of linear transformations of V: $\rho : G \to GL(V)$. The *degree* of the representation is the dimension of V. The the kernel $ker(\rho)$ of the representation ρ is the set of elements $s \in G$ such that $\rho(s)$ is the identity transformation of $GL(V)$. The representation ρ is *faithful* if $ker(\rho)$ reduces to the identity element of G. Using the isomorphism between the linear group and the group of matrices the elements of $\rho(G)$ may be expressed by non singular $n \times n$ matrices of $GL(n, K)$.

We write ρ_s for the image $\rho(s)$ of the element s of G.

The representations of a group G are related to the actions of G on a set Ω as follows. Consider the formal linear combinations of the elements of Ω with coefficients in a field K, $v = c_1\alpha_1 + c_2\alpha_2 + \cdots + c_n\alpha_n$, $\alpha_i \in \Omega$. Then we may extend the action of G to these by defining, for $s \in G$, $v^s = c_1\alpha_1^s + c_2\alpha_2^s + \cdots + c_n\alpha_n^s$. These elements form a vector space V with basis Ω, and the action of G on Ω is given by the permutation matrix P_s defined by s (Theorem 3.1). The mapping $s \to P_s$ is a homomorphism of G into the group $GL(n, K)$, the *permutation representation* of G. In turn, a representation $\rho : G \to GL(V)$ defines an action of G on V given by $v^s = \rho_s(v)$.

Machì A.: Groups. An Introduction to Ideas and Methods of the Theory of Groups.
DOI 10.1007/978-88-470-2421-2_6, Springer-Verlag Italia 2012

Fixing a basis $e_1, e_2, \dots, e_n$ of V, let R_s be the matrix of ρ_s in this basis. Then $\det(R_s) \neq 0$; ρ being a homomorphism, $R_{st} = R_s R_t$. If $r_{i,j}(s)$ is the (i,j) entry of R_s, the last equality means $r_{i,k}(st) = \sum_j r_{i,j}(s) r_{j,k}(t)$.

Examples 6.1. 1. If the space is of dimension 1, the representations are homomorphism in K^*. If $s \in G$ has period m, $\rho(s)^m = \rho(s^m) = \rho(1) = 1$, so that if G is finite, the images of the elements of G are roots of unity of K. In particular, over the complex field these images are of modulus 1. A non trivial example of a 1-dimensional representation is the *alternating representation* of the symmetric group S^n: $\rho(s) = 1$, if s is even, and $\rho(s) = -1$, if s is odd.

2. If $\rho_s = I$, the identity transformation, for all $s \in G$, ρ is the *trivial* representation. If ρ is trivial, and of degree 1 (i.e. ρ_s equals the 1-element of the field K) then ρ is the *principal* representation; it is denoted 1_G.

3. Let $\mathbf{R}$ be the additive group of the reals, and ρ the mapping of $\mathbf{R}$ into $GL(2, \mathbf{R})$ given by $\alpha \to \begin{pmatrix} \cos\alpha & -\sin\alpha \\ \sin\alpha & \cos\alpha \end{pmatrix}$. Then ρ is a representation of $\mathbf{R}$.

4. Let G be finite, and let e_t be a basis of V indexed by the elements of G. By defining $\rho_s(e_t) = e_{ts}$, we obtain the right regular representation, and by defining $\rho_s(e_t) = e_{s^{-1}t}$ the left regular representation (see Definition 1.9). Note that $e_s = \rho_s(e_1)$; in other words, the images of e_1, where 1 is the identity of G, under the various ρ_s, make up a basis of V.

Conversely, if the space V contains a vector v such that its images under ρ_s, $s \in G$ make up a basis of V, then ρ is equivalent to the regular representation according to the following definition.

Definition 6.2. (cf. Definition 3.5). Let ρ and ρ' two representations of a group G with spaces V and W over a field K. Then ρ and ρ' are *equivalent* (or *similar*) if there exists an isomorphism $\tau : V \to W$ commuting with the action of G:

$$\tau\rho_s = \rho'_s\tau, \; s \in G.$$

If R_s is the matrix expressing ρ_s in a given basis, then equivalence means that there exists a matrix T such that $TR_sT^{-1} = R'_s$, for all $s \in G$.

In the case of *Ex.* 4 above, the equivalence is obtained by defining $\tau(e_s) = \rho_s(v)$.

Let ρ and ρ' be two representations of G on two spaces V and W over K. The *direct sum* $\rho \oplus \rho'$ is the representation on $V \oplus W$ defined by:

$$(\rho \oplus \rho')_s(u + v) = \rho_s(u) + \rho'_s(v).$$

In terms of matrices, $(\rho \oplus \rho')_s$ is represented by $\begin{pmatrix} R_s & 0 \\ 0 & R'_s \end{pmatrix}$, where R_s and R'_s are relative to ρ_s and ρ'_s, respectively.

A subspace W of V is *invariant* under the action of G, or G-invariant, if $\rho_s(w) \in W$, for all $s \in G$ and $w \in W$. If W is G-invariant, then ρ induces two representations, one on W and one on the quotient space V/W. The first one is simply the restriction of ρ to W, the second is obtained by setting $\rho_s(v+W) = \rho_s(v) + W$ (the invariance of W implies that this action is well defined). These representations on W and V/W are the *constituents* of ρ.

Definition 6.3. A representation ρ on a space V is *irreducible* if the only subspaces invariant under the action of G are $\{0\}$ and V; otherwise ρ is reducible. The representation ρ is *completely reducible* if every invariant subspace W has an invariant complement W', that is $V = W \oplus W'$. (Note that an irreducible representation is completely reducible.)

In terms of matrices, the representation ρ is reducible if there exists an invertible matrix T over K, independent of s, such that $TR_sT^{-1} = \begin{pmatrix} R'_s & 0 \\ * & R''_s \end{pmatrix}$; it is completely reducible if there exists T such that:

$$TR_sT^{-1} = \begin{pmatrix} R_s^{(1)} & & & \\ & R_s^{(2)} & & 0 \\ & 0 & \ddots & \\ & & & R_s^{(m)} \end{pmatrix}$$

for some m, and where the matrices $R_s^{(i)}$, $s \in G$, $i = 1, 2, \ldots, m$, are irreducible.

Reducibility depends on the ground field as the following examples show.

Examples 6.2. **1.** Let $G = \{1, x, x^2\}$ be the cyclic group of order 3, and let ρ be the 2-dimensional representation:

$$\rho_1 = \begin{pmatrix} 1 & 0 \\ 0 & 1 \end{pmatrix}, \ \rho_x = \begin{pmatrix} -1 & 1 \\ -1 & 0 \end{pmatrix}, \ \rho_{x^2} = \begin{pmatrix} 0 & -1 \\ 1 & -1 \end{pmatrix}.$$

Over the real field this representation is irreducible. Indeed, if

$$T\rho_xT^{-1} = \begin{pmatrix} a & 0 \\ * & b \end{pmatrix}$$

then taking traces and determinants we have $a + b = -1$ and $ab = 1$, and so $a^2 + a + 1 = 0$ which has no real roots. Over the complex numbers, the matrices can be reduced to the form:

$$\rho_1 = \begin{pmatrix} 1 & 0 \\ 0 & 1 \end{pmatrix}, \ \rho_x = \begin{pmatrix} w & 0 \\ 0 & w^2 \end{pmatrix}, \ \rho_{x^2} = \begin{pmatrix} w^2 & 0 \\ 0 & w \end{pmatrix}$$

where w is a primitive third root of unity, and the representation is even completely reducible. It may happen that a representation remains irreducible for every extension of the base field (see *Ex.* below).

2. Consider the permutation representation associated with the action of a group on a set Ω defined at the outset of the present chapter. If Ω is split into orbits, the space V splits into the direct sum of subspaces, each having as basis the elements of an orbit Δ. These subspaces are never irreducible if $|\Delta| > 1$: transitivity does not coincide with irreducibility. Indeed, linearity allows a further decomposition besides that into orbits. More precisely, if $\Delta = \{e_1, e_2, \ldots, e_t\}$, then the subspace V' having the elements of Δ as basis always has an invariant subspace, that is, $\langle v \rangle$, where $v = e_1 + e_2 + \ldots + e_t$, and on which the representation is trivial. This subspace of dimension 1 has a complement, the hyperplane W of equation $x_1 + x_2 + \cdots + x_t = 0$ of dimension $t-1$, and we have $V' = \langle v \rangle \oplus W$.

Assume now $G = S^n$; let $\Omega = \{e_1, e_2, \ldots, e_n\}$ and let V be a vector space with basis Ω. In the representation of S^n defined by $e_i^\sigma = e_{\sigma(i)}$, the hyperplane W of equation $x_1 + x_2 + \cdots + x_n = 0$ is invariant, and on W the representation is irreducible. Indeed, let $U \neq \{0\}$ be an invariant subspace of W. If $0 \neq u = \sum_i x_i e_i \in U$, then at least two coefficients are different, $x_1 \neq x_2$ say. Consider the transposition $\tau = (1,2)$ and the difference $u - u^\tau$; we have:

$$\begin{aligned} u - u^\tau &= x_1e_1 + x_2e_2 + x_3e_3 + \cdots + x_ne_n - x_2e_1 - x_1e_2 - x_3e_3 - \cdots - x_ne_n \\ &= (x_2 - x_1)(e_2 - e_1), \end{aligned}$$

so U is invariant, $(x_2 - x_1)(e_2 - e_1) \in U$, and being $x_1 \neq x_2$, $e_2 - e_1 \in U$ too. With $\tau_i = (2, i)$, $i = 3, \ldots, n$, invariance also implies $(e_2 - e_1)^{\tau_i} = e_i - e_1 \in U$. But these vectors generate W: if $w \in W$, then

$$\begin{aligned} w &= x_1e_1 + x_2e_2 + \cdots + x_ne_n = -(x_2 + \cdots + x_n)e_1 + x_2e_2 + \cdots + x_ne_n \\ &= x_2(e_2 - e_1) + x_3(e_3 - e_1) + \cdots + x_n(e_n - e_1), \end{aligned}$$

and therefore $U = W$. Hence, on the space V, S^n admits the trivial representation on the subspace $\langle v \rangle$, where $v = e_1 + e_2 + \cdots + e_n$, and an irreducible representation of degree $n-1$ on the complement of $\langle v \rangle$.

Let us see an example with the group S^3. Let $S^3 = \langle s = (12), t = (132) \rangle$ and its 3-dimensional representation with the permutation matrices:

$$\rho_s = \begin{pmatrix} 0 & 1 & 0 \\ 1 & 0 & 0 \\ 0 & 0 & 1 \end{pmatrix}, \ \rho_t = \begin{pmatrix} 0 & 1 & 0 \\ 0 & 0 & 1 \\ 1 & 0 & 0 \end{pmatrix},$$

(it is sufficient to define ρ on the generators and verify that $\rho_s^2 = \rho_t^3 = (\rho_s\rho_t)^2 = 1$). ρ is reducible: with the matrix $T = \begin{pmatrix} 1 & 1 & 1 \\ -1 & 1 & 0 \\ -1 & 0 & 1 \end{pmatrix}$, which allows passing from the basis $\{e_1, e_2, e_3\}$ to the basis $v = e_1 + e_2 + e_3, e_2 - e_1, e_3 - e_1$ corresponding to the decomposition into invariant subspaces $V = \langle v \rangle \oplus W$, we have:

$$T\rho_sT^{-1} = \begin{pmatrix} 1 & 0 & 0 \\ 0 & -1 & 0 \\ 0 & -1 & 1 \end{pmatrix}, \ T\rho_tT^{-1} = \begin{pmatrix} 1 & 0 & 0 \\ 0 & -1 & 1 \\ 0 & -1 & 0 \end{pmatrix}.$$

Let us verify that the matrices of the representation on W make up an irreducible system. Indeed, let S be such that

$$S\begin{pmatrix}-1 & 0\\ -1 & 1\end{pmatrix}S^{-1} = \begin{pmatrix}r & 0\\ * & s\end{pmatrix}, \quad S\begin{pmatrix}-1 & 1\\ -1 & 0\end{pmatrix}S^{-1} = \begin{pmatrix}p & 0\\ * & q\end{pmatrix},$$

by equating traces and determinants we have $r+s=0, rs=1$, i.e. $r=\pm 1, s=\mp 1$; and $p+q=-1$, $pq=1$, i.e. $p=w, w^2$ and $q=w^2, w$ (third roots of unity). Summing memberwise we have

$$S\begin{pmatrix}-2 & 1\\ -2 & 1\end{pmatrix}S^{-1} = \begin{pmatrix}\pm 1 \pm a & 0\\ * & \mp 1 + b\end{pmatrix},$$

$a=w, w^2$ and $b=w^2, w$, which is impossible over any field (the determinant on the left is zero, that on the right is not zero.) Hence this representation is absolutely irreducible.

6.1.1 Maschke's Theorem

If $V = W \oplus W'$, and $v = w + w'$, the *projector* $\pi : V \to W$ makes correspond to the vector v its component on W. If $v \in W$, then $\pi(v) = v$. Conversely, if π is a linear mapping of V in itself such that $Im(\pi) = W$ and $\pi(w) = w$ if $w \in W$, then $V = W \oplus ker(\pi)$, a direct sum. Indeed, if $v \in W \cap ker(\pi)$, then $\pi(v) = v$ and $\pi(v) = 0$, so $v = 0$. Moreover, $v = \pi(v) + (v - \pi(v))$, $\pi(\pi(v)) = \pi(w) = w = \pi(v) \in W$ and $\pi(v - \pi(v)) = \pi(v) - \pi^2(v) = \pi(v) - \pi(v) = 0$, i.e. $v - \pi(v) \in ker(\pi)$. In other words, given a subspace W and a complement W', $V = W \oplus W'$, the subspace W' determines the projector π that associates with a vector v its component in W; then W' is a kernel of π. Conversely, if π is such that $\pi(v) \in W$ and $\pi(w) = w$, we have seen that $V = W \oplus ker(\pi)$, so π determines a complement of W (its kernel). Therefore, we have a one-to-one correspondence between the projectors of V on W and the complements of W in V.

Theorem 6.1 (Maschke). *A representation ρ of a finite group G over a field whose characteristic does not divide the order of the group is completely reducible.*

Proof. If ρ is irreducible, there is nothing to prove. Hence assume there exists an invariant subspace W and let us prove that there is a complementary subspace of W which is also invariant. Let V' be a complement of W, $V = W \oplus V'$, let π be the corresponding projector, and consider the average over the group G:

$$\pi^\circ = \frac{1}{|G|}\sum_{t\in G}\rho_t \pi \rho_{t^{-1}}. \tag{6.1}$$

We show that π° is a projector of V on W. Let $v \in V$, $t \in G$; then

$$\rho_t \pi \rho_{t^{-1}}.v = \rho_t \pi(\rho_{t^{-1}}.v) = \rho_t.w \in W,$$

where $w = \pi(\rho_{t^{-1}}.v) \in W$, and $\rho_t.w \in W$ because of the invariance of W. Hence $\pi^\circ(v)$ is a sum of elements of W multiplied by the scalar $\frac{1}{|G|}$, and therefore belongs to W.

Let now $w \in W$, $t \in G$. We have:

$$\rho_t \pi \rho_{t^{-1}}.w = \rho_t \pi(\rho_{t^{-1}}.w) = \rho_t \rho_{t^{-1}}.w = w,$$

because $\rho_{t^{-1}}.w \in W$ and therefore is fixed by π, for all $t \in G$. It follows $\pi^\circ(w) = \frac{1}{|G|}(|G|w) = w$, so π° is a projector; to it there corresponds a complement of W_0 (its kernel): $V = W \oplus W_0$. Let us show that W_0 is $G-$invariant, i.e. that $\pi^\circ(v) = 0$ implies $\pi^\circ(\rho_s.v) = 0,\ s \in G$; then W_0 will be the required subspace. To this end, we prove that π° commutes with the action of G: from $\pi^o(v) = 0$ it will follow $\pi^\circ(\rho_s.v) = \rho_s\pi^\circ(v) = \rho_s.0 = 0$. Now,

$$\begin{aligned}\rho_s \pi^\circ \rho_{s^{-1}} &= \rho_s[\frac{1}{|G|}\sum_{t\in G}\rho_t\pi\rho_{t^{-1}}]\rho_{s^{-1}} = \frac{1}{|G|}\sum_{t\in G}\rho_s\rho_t\pi\rho_{t^{-1}}\rho_{s^{-1}}\\ &= \frac{1}{|G|}\sum_{t\in G}\rho_{st}\pi\rho_{(st)^{-1}}.\end{aligned}$$

But as t varies in G, st runs over all elements of G, and therefore the last term is $\frac{1}{|G|}\sum_{t\in G}\rho_t\pi\rho_{t^{-1}}$, that is π°, as required. Then $\rho = \rho' \oplus \rho''$, where ρ' e ρ'' are the restrictions of ρ to W and W_0, respectively. If these are irreducible, then stop. Otherwise, by repeatedly applying the procedure above, V decomposes into the sum of G-invariant subspaces and ρ into the sum of irreducible representations. ◇

The assumption on the characteristic of the field was used in the above proof because $1/|G|$ is meaningless if the characteristic divides $|G|$. However, the assumption cannot be dropped as *Ex.* 1 and 2 below show.

Examples 6.3. 1. The group S^3 acts on the space $\mathbf{F}_2 \oplus \mathbf{F}_2 = \{0, u, v, u+v\}$ over $\mathbf{F}_2$; ρ_s, $s = (1,2)$, fixes $\{0, u\}$ and interchanges $\{0, v\}$ and $\{0, u+v\}$. The subspace $\{0, u\}$ admits the latter as complements, neither of which is invariant under ρ_s.

2. The cyclic group $G = \langle s\rangle$ of prime order p admits over $\mathbf{F}_p$ the reducible representation:

$$\rho(s^k) = \begin{pmatrix}1 & 0\\ k & 1\end{pmatrix},\ k = 0, 1, \ldots, p-1,$$

(if $V = \langle v_1, v_2\rangle$, the subspace $\langle v_1\rangle$ is invariant). But if $T\rho(s)T^{-1} = \begin{pmatrix}a & 0\\ 0 & b\end{pmatrix}$, by equating traces and determinants we have $a + b = 2$ and $ab = 1$, from which $a = b = 1$, a contradiction.

3. An argument like the one of *Ex.* 2 above shows that the result does no longer hold in the case of an infinite group. The matrices: $\begin{pmatrix}1 & 0\\ k & 1\end{pmatrix}$, $k \in \mathbf{Z}$, form a group isomorphic to **Z**, and therefore they afford a faithful representation of this group in dimension 2. The representation is reducible, but not completely (here too the existence of a matrix T would give $a = b = 1$).

Let ρ be a representation of a group G with space V over the complex field **C**. We recall that a Hermitian form is a function $V \times V \to \mathbf{C}$ such that:

i) $(u, v) = \overline{(v, u)}$;
ii) $(\alpha u + \beta v, w) = \alpha(u, w) + \beta(v, w)$;
iii) $(v, v) > 0$, when $v > 0$ (non degenerate).

The form is *G-invariant* if $(\rho_s(u), \rho_s(v)) = (u, v)$, $s \in G$. Starting with a given form it is always possible to determine a G-invariant form. Indeed, by defining $(u|v) = \sum_s(\rho_s(u), \rho_s(v))$, we have, for $t \in G$, $(\rho_t(u), \rho_t(v)) = \sum_s(\rho_s\rho_t(u), \rho_s\rho_t(v)) = \sum_s(\rho_{st}(u), \rho_{st}(v)) = \sum_s(\rho_s(u), \rho_s(v)) = (u|v)$. This invariance means that in an orthonormal basis of V the matrix of ρ_s is *unitary* (the inverse coincides with the conjugate transpose), and in the real case *orthogonal*.

The orthogonal subspace $W^\perp$ of a G-invariant subspace W is G-invariant. Indeed, if $v \in W^\perp$, then $(w, v) = 0$, $w \in W$, so $(\rho_s w, v) = 0$ by the invariance of W. It follows $0 = (\rho_s w, 1.v) = (\rho_s w, \rho_s\rho_{s^{-1}}v) = (w, \rho_{s^{-1}}v)$, where the second equality follows from the invariance of the form under ρ_s and the third from that of W under $\rho_{s^{-1}}$. Hence $V = W \oplus W^\perp$ with both subspaces invariant under G; in this way, a new proof of Maschke's theorem is obtained.

6.2 Characters

In the case of an action of a group on a set Ω, we have defined the character of the action as the function χ which associates with an element $g \in G$ the number of points it fixes. If g is represented with a permutation matrix, to a point fixed by g there corresponds a 1 on the main diagonal, so that $\chi(g)$ is the sum of the elements on the diagonal, i.e. the trace of the matrix representing g.

Definition 6.4. Let ρ be a representation of the finite group G over **C**. The function $\chi : G \to \mathbf{C}$, which associates with $s \in G$ the trace of the matrix ρ_s, $\chi(s) = \mathrm{tr}(\rho_s)$ is the *character of G afforded by ρ*. The degree of χ is the degree of ρ. If ρ is of degree 1, the character is *linear* and $\chi = \rho$. The character afforded by the principal representation, of constant value 1, is the *principal* character, and is denoted χ_1 or 1_G. Two equivalent representations afford the same character: the matrices representing the same element in the two representations are conjugate, so they have the same trace. The character of the alternating representation of S^n will be referred to as the *alternating* character.

Theorem 6.2. *Let χ be the character of a representation of a group G on a space V of dimension n. Then:*

i) $\chi(1) = n$, *the degree of the representation;*
ii) *if* $o(s) = m$, *then* $\chi(s)$ *is a sum of* m*-th roots of unity, so* $|\chi(s)| \leq n$;
iii) $|\chi(s)| = n$ *if and only if* ρ_s *is scalar;*
iv) $\chi(s) = n$ *if and only if* ρ_s *is the identity;*
v) $\chi(s^{-1}) = \overline{\chi(s)}$;
vi) $\chi(tst^{-1}) = \chi(s)$.

Proof. *i)* $\chi(1)$ is the trace of the identity matrix;
ii) the eigenvalues ρ_s are roots of unity. Indeed, if $s^m = 1$, and $\rho_s(v) = \lambda v$, $v \neq 0$, then $\rho_s^m(v) = \lambda^m v$. But $\rho_s^m = \rho_{s^m} = \rho_1 = I$, and therefore $v = Iv = \lambda^m v$, and in particular $\lambda^m = 1$ (observe that this also follows from the fact that ρ_s may be given by a unitary matrix). Hence $\lambda = \epsilon$, a root of unity. From $\chi(s) = \sum_{i=1}^n \epsilon_i$ it follows $|\chi(s)| \leq \sum_{i=1}^n |\epsilon_i| = n$.
iii) If the ϵ_i are all equal, $\epsilon_i = \epsilon$, then $\chi(s) = n\epsilon$ and so $|\chi(s)| = n$. Conversely, if $|\chi(s)| = n$, then since $|\epsilon_i| = 1$, all i, the ϵ_i are all equal, $\epsilon_i = \epsilon$, say. The characteristic polynomial of ρ_s is $(x - \epsilon)^n$, and since ρ_s satisfies $x^m - 1$ it also satisfies the gcd of the two polynomials, i.e. $x - \epsilon$. This means that ρ_s is the multiplication by ϵ.
iv) If $\chi(s) = n$, then $|\chi(s)| = n$, so ρ_s is scalar by *iii)* and $\chi(s) = n\epsilon$. It follows $\epsilon = 1$ and $\rho(s)$ is the identity transformation.
v) From $\lambda\overline{\lambda} = 1$ it follows:

$$\overline{\chi(s)} = \overline{\mathrm{tr}(\rho_s)} = \sum_i \overline{\lambda_i} = \sum_i \lambda_i^{-1} = \mathrm{tr}(\rho_s^{-1}) = \mathrm{tr}(\rho_{s^{-1}}) = \chi(s^{-1}).$$

vi) Conjugate matrices have the same trace. ◇

From *v)* of the above theorem it follows that s is conjugate to its inverse s^{-1} in the group G then $\chi(s)$ is a real number for all characters χ, and conversely. A *real character* is a character that only takes real values.

Theorem 6.3. *Let χ be the character of a representation ρ of degree n of a group G. Then the set of $s \in G$ such that $\chi(s) = n$ is the kernel of the representation ρ and therefore is a normal subgroup of G (one also says that this set is the kernel $ker(\chi)$ of χ).*

Proof. If $\chi(s) = n$, then by *iv)* of the previous theorem ρ_s is the identity transformation, i.e. $s \in ker(\rho)$. Conversely, if $s \in ker(\rho)$, then ρ_s is the identity, hence $\chi(s) = \chi(1) = n$. ◇

We recall that the tensor (or Kronecker) product $T = V \otimes_K W$ of two vector spaces over the same field K is the space over K with basis the pairs of basis vectors of V and W; the notation $v_i \otimes w_j$, $i = 1, 2, \ldots$, $j = 1, 2, \ldots$ is used. If $v = \sum_i \alpha_i v_i$ and $w = \sum_j \beta_j w_j$, the tensor product of v and w is $v \otimes w = \sum_{i,j} \alpha_i \beta_j (v_i \otimes w_j)$; an element of T is a sum of elements of this type.

It is easily seen that the product thus defined is bilinear. If $A = (a_{i,j})$ and $B = (b_{i,j})$ are the matrices of two linear transformations of V and W, respectively, their tensor product is defined as the linear transformation of $V \otimes_K W$ given by$(A \otimes B)(v_i \otimes w_j) = Av_i \otimes Bw_j$. If the dimensions of V and W are finite, n and m say, by choosing the lexicographic order according to the first index of the basis vectors of $V \otimes_K W$ $(v_1 \otimes w_1, v_1 \otimes w_2, \ldots, v_1 \otimes w_m, v_2 \otimes w_1, \ldots, v_n \otimes w_m)$, the matrix $A \otimes B$ has as entry (i, j) the entire matrix B multiplied by $a_{i,j}$. It is immediate to verify that the trace of the tensor product $A \otimes B$ is the product of the traces: $\mathrm{tr}(A \otimes B)=\mathrm{tr}(A)\mathrm{tr}(B)$, whatever the order chosen for the basis of $V \otimes_K W$.

The tensor product of two spaces allows the definition of the product of two representations over these spaces: $\rho' \otimes \rho''$ is defined on $V \otimes_K W$ as $(\rho' \otimes \rho'')(v \otimes w) = \rho'(v) \otimes \rho''(w)$. As seen above, the trace of the tensor product of two matrices is the product of the traces, so taking the product of two characters we obtain again a character, that of the tensor product of the two representations: $\chi' \cdot \chi'' = \chi(\rho' \otimes \rho'')$.

Lemma 6.1 (Schur). *Let ρ' and ρ'' be two irreducible representations with spaces V_1 and V_2, and let φ be a linear transformation $V_1 \to V_2$ commuting with the action of G: $\varphi\rho'_s = \rho''_s\varphi$, $s \in G$. Then:*

i) *if ρ' and ρ'' are not equivalent, then φ is the zero map $V_1 \to \{0\}$;*
ii) *if the field is the complex field, $V_1 = V_2$, and $\rho' = \rho''$, then φ is the multiplication by a scalar.*

Proof. i) Let $\varphi \neq 0$ with kernel W. Then, for $w \in W$, we have $\varphi(\rho'_s(w)) = \rho''_s\varphi(w) = \rho''_s(0) = 0$; it follows $\rho'_s(w) \in W$, and therefore W is G-invariant. ρ'_s being irreducible, either $W = V_1$ or $W = \{0\}$. In the first case, $ker(\varphi) = V_1$ and therefore $\varphi = 0$, which is excluded. Hence φ is injective. Let us prove that it is surjective. From $\rho''_s(\varphi(v)) = \varphi(\rho'_s(v)) \in Im(\varphi)$ we see that $Im(\varphi)$ is G-invariant, and since $Im(\varphi) \neq \{0\}$ because $\varphi \neq 0$, we have $Im(\varphi) = V_2$. Therefore φ is an isomorphism, and the assumption that it commutes with the action of G implies the equivalence.

ii) Let λ be an eigenvalue of φ: $\varphi' = \varphi - \lambda I$. Then the kernel of φ' is not zero, and by *i)* (φ' also commutes with ρ) this kernel is the whole of V, i.e. $\varphi' = 0$ and $\varphi = \lambda I$. ◇

In terms of matrices, Schur's lemma may be stated as follows:

i) Let $A(s), B(s)$ two irreducible systems of matrices depending on the same parameter s and relative to two vector spaces of dimension n and m, respectively. Assume there exists an $n \times m$ matrix Φ (rectangular, in general) such that $\Phi A(s) = B(s)\Phi$. Then either $\Phi = 0$, or Φ is invertible, and in the latter case $n = m$ and $A(s)$ and $B(s)$ are equivalent.

ii) Over **C**, if a matrix Φ commutes with the matrix of an irreducible system, then it is a scalar matrix.

Corollary 6.1. *Let* $\mathbf{C}$ *be the complex field. Then:*

i) *the irreducible representations of an abelian group are of dimension* 1*;*
ii) *if a finite group admits a faithful irreducible representation, then the center is cyclic.*

Proof. i) The elements of the group pairwise commute, and this is also the case for the matrices of an irreducible representation. But each matrix commutes with all the others, so by *ii)* of Schur's lemma every matrix of the group is scalar. Irreducibility then implies that this is only possible if the matrices are 1×1, i.e. if the space is 1-dimensional.

ii) The representation being irreducible, the matrices representing the center of the group are scalar matrices, and the representation being faithful, these matrices form a group isomorphic to a finite subgroup of the complex field, and therefore is cyclic. ◊

Corollary 6.2. *Let* $\varphi : V_1 \to V_2$ *be a linear transformation, and set:*

$$\varphi' = \frac{1}{|G|}\sum_{t\in G} \rho'_t \varphi (\rho''_t)^{-1}.$$

Then:

i) *if* ρ' *and* ρ'' *are not equivalent, then* $\varphi' = 0$*;*
ii) *if* $V_1 = V_2$ *and* $\rho' = \rho''$*,* φ' *is the multiplication by* $\frac{1}{n}tr(\varphi)$*, where* n *is the dimension of* V_1*.*

Proof. By Schur's lemma, it suffices to show that, for $s \in G$, $\rho'_s\varphi' = \varphi'\rho''_s$. We have:

$$\rho'_s\varphi'(\rho''_s)^{-1} = \frac{1}{|G|}\sum_{t\in G} \rho'_s\rho'_t\varphi(\rho''_t)^{-1}(\rho'_s)^{-1} = \frac{1}{|G|}\sum_{t\in G} \rho'_{st}\varphi\rho''_{st^{-1}} = \varphi. \quad ◊$$

Let us now consider this corollary in matrix form. Let ρ'_t be given by $A = (a_{i,j}(t))$ and ρ''_t by $B = (b_{i,j}(t))$, and let φ be the transformation represented by the matrix $E_{h,k} = (\delta_{h,k})$. Then $AE_{h,k}B = (a_{i,h}b_{k,j})$; in case *i)*, $\sum_t \rho'_t\varphi\rho''^{-1}_t = 0$, so that, in terms of matrices, $\sum_t A(t)E_{h,k}B(t^{-1}) = 0$, or $\sum_t a_{i,h}(t)b_{k,j}(t^{-1}) = 0$, for all i, h, k, j. In case *ii)*, $\frac{1}{|G|}\sum_t \rho(t)\varphi\rho(t^{-1}) = \frac{1}{n}\mathrm{tr}(\varphi)I$, where I is the identity transformation. With φ in the form $E_{h,k}$ we have $\sum_t A(t)E_{h,k}A(t^{-1}) = \frac{|G|}{n}\mathrm{tr}(\varphi)I$, where now I is the identity matrix. The matrix being scalar, the entries off the main diagonal are zero: $\sum_t a_{i,h}(t)a_{k,j}(t^{-1}) = 0$, if $i \neq j$. There remain the entries $\sum_t a_{i,h}(t^{-1})a_{k,i}(t)$, which, if $h \neq k$, are zero, since $\mathrm{tr}(E_{h,k}) = 0$. With $h = k$, $\mathrm{tr}(E_{h,k}) = 1$, and $\sum_t A(t)E_{h,k}A(t^{-1}) = \mathrm{diag}(\frac{|G|}{n})$, that is $\sum_t a_{i,h}(t)a_{k,j}(t^{-1}) = \frac{|G|}{n}$, for all i and h.

Summing up:

Corollary 6.3. *Let ρ' and ρ'' be represented by the matrices $A(t)$ and $B(t)$, respectively. Then:*

i) if ρ' is not equivalent to ρ'',

$$\sum_t a_{i,h}(t)b_{k,j}(t^{-1}) = 0, \tag{6.2}$$

for all i, h, k, j;

ii) if $\rho' = \rho''$,

$$\sum_t a_{i,h}(t)a_{k,j}(t^{-1}) = \begin{cases} 0, & \text{if } i \neq j \text{ or } h \neq k, \\ \frac{|G|}{n}, & \text{otherwise.} \end{cases} \tag{6.3}$$

Let us suppose that the matrices A and B are unitary (we have seen that this is always possible). Then $a_{i,j}(t^{-1}) = \overline{a_{j,i}(t^{-1})}$ (because $A^{-1} = \bar{A}^t$). Formulae (6.2) and (6.3) express orthogonality relations w.r.t. a scalar product that we now define.

Let φ and ψ be two complex functions defined on a finite group G, and set

$$(\varphi, \psi) = \frac{1}{|G|}\sum_{s \in G} \varphi(s)\overline{\psi(s)}.$$

This is a scalar product with the following properties:

i) it is linear in φ;

ii) it is is semilinear in ψ:

$$(\varphi, \alpha(\psi_1 + \psi_2)) = \overline{\alpha}(\varphi, \psi_1) + \overline{\alpha}(\varphi, \psi_2);$$

iii) $(\varphi, \varphi) > 0$ for $\varphi \neq 0$.

(*i*) and *ii*) follow easily by calculation; as for *iii*) we have

$$(\varphi, \varphi) = \frac{1}{|G|}\sum_{t \in G} \varphi(t)\overline{\varphi(t)} = \frac{1}{|G|}\sum_{t \in G} |\varphi(t)|^2,$$

a sum of squares of real numbers that are not all zero because $\varphi \neq 0$, so for at least one $t \in G$ $\varphi(t) \neq 0$).

We call φ and ψ *orthogonal* if $(\varphi, \psi) = 0$. The *norm* of φ is $\sqrt{(\varphi, \varphi)}$.

The importance of the product just defined in representation theory lies in the fact that w.r.t. it the irreducible characters have norm 1 and are pairwise orthogonal, as the following result shows.

Theorem 6.4. *i) If χ is the character of an irreducible representation of a finite group G, then $(\chi,\chi)=1$;*

ii) if χ and χ' are the characters of two non equivalent irreducible representations of G, then $(\chi,\chi')=0$.

Proof. *i)* We have

$$(\chi,\chi)=\frac{1}{|G|}\sum_{t\in G}\chi(t)\overline{\chi(t)}=\frac{1}{|G|}\sum_{t\in G}\chi(t)\chi(t^{-1}).$$

If ρ is given in matrix form $\rho_t=(a_{i,j}(t))$, then $\chi(t)=\sum_i a_{i,i}(t)$, so that

$$(\chi,\chi)=\frac{1}{|G|}\sum_t(\sum_{i=1}^n(a_{i,i}(t)\sum_{i=1}^n(a_{i,i}(t)^{-1})=\frac{1}{|G|}\sum_t(\sum_{i=1}^n(a_{i,i}(t)a_{i,i}(t^{-1})),$$

because, by (6.3), $\sum_{i=1}^n a_{i,i}(t)a_{j,j}(t^{-1})=0$ for $i\neq j$. By (6.3) again, the summands are equal to $\frac{1}{n}$, so the sum is $n\frac{1}{n}=1$.

ii) The proof is similar by applying (6.2). ◇

The result of the theorem may be summarized as follows:

$$\delta_{i,j}=\frac{1}{|G|}\sum_{s\in G}\chi_i(s)\chi_j(s^{-1})=\frac{1}{|G|}\sum_{k=1}^r\chi_i(s_k)\overline{\chi_j(s_k)} \tag{6.4}$$

($\delta_{i,j}$ is the Kronecker symbol). (6.4) is the *first orthogonality relation.*

Let us now consider a few applications of the above.

Theorem 6.5. *Let ρ be a representation of a finite group G on a space V, let φ be its character, and suppose that V splits into the direct sum of irreducible subspaces $V=W_1\oplus W_2\oplus\ldots\oplus W_k$. If ρ_i are the representations on the W_i, with character χ_i, the number of equivalent ρ_i is (φ,χ_i), where χ_i is the character of ρ_i.*

Proof. By assumption, $\varphi=\chi_1+\chi_2+\cdots+\chi_k$, from which $(\varphi,\chi_i)=(\chi_1,\chi_i)+(\chi_2,\chi_i)+\cdots+(\chi_k,\chi_i)$. If ρ_j is not equivalent to ρ_i, then $(\chi_j,\chi_i)=0$ (Theorem 6.4), so the sum contains as many 1 as there are ρ_j equivalent to ρ_i. ◇

The product (φ,χ) is independent of the chosen decomposition of V (for example, it does not depend on the chosen basis). Hence:

Corollary 6.4. *The number of ρ_i equivalent to a given irreducible representation does not depend on the decomposition of V.*

One speaks in this case of the *number of times W appears in V*, or of the *multiplicity of W in V*, or of *ρ_i in ρ*, or of *χ_i in φ*.

If an irreducible subspace W_i appears m_i times in V as a direct summand, one writes $V = m_1W_1 \oplus m_2W_2 \oplus \cdots \oplus m_rW_r$, $\varphi = m_1\chi_1 + m_2\chi_2 + \cdots + m_r\chi_r$, and one has

$$m_i = (\varphi, \chi_i). \tag{6.5}$$

Moreover,

$$(\varphi, \varphi) = (\sum_i m_i\chi_i, \sum_i m_i\chi_i) = \sum_i m_i m_j(\chi_i, \chi_j),$$

and since $(\chi_i, \chi_j) = 0$ if $i \neq j$ and $(\chi_i, \chi_i) = 1$,

$$(\varphi, \varphi) = \sum_{i=1}^{r} m_i^2. \tag{6.6}$$

Theorem 6.6. *Let φ be a character. Then:*
i) (φ, φ) *is an integer;*
ii) $(\varphi, \varphi) = 1$ *if and only if φ is irreducible.*

Proof. i) Follows from (6.6).

ii) The sum in (6.6) equals 1 if and only if the m_i are all zero except one, i.e. if and only if the space V coincides with one of the irreducible W_i. ◇

Theorem 6.7. *Two representations ρ' and ρ'' are equivalent if and only if they have the same character.*

Proof. Necessity has already been seen. As to sufficiency, if the two representations have the same character they contain a given irreducible representation the same number of times. It follows that if $V' = \sum_{i,j}^{\oplus} W_{i,j}$ and $V'' = \sum_{i,j}^{\oplus} W'_{i,j}$, then for each $W_{i,j}$ there exists an isomorphism $f_{i,j}$ such that $f_{i,j}\rho'_t = \rho''_t f_{i,j}$. The mapping $f = \sum_{i,j} f_{i,j}$ sending $v = \sum_{i,j} w_{i,j}$ to the sum $\sum_{i,j} f_{i,j}(w_{i,j})$ is such that $\rho'_t f = f\rho''_t$, for all $t \in G$. ◇

Remark 6.1. In Chapter 3, *Ex.* 3.8 and *ex.* 79, we have seen that two actions of a group G on a set Ω may have the same character without being equivalent. This means that there may not exist a permutation φ of Ω such that $\varphi(\alpha^g) = \varphi(\alpha)^g$; in terms of matrices, that there may not exist a permutation matrix B such that $BA(g)B^{-1} = A'(g)$, where $A(g)$ and $A'(g)$, $g \in G$, are the matrices of the permutations of Ω induced by the elements of G in the two actions. The above theorem states that if one allows the more general matrices with complex entries, then the equality of the characters is sufficient for the existence of a matrix that establishes the equivalence. In other words, two actions having the same character may not be equivalent as actions, but they are equivalent as representations over the complex numbers.

Let V be a vector space with basis e_t indexed by the element of a group G, $t \in G$. The mapping $\rho_s : e_t \to e_{ts}$ defines the *regular representation* of G. If $s \neq 1$, $st \neq t$, so that the diagonal entries of the matrix are all zero (no e_t is fixed). In particular, $\mathrm{tr}(\rho_s) = 0$. If $s = 1$, $\mathrm{tr}(\rho_1)$ is the trace of the identity matrix, and therefore equals the dimension of V, i.e. $|G|$.

Theorem 6.8. *Let φ be the character of the regular representation. Then:*
i) $\varphi(1) = |G|$;
ii) $\varphi(s) = 0,\ s \neq 1$.

Theorem 6.9. *Every irreducible representation of a finite group G is contained in the regular representation a number of times equal to its degree. In terms of characters,*

$$\varphi = \chi_1(1)\chi_1 + \chi_2(1)\chi_2 + \cdots \chi_r(1)\chi_r.$$

Proof. If φ is the character of the regular representation and χ that of the irreducible representation under consideration of degree $\chi(1)$, from (6.5) the sought-for number is (φ, χ). Now,

$$(\varphi, \chi) = \frac{1}{|G|}\sum_s \varphi(s)\overline{\chi(s)}.$$

But $\varphi(s) = 0$ if $s \neq 1$, and $\varphi(1) = |G|$, and therefore

$$(\varphi, \chi) = \frac{1}{|G|}\sum_s \varphi(1)\overline{\chi(1)} = \frac{1}{|G|}|G| \cdot \chi(1) = \chi(1),$$

as required. ◇

In particular, there are only a finite number of irreducible representations for a finite group.

Corollary 6.5. *The degrees n_i of the irreducible representations of a finite group G satisfy the relation:*

$$\sum_{i=1}^{r} n_i^2 = |G| \tag{6.7}$$

where r is the number of conjugacy classes of G.

Proof. By the previous theorem, and with the same notation, $\varphi = \sum_{i=1}^r n_i\chi_i$. Evaluating at $1 \in G$,

$$|G| = \varphi(1) = \sum_{i=1}^{r} n_i\chi_i(1) = \sum_{i=1}^{r} n_i n_i = \sum_{i=1}^{r} n_i^2.$$ ◇

From this corollary it follows that *a group G is abelian if, and only if, every irreducible character is linear*. Indeed, G is abelian if, and only if, $r = |G|$, i.e if, and only if, the sum (6.7) equals r. But $n_i \geq 1$ all i, so $r = |G|$ if, and only $n_i = 1$, all $i = 1, 2, \ldots, r$.

We shall see another property of the degrees the irreducible representations of a finite group: they divide of the order of the group.

In Chapter 3 we have called class function (or central function) a function f defined on a group which takes the same value on conjugate elements. By Theorem 6.2, *vi*), characters are class functions. In the space of complex valued functions defined on a group G consider a class function f; we define its *Fourier transform* $\widehat{f}$ *w.r.t. the representation* ρ, with representation space V, as the linear transformation $V \to V$:

$$\widehat{f}_\rho = \sum_{t \in G} f(t)\rho_t.$$

Theorem 6.10. *If ρ is an irreducible representation of degree n and character χ, then $\widehat{f}_\rho$ is the multiplication by λ, where*

$$\lambda = \frac{1}{n}\sum_t f(t)\chi(t) = \frac{|G|}{n}(f, \overline{\chi}).$$

Proof. (We write $\widehat{f}$ for $\widehat{f}_\rho$.) Let us show that $\widehat{f}\rho_s = \rho_s\widehat{f}$, $s \in S$:

$$\rho_s\widehat{f}\rho_s^{-1} = \sum_t f(t)\rho_s\rho_t\rho_s^{-1} = \sum_t f(t)\rho_{sts^{-1}}.$$

Setting $u = s^{-1}ts$ gives:

$$\rho_s^{-1}\widehat{f}\rho_s = \sum_u f(sus^{-1})\rho_u = \sum_u f(u)\rho_u = \widehat{f}.$$

By Schur's lemma, $\widehat{f}$ is the multiplication by a scalar λ, $\widehat{f} = \lambda I$, with trace $n\lambda$. It follows:

$$n\lambda = \mathrm{tr}(\widehat{f}) = \sum_t f(t)\mathrm{tr}(\rho_t) = \sum_t f(t)\chi(t),$$

from which

$$\lambda = \frac{1}{n}\sum_t f(t)\chi(t) = \frac{|G|}{n}(f, \overline{\chi}),$$

as required. 

The central functions with values in $\mathbf{C}$ form a vector space $\mathcal{H}$ over $\mathbf{C}$ for which the following result holds.

Theorem 6.11. *The characters $\chi_i, i = 1, 2, \ldots, r$, of the irreducible representations form an orthonormal basis of $\mathcal{H}$.*

Proof. We already know that these characters form an orthonormal system; we must show that this system is complete, that is, that an element f of $\mathcal{H}$ orthogonal to the χ_i is zero. For this purpose, consider $\widehat{\overline{f}} = \sum_t \overline{f(t)}\rho_t$, where

ρ is a representation. If ρ is irreducible, by the preceding theorem $\widehat{\overline{f}}$ is the multiplication by $\frac{|G|}{n}(f,\chi)$; by assumption, $(f,\chi)=0$, so that $\widehat{\overline{f}}=0$. If ρ is reducible, then $\rho=\rho'+\rho''+\cdots+\rho^{(h)}$, with the $\rho^{(i)}$ irreducible, so

$$\sum_t \overline{f(t)}\rho_t = \sum_t \overline{f(t)}(\rho'_t+\rho''_t+\cdots+\rho_t^{(h)}) = \sum_{t,i} \overline{f(t)}\rho_t^{(i)} = 0.$$

In any case, $\widehat{\overline{f}}=0$. Let ρ be the regular representation, and let us evaluate $\widehat{\overline{f}}$ at e_1:

$$0=\widehat{\overline{f}}.e_1=\sum_t \overline{f(t)}\rho_t.e_1=\sum_t \overline{f(t)}e_t.$$

The independence of the e_t now implies $\overline{f(t)}=0$, all t, so $\overline{f}=0$ and $f=0$.◊

Theorem 6.12. *The number of non equivalent irreducible representations of a group G equals the number of conjugacy classes of G.*

Proof. By the previous theorem, the dimension of the space $\mathcal{H}$ equals the number of irreducible characters of G. Since a central function is determined once its values on the conjugacy classes are arbitrarily assigned, the dimension of $\mathcal{H}$ also equals the number of these classes. ◊

Write C_s for be the conjugacy class of the element s, and let f_s be the function whose value is 1 on C_s and 0 on the other classes. It is a class function, so (Theorem 6.11) it is a linear combination of the characters χ_i: $f_s=\sum_{i=1}^r \alpha_i\chi_i$,

$$\alpha_i=(f_s,\chi_i)=\frac{1}{|G|}\sum_{t\in G} f_s(t)\overline{\chi_i(t)}=\frac{1}{|G|}\sum_{t\in C_s}\overline{\chi_i(t)}=\frac{1}{|G|}|C_s|\,\overline{\chi_i(t)},$$

and therefore $f_s(t)=\frac{|C_s|}{|G|}\sum_{i=1}^h \overline{\chi_i(s)}\chi_i(t)$. It follows:

$$\sum_{i=1}^r \chi_i(s)\overline{\chi_i(t)}=\begin{cases}\frac{|G|}{|C_s|}, & \text{if } t=s,\\ 0, & \text{if } t \text{ and } s \text{ are not conjugate.}\end{cases} \tag{6.8}$$

(6.8) is the *second orthogonality relation.*

We recall that an algebraic integer is the root of a monic polynomial with integer coefficients, and that sums and products of algebraic integers are again algebraic integers. We have:

Theorem 6.13. *Let χ be a character of a group G. Then $\chi(s)$ is an algebraic integer for all $s\in G$.*

Proof. $\chi(s)$ is the trace of the matrix ρ_s, and therefore is the sum of its eigenvalues. These are roots of unity (see the proof of Theorem 6.2, *ii*)) and therefore algebraic integers. ◊

Theorem 6.14. *Let $C_i, i = 1, 2, \ldots, r$, of order k_i, the conjugacy classes of a group G. Let χ be the character of an irreducible representation ρ, and let $\chi(x_i)$ be its value on C_i. Then,*

$$\frac{k_l\chi(x_l)}{n} \cdot \frac{k_p\chi(x_p)}{n} = \sum_{i=1}^{r} c_{l,p,i}\frac{k_i\chi(x_i)}{n}, \tag{6.9}$$

where n is the degree of χ and the $c_{l,p,i}$ are nonnegative integers. Moreover, $c_{l,p,i}$ gives the number of times an element of the class C_i can be written as a product of an element of C_l and one of C_p.

Proof. Let $M_i = \sum_{s\in C_i}\rho(s)$. For each $t \in G$ we have $\rho(t^{-1})\rho(s)\rho(t) = \rho(t^{-1}st) = \rho(s)$, and therefore $\rho(t^{-1})M_i\rho(t) = M_i$. The matrix M_i commutes with the matrices $\rho(t), t \in G$; this set being irreducible, M_i is scalar (Schur): $M_i = \text{diag}(\alpha_i)$, and $\text{tr}(M_i) = n\alpha_i$. But $\text{tr}(M_i) = \sum_{s\in K_i}\text{tr}(\rho(s)) = \sum_{i=1}^{k_i}\chi(x_i) = k_i\chi(x_i)$, so

$$\alpha_i = \frac{k_i\chi(x_i)}{n}. \tag{6.10}$$

Similarly, let $M_l = \sum_{s\in C_l}\rho(s)$, $M_p = \sum_{t\in C_p}\rho(s)$, and

$$M_lM_p = \sum_{s\in C_l, t\in C_p}\rho(st). \tag{6.11}$$

But for all $u \in G$,

$$\rho(u^{-1})M_lM_p\rho(u) = \rho(u^{-1})M_l\rho(u)\cdot\rho(u^{-1})M_p\rho(u) = M_lM_p,$$

and therefore in the sum (6.11), together with the $\rho(s)$ there also appear all the $\rho(u^{-1}su)$. It follows that the summands of (6.11) appear in blocks, each of which corresponds to a conjugacy class, and whose sum is an M_i. Let $c_{l,p,i} \geq 0$ denote the number of times in which the blocks corresponding to the class C_i appear; then $M_lM_p = \sum_{i=1}^{r} c_{l,p,i}M_i$. These matrices being scalar, $\alpha_l\alpha_p = \sum_{i=1}^{r} c_{l,p,i}\alpha_i$, i.e. (6.9). The number of ways in which $g \in C_i$ is a product st, $s \in C_l$, $t \in C_p$ does not depend on g but only on its conjugacy class C_i, and this is precisely the number of times in which the blocks corresponding to C_i appear in (6.11), i.e $c_{l,p,i}$. ♢

Corollary 6.6. *The $\alpha_i = k_i\chi(x_i)/n$ are algebraic integers.*

Proof. Consider the equalities $\alpha_l\alpha_p = \sum_{i=1}^{r} c_{l,p,i}\alpha_i$ as p varies, $p = 1, 2, \ldots, r$, for a fixed α_l:

$$\begin{aligned}
\alpha_l\alpha_1 &= c_{l,1,1}\alpha_1 + c_{l,1,2}\alpha_2 + \cdot + c_{l,1,r}\alpha_r,\\
\alpha_l\alpha_2 &= c_{l,2,1}\alpha_1 + c_{l,2,2}\alpha_2 + \cdot + c_{l,2,r}\alpha_r,\\
&\vdots\\
\alpha_l\alpha_r &= c_{l,r,1}\alpha_1 + c_{l,r,2}\alpha_2 + \cdot + c_{l,r,r}\alpha_r.
\end{aligned}$$

Thus we have a system of homogeneous linear equations in the α_p, with matrix

$$\begin{pmatrix} c_{l,1,1}-\alpha_l & c_{l,1,2} & \dots & c_{l,1,r} \\ c_{l,2,1} & c_{l,2,2}-\alpha_l & \dots & c_{l,2,r} \\ \vdots & \vdots & \vdots & \vdots \\ c_{l,r,1} & c_{l,r,2} & \dots & c_{l,r,r}-\alpha_l \end{pmatrix}.$$

The system admits a solution $\alpha_1, \alpha_2, \dots, \alpha_r$, where the α_i are not all zero (for instance $\alpha_1 = 1$) and therefore the determinant is zero. Setting $x = \alpha_l$, this determinant is a polynomial in x with integer coefficients (the $c_{l,p,i}$ are integers), with leading coefficient ± 1 and admitting the root α_l. This proves that α_l is an algebraic integer. ♢

Corollary 6.7. *The degree n of an irreducible representation divides the order of the group.*

Proof. With χ irreducible, $\sum_s \chi(s)\overline{\chi(s)} = |G|$, that we write

$$\sum_i \frac{k_i\chi(x_i)}{n} \cdot \overline{\chi(x_i)} = \frac{|G|}{n}.$$

By the previous corollary, the $\frac{k_i\chi(x_i)}{n}$ are algebraic integers, and so are the $\overline{\chi(x_i)}$. Hence $\frac{|G|}{n}$, being a sum of products of algebraic integers is an algebraic integer, and being a rational number is an integer; hence n divides $|G|$. ♢

Theorem 6.15. *The number of linear characters of a group G equals the index in G of the derived group G′.*

Proof. The representations of the abelian group G/G' are all linear; composition with the projection $\pi : G \to G/G'$ yields linear representations of G. Conversely, if $\rho : G \to \mathbf{C}^*$ is a linear representation of G, then ρ factorizes through $\rho' : G/G' \to \mathbf{C}^*$: $\rho = \rho'\pi$. The mapping $\rho \to \rho'$ is a bijection. ♢

We resume the relations existing between the degrees n_i of the irreducible representations and the order of the group G:

i) n_i divides $|G|$, $i = 1, 2, \dots, r$ (Corollary 6.7);

ii) $\sum_{i=1}^r n_i^2 = |G|$ (Corollary6.5),

where r is the number of conjugacy classes of G.

Examples 6.4. 1. The group S^3 has three conjugacy classes, and therefore three irreducible representations. There always is the principal one, of degree 1, and the alternating one, also of degree 1. We have seen one of degree 2 (*Ex.* 6.2, 2); its existence also follows from the equality $n_1^2 + n_2^2 + n_3^2 = 6$, because, being $n_1 = n_2 = 1$, the only possibility is $n_3 = 2$. The group S^4 has five conjugacy classes; the derived group of index 2, hence S^4 has two linear characters. The non principal one is the alternating character. From *ii*) above we see that the degrees of the remaining characters are 2, 3 and 3.

2. The group D_4 has five conjugacy classes, and so does the quaternion group $\mathcal{Q}$. For both groups the only possibility for the degrees is $1, 1, 1, 1, 2$.

3. Let C_a and C_b be the classes of the elements a and b, and let $c^x_{a,b}$ the number of times in which a given element x can be expressed as a product $a_i b_j$, $a_i \in C_a$, $b_j \in C_b$. Then $\sum_{i,j} \chi(a_i b_j) = \sum_{x \in G} c^x_{a,b}\chi(x)$. Let us consider $\chi(a_i b_j)$; since $a_i = y^{-1}ay$ for some y, $\chi(a_i b_j) = \chi(y^{-1}ayb_j) = \chi(a \cdot yb_jy^{-1}) = \chi(ab_k)$. If $h \neq j$, $\chi(a_i b_h) = \chi(y^{-1}ayb_h) = \chi(ayb_hy^{-1}) = \chi(ab_{h'})$, with $h' \neq k$. Hence in the following table the values on each row are equal to those on the first one, up to the order (set $a_1 = a$):

$$\begin{array}{cccc}
\chi(ab_1), & \chi(ab_2), & \ldots, & \chi(ab_s), \\
\chi(a_2b_1), & \chi(a_2b_2), & \ldots, & \chi(a_2b_s), \\
\vdots & \vdots & \ldots & \vdots \\
\chi(a_tb_1), & \chi(a_tb_2), & \ldots, & \chi(a_tb_s).
\end{array}$$

It follows that the sum of all the elements of the table equals that of the first row times $|C_a| = k_a$:

$$\sum_{i,j} \chi(a_i b_j) = k_a \sum_j \chi(ab_j). \tag{6.12}$$

Let us now consider the sum $\sum_{y \in G} \chi(ab^y)$. Two elements $y, y' \in G$ give rise to the same conjugate b if and only if they belong to the same coset of $\mathbf{C}_G(b)$. In other words, if $C_b = \{b_1, b_2, \ldots, b_s\}$, each b_i is obtained $|\mathbf{C}_G(b)| = \frac{|G|}{k_b}$ times:

$$\sum_{y \in G} \chi(ab^y) = \sum_j |\mathbf{C}_G(b)| \chi(ab_j) = \frac{|G|}{k_b} \sum_j \chi(ab_j).$$

Equality (6.12) then becomes:

$$\sum_{i,j} \chi(a_i b_j) = \frac{k_a k_b}{|G|} \sum_y \chi(ab^y),$$

and therefore $\sum_x c^x_{ab}\chi(x) = \frac{k_a k_b}{|G|} \sum_y \chi(ab^y)$. Hence (6.9) is equivalent to

$$\chi(a)\chi(b) = \frac{n}{|G|} \sum_y \chi(ab^y).$$

In particular, for $b = a^{-1}$,

$$\chi(a)\overline{\chi(a)} = \frac{n}{|G|} \sum_y \chi([a, y]).$$

4. From equality (6.9) an explicit expression of the number of ways in which an element x_q of a class C_q may be written as a product of an element of C_l and one of C_p can be drawn. Set $\chi = \chi_m$, multiply both sides of (6.9) by $\overline{\chi_m(x_q)}n_m$, where n_m is the degree of χ_m, and sum over m. Reversing the order of summation on the right hand side we have:

$$\sum_{i=1}^{r} c_{l,p,i}k_i \sum_{m=1}^{r} \chi_m(x_i)\overline{\chi_m(x_q)}.$$

By (6.8), if $i \neq q$ the terms of sum are zero, otherwise the sum equals $|G|/k_q$. The required expression:

$$c_{l,p,q} = \frac{k_l k_p}{|G|} \sum_{m=1}^{r} \chi_m(x_l)\chi_m(x_p)\overline{\chi_m(x_q)}$$

follows.

As in the case of the action of a group on a set, we now consider induced matrix representations. Let H be a subgroup of a group G, ρ a representation of H. One may think of obtaining a representation of G by simply defining ρ to be zero outside H. By this would give rise to singular matrices, and therefore not a representation of G. The right idea, due to Frobenius, goes as follows[1]. Let $t_1, t_2, \ldots, t_m$ be representatives of the cosets of H, and let $\rho^G(g)$ be the matrix whose (i,j) entry is the block $\rho(t_i g t_j^{-1})$, and 0 if $t_i g t_j^{-1} \notin H$.

Theorem 6.16. *The mapping $g \to \rho^G(g)$ is a matrix representation of G.*

Proof. We have to prove that $\rho^G(x)\rho^G(y) = \rho^G(xy)$, i.e.

$$\sum_{i=1}^{m} \rho(t_i x t_k^{-1})\rho(t_k y t_j^{-1}) = \rho(t_i x y t_j^{-1}). \tag{6.13}$$

If $t_i x y t_j^{-1} \notin H$, then the right-hand side of (6.13) is zero; we prove that the left-hand side is also zero. Indeed, for each k, either $t_i x t_k^{-1}$ or $t_k y t_j^{-1}$ does not belong to H, otherwise the product $t_i x t_k^{-1} \cdot t_k y t_j^{-1} = t_i x y t_j^{-1}$ belongs to H, contrary to assumption. Hence each summand is zero, and therefore the sum is zero. If $v = t_i x y t_j^{-1} \in H$, let $t_i x$ belong to the coset Ht_r, so $u = t_i x t_r^{-1} \in H$. It follows that $t_i x t_k^{-1} \notin H$ if $k \neq r$, so the left-hand side of (6.13) reduces to the summand for which $k = r$, i.e. $\rho(t_i x t_r^{-1})\rho(t_r x t_j^{-1})$. Now $t_r y t_j^{-1} = u^{-1}v$, an element of H; hence (6.13) becomes $\rho(u)\rho(u^{-1}v) = \rho(uv)$, which is true because ρ is a representation. ◇

The representation ρ^G of G is said to be *induced* by the representation ρ of H. Moreover, the matrix $\rho^G(g)$ is a *block-permutation* matrix (every

[1] See Section 3.4.

row and column contains only one non zero block $\rho(t_i g t_j^{-1})$). Indeed, $t_i x$ belongs to a unique coset Ht_r, so there is a unique element of the list $\rho(t_i g t_1^{-1}), \rho(t_i g t_2^{-1}), \ldots, \rho(t_i g t_m^{-1})$ of entries of row i for which $t_i g t_j^{-1}$ belongs to H, and this element is $t_i g t_r^{-1}$. Similarly for the other cases.

If χ is the character of the representation ρ of H, we denote χ^G the induced character of ρ^G. For $g \in G$, $\chi^G(g)$ is the trace of the matrix $\rho^G(g)$:

$$\chi^G(g) = \sum_{i=1}^{m} \chi(t_i g t_i^{-1}), \tag{6.14}$$

with the convention that $\chi(x) = 0$ if $x \notin H$. As proved in Theorem 3.23, *ii*), this formula can be written in the following form:

$$\chi^G(g) = \frac{1}{|H|} \sum_{x \in G} \chi(xgx^{-1}), \tag{6.15}$$

with $g \in G$ and $\chi(xgx^{-1}) = 0$ if $xgx^{-1} \notin H$. More generally, if φ is a class function on H, then (6.15) defines the induced class function φ^G.

To illustrate the theorem, let $G = S^3$, $H = \{I, (2,3)\}$, and let $t_1 = 1, t_2 = (1,2), t_3 = (1,3)$ be representatives of the cosets of H. Let ρ be a matrix representation of H of degree r, and let $R = \rho((2,3))$ be the $r \times r$ matrix representing the element (2,3) of H. Then the matrices:

$$\rho^G(I) = \begin{pmatrix} I & 0 & 0 \\ 0 & I & 0 \\ 0 & 0 & I \end{pmatrix}, \quad \rho^G((1,2)) = \begin{pmatrix} 0 & I & 0 \\ I & 0 & 0 \\ 0 & 0 & R \end{pmatrix}, \quad \rho^G((1,3)) = \begin{pmatrix} 0 & 0 & I \\ 0 & R & 0 \\ I & 0 & 0 \end{pmatrix}$$

$$\rho^G((2,3)) = \begin{pmatrix} I & 0 & 0 \\ 0 & 0 & R \\ 0 & R & 0 \end{pmatrix}, \quad \rho^G((1,2,3)) = \begin{pmatrix} 0 & 0 & R \\ I & 0 & 0 \\ 0 & I & 0 \end{pmatrix} \quad \rho^G((1,3,2)) = \begin{pmatrix} I & R & 0 \\ 0 & 0 & I \\ R & 0 & 0 \end{pmatrix},$$

yield a representation of S^3 of degree $3r$.

If ρ is a representation of a group G, then its restriction ρ_H to a subgroup H is a representation of H (but if ρ is irreducible, ρ_H need not be irreducible). If χ is the character of ρ, we denote χ_H the character of ρ_H. More generally, if φ is a class function defined on G we denote φ_H the restriction of φ to H.

Induction and restriction are related by the following theorem.

Theorem 6.17 (Frobenius reciprocity). *Let $H \leq G$, φ a class function on H and ψ a class function on G. Then:*

$$(\varphi^G, \psi) = (\varphi, \psi_H).$$

In particular, if φ and ψ are irreducible characters, then the multiplicity of ψ_H in φ equals that of ψ in φ^G.

Proof. From the definitions we have:

$$(\varphi^G, \psi) = \frac{1}{|G|} \sum_{g \in G} \varphi^G(g)\overline{\psi(g)} = \frac{1}{|G|}\frac{1}{|H|} \sum_{g \in G} \sum_{x \in G} \varphi(xgx^{-1})\overline{\psi(g)}.$$

Setting $y = xgx^{-1}$, and since $\psi(g) = \psi(y)$, we have:

$$\frac{1}{|G|}\frac{1}{|H|} \sum_{y \in G} \sum_{x \in G} \varphi^G(y)\overline{\psi(y)} = \frac{1}{|G|}\frac{1}{|H|}|G| \sum_{y \in H} \varphi(y)\overline{\psi(y)} = (\varphi, \psi_H). \qquad \diamondsuit$$

We give an application of the above result. Recall that for a group G acting on a set Ω the character of the permutation representation corresponding to the action is the function which associates with an element $g \in G$ the number of points of Ω fixed by g.

Lemma 6.2. *Let G act on Ω with permutation character χ. Then $(\chi, 1_G)$ equals the number of orbits of G on Ω.*

Proof. $(\chi, 1_G) = \frac{1}{|G|}\sum_{g \in G} \chi(g)\overline{1_G(g)} = \sum_{g \in G} \chi(g)$, which is the number of orbits by Burnside's counting formula (Theorem 3.18). $\diamondsuit$

Lemma 6.3. *Let G act transitively on a set Ω, and let $H = G_\alpha$ be the stabilizer of the point α. Then the induced character 1_H^G is the (permutation) character of G (i.e., $1_H^G(g)$ equals the number of points of Ω fixed by $g \in G$).*

Proof. $1_H(xgx^{-1}) = 1$ each time $xgx^{-1} \in H$, and this happens if and only if $\alpha^{xgx^{-1}} = \alpha$, i.e. $(\alpha^x)^g = \alpha^x$. Since $\alpha^{hx} = \alpha^x$, for all $h \in H$, each point fixed by g occurs with multiplicity $|H|$, and so does each conjugate $xgx^{-1} \in H$. Now $1_H^G(g) = \frac{1}{|H|}\sum_{x \in G} 1_H(xgx^{-1})$; if g fixes m points, then each term of the sum appears $|H|m$ times, so $1_H^G(g) = m$. $\diamondsuit$

By (6.14), if χ_1 is the principal character of a subgroup H, then $\chi_1^G(g) = 1$ each time $t_i g t_i^{-1} \in H$, i.e., each time $t_i g \in Ht_i$, i.e. g fixes the coset Ht_i under multiplication. Hence χ_1^G is the character of the representation of G on the right cosets of H (see *Ex.* 3.9, 2). We recall that the rank of a group G acting transitively on a set Ω is the number of orbits of the stabilizer of a point. If $H = \{1\}$, χ_1^G is the character of the regular representation (see *Ex.* 3.9, 1). This example shows that an induced character χ^G may be reducible even if χ is irreducible.

Theorem 6.18. *Let G be transitive on Ω with permutation character χ. Then the rank r of G is given by (χ, χ).*

Proof. Let H be the stabilizer of a point. The restriction χ_H is the character of H on Ω, and $(\chi_H, 1_H)$ the number of orbits of H by Lemma 6.2, i.e. the rank r of G. By Frobenius reciprocity the latter product equals $(\chi, 1_H^G)$ which, again by Lemma 6.2, equals the number of orbits of G. $\diamondsuit$

Corollary 6.8. *Let G be 2-transitive, with permutation character χ. Then*

$$\chi = 1_G + \varphi,$$

where φ is an irreducible character of G.

Proof. By Theorem 6.18 and formula (6.6), $2 = (\chi, \chi) = \sum_{i=1}^r m_i^2$. The m_i being integers, the only possibility is $m_i = 1$ for two values of i, and $m_i = 0$ for the other values, so χ is the sum of two irreducible characters. By Lemma 6.2, $(\chi, 1_G)$ equals the number of orbits of G, so $(\chi, 1_G) = 1$, i.e. 1_G occurs in χ (with multiplicity 1). Hence one of the two characters occurring in χ is 1_G, and the result follows. ◇

6.3 The Character Table

The irreducible characters of a group G may be presented in a square array of complex numbers, the *character table*, whose rows correspond to the characters (starting with the principal character) and the columns to the conjugacy classes (starting with the class $\{1\}$). The rows are pairwise orthogonal, and so are the columns, as follows from the first and second orthogonality relations ((6.4) and (6.8), respectively).

Examples 6.5. 1. *The table of S^3.* S^3 has two linear characters (*Ex.* 6.4, 1). The irreducible representation of degree 2 (*Ex.* 6.2, 2) affords a character whose value is zero on the class of transpositions and -1 on the class of 3-cycles. Hence the table:

	(1)	(1,2)	(1,2,3)
χ_1	1	1	1
χ_2	1	-1	1
χ_3	2	0	-1

If φ is the character of the regular representation, then $\varphi(1) = 6$ and $\varphi(g) = 0$ if $g \neq 1$. By Theorem 6.9, the latter contains once the two linear representations and twice that of degree 2: $\varphi = \chi_1 + \chi_2 + 2\chi_3$. For instance, on $g = (1, 2, 3)$ we have $\varphi(g) = \chi_1(g) + \chi_2(g) + 2\chi_3(g) = 1 + 1 + 2 \cdot -1 = 0$.

We now consider the induced characters on $G = S^4$ of the above characters of S^3. The elements of S^4 having conjugates in S^3 are the identity, the transpositions and the 3-cycles. Hence $\chi_i^G(g) = 0$ for the elements g with cycle structure of type (1,2)(3,4) and (1,2,3,4). As for the other elements, using formula (6.14) with $t_1 = 1, t_2 = (1, 4), t_3 = (2, 4)$ and $t_4 = (3, 4)$, we see that the transpositions admit two conjugates contained in S^3: for example, for (1,2) we obtain (1,2) (conjugating with $t_1 = 1$) and (1,3) (conjugating with $t_4 = (3, 4)$); for (2,4), we obtain (1,2) (conjugating with $t_2 = (1, 4)$) and (2,3) (conjugating with $t_2 = (1, 4)$). Hence for g a transposition we have $\chi_i^G(g) = 2\chi_i(g)$. The 3-cycles have only one conjugate in S^3 (e.g. (1,2,3) only has itself, and

(1,4,3) has (1,3,2)); it follows that for g a 3-cycle, $\chi_i^G(g) = \chi_i(g)$. Finally, $\chi_i(1) = 4\chi_i(1)$. Summing up,

	$\{1\}$	$(1,2)$	$(1,2,3)$	$(1,2)(3,4)$	$(1,2,3,4)$
χ_1^G	4	2	1	0	0
χ_2^G	4	−2	1	0	0
χ_3^G	8	0	−1	0	0

Note that $(\chi_1^G(1)^2 = 16$, $\chi_1^G((1,2))^2 = 4$, $(\chi_1^G((1,2,3))^2 = 1$, and that $\chi_1^G((1,2)(3,4)) = \chi_1^G((1,2,3,4)) = 0$. It follows $(\chi_1^G, \chi_1^G) = \frac{1}{24}(16 + 6 \cdot 4 + 8 \cdot 1) = 2$, so χ_1^G is not irreducible. Similarly, $(\chi_2^G, \chi_2^G) = 2$ and $(\chi_3^G, \chi_3^G) = 3$, so none of the induced character is irreducible. However, we can obtain an irreducible character of S^4 if we subtract from χ_1^G the principal character 1_G. Indeed, let $\chi = \chi_1^G - 1_G$ (it is a character by *ex.* 12), with values 3,1,0,–1,–1; we have $(\chi, \chi) = \frac{1}{24}(1 \cdot 3 \cdot 3 + 6 \cdot 1 \cdot 1 + 8 \cdot 0 \cdot 0 + 3 \cdot -1 \cdot -1 + 6 \cdot -1 \cdot -1) = 1$. It is the character of the permutation representation of S^4 on the space of dimension 3 (it is the character φ of Corollary 6.8).

The characters of S^3, and more generally those of S^n, are real because an element and its inverse are conjugate, hence $\overline{\chi(s)} = \chi(s^{-1}) = \chi(s))$. But there is more.

Let $s \in G$, $o(s) = m$, and consider the field $\mathbf{Q}(\lambda)$ obtained by adjoining a primitive m-th root of unity λ to $\mathbf{Q}$. The automorphism group $\mathcal{G}$ of $\mathbf{Q}(\lambda)$ over $\mathbf{Q}$ (the Galois group) is isomorphic to the group of integers less than m and coprime to m. For $\sigma \in \mathcal{G}$, $\sigma(\lambda) = \lambda^t$, $(t, m) = 1$, and the mapping $\sigma \to t$ is an isomorphism. Write $\sigma = \sigma_t$. For $s \in G$, $\chi(s)$ is a sum of m-th roots of unity, so $\chi(s) \in \mathbf{Q}(\lambda)$.

Lemma 6.4. *Let χ be a character of G. With the above notation, and $\sigma_t \in \mathcal{G}$,*

$$\sigma_t(\chi(s)) = \chi(s^t), \quad \forall s \in G.$$

Moreover, $\chi(s)$ is rational if, and only if, $\chi(s^t) = \chi(s)$, for all integers t coprime to m.

Proof. If λ_i are the eigenvalues of the matrix $\rho(s)$, those of $\rho(s^t)$ are the powers λ_i^t. Hence,

$$\sigma_t(\chi(s)) = \sigma_t(\sum \lambda_i) = \sum \sigma_t(\lambda_i) = \sum \lambda_i^t = \chi(s^t).$$

If $\chi(s)$ is rational, then it is invariant under all $\sigma_t \in \mathcal{G}$, hence $\chi(s) = \sigma_t(\chi(s))$ which equals $\chi(s^t)$. Conversely, if $\chi(s^t) = \chi(s)$, for all $\sigma_t \in \mathcal{G}$ then $\chi(s)$ is invariant for all $\sigma_t \in \mathcal{G}$ and therefore is rational. ◇

Theorem 6.19. *The characters of the symmetric group S^n are rational valued, and therefore integer valued.*

Proof. For $s \in S^n$, and t coprime to $o(s)$, s^t and s have the same cyclic structure, so they are conjugate, and therefore $\chi(s^t) = \chi(s)$. By the lemma, χ is rational valued. But the values of the characters are algebraic integers, and rational algebraic integers are ordinary integers. ◇

2. *The table of* A^4. The derived group of A^4 is the Klein group, hence there are three linear characters. Moreover, we have four conjugacy classes: C_1; C_2, which contains the elements of order 2; C_3 and C_4, both with elements of order 3. The equality $1 + 1 + 1 + n_4^2 = 12$ implies $n_4 = 3$. The elements of order 2 belong to the Klein group, and therefore also to the kernel of every representation of degree 1. Hence the first column of the table is 1,1,1,3. The second one has 1,1,1 in the first three entries; by the orthogonality with the first the fourth entry is –1:

	1	(1,2)(3,4)	(1,2,3)	(1,3,2)
χ_1	1	1	1	1
χ_2	1	1	x	
χ_3	1	1	y	
χ_4	3	-1	z	

The orthogonality of the third column with the first two yields

$$1 + x + y + 3z = 0,\ 1 + x + y - z = 0,$$

from which $z = 0$. The same argument applied to the fourth column yields $\chi_4(C_4) = 0$. The orthogonality between columns 1 and 3 and between 2 and 3 now yields $1+x+y = 0$. Next, the elements of C_3 are of order 3, and therefore x and y are third roots of unity, and they are distinct; set $x = w$ and $y = w^2$:

	1	(1,2)(3,4)	(1,2,3)	(1,3,2)
χ_1	1	1	1	1
χ_2	1	1	w	
χ_3	1	1	w^2	
χ_4	3	-1	0	0

Now let x and y the values of χ_2 and χ_3 on C_4, and consider the orthogonality between the first two rows:

$$1 \cdot 1 + 3(1 \cdot 1) + 4(1 \cdot w) + 4(1 \cdot x) = 0,$$

from which $1 + w + x = 0$, i.e. $x = w^2$. Similarly, $y = w$. Summing up:

	1	(1,2)(3,4)	(1,2,3)	(1,3,2)
χ_1	1	1	1	1
χ_2	1	1	w	w^2
χ_3	1	1	w^2	w
χ_4	3	-1	0	0

Consider the permutation representation of degree 4 of A^4. Since A^4 is 2-transitive, its character χ equals $1_{A^4} + \varphi$, with φ irreducible. Hence $4 = \chi(1) = 1_{A^4}(1) + \varphi(1) = 1 + \varphi(1)$, i.e. φ is of degree 3. It is character χ_4 of the above table, and is the character of the irreducible representation on the 3-dimensional space of *Ex.* 6.2, 2.

If we induce the character χ_1 on S^4 we find an irreducible character of S^4 (we have $\frac{1}{24}(1{\cdot}2^2{+}3{\cdot}2^2{+}4{\cdot}1^2{+}4{\cdot}1^2) = 1$). The same holds for χ_2 and χ_3 but not for χ_4. Note that $\chi_2((1,2,3))$ being different from $\chi_2((1,2,3)^2) = \chi_2((1,3,2))$, is irrational (Lemma 6.4).

3. $G = A^5$ has five conjugacy classes:

	1	(1,2)(3,4)	(1,2,3)	(1,2,3,4,5))	(1,3,5,2,4)
χ_1	1	1	1	1	1

In the representation as a permutation group, the characters (number of fixed points) are respectively, 5,1,2,0,0. A^5 being 2-transitive, the representation of degree 4 of *Ex.* 6.2, 2 is irreducible, and the character has values 4,0,1,−1,−1 obtained by the above by subtracting the trivial representation on the subspace of dimension 1. Hence we may add a second row to the above table:

	1	(1,2)(3,4)	(1,2,3)	(1,2,3,4,5))	(1,3,5,2,4)
χ_1	1	1	1	1	1
χ_2	4	0	1	−1	−1

Let χ be a non principal linear character of A^4. From the character table of $H = A^4$ we see that $\chi(g) = 1$ if $o(g) = 1$ or 2, $\chi((1,2,3)) = w$, $\chi((1,3,2)) = w^2$. If g is a 3-cycle, then $|\mathbf{C}_G(g)| = |\mathbf{C}_H(g)| = 3$ and $\chi^G(g) = 3\cdot(\frac{w}{3} + \frac{w^2}{3}) = w + w^2 = -1$. Since $(\chi^G, \chi^G) = \frac{1}{60}(25 + 15 + 20) = 1$, χ^G is irreducible. We get one more row of the table:

	1	(1,2)(3,4)	(1,2,3)	(1,2,3,4,5))	(1,3,5,2,4)
χ_1	1	1	1	1	1
χ_2	4	0	1	−1	−1
χ_3	5	1	−1	0	0

The sum of the squares of the degrees must equal the order of the group. Hence $1^2 + 4^2 + 5^2 + a^2 + b^2 = 60$, $a^2 + b^2 = 18$, and the only possibility in integers is $a = b = 3$. Hence we have two irreducible representations of degree 3, χ_4 and χ_5. Let a and b be the values of χ_4 and χ_5 on (1,2)(3,4). By the second orthogonality relation, $1\cdot 1 + 4\cdot 0 + 5\cdot 1 + 3a + 3b = 0$, and hence $a + b = -2$. Again by the the second orthogonality relation, $\sum_{i=1}^5 |\chi_i((1,2)(3,4))|^2 = 1 + 0 + 1 + a^2 + b^2 = \frac{|G|}{|\mathbf{Cl}_G((1,2)(3,4))|} = \frac{|G|}{15} = 4$, from which $a^2 + b^2 = 2$. Hence $a = b = -1$. Similarly, if c and d are the values of χ_4 and χ_5 on (1,2,3) we have $\sum_{i=1}^5 |\chi_i((1,2,3))| = 1 + 1 + 1 + c^2 + d^2 = \frac{|G|}{|\mathbf{Cl}_G((1,2,3))|} = \frac{|G|}{20} = 3$. Hence $c^2 + d^2 = 0$, and $c = d = 0$. At this point:

	1	(1,2)(3,4)	(1,2,3)	(1,2,3,4,5))	(1,3,5,2,4)
χ_1	1	1	1	1	1
χ_2	4	0	1	−1	−1
χ_3	5	1	−1	0	0
χ_4	3	−1	0		

Let $c = (1,2,3,4,5)$ and $d = (1,3,5,2,4)$. Let $a = \chi_4(c)$, and $b = \chi_4(d)$; row-orthogonality yields

$$0 = \sum_{i=1}^{5} k_i \chi_4(s_i) = 1 \cdot 3 + 15 \cdot -1 + 20 \cdot 0 + 12a + 12b,$$

i.e. $a + b = 1$. Also,

$$60 = \sum_{i=1}^{5} k_i \chi_4(s_i)^2 = 1 \cdot 9 + 15 \cdot 1 + 20 \cdot 0 + 12a^2 + 12b^2$$

i.e. $a^2 + b^2 = 3$ (the characters are real because every element is conjugate to its inverse). Hence $a^2 - a - 1 = 0$ so $a = (1+\sqrt{5})/2$, and $b = (1-\sqrt{5})/2$ (or conversely). As in the case of A^4, $\chi_4((1,2,3,4,5))$ is different from $\chi_4((1,3,5,2,4))$ and so is irrational. Now see *ex.* 14.

4. *The table of D_4.* $D_4 = \{1, a, a^2, a^3, b, ab, a^2, a^3b\}$ has five conjugacy classes, and the derived group C_2 has order 2, so D_4 has four linear characters, and since the sum of the squares must be 8, the fifth one has degree 2. Moreover, D_4/C_2 is a Klein group, so its elements have order 2 or 1. But in the group $\mathbf{C}^*$ only −1 has order 2, so the image of a linear representation (not ρ_1) is $\{1, -1\}$, and therefore the kernel has order 4. The three subgroups of order 4 are possible kernels, and in fact we have three homomorphism:

$$\rho_2 : \{1, a, a^2, a^3\} \to 1, \ \{b, ab, a^2b, a^3b\} \to -1,$$

$$\rho_3 : \{1, a^2, b, a^2b\} \to 1, \ \{a, a^3, ab, a^3b\} \to -1,$$

$$\rho_4 : \{1, a^2, ab, a^3b\} \to 1, \ \{a, a^3, b, a^2b\} \to -1,$$

that allow us to complete the first four rows of the character table:

	$\{1\}$	$\{a^2\}$	$\{a, a^3\}$	$\{b, a^2b\}$	$\{ab, a^3b\}$
χ_1	1	1	1	1	1
χ_2	1	1	1	−1	−1
χ_3	1	1	−1	1	−1
χ_4	1	1	−1	−1	1
χ_5	2	−2	0	0	0

where the last row follows from the orthogonality relations. With the same proof it can be shown that the quaternion group has the same character table of D_4. Hence, the character table does not determine the group.

In the dihedral group D_n, the commutator subgroup has order $n/2$ or n, according as n is even or odd. Hence D_n has four or two representations of degree 1, respectively. All non linear irreducible representations have degree 2; there are $n/2 - 1$ (n even) or $(n-1)/2$ (n odd) of them[2].

6.3.1 Burnside's Theorem and Frobenius Theorem

Let us now see two important application of what we have seen above. The first is a famous theorem due to Burnside, know as the "$p^a q^b$ theorem",

First two lemmas.

Lemma 6.5. *Let χ be an irreducible character of degree n, C a conjugacy class of cardinality coprime to n. Then for $s \in C$ we have either $\chi(s) = 0$ or $|\chi(s)| = n$, and in the latter case ρ_s is a scalar transformation.*

Proof. First we prove that $\frac{\chi(s)}{n}$ is an algebraic integer. Let a and b be integers such that $a|C| + bn = 1$; multiplying this equality by $\chi(s)$ and dividing by n yields $\frac{a|C|\chi(s)}{n} + b\chi(s) = \frac{\chi(s)}{n}$. The first summand is an algebraic integer (Corollary 6.6), and so is $b\chi(s)$, and therefore also $\frac{\chi(s)}{n}$. Moreover, $\chi(s)$ is a sum of $|G|$-th roots of unity:

$$\lambda = \frac{\chi(s)}{n} = \frac{w_1 + w_2 + \cdots + w_n}{n}.$$

Let $\lambda_i = \frac{w_1^i + w_2^i + \cdots + w_n^i}{n}$, $i = 2, \ldots, |G|$, $\lambda_1 = \lambda$; then:

$$|\lambda_i| = |\frac{w_1^i + w_2^i + \cdots + w_n^i}{n}| \leq \frac{|w_1^i| + |w_2^i| + \cdots + |w_n^i|}{n} = \frac{n}{n} = 1.$$

Consider the polynomial $f(x) = \prod_{i=1}^{|G|}(x - \lambda_i)$. This polynomial has coefficients that are symmetric functions of the λ_i, and therefore polynomials with rational coefficients in the elementary symmetric functions of $w_1, w_2, \ldots, w_{|G|}$, i.e. in the coefficients of $x^{|G|} - 1$, that are equal to 0, 1 or –1. Hence the coefficients of $f(x)$ are rational numbers. $f(x)$ has the root λ_1 in common with $p(x)$, the minimum polynomial of λ; the latter being irreducible, $p(x)$ divides $f(x)$. The other roots of $p(x)$ also are roots of $f(x)$, so they are among the λ_i and therefore have absolute value less than or equal to 1. But the product of all these roots is the constant term of $p(x)$, an integer, a_m say[3]. If $a_m = 0$, irreducibility implies $p(x) = x$ and therefore $\lambda = 0$. This proves the first part. If $|a_m| = 1$, then $|\lambda| = 1$ (and the other $|\lambda_i| = 1$ as well), so that $\frac{|\chi(s)|}{n} = 1$ and $|\chi(s)| = n$. By Theorem 6.2, *iii*), ρ_s is scalar. ◇

---

[2] Ledermann, p. 65–66.

[3] It follows from a theorem of Gauss that the minimum polynomial of an algebraic integer has integer coefficients and is monic.

Lemma 6.6. *If in a group G a conjugacy class C contains a number of elements which is a power of a prime p, then the group is not simple.*

Proof. Since $|G| = 1 + n_2^2 + \cdots + n_r^2$, and $p||G|$, there exists n_k such that $(p, n_k) = 1$, so that, by the assumption on C, $(|C|, n_k) = 1$. By the previous lemma, for $1 \neq s \in C$ we have either $\chi(s) = 0$ or $|\chi(s)| = n_k$, and, in the latter case ρ_s is scalar, it commutes with all $\rho_t, t \in G$, and therefore belongs to the center of $\rho(G)$. If ρ is faithful, s belongs to the center of G; otherwise ρ has a nontrivial kernel. In either case G is not simple.

If $\chi(s) = 0$, let φ be the regular representation of G; we know that $\rho(1) = |G|$, and $\rho(s) = 0, s \neq 1$. Then $\varphi(s) = \sum_i n_i\chi_i(s) = 0$ and $1 + \sum_{i \neq 1} n_i\chi_i(s) = 0$. If $p \nmid n_i$, then either $|\chi_i(s)| = n_i$, and for what we have seen above G is not simple, or $\chi_i(s) = 0$. If $\chi_i(s) = 0$ for all i, then we would have $1 = 0$. The only non zero terms of the sum come from the n_i such that $p|n_i$. Let $n_i = pk_i$; then $0 = 1 + \sum pk_i\chi_i(s) = 1 + p(\sum k_i\chi_i(s))$. Now, the $\chi_i(s)$ are algebraic integers, the h_i are integers, and therefore the above sum is a sum, a say, of products of algebraic integers, and so is an algebraic integer. From $1 + pa = 0$ it follows that $a = -\frac{1}{p}$ is rational, and therefore integer, a contradiction. ◇

Theorem 6.20 (Burnside). *A group of order $p^a q^b$ is solvable.*[4]

Proof. If G is a p-group there is nothing to prove. Let $a, b > 0$, S a Sylow p-subgroup of G, and let $1 \neq x \in \mathbf{Z}(S)$. Then $\mathbf{C}_G(x)$ contains S and therefore $|\mathbf{cl}(x)| = [G : \mathbf{C}_G(x)]$ divides q^b. By the lemma above, G is not simple. If N is a normal subgroup of G, then by induction, N and G/N are solvable, and therefore so is G. ◇

If a character is not irreducible, then it is a linear combination of irreducible characters, with positive integer coefficients. If we allow negative or zero coefficients, we have a class function (but not the character of a representation if some of the coefficients are negative) called a *virtual* or a *generalized* character. Formula (6.6) also holds for virtual characters. However, Theorem 6.6, *ii*), should be modified as follows:

Lemma 6.7. *Let α be a virtual character such that $(\alpha, \alpha) = 1$. Then $\alpha = \pm\chi$, where χ is an irreducible character. Hence, if $\alpha(1) > 0$, then $\alpha = \chi$.*

We now prove the important result due to Frobenius announced in Section 5.4. Unlike Burnside's theorem, so far no proof is known that does not make use of character theory.

Theorem 6.21 (Frobenius). *The elements fixing no point in a Frobenius group G acting on a set of n points form, together with the identity, a normal subgroup K.*

---

[4] For a long time this results has defeated many attempts to prove it without character theory. In the 1970's, proofs using deep group-theoretic results have been found.

Proof. (Clearly the issue is not on normality, but rather the fact that K is a subgroup.) The idea of the proof is to construct a character of G and to show that K is the set of elements of G on which the value of this character equals the degree. The result will then follow from Theorem 6.3. Let H be the stabilizer of a point, and let $t_1 = 1, t_2, \ldots, t_m$ be a transversal of the right cosets of H. Then the stabilizers of the other points are the conjugates $t_2^{-1}Ht_2, \ldots, t_m^{-1}Ht_m$ of H.

Let ψ be an irreducible character of H, and let ψ^G be a transversal; we have $\psi^G(1) = n\psi(1)$. Consider now the permutation representation of G; it is the sum of the trivial representation and a representation of degree $n-1$ (*Ex.* 6.2, 2). If φ is the character of the latter, then $\varphi(1) = n-1$. The elements outside H and its conjugates fix no point, so in the permutation representation their character is zero; hence the value of φ at these points is -1, and is 0 elsewhere.

If we subtract $\psi(1)$ times the above character φ from ψ^G we obtain the virtual character

$$\chi = \psi^G - \psi(1)\varphi. \tag{6.16}$$

Note that from $\psi^G(1) = n\psi(1)$ it follows:

$$\chi(1) = \psi^G(1) - \psi(1)(n-1) = \psi(1),$$

i.e. the degree of χ equals that of ψ. Moreover, for $x \in K$,

$$\chi(x) = \psi^G(x) - \psi(1)\varphi(x) = 0 - \psi(1) \cdot -1 = \psi(1).$$

Hence, for $x \in K$,

$$\chi(x) = \chi(1).$$

We now show that χ is an actual character, and is irreducible. Let us calculate

$$\sum_{x \in G} (\psi^G(x) - \psi(1)(\varphi(x))(\overline{\psi^G(x)} - \psi(1)\overline{\varphi(x)}). \tag{6.17}$$

We first sum over the elements of H; recalling that $\varphi(h) = 0$ for $1 \neq h \in H$,

$$\sum_{1 \neq h \in H} \psi(h)\overline{\psi(h)} = |H| - \psi^2(1),$$

because ψ is irreducible. Now consider (6.14), and let x be an element of a conjugate $t_j^{-1}Ht_j$ of H; then $x = t_j^{-1}ht_j$, so (6.14) becomes:

$$\psi^G(x) = \sum_i \psi(t_i t_j^{-1} h t_j t_i^{-1}). \tag{6.18}$$

The i-th term of the sum is nonzero only if $t_i t_j^{-1} h t_j t_i^{-1} \in H$, i.e. $t_j^{-1}ht_j \in t_i^{-1}Ht_i$, which is possible only if $j = i$. Hence (6.18) reduces to

$$\psi^G(t_j^{-1}ht_j) = \psi(h),\ j = 1, 2, \ldots, m.$$

For the elements x outside H and its conjugates $\psi^G(x) = 0$ and $\varphi(x) = -1$, so their contribution to the sum (6.17) is $\psi(1)^2$. Hence,

$$\psi(1)^2 + n(|H| - \psi(1)^2) + (n-1)\psi(1)^2 = n|H| = |G|,$$

so $(\chi, \chi) = 1$, and χ is an irreducible character of G (Lemma 6.7).

Hence χ is a character, and we have seen that its value on the elements of K is $\chi(1)$. However we cannot conclude that this is the character we seek because there might be other elements y of G, not in K, for which $\chi(y) = \chi(1)$. To overcome this difficulty, we consider characters of G like (6.16), this time taking into account all the irreducible characters χ_i of H:

$$\chi^{(i)} = \chi_i^G - \chi_i(1)\varphi, \quad i = 1, 2, \ldots, n.$$

Like the above χ the $\chi^{(i)}$ are irreducible; we construct the character:

$$\eta = \sum_i \chi_i(1)\chi^{(i)}. \tag{6.19}$$

We have:

$$\chi^{(i)}(1) = \chi_i(1), \ \ \chi_i^{(i)}(t_j^{-1}ht_j) = \chi_i(h), \ \ \chi^{(i)}(k) = \chi_i(1),$$

where $h \in H$, and $k \in K$. Substituting in (6.19) we have:

$$\eta(1) = \eta(k) = \sum \chi_i(1)^2 = |H|,$$

(Corollary 6.5) and

$$\eta(h) = \sum_i \chi_i(1)\chi_i(h) = \rho(h) = 0,$$

because the sum is the regular representation ρ of H (Theorem 6.8). Hence K is the set of elements of G for which the value of η is $\eta(1)$, so is the required normal subgroup. $\diamondsuit$

Remark 6.2. A different virtual character is considered by M. Hall, p. 292–294, that is, $\omega = \rho - h\varphi$, $h = |H|$, where ρ is the character of the regular representation of G and φ is as in the theorem above. Then $\omega(1) = \rho(1) - h\varphi(1) = nh - (n-1) = h$; for $x \in H$ and its conjugates, $\omega(x) = \rho(x) - h\varphi(x) = 0 - h \cdot 0 = 0$ and, for $y \in K$, $\omega(y) = \rho(y) - h\varphi(y) = 0 - h \cdot -1 = h$. Hence, if ω is a character, it as the required properties, so what one has to prove is that this virtual character is an actual character. Isaacs, pp. 100–101, considers for an irreducible non principal character φ of H, the virtual character $\theta = \varphi - \varphi(1)1_H$, and proves that $\varphi^* = \theta^G + \varphi(1)1_G$ is an irreducible character of G. The required normal subgroup will be the intersection of the kernels of the φ^* as φ runs over all the irreducible characters of H. Our treatment follows Ledermann, pp. 153–157.

Exercises

1. If ρ is a representation of G on V, prove that ρ^* defined by $\rho_s^*(f).v = f(\rho_{s^{-1}}.v)$, where $f \in V^*$, the dual of V, is a representation $G \to GL(n, V^*)$. If A is the matrix of ρ_s, then the matrix of ρ^* is the inverse transpose of A (*contragredient* representation).

2. Prove that if V is of dimension n, and ρ is irreducible, then ρ_s^* is also irreducible. [*Hint*: consider an invariant subspace W of V^*, and prove that its annihilator $W^0 = \{v \in V \mid f(v) = 0, f \in W\}$ is also invariant, and use the equality $\dim W + \dim W^0 = n$.]

3. The representation of the reals given in *Ex.* 6.1, 3, is not faithful. Determine its kernel.

4. Let V be the space of polynomials with real coefficients, and let ρ_s be the linear transformation of V defined by $\sigma_s(f).v = f(v - s)$. Prove that ρ_s is a linear transformation of the additive group of the reals.

5. *i)* Let χ be a character of a group G such that $\chi = \sum n_i\chi_i$, with the χ_i irreducible. Then $\ker(\chi)$ (Theorem 6.3) equals $\cap ker(\chi_i)$, $n_i > 0$. [*Hint*: Theorem 6.2 *ii*)).]

ii) The intersection $\cap ker(\chi_i)$ of all the irreducible characters of G is the identity. [*Hint*: consider the character of the regular representation.]

6. Prove that a finite p-group admits a faithful irreducible representation if, and only, if the center is cyclic. [*Hint*: there is a unique subgroup of order p in the center.]

7. Using the results of this chapter prove that a group of order p^2 is abelian.

8. *i)* A class function φ is a character if and only if $\varphi \neq 0$ and the scalar products with the irreducible characters are non negative integers;

ii) use *i)* and Frobenius reciprocity to prove that if φ is a character of a subgroup H of a group G, then φ^G, as defined by formula (6.15), is a character of G. [*Hint*: $[\varphi, \chi_H]$ is a nonnegative integer.]

9. Let G be a permutation group, and let $\chi(g)$ be the number of fixed points of $g \in G$. Prove that the function $\nu(g) = \chi(g) - 1$ is a character of G.

10. Let λ be a linear character of G, and let χ be an irreducible character. Prove that the product $\chi\lambda$ defined by $\chi\lambda.(g) = \chi(g)\lambda(g)$ is also an irreducible character. [*Hint*: recall that if α is a root of unity, then $\alpha\overline{\alpha} = 1$.]

11. Determine the character table of S^4.

12. Let $H \leq G$. Prove that:

i) $(1_H^G, 1_G) = 1$;

ii) $1_H^G - 1_G$ is a character.

13. Prove that with $H = A^4$ and $G = A^5$, the character $1_H^G - 1_G$ is the same as the character of the permutation representation of A^5 on the space of dimension 4.

14. Complete the character table of A^5.

15. (Solomon) Prove that the sum of the elements on a row of the character table is a non negative integer. [*Hint*: as in the regular representation, let v_s, $s \in G$, be the vectors of a basis for V and let ρ be the representation of the group induced by conjugation, $\rho_t : v_s \to v_{t^{-1}st}$; then $\chi(t) = |\mathbf{C}_G(t)|$. Consider the number of times the representation ρ_i with character χ_i appears in ρ.]

6.3.2 Topological Groups

In order to prove Maschke's theorem on complete reducibility we have shown that there exists a G-invariant complement W_0 of the G-invariant subspace $W \subset V$ by constructing a projection $\pi : V \to W$ that commutes with the action of the group G. This is obtained by averaging over the group the projection π associated with the decomposition $W \oplus V'$, where V' is any complement of W. In the case of infinite groups complete reducibility may not be achieved, as the example of the integers shows (*Ex.* 6.3, 3). In this section we sketch the possibility of extending the mentioned results to infinite groups. Instead of finite groups, we shall be considering compact topological groups. If G is such a group, then a measure μ can be defined a on the subsets of G such that $\mu(G)$ is finite. In this way, the finiteness of G is replaced by the finiteness of $\mu(G)$. In the discrete case, compactness means that the group is finite, the invariant measure is the counting measure, and $\mu(G)$ is simply the cardinality of G. First some definitions and examples.

Definition 6.5. A *topological group* G is a group endowed with a structure of a Hausdorff topological space with respect to which the mappings from the product spaces $G \times G$ to G given by $(g, h) \to gh$, and from G to G given by $g \to g^{-1}$ are continuous.

Any group becomes a topological group if endowed with the discrete topology.

Examples 6.6. **1.** The group $\mathbf{R}$ of the real numbers with the topology of the real line is a topological group. The two functions $(x, y) \to x + y$ and $x \to -x$ are continuous. Similarly, the set $\mathbf{R}^n$ of the n-tuples $(x_1, x_2, \ldots, x_n)$ with the usual topology is a topological group, and so is the set $\mathbf{C}^n$ of the n-tuples of complex numbers $(z_1, z_2, \ldots, z_n)$, $z_k = x_k + iy_k$, where x_k, y_k are real numbers, with componentwise sum. With such an n-tuple there is associated the $2n$-tuple $(x_1, y_1, x_2, y_2, \ldots, x_n, y_n)$ of $\mathbf{R}^{2n}$; this correspondence is bijective, and allows one to consider as open sets of $\mathbf{C}^n$ the inverse images of the open sets of $\mathbf{R}^{2n}$. These are the *natural* topologies of $\mathbf{R}^n$ and $\mathbf{C}^n$.

2. The groups $GL(n, \mathbf{R})$ and $GL(n, \mathbf{C})$. A matrix $A = (a_{i,j})$ of $GL(n, \mathbf{R})$ may be considered as a point $(a_{1,1}, \ldots, a_{1,n}, a_{2,1}, \ldots, a_{2,n}, \ldots, a_{n,1}, \ldots, a_{n,n})$ of $\mathbf{R}^{n^2}$, and therefore one can endow $GL(n, \mathbf{R})$ with the topology induced by that of $\mathbf{R}^{n^2}$. In this way, $GL(n, \mathbf{R})$ becomes a topological space, and indeed a topological group. The inverse of a matrix $A = (a_{i,j})$ is $\frac{1}{\det(A)}(\Delta_{i,j})$, where $\Delta_{i,j}$ is the matrix of the cofactors, which are determinants and hence polynomials in the entries of the matrix, and therefore continuous functions of the latter. Similarly, the row-column products of the product matrices are polynomials in the entries; hence $GL(n, \mathbf{R})$ is a topological group w.r.t. the natural topology. A basis of the neighborhoods of a matrix $A = (a_{i,j})$ is given by the invertible matrices $B = (b_{i,j})$ for which, given $\epsilon > 0$, $|a_{i,j} - b_{i,j}| < \epsilon$, $i, j = 1, 2, \ldots, n$. Note that $GL(n, \mathbf{R})$ is an open set of $\mathbf{R}^{n^2}$; indeed, the function that associates with the matrix its determinant is a continuous function $GL(n, \mathbf{R}) \to \mathbf{R}$ and $GL(n, \mathbf{R})$ is the inverse image of the complement of the closed set $\{0\}$ of $\mathbf{R}$. This also holds, with obvious modifications, for $GL(n, \mathbf{C})$.

3. A subgroup of a topological group becomes a topological group by endowing it with the induced topology. With this topology, $SL(n, \mathbf{R})$ is a closed subgroup of $GL(n, \mathbf{R})$ because is the counter-image of the closed subgroup $\{1\}$ under the continuous mapping determinant. The *orthogonal* group $O(n)$ consists of the matrices $A = (a_{i,j})$ of $GL(n, \mathbf{R})$ such that $AA^t = I$, that is, $\sum_k a_{i,k}a_{j,k} = \delta_{i,j}$ (rows are orthonormal). These sums are continuous, and $O(n)$ is a closed subgroup of $GL(n, \mathbf{R})$; its intersection with $SL(n, \mathbf{R})$ yields the closed subgroup $SO(n)$, the *special orthogonal group*. Similarly, the *unitary group* $U(n)$[5] (matrices such that $A\bar{A}^t = I$) is a closed subgroup of $GL(n, \mathbf{C})$, as well as its subgroup $SU(n)$ obtained by intersection with $SL(n, \mathbf{C})$.

4. The group $O(n)$ is compact. Indeed, due to the orthonormality condition, the sum of the squares of the elements of each row of a matrix of $O(n)$ is 1, and therefore the sum of the squares of the elements of the matrix is n. Hence a matrix of $O(n)$ is a point on the surface of the sphere centered at the origin with radius $\sqrt{n}$, so that $O(n)$ is a bounded set of $\mathbf{R}^{n^2}$ and therefore compact. Similarly, the unitary group is compact.

5. The group $SO(n)$, as a closed subset of the compact space $O(n)$, is itself compact. Let us consider the case $n = 2$. Let $\begin{pmatrix} a_{1,1} & a_{1,2} \\ a_{2,1} & a_{2,2} \end{pmatrix} \in SO(n)$. The orthogonality condition implies the equalities:

$$a_{1,1}^2 + a_{1,2}^2 = 1,\ a_{2,1}^2 + a_{2,2}^2 = 1,\ a_{1,1}a_{2,1} + a_{1,2}a_{2,2} = 0;$$

we have the further condition $a_{1,1}a_{2,2} - a_{1,2}a_{2,1} = 1$ given by the fact the the determinant is 1. The first condition allows one to set $a_{1,1} = \cos\alpha$ and $a_{1,2} = -\operatorname{sen}\alpha$, $\alpha \in [0, 2\pi]$; from the other conditions it follows that the matrix is $\begin{pmatrix} \cos\alpha & -\sin\alpha \\ \sin\alpha & \cos\alpha \end{pmatrix}$. Hence the group $SO(2)$ consists of the rotation matrices (rotations around the origin of the plane). If we make correspond with the matrix of a rotation of an angle θ the complex number $e^{i\theta}$ we have an isomorphism φ between $SO(2)$ and the group $S^1 = \mathbf{R}/2\pi\mathbf{Z}$ of the circle with center at the origin and radius 1. By defining the open sets of S^1 as the images under φ of the open sets of $SO(2)$ we have a homeomorphism.

---

[5] This symbol has already been used with a different meaning (see p. 21).

Definition 6.6. A finitely dimensional (complex) representation of a topological group is a continuous homomorphism into the topological group $GL(n, \mathbf{C})$.

Example 6.7. In this example we see how one can define an average operation similar to (6.1) of Maschke's theorem in the case of a representation of an infinite group. Consider the group S^1, and let f be a continuous function defined on S^1 with values on a space V. Define the *mean* of f over S^1 to be the integral

$$\langle f \rangle = \frac{1}{2\pi} \int_0^{2\pi} f(e^{i\theta}) d\theta.$$

Due to the translation invariance of the Lebesgue measure over $\mathbf{R}$, and the periodicity of the function $\theta \to f(e^{i\theta})$, we have that $\langle f \circ L_g \rangle = \langle f \rangle$, for all $g \in S^1$, where L_g denotes left multiplication by g. If $\rho : S^1 \to GL(V)$ is a representation of S^1 and $f(e^{i\theta}) = \rho(e^{i\theta}).v$, some $v \in V$, then f also satisfies the property

$$\rho(e^{i\alpha})\langle f \rangle = \langle f \rangle,$$

as the following calculation shows:

$$\begin{aligned}\rho(e^{i\alpha})\langle f \rangle &= \rho(e^{i\alpha})\frac{1}{2\pi} \int_0^{2\pi} \rho(e^{i\theta}) v d\theta \\ &= \frac{1}{2\pi} \int_0^{2\pi} \rho(e^{i(\alpha+\theta)}) v d\theta = \frac{1}{2\pi} \int_0^{2\pi} \rho(e^{i\theta'}) v d\theta'.\end{aligned}$$

Hence the mean $\langle f \rangle$ of the function $f(e^{i\theta}) = \rho(e^{i\theta}).v$ provides an S^1-invariant element of V. Note that the passage of ρ under the integral sign is guaranteed by the continuity of ρ. By making use of this operation of mean on the topological group S^1 the proof of Maschke's theorem holds for S^1 in zero characteristic: the complex linear representations of S^1 are completely reducible. The possibility of constructing an invariant mean on the functions defined on S^1 is made possible by the property of the measure induced on S^1 by the Lebesgue measure on $\mathbf{R}$, which is translation invariant and finite on compact sets.

As for the Lebesgue measure on the reals, in any locally compact group it is possible to define a measure which is translation invariant and finite on the compact sets.

Definition 6.7. Let G be a locally compact topological group. A *left (right) invariant Haar measure* for G is a measure μ defined on Borel σ-algebra[6] of G and such that:

i) $\mu(U) = \mu(gU)$ $(\mu(U) = \mu(Ug))$ for all subsets U of G and elements g of G;

[6] The Borel $\sigma-$algebra of a topological space is the smallest σ-algebra among those that contain all the open sets of the space.

ii) $\mu(U) > 0$ if U is open;

iii) $\mu(U)$ is finite if U is compact.

Theorem 6.22 (Haar). *A locally compact topological group has a left invariant measure which is unique up to scalars. If the group is compact (or if the group is abelian, in which case left and right invariance coincide) such a measure is also right invariant.*

If G is a compact group, and μ is the Haar measure on G, then $\mu(G)$ is finite. If $\rho : G \to GL(n, \mathbf{C})$ is a continuous linear representation of G, and v is a vector of the representation space V, then the mean

$$\langle v \rangle = \frac{1}{\mu(G)} \int_G \rho(g) v \, d\mu,$$

is a G-invariant element of V. This fact allows the construction of G-invariant projectors. Hence the existence of the Haar measure for a compact group implies, as seen above, the complete reducibility of the continuous representations of G.

7

Extensions and Cohomology

Among the subjects we shall touch upon in this chapter there is that of extensions. We have already mentioned this problem in Chapter 2 when talking about Hölder's program for the classification of finite groups. As we shall see, the solution proposed by Schreier in the years 1920's allows a classification of the groups that are extensions of an abelian group A by a group π by means of equivalence classes of functions $\pi \times \pi \to A$. Unfortunately, one does not obtain a system of invariants for the isomorphism classes of the groups obtained in this way, because nonequivalent functions can yield isomorphic groups. No characterization of the isomorphism classes is known.

7.1 Crossed Homomorphisms

Definition 7.1. Given two groups π and G, and a homomorphism $\varphi : \pi \to \mathbf{Aut}(G)$, the group π acts on G by $g^\sigma = g^{\varphi(\sigma)}$. We shall then say that G is a *π-group* and, if it is abelian, a *π-module.*

Let G be a π-group; we denote by G^π the set

$$G^\pi = \{x \in G \mid x^\sigma = x, \ \forall \sigma \in \pi\},$$

that is, the set of fixed points of π on G. It is a subgroup, the largest subgroup of G on which π acts trivially. In the semidirect product $\pi \times_\varphi G$, after the usual identifications, G^π is the centralizer in G of π (for this reason, G^π is also denoted by $\mathbf{C}_G(\pi)$).

Given an action of π on G, we know how to construct the semidirect product $\pi \times_\varphi G$. In general, together with π there exist in this group other complements of G. The purpose of the present section is to classify these complements by means of certain functions $f : \pi \to G$.

If ω is a complement for G and $\sigma \in \pi$, then there exists a unique element $g \in G$ such that $\sigma g \in \omega$. Indeed, if σg and σh are two elements of ω, then

Machì A.: Groups. An Introduction to Ideas and Methods of the Theory of Groups.
DOI 10.1007/978-88-470-2421-2_7, Springer-Verlag Italia 2012

$\sigma g(\sigma h)^{-1} = \sigma(gh^{-1})\sigma^{-1}$ belongs to ω and, G being normal, also to G, and therefore equals 1. It follows $h = g$. Therefore, it is possible to define a function $f : \pi \to G$ that associates with $\sigma \in \pi$ the element $g \in G$ such that σg belongs to ω.

The function f enjoys the property:

$$f(\sigma\tau) = f(\sigma)^\tau f(\tau). \tag{7.1}$$

Indeed, on the one hand $\sigma g \cdot \tau h = \sigma\tau g^\tau h = \sigma\tau f(\sigma\tau)$, and on the other hand $\sigma g \cdot \tau h = \sigma f(\sigma)\tau f(\tau) = \sigma\tau f(\sigma)^\tau f(\tau)$, recalling that in $\pi \times_\varphi G$ conjugacy by means of an element $\sigma \in \pi$ coincides with the action of σ. Equality (7.1) follows. Observe that $f(1) = 1$, as can be seen by setting $\sigma = \tau = 1$ in (7.1), and that if the action is trivial then f is a homomorphism.

A function $f : \pi \to G$ satisfying (7.1) is a *crossed homomorphism*. Conversely, given a crossed homomorphism, the pairs $(\sigma, f(\sigma))$ make up a subgroup of $\pi \times_\varphi G$:

i) $\sigma f(\sigma)\tau f(\tau) = \sigma\tau f(\sigma)^\tau f(\tau) = \sigma\tau f(\sigma\tau)$;

ii) $1 = 1 \cdot 1 = 1 \cdot f(1)$;

iii) $(\sigma f(\sigma))^{-1} = \sigma^{-1} f(\sigma^{-1})$. Indeed,

$$\sigma f(\sigma) \cdot \sigma^{-1} f(\sigma^{-1}) = \sigma\sigma^{-1} f(\sigma)^{\sigma^{-1}} f(\sigma^{-1}) = 1 \cdot f(\sigma\sigma^{-1}) = f(1) = 1.$$

We denote by ω this subgroup. If $x \in \pi \times_\varphi G$, then $x = \sigma g$, and since $f(\sigma) \in G$ there exists $g_1 \in G$ such that $g = f(\sigma)g_1$, and therefore $x = \sigma g = \sigma f(\sigma)g_1 \in \omega G$, so that $\omega G = \pi \times_\varphi G$. If $\sigma f(\sigma) = g \in G$, then $f(\sigma) = 1$ (uniqueness of the writing) so that $f(\sigma) = 1$ and $g = 1$. In other words, the intersection $\omega \cap G = \{1\}$. Thus, ω is a complement of G in the semidirect product $\pi \times_\varphi G$: like the elements of π, those of ω also form a complete system of representatives of the cosets of G in $\pi \times_\varphi G$.

In addition, it is immediately seen that the function $\pi \to G$ to which ω gives rise is the function f used to construct it, and that the complement $\omega = \pi$ corresponds to the f such that $f(\sigma) = 1$ for all $\sigma \in \pi$. We have the following theorem.

Theorem 7.1. *The crossed homomorphisms $f : \pi \to G$ are in a one-to-one correspondence with the complements of G in the semidirect product $\pi \times_\varphi G$.*

A complement ω of G corresponding to a crossed homomorphism f is the image of π under the mapping $s : \pi \to \pi \times_\varphi G$, given by $s(\sigma) = (\sigma, f(\sigma))$. This mapping s is a homomorphism because f is a crossed homomorphism, and viceversa. If ψ is the homomorphism $\psi : \pi \times_\varphi G \to \pi$, given by $\psi(\sigma, g) = \sigma$, then $\psi s = id_\pi$. A homomorphism s with this property is called a *splitting*. Thus the crossed homomorphisms correspond bijectively to the splittings.

Let now ω_1 and ω_2 be two complements for G in $\pi \times_\varphi G$, and assume they are conjugate, $\omega_1^x = \omega_2$. If $x = yg_1$, $y \in \omega_1$ and $g_1 \in G$, then $\omega_1^x = \omega_1^{g_1} = \omega_2$.

Let f_1 and f_2 be the crossed homomorphisms corresponding to ω_1 e ω_2, respectively. Then $(\sigma f_1(\sigma))^{g_1} = \tau f_2(\tau)$, for some $\tau \in \pi$, and $\sigma[\sigma^{-1}\sigma^{g_1} f_1(\sigma)^{g_1}] = \tau f_2(\tau)$, with $\sigma^{-1}\sigma^{g_1} \in G$, because of the normality of G. Therefore, the expression in brackets is an element of G. By uniqueness, it follows $\sigma = \tau$ and $(g_1^{-1})^\sigma f_1(\sigma) g_1 = f_2(\sigma)$, and setting $g_1^{-1} = g$,

$$f_2(\sigma) = g^\sigma f_1(\sigma) g^{-1}, \ \forall \sigma \in \pi. \tag{7.2}$$

Conversely, it is clear that if f_1 and f_2 are as in (7.2) for a given $g \in G$, then ω_1 and ω_2 are conjugate by means of g^{-1}, and the relation between f_1 and f_2 is an equivalence relation, $f_1 \sim f_2$.

Theorem 7.2. *The conjugacy classes of the complements of G in the semidirect product $\pi \times_\varphi G$ are in a one-to-one correspondence with the classes $[f]$ of the equivalence given by* (7.2).

We have seen that the crossed homomorphism f corresponding to the complement π is the one for which $f(\sigma) = 1$ for all $\sigma \in \pi$; the crossed homomorphisms f equivalent to it are those for which there exists a fixed element $g \in G$ such that:

$$f(\sigma) = g^\sigma g^{-1}, \ \ \forall \sigma \in \pi. \tag{7.3}$$

A crossed homomorphism f satisfying (7.3) is called *principal.* Thus:

Corollary 7.1. *If any two complements of G in the semidirect product $\pi \times_\varphi G$ are conjugate, then every crossed homomorphism is principal, and conversely.*

If $H \leq G$ is π-invariant, π acts on the cosets of H: $(Hg)^\sigma = Hg^\sigma$. Clearly, if π fixes an element of a coset, then it fixes that coset: just take as representative the fixed element. Conversely, let Hg be fixed by all the elements of π; then $g^\sigma g^{-1} \in H$, for all $\sigma \in \pi$, and the correspondence $f : \sigma \to g^\sigma g^{-1}$ is a crossed homomorphism of π in H:

$$f(\sigma\tau) = g^{\sigma\tau} g^{-1} = g^{\sigma\tau} \cdot g^{-\tau} g^\tau \cdot g^{-1} = (g^\sigma g^{-1})^\tau g^\tau g^{-1} = f(\sigma)^\tau f(\tau).$$

If two complements of H in the semidirect product π by H are conjugate, every crossed homomorphism π in H is principal, and therefore there exists $h \in H$ such that $g^\sigma g^{-1} = h^\sigma h^{-1}$, $(h^{-1}g)^\sigma = h^{-1}g$.

Corollary 7.2. *If $H \leq G$ is π-invariant, and any two complements of H in the semidirect product of $\pi \times_\varphi H$ are conjugate, then π fixes a coset of H if, and only if, it fixes an element of that coset.*

If $H \trianglelefteq G$ and is π-invariant, the image of the fixed points G^π in the canonical homomorphism $G \to G/H$ is obviously contained in $(G/H)^\pi$, but it may not coincide with it, as the following examples show.

Examples 7.1. **1.** Consider the Klein group $G = \{1, a, b, c\}$, and let σ be the automorphism of G that interchanges a and b and fixes c. If $\pi = \{1, \sigma\}$, the subgroup $H = \{1, c\}$ is normal and π-invariant, and we have $G^\pi = H$, the image of G^π in G/H is the identity, whereas $(G/H)^\pi = G/H$.

2. Let G be the group of the integers, $\pi = \{1, \sigma\}$, where $\sigma : n \to -n$. We have $G^\pi = \{0\}$, and if $H = \langle 2 \rangle$, then $(G/H)^\pi = G/H$.

In the next section we will introduce a group which, in the case of an abelian group acted upon by a group π, measures the extent to which $(G/H)^\pi$ differs from $G^\pi H/H$ (see *ex.* 6 to 8).

Corollary 7.3. *If $H \trianglelefteq G$ is as in* Corollary 7.2, *then the image of the fixed points of π in G is the set of the fixed points of the image: $G^\pi H/H = (G/H)^\pi$.*

In the first of the above examples the semidirect product of π by H is the Klein group; this group being abelian, there are two conjugacy classes, each with a single element, of complements of H in the group. In the second example also there are two conjugacy classes (see *Ex.* 7.2, 2).

Remark 7.1. Our action is a right action; with the left action (7.1) becomes $f(\sigma\tau) = (\sigma f(\tau))f(\sigma)$.

7.2 The First Cohomology Group

Let now A be a π-*module*, written additively. The crossed homomorphisms of π in A are called 1-*cocycles*; the set of 1-cocycles is denoted by $Z^1(\pi, A)$. A structure of an abelian group can be given to this set by defining $(f_1 + f_2)(\sigma) = f_1(\sigma) + f_2(\sigma)$, $\sigma \in \pi$; the zero element is the 1-cocycle f such that $f(\sigma) = 0$ for all $\sigma \in \pi$, and the opposite of a 1-cocycle f is $-f : \sigma \to -f(\sigma)$. The principal crossed homomorphisms are called 1-*cobords*; they form a subgroup $B^1(\pi, A)$ of $Z^1(\pi, A)$. Indeed, if $f_1(\sigma) = x^\sigma - x$, $f_2(\sigma) = y^\sigma - y$ then $(f_1 + f_2)(\sigma) = (x + y)^\sigma - (x + y)$, and $-f(\sigma) = (-x)^\sigma - (-x)$.

Definition 7.2. The quotient $H^1(\pi, A) = Z^1(\pi, A)/B^1(\pi, A)$ is the *first cohomology group of π with coefficients in A.*

Theorem 7.2 can be rephrased as follows: *the order of $H^1(\pi, A)$ equals the number of conjugacy classes of the complements of A in the semidirect product of π by A.*

Examples 7.2. **1.** Let π be finite, acting trivially on the integers. The action being trivial, every 1-cocycle is a homomorphism, and since the only homomorphism between a finite group and $\mathbf{Z}$ is the one sending everything to $\{0\}$, we have $Z^1(\pi, \mathbf{Z}) = \{0\}$, and, *a fortiori*, $H^1(\pi, \mathbf{Z}) = \{0\}$.

2. If $\pi = \{1, \sigma\}$ is the automorphism group of $\mathbf{Z}$, where $\sigma(n) = -n$, the semidirect product is the group D_∞, and the complements of $\mathbf{Z}$ are the groups

$\{(1,0),(\sigma,n)\}$, of order 2. Two complements $\{(1,0),(\sigma,n)\}$ and $\{(1,0),(\sigma,m)\}$ are conjugate if, and only if, m and n have the same parity, as it is easily seen; it follows $|H^1(\pi,\mathbf{Z})|=2$.

3. If π is finite, $|\pi|=n$, and D is an *n-divisible* group (every element of D is divisible by n) and torsion-free, then:

$$H^1(\pi,D)=\{0\}.$$

For $f\in Z^1(\pi,D)$ we have $nf\in B^1(\pi,D)$, that is, $nf(\sigma)=d^\sigma-d$, for all σ and some $d\in D$. D being n-divisible, $d=nd_1$, and therefore $nf(\sigma)=(nd_1)^\sigma-nd_1=n(d_1^\sigma-d_1)=nf'(\sigma)$, where $f'\in B^1$ is defined by $f'(\sigma)=d_1^\sigma-d_1$. $n(f(\sigma)-f'(\sigma))=0$ and therefore, D being torsion-free, $f(\sigma)-f'(\sigma)=0$, $f(\sigma)=f'(\sigma)$; this equality holding for all σ, $f=f'$, and $f\in B^1$.

4. Let us determine $H^1(\pi,A)$ in case π is a cyclic group. Firstly, a 1-cocycle f is determined once the image $a=f(\sigma)$ of the generator σ of π is known. Indeed, $f(\sigma^2)=f(\sigma\cdot\sigma)=f(\sigma)^\sigma+f(\sigma)=a^\sigma+a$, and for $k\geq 0$, $f(\sigma^k)=a+a^\sigma+\cdots+a^{\sigma^{k-1}}$. Moreover, from $0=f(1)=f(\sigma\cdot\sigma^{-1})$ it follows $0=f(\sigma)^{\sigma^{-1}}+f(\sigma^{-1})=a^{\sigma^{-1}}+f(\sigma^{-1})$, from which $f(\sigma^{-1})=-a^{\sigma^{-1}}$, and in general for negative powers of σ, $f(\sigma^{-k})=-a^{\sigma^{-k}}-a^{\sigma^{-(k-1)}}-\cdots-a^{\sigma^{-1}}$. By associating f with $a=f(\sigma)$ we obtain an injective homomorphism

$$Z^1(\pi,A)\to A. \tag{7.4}$$

If $f(\sigma)=a^\sigma-a$, then f is a 1-cobord: for all k,

$$f(\sigma^k)=(a^\sigma-a)+(a^\sigma-a)^\sigma+\cdots+(a^\sigma-a)^{\sigma^{k-1}}=a^{\sigma^k}-a.$$

Conversely, for a 1-cobord we have, by definition, $f(\sigma^k)=a^{\sigma^k}-a$, and therefore, in particular, $f(\sigma)=a^\sigma-a$.

Let us now distinguish the two cases π finite and π infinite.

i) π *finite*, $|\pi|=n$. In this case, let us define the *norm* of $a\in A$ as the sum of the images of a under the various elements of π:

$$N(a)=\sum_{\sigma\in\pi}a^\sigma.$$

As a varies in A, the elements of the form $N(a)$ make up a subgroup (because A is abelian). Moreover, $N(a)$ is fixed by every element of π, because when $\tau\in\pi$ varies, the products $\sigma\tau$ run over π. Therefore, the subgroup of the elements of the form $N(a)$ is contained in A^π. The elements of A of zero norm also form a subgroup of A^π.

If π is cyclic, f is a 1-cocycle and $f(\sigma)=a$ then, by (7.2), $0=f(1)=f(\sigma^{n-1})=N(a)$, and the image a of σ under f has zero norm. Conversely, if $N(a)=0$, setting $f(\sigma)=a$ a 1-cocycle is defined. Therefore, the image of

$Z^1(\pi, A)$ in (7.4) is $\{a \in A \mid N(a) = 0\}$. The elements of A of the form $a^\sigma - a$ have zero norm, and constitute the image of $B^1(\pi, A)$; thus:

$$H^1(\pi, A) = \frac{Z^1(\pi, A)}{B^1(\pi, A)} \simeq \frac{\{a \in A \mid N(a) = 0\}}{\{a^\sigma - a,\ a \in A\}}.$$

ii) π infinite. In this case, letting $f(\sigma) = a$ for all $a \in A$ a cocycle is defined: the mapping (7.4) is surjective, and therefore is an isomorphism; we have:

$$H^1(\pi, A) \simeq \frac{A}{\{a^\sigma - a,\ a \in A\}}.$$

Lemma 7.1. *Let π be finite, $|\pi| = n$. Then $nH^1(\pi, A) = 0$. Therefore, the order of every element of $H^1(\pi, A)$ divides n.*

Proof. In (7.1) let us fix τ and sum over σ: $\sum_\sigma f(\sigma\tau) = \sum_\sigma f(\sigma)^\tau + \sum_\sigma f(\tau)$. As σ varies, the product $\sigma\tau$ runs over all elements of π; the above becomes $\sum_\sigma f(\sigma) = (\sum_\sigma f(\sigma))^\tau + nf(\tau)$. Setting $\sum_\sigma f(\sigma) = -g$ we have $nf(\tau) = g^\tau - g$, for all $\tau \in \pi$, that is $nf \in B^1(\pi, A)$. ◇

Theorem 7.3. *If $(|\pi|, |A|) = 1$, any two complements of an abelian group A in the semi-direct product of π by A are conjugate.*

This theorem also holds in the more general case in which A is a solvable group.

Lemma 7.2. *Let G be a π-group, $H \trianglelefteq G$ and π-invariant. If every crossed homomorphism of π in H and of π in G/H is principal, then every crossed homomorphism of π in G is principal.*

Proof. Let f be a crossed homomorphism of π in G. The mapping $g : \pi \to G/H$ given by $\sigma \to Hf(\sigma)$ is easily seen to be a crossed homomorphism of π in G/H, and as such is principal. Then there exists $Hx \in G/H$ such that $g(\sigma) = (Hx)^\sigma(Hx)^{-1} = Hx^\sigma x^{-1}$, $\sigma \in \pi$. It follows $Hf(\sigma) = Hx^\sigma x^{-1}$ and $f(\sigma)xx^{-\sigma} \in H$, and therefore also $x^{-\sigma}f(\sigma)x \in H$. The mapping $g_1 : \sigma \to f(\sigma)xx^{-\sigma}$ is a crossed homomorphism of π in H and therefore principal. Then there exists $h \in H$ such that $x^{-\sigma}f(\sigma)x = h^\sigma h^{-1}$. But then $f(\sigma) = x^\sigma h^\sigma h^{-1}x^{-1} = (xh)^\sigma(xh)^{-1}$, $\sigma \in \pi$, and so f is principal. ◇

Theorem 7.4. *If G is a solvable π-group and $(|\pi|, |G|) = 1$, then the complements of G in the semidirect product of π by G are all conjugate.*

Proof. By induction on $|G|$. If G is abelian, the result has already been seen. Assume $G' \neq \{1\}$. G' being π-invariant, and since $G' < G$ and $|G/G'| < |G|$, by induction the complements in their respective semidirect products are all conjugate, and therefore the crossed homomorphisms of π in G' and in G/G' are all principal (Corollary 7.1). By the preceding lemma, those of π in G are all principal, and again by Corollary 7.1 the result follows. ◇

Example 7.3. As an application of this theorem let us prove Corollary 5.11 in the following form: *if an automorphism σ of a finite p-group P induces the identity on $P/\mathbf{\Phi}$, where $\mathbf{\Phi}$ is the Frattini subgroup of P, and $p \nmid o(\sigma) = 1$, then σ and is identity on P.* With $\pi = \langle\sigma\rangle$ we have $(|P|, |\pi| = 1)$, and by Theorem 7.3 and Corollary 7.3 $P^\pi\mathbf{\Phi}/\mathbf{\Phi} = (P/\mathbf{\Phi})^\pi = (P/\mathbf{\Phi})$, from which $P^\pi\mathbf{\Phi} = P$, and therefore $P^\pi = P$.

The conclusion of Theorem 7.4 also holds when instead of G, the group π is solvable, as will be shown in a moment. Observe that if π_1 is a normal subgroup of π, then the quotient π/π_1 acts on G^{π_1}, the fixed points of π_1, by $g^{\pi_1\sigma} = g^\sigma$. We have a lemma "dual" to Lemma 7.2:

Lemma 7.3. *If $\pi_1 \trianglelefteq \pi$, and every crossed homomorphism of G^{π_1} in G and of π/π_1 in G_1^π is principal, then every crossed homomorphism of π in G is principal.*

Proof. If f is a crossed homomorphism of π in G, its restriction to π_1 is a crossed homomorphism of π_1 in G, and therefore principal. Hence there exists $x \in G$ such that $f(\tau) = x^\tau x^{-1}$, for all $\tau \in \pi_1$. For $\sigma \in \pi$ set $g(\sigma) = x^{-\sigma}f(\sigma)x$, thus obtaining a crossed homomorphism of π in G and such that $g(\sigma) = 1$ for $\sigma \in \pi_1$. For $\eta, \eta' \in \pi_1$ we have $g(\eta\sigma) = g(\eta)^\sigma g(\sigma) = g(\sigma)$, and similarly for $g(\eta'\sigma)$, so that g is constant on the cosets of π_1. Moreover, if $\tau \in \pi_1$ and $\sigma \in \pi$, then $g(\sigma\tau) = g(\sigma)^\tau g(\tau) = g(\sigma^\tau)$, and $\sigma\tau = \tau'\sigma$, by the normality of π_1, $\tau' \in \pi_1$, $g(\tau'\sigma) = g(\tau')^\sigma g(\sigma) = g(\sigma)$, and therefore $g(\sigma)^\tau = g(\sigma)$ for all $\sigma \in \pi$ and $\tau \in \pi_1$, that is, $g(\sigma) \in G^{\pi_1}$ for all $\sigma \in \pi$. The function $\overline{g} : \pi/\pi_1 \to G^{\pi_1}$, given by $\pi_1\sigma \to g(\sigma)$ is well defined, and clearly is a crossed homomorphism of π/π_1 in G^{π_1}. As such, it is principal: there exists $y \in G^{\pi_1}$ such that $\overline{g}(\pi_1\sigma) = y^{\pi_1\sigma}y^{-1} = y^\sigma y^{-1}$. It follows $g(\sigma) = y^\sigma y^{-1}$ and therefore $f(\sigma) = x^\sigma g(\sigma)x^{-1} = x^\sigma y^\sigma y^{-1}x^{-1} = (xy)^\sigma(xy)^{-1}$, and f is principal. ◊

Theorem 7.5. *If G is a π-group with π solvable and $(|\pi|, |G|) = 1$, then the complements of G in the semidirect product of π by G are all conjugate.*

Proof. If π is not simple, the claim follows by induction from the preceding lemma and from Corollary 7.1. If it is simple, being solvable is of prime order p, and from $(p, |G|) = 1$ it follows that π is Sylow in the semidirect product of π by G. The result follows from the Sylow theorem. ◊

Exercises

1. If $H \leq G$ is π-invariant, its centralizer and normalizer are also invariant.

2. Define $[G, \pi] = \langle g^{-1}g^\sigma,\ g \in G,\ \sigma \in \pi\rangle$, and show that this subgroup

i) is normal;

ii) is π-invariant;

iii) is the smallest normal π-invariant subgroup of G such that π acts trivially on the quotient (therefore, $[G, \pi]$ is the subgroup"dual" to G^π).

3. Let G be a group. For a fixed $y \in G$, the function $f_y : x \to [x, y]$, is a crossed homomorphism $G \to G'$.

4. If $V = \{1, a, b, c\}$ is the Klein group and $\pi = \{1, \sigma\}$ acts on V according to $a^\sigma = b$, $b^\sigma = a$, $c^\sigma = c$, show that $H^1(\pi, V) = \{0\}$. If the action of π on V is trivial, then $H^1(\pi, V) \simeq V$.

5. Show that if any two complements of G in the semidirect product of π by G are conjugate, then
i) $G = G^\pi[G, \pi]$;
ii) $[G, \pi, \pi] = [G, \pi]$.

We know that G is a π-group and A is a subgroup of G which is normal and π-invariant, the inclusion

$$G^\pi A/A \subseteq (G/A)^\pi \tag{7.5}$$

may be strict (see *Ex.* 7.1, 1), and that if any two complements of A in the semidirect product $\pi \times_\varphi A$ are conjugate, then equality holds. If A is abelian, the condition that any two complements are conjugate becomes $H^1(\pi, A) = \{0\}$.

If A and B are two π-modules, a *π-morphism* $\alpha : A \to B$ is a homomorphism commuting with the action of π, $\alpha(x^\sigma) = \alpha(x)^\sigma$, $x \in A$, $\sigma \in \pi$. We say that a sequence

$$\cdots \to A_{i-1} \stackrel{\alpha_{i-1}}{\to} A_i \stackrel{\alpha_i}{\to} A_{i+1} \to \cdots$$

of groups A_i and homomorphisms α_i is *exact at point* A_i if $ker(\alpha_i) = Im(\alpha_{i-1})$; it is *exact* if it is exact at each point. Exactness of the sequence:

$$\{0\} \to A \stackrel{\alpha}{\to} B \stackrel{\beta}{\to} C \to \{0\} \tag{7.6}$$

means that α is injective and β is surjective. Such a sequence is called a *short exact sequence.*

6. If in (7.6) α and β are π-morphisms, show that, by defining $\alpha^\pi : A^\pi \to B^\pi$ according to $\alpha^\pi : a \to \alpha(a)$, $a \in A^\pi$, and similarly β^π, we have $\alpha^\pi(A^\pi) \subseteq B^\pi$ and $\beta^\pi(B^\pi) \subseteq C^\pi$, and that the sequence of abelian groups (trivial π-modules)

$$\{0\} \to A^\pi \stackrel{\alpha^\pi}{\to} B^\pi \stackrel{\beta^\pi}{\to} C^\pi$$

is exact. (Note that $B^\pi/A^\pi = B^\pi/B^\pi \cap A \simeq B^\pi A/A^\pi$.)

The fact that the inclusion (7.5) may be strict shows that, in general, the sequence obtained by the previous one by adding $\to \{0\}$ is not exact at C^π. In other words, taking fixed points does not respect exact sequences. Let us show what can be add in order to restore the exactness lost when passing from $(-)$ to $(-)^\pi$. For the sake of simplicity, let us consider the case in which α is injective, i.e. $A \subseteq B$.

Let $c \in C^\pi$; β being surjective, there exists b such that $\beta(b) = c$. If $\sigma \in \pi$, then $c^\sigma = c$, that is $\beta(b)^\sigma = \beta(b)$, and β being a π-morphism, $\beta(b^\sigma) = \beta(b)$, $b^\sigma - b \in ker(\beta) = A$. The function $f : \pi \to A$, given by $\sigma \to b^\sigma - b$ is a 1-cocycle of π in A, and in fact a 1-cobord of π in B.

If $c = \beta(b')$, we obtain in the same way a 1-cocycle f' of π in A given by $f'(\sigma) = b'^\sigma - b'$. Now $c = \beta(b') = \beta(b)$ implies $b' - b \in ker(\beta) = A$, and therefore

$b' - b = a$ for some $a \in A$, from which $b' = a + b$ and

$$f'(\sigma) = b'^{\sigma} - b' = (a+b)^{\sigma} - (a+b) = (a^{\sigma} - a) + (b^{\sigma} - b) = f(\sigma) + a^{\sigma} - a$$

so that $f' \sim f$.

7. Show that the mapping $\delta : C^{\pi} \to H^1(\pi, A)$, defined in the way seen above by $c \to [f] = f + B^1(\pi, A)$, is a homomorphism that makes the sequence

$$\{0\} \to A^{\pi} \xrightarrow{\alpha^{\pi}} B^{\pi} \xrightarrow{\beta^{\pi}} C^{\pi} \xrightarrow{\delta} H^1(\pi, A)$$

exact.

A π-morphism $\alpha : A \to B$ induces a homomorphism

$$\alpha' : Z^1(\pi, A) \to Z^1(\pi, B),$$

given by $f \to \alpha f$. Moreover, if $f_1 \sim f_2$, then $f_2(\sigma) = f_1(\sigma) + a^{\sigma} - a$ so that

$$\alpha f_2(\sigma) = \alpha(f_1(\sigma) + a^{\sigma} - a) = \alpha(f_1(\sigma)) + \alpha(a)^{\sigma} - \alpha(a),$$

and αf_1 e αf_2 differ by the 1-cobord $\sigma \to \alpha(a)^{\sigma} - \alpha(a)$.

Thus, we can consider the two morphisms:

$$H^1(\pi, A) \xrightarrow{\overline{\alpha}} H^1(\pi, B) \quad \text{and} \quad H^1(\pi, B) \xrightarrow{\overline{\beta}} H^1(\pi, C),$$

induced by α and β according to $\overline{\alpha}[f] = [\alpha f]$ and $\overline{\beta}[f] = [\beta f]$.

8. Show that the sequence

$$\{0\} \to H^0(\pi, A) \xrightarrow{\alpha^{\pi}} H^0(\pi, B) \xrightarrow{\beta^{\pi}} H^0(\pi, C) \xrightarrow{\delta} H^1(\pi, A) \xrightarrow{\overline{\alpha}} H^1(\pi, B) \xrightarrow{\overline{\beta}} H^1(\pi, C),$$

where we have set $(-)^{\pi} = H^0(\pi, -)$, is exact.

This sequence is not necessarily exact at the point $H^1(\pi, C)$, as it can be seen by considering the short exact sequence $0 \to \mathbf{Z} \xrightarrow{\alpha} \mathbf{Z} \xrightarrow{\beta} \mathbf{Z}_2 \to 0$ where $\pi = C_2$ is the cyclic group of order 2, $\alpha : n \to 2n$, β is the canonical homomorphism and the action of π is trivial. Then $H^1(\pi, \mathbf{Z}) = 0$, while $H^1(\pi, \mathbf{Z}_2)$ has two elements since there are two homomorphisms $C_2 \to \mathbf{Z}_2$. This means that if $\beta : B \to C$ is a surjective π-morphism, the morphism induced by $\overline{\beta} : H^1(\pi, B) \to H^1(\pi, C)$ is not necessarily surjective.

The sequence cannot be extended by adding "$\to \{0\}$". We will see how it can be extended (*ex.* 15).

7.2.1 The Group Ring $\mathbf{Z}\pi$

For $\mathbf{Z}$ the ring of integers and π a group, let us consider the set of functions $u : \pi \to \mathbf{Z}$ with finite support ($u(\sigma) \neq 0$ for at most a finite number of elements σ of π). Denote by $\mathbf{Z}\pi$ this set. It can be endowed with the structure of a free abelian group with addition $(u+v)(\sigma) = u(\sigma) + v(\sigma)$, a basis been given by the functions u_σ defined as follows:

$$u_\sigma(\tau) = \begin{cases} 1, \text{ if } \tau = \sigma, \\ 0, \text{ otherwise.} \end{cases}$$

Indeed, if $u \in \mathbf{Z}\pi$ and $\{\sigma_1, \sigma_2, \ldots, \sigma_n\}$ is the support of u, then $u = u(\sigma_1)u_{\sigma_1} + u(\sigma_2)u_{\sigma_2} + \cdots + u(\sigma_n)u_{\sigma_n}$. Thus, every element of $\mathbf{Z}\pi$ has a unique expression as

$$u = \sum_{\sigma \in \pi} u(\sigma)u_\sigma, \tag{7.7}$$

so that $\mathbf{Z}\pi$ is a free group. $\mathbf{Z}\pi$ also has a ring structure, with the product defined by extending by linearity the group product. This product is the *convolution product*:

$$u \circ v(\sigma) = \sum_{\tau \in \pi} u(\tau)v(\tau^{-1}\sigma) \tag{7.8}$$

(we shall simply write uv for $u \circ v$). It is easily verified that this product is associative and that the distributive property with respect to addition holds.

Consider now the mapping $\pi \to \mathbf{Z}\pi$ given by $\sigma \to u_\sigma$. Observe that $u_\sigma u_\tau(\eta) = \sum_{\gamma \in \pi} u_\sigma(\gamma)u_\tau(\gamma^{-1}\eta)$, and since if $\gamma \neq \sigma$ or if $\gamma^{-1}\eta \neq \tau$ the terms of the sum are zero, only the term for $\gamma = \sigma$ and $\gamma^{-1}\eta = \tau$, i.e. $\eta = \sigma\tau$ remains. In other words:

$$u_\sigma u_\tau(\eta) = \begin{cases} 1, \text{ if } \eta = \sigma\tau, \\ 0, \text{ otherwise,} \end{cases}$$

and therefore, by definition, $u_\sigma u_\tau = u_{\sigma\tau}$. Thus, the elements u_σ combine as those of π, and this fact allows us to embed π in $\mathbf{Z}\pi$ by identifying σ with u_σ, and (7.7) can be written more simply as

$$u = \sum_{\sigma \in \pi} u(\sigma)\sigma\ . \tag{7.9}$$

In this way $\mathbf{Z}\pi$ becomes the set of formal linear combinations with integer coefficients of the elements of π. The product of two elements $u = \sum_{\sigma \in \pi} u(\sigma)\sigma$ and $v = \sum_{\tau \in \pi} u(\tau)\tau$ is $uv = \sum_{\sigma,\tau \in \pi} u(\sigma)v(\tau)\sigma\tau$. Writing uv in the form (7.9) the coefficient of $\eta \in \pi$ is the one obtained for $\sigma\tau = \eta$, i.e. $\tau = \sigma^{-1}\eta$, so that this coefficient is $\sum_{\sigma \in \pi} u(\sigma)v(\sigma^{-1}\eta)$. Therefore $uv = \sum_{\eta \in \pi}(\sum_{\sigma \in \pi} u(\sigma)v(\sigma^{-1}\eta))\eta$. (Observe that from $\sigma\tau = \eta$ it follows $\sigma = \eta\tau^{-1}$, and $\sum_{\sigma \in \pi} u(\sigma)v(\sigma^{-1}\eta) = \sum_{\tau \in \pi} u(\eta\tau^{-1})v(\tau)$).

A π-module A becomes a $\mathbf{Z}\pi$-module by defining,

$$a.u = \sum_\sigma u_\sigma a^\sigma,$$

for $a \in A$ and u as in (7.7). Conversely, if A is a $\mathbf{Z}\pi$-module, A becomes a π-module by embedding π in $\mathbf{Z}\pi$ ($\sigma \to 1\sigma$) and defining $a^\sigma = a.1\sigma$.

As we know, to say that A is a π-module amounts to saying that it has been assigned a homomorphism of π in the automorphism group $\mathbf{Aut}(A)$ of the abelian group A. By composing this homomorphism with the inclusion $\mathbf{Aut}(A) \subseteq \mathbf{End}(A)$ we have a mapping $\varphi : \pi \to \mathbf{End}(A)$ such that $\varphi(\sigma\tau) = \varphi(\sigma)\varphi(\tau)$ e $\varphi(1) = I_A$. This mapping φ extends to a homomorphism φ' between the rings $\mathbf{Z}\pi$ and $\mathbf{End}(A)$[1]: $\varphi'(\sum_\sigma u_\sigma \sigma) = \sum_\sigma u_\sigma \varphi(\sigma)$. Thus, by defining: $a.\sum_\sigma u_\sigma\sigma = \varphi'(\sum_\sigma u_\sigma\sigma)(a)$, A becomes a $\mathbf{Z}\pi$–module. Conversely, if A is a $\mathbf{Z}\pi$–module, by definition we have a homomorphism $\mathbf{Z}\pi \to \mathbf{End}(A)$ which induces a homomorphism $\pi \to \mathbf{Aut}(A)$ (invertible elements of $\mathbf{Z}\pi$ map into invertible elements of $\mathbf{End}(A)$, that is into elements of $\mathbf{Aut}(A)$). Thus, the two notions of a π–module and a $\mathbf{Z}\pi$–module are equivalent, so that we can speak of a module, a notion belonging to ring theory, for an abelian group acted upon by a group .

The map $\pi \to \mathbf{Z}$ which sends all the elements $\sigma \in \pi$ to 1 (*augmentation map*) induces the homomorphism between the rings $\mathbf{Z}\pi$ and $\mathbf{Z}$ which sends an element of $\mathbf{Z}\pi$ to the sum of its coefficients $\sum u_\sigma\sigma \to \sum u_\sigma$, which is surjective. Its kernel, denoted $\mathbf{I}\pi$, is the ideal of the elements of $\mathbf{Z}\pi$ for which $\sum u_\sigma = 0$ (*augmentation ideal*).

Lemma 7.4. *i)* $\mathbf{I}\pi$ *is a free abelian group with basis* $\sigma - 1$, $1 \neq \sigma \in \pi$;

ii) if $\{\sigma_i\}$ *is a system of generators for* π, *then* $\{\sigma_i - 1\}$ *is a system of generators for* $\mathbf{I}\pi$ *as an ideal.*

Proof. i) If $\sum_\sigma u_\sigma\sigma \in \mathbf{I}\pi$, then $\sum_\sigma u_\sigma = 0$ and $\sum_\sigma u_\sigma\sigma = \sum_\sigma u_\sigma(\sigma - 1)$. Hence the elements $\sigma - 1$ generate $\mathbf{I}\pi$. From this equality it also follows that they are free generators: $0 = \sum u_\sigma(\sigma - 1) = \sum_\sigma u_\sigma$ implies $u_\sigma = 0$ since the σ's are free generators for $\mathbf{Z}\pi$.

ii) Let us show that $\sigma - 1 \in \mathbf{I}\pi$ for all $\sigma \in \pi$. This holds for the words of length 1 in the σ_i: $\sigma_i - 1 \in \mathbf{I}\pi$ and $\sigma_i^{-1} = -\sigma_i^{-1}(\sigma_i - 1) \in \mathbf{I}\pi$. Assume the result true for words of length m. A word of length $m+1$ has the form $\sigma_i^{\pm 1}\sigma$, with σ of length m, and we have $\sigma_i^{\pm 1}\sigma - 1 = \sigma_i^{\pm 1}(\sigma - 1) + (\sigma_i^{\pm 1} - 1)$, with both terms in $\mathbf{I}\pi$. ◇

Note that $\mathbf{Z}\pi = \mathbf{I}\pi \oplus \mathbf{Z}$ as abelian groups: indeed, $\sum u_\sigma\sigma = \sum u_\sigma \cdot (\sigma - 1) + \sum u_\sigma \cdot 1$.

If A and B are two π-modules, it is possible to endow the set $\mathbf{Hom}(A,B)$ of morphisms from A to B with the structure of a π-module, with the addition $(f+g)(a) = f(a) + g(a)$ and the action of π given by $f^\sigma(a) = (f(a^{\sigma^{-1}}))^\sigma, \sigma \in \pi$. The set $\mathbf{Hom}_\pi(A,B)$ of π–morphisms preserving the module structure $(f(a^\sigma) = f(a)^\sigma)$ is an abelian group with the obvious addition. If $\varphi : \mathbf{Z}\pi \to A$ is a π-morphism and $\varphi(1) = a$, then $\varphi(\sigma) = \varphi(1\sigma) = \varphi(1^\sigma) = \varphi(1)^\sigma = a^\sigma$, and therefore $\varphi(\sum u_\sigma\sigma) = \sum u_\sigma\varphi(\sigma) = \sum u_\sigma a^\sigma = (\sum u_\sigma a)^\sigma$. Hence a π-morphism φ is determined by the image of 1; if $\varphi(1) = a$ we write $\varphi = \varphi_a$. It

[1] It is well known that the endomorphisms of an abelian group form a ring.

is easily seen that an element $a \in A$ determines a morphism $\varphi(\sigma) = a^\sigma$ which sends 1 to a, that is φ_a. Thus we have an isomorphism $\mathbf{Hom}_\pi(\mathbf{Z}\pi, A) \simeq A$.

The inclusion $\iota : \mathbf{I}\pi \to \mathbf{Z}\pi$ induces a homomorphism

$$\iota^* : \mathbf{Hom}_\pi(\mathbf{Z}\pi, A) \to \mathbf{Hom}_\pi(\mathbf{I}\pi, A) \tag{7.10}$$

obtained by sending a π–morphism $\varphi_a : \mathbf{I}\pi \to A$ to φ_a composed with the inclusion: $\varphi_a \to \varphi_a \circ \iota$. As to the elements of $\mathbf{I}\pi$ we have $\iota^*(\sigma - 1) = \varphi_a \circ \iota(\sigma - 1) = \varphi_a(\sigma - 1) = \varphi_a(\sigma) - \varphi_a(1) = a^\sigma - a$.

Lemma 7.5. *i)* $\mathbf{Hom}_\pi(\mathbf{I}\pi, A) \simeq Z^1(\pi, A)$*;*

ii) if π is a free group with basis $\{\sigma_i\}$, then $\mathbf{I}\pi$ is a free π–module with basis $\{\sigma_i - 1\}$.

Proof. i) If $g \in \mathbf{Hom}_\pi(\mathbf{I}\pi, A)$ the function $f : \sigma \to g(\sigma - 1)$ is a crossed homomorphism of π in A. Indeed, $f(\sigma\tau) = g(\sigma\tau - 1) = g(\sigma\tau - \tau + \tau - 1) = g((\sigma - 1)\tau + \tau - 1) = g(\sigma - 1)^\tau + g(\tau - 1) = f(\sigma)^\tau + f(\tau)$. Moreover, it is clear that the mapping $g \to f$ is a homomorphism. Conversely, for $f \in Z^1(\pi, A)$ define $g : \sigma - 1 \to f(\sigma)$; let us show that $g \in \mathbf{Hom}_\pi(\mathbf{I}\pi, A)$. First, $\mathbf{I}\pi$ being free abelian on $\{\sigma - 1,\ \sigma \in \pi\}$, g extends to a homomorphism of abelian groups $\mathbf{I}\pi \to A$. This extension is a π–morphism: $g((\sigma - 1)\tau) = g(\sigma\tau - \tau) = g((\sigma\tau - 1) - (\tau - 1)) = g(\sigma\tau - 1) - g(\tau - 1) = f(\sigma\tau) - f(\tau) = f(\sigma)^\tau + f(\tau) - f(\tau) = f(\sigma)^\tau = g(\sigma - 1)^\tau$.

ii) We must show that every mapping φ of $\{\sigma_i - 1\}$ into a π–module A extends to a π-morphism of $\mathbf{I}\pi$ in A. If E is the semidirect product of π by A, φ extends to a mapping of $\{\sigma_i\}$ into E given by $\sigma_i \to (\sigma_i, \varphi(\sigma_i - 1))$, and since π is free, this extends to a homomorphism φ' of π into E: $\varphi'(\sigma) = (\sigma, f(\sigma))$, for some f of π into A. The fact that φ' is a homomorphism implies that f is a crossed homomorphism (and conversely): $\varphi'(\sigma)\varphi'(\tau) = (\sigma, f(\sigma))(\tau, f(\tau)) = (\sigma\tau, f(\sigma)^\tau f(\tau))$, and $\varphi'(\sigma\tau) = (\sigma\tau, f(\sigma\tau))$. Thus the isomorphism *i)* yields a π–morphism of $\mathbf{I}\pi$ into A, that is $\widetilde{\varphi}(\sigma - 1) = f(\sigma)$. Now, $\varphi(\sigma_i) = \varphi'(\sigma_i) = (\sigma_i, f(\sigma_i))$, from which $\widetilde{\varphi}(\sigma - 1) = \varphi(\sigma - 1)$, hence $\widetilde{\varphi}$ extends φ. ◇

If f is principal, $f(\sigma) = a^\sigma - a$, and the corresponding $g : \mathbf{I}\pi \to A$ is given by $g(\sigma - 1) = a^\sigma - a$. Therefore, we can extend (7.10) by means of "$\to H^1(\pi, A)$", and we have:

Theorem 7.6. *The sequence:*

$$\mathbf{Hom}_\pi(\mathbf{Z}\pi, A) \xrightarrow{\iota^*} \mathbf{Hom}_\pi(\mathbf{I}\pi, A) \xrightarrow{\varphi^*} H^1(\pi, A) \to \{0\} \tag{7.11}$$

is exact, that is:

$$H^1(\pi, A) \simeq \mathbf{Hom}_\pi(\mathbf{I}\pi, A)/\iota^*(\mathbf{Hom}_\pi(\mathbf{Z}\pi, A)).$$

Proof. The kernel of φ^* consists of the elements g such that the corresponding f is principal. From what it has been observed before Lemma 7.5 the g's are the elements of the image of ι^*. ◇

As we have seen, the principal crossed homomorphisms of π into A correspond to the images of $\mathbf{Hom}_\pi(\mathbf{Z}\pi, A)$, hence to the morphisms of $\mathbf{I}\pi$ into A that lift to $\mathbf{Z}\pi$. If φ lifts to φ', then $\varphi(\sigma-1) = \varphi'(\sigma-1) = \varphi'(\sigma) - \varphi'(1)$ ($\varphi'(\sigma)$ makes sense because φ' is defined on the whole of $\mathbf{Z}\pi$; hence the second equality is legitimate). Now, $\varphi'(\sigma) = \varphi'(1\cdot\sigma) = \varphi'(1)^\sigma$, and setting $\varphi'(1) = a$, we have $\varphi(\sigma-1) = \varphi'(\sigma-1) = a^\sigma - a$. In conclusion:

Theorem 7.7. *Let A be a π–modulo. If every π–morphism of $\mathbf{I}\pi$ in A lifts to one of $\mathbf{Z}\pi$ in A, then $H^1(\pi, A) = \{0\}$.*

As in the case of abelian groups, one can define injective π-modules: A is injective if for every module C and every submodule B of C, a morphism of π–modules of B in A lifts to one of C in A.

Corollary 7.4. *The first cohomology group of a group π with coefficients in an injective π-module is trivial.*

Exercises

9. Let π be a cyclic group of order n and let $\mathbf{Z}[x]$ be the ring of polynomials in x with integer coefficients. Show that $\mathbf{Z}\pi \simeq \mathbf{Z}[x]/(x^n - 1)$.

10. Let $\pi_1 \leq \pi$, and let T be a transversal for the left cosets of π_1. Show that $\mathbf{Z}\pi$ is a free right $\mathbf{Z}\pi_1$ module with basis T.

11. Let $\{C_\lambda\}_{\lambda\in\Lambda}$ be the set of finite conjugate classes of π, and let $\{\overline{C}_\lambda\} = \{\sum_{\sigma\in \mathbf{C}_\lambda}\sigma\}$. Show that the center $\mathbf{Z}\pi$ is a free abelian group with basis $\{\overline{C}_\lambda\}_{\lambda\in\Lambda}$.

12. Prove the equality $\mathbf{Hom}_\pi(A,B) = (\mathbf{Hom}(A,B))^\pi$.

13. If A is a π-module and $\mathbf{Z}$ is a trivial π-module, show that $\mathbf{Hom}_\pi(\mathbf{Z}, A) \simeq A^\pi$. [*Hint*: consider the mapping $f \to f(1)$.]

7.3 The Second Cohomology Group

Let A be an abelian group, E an extension of A. Then A becomes an E/A-module by defining

$$a^{Ax} = x^{-1}ax,$$

A being abelian, this is well defined, because if $Ax = Ay$, then $y = a_1x$, and $y^{-1}ay = x^{-1}a_1^{-1}aa_1x = x^{-1}ax$. Hence we have a homomorphism $\psi : E/A \to \mathbf{Aut}(A)$. Observe that A is a trivial E/A-module if, and only if, A is contained in the center of E; in this case, the extension is *central*.

If A is a π-module according to a homomorphism $\varphi : \pi \to \mathbf{Aut}(A)$, and E is an extension of A (as an abelian group) with $E/A \simeq \pi$, the structure of $E/A \simeq \pi$-module does not necessarily concide with the original one (in other words, the two homomorphisms φ and ψ are not necessarily the same). If they do coincide, we say that the extension E of A by π *realizes* the group of operators π. The *extension problem for the π-module A* can then be phrased as follows: determine all the extensions E of the abelian group A realizing π.

This problem always admits at least one solution, i.e. the semidirect product $E = \pi \times_\varphi A$, for which $(\sigma, 1)^{-1}(1, a)(\sigma, 1) = (1, a^{\varphi(\sigma)})$. With the usual identifications, we have $\sigma^{-1}a\sigma = a^\sigma$, so that E realizes π.

Let us now try to determine other extensions, if any. By "determine" we mean "to specify what is the product" in the groups E, and this will be done by means of certain functions $\pi \times \pi \to A$ that arise in a way that we will see shortly. As we observed at the outset of this chapter, Schreier's solution of the extension problem that we propose does not provide a system of invariants for the isomorphism classes. The above mentioned functions, up to a certain equivalence, allow us to construct the multiplication table of the groups, but nonequivalent functions may give rise to isomorphic groups.

Let E be an extension realizing π, and let λ be a choice function of the representatives of the cosets $\sigma \in \pi$ of A in E. If $\lambda(\sigma)$ and $\lambda(\tau)$ are representatives of σ and τ, we have $\sigma\tau = \lambda(\sigma)A \cdot \lambda(\tau)A = \lambda(\sigma)\lambda(\tau)A$, so that, for a certain element $g(\sigma, \tau)$ of A depending on σ and τ, $\lambda(\sigma)\lambda(\tau) = \lambda(\sigma\tau)g(\sigma, \tau)$. In this way, the choice function λ determines a function $g : \pi \times \pi \to A$. Now,

$$(\lambda(\sigma)\lambda(\tau))\lambda(\eta) = \lambda(\sigma\tau)g(\sigma, \tau)\lambda(\eta) = \lambda(\sigma\tau)\lambda(\eta) \cdot \lambda(\eta)^{-1}g(\sigma, \tau)\lambda(\eta),$$

and since E realizes π, we have $\lambda(\eta)^{-1}g(\sigma, \tau)\lambda(\eta) = g(\sigma, \tau)^\eta$, so that

$$(\lambda(\sigma)\lambda(\tau))\lambda(\eta) = \lambda(\sigma\tau)\lambda(\eta)g(\sigma, \tau)^\eta = \lambda(\sigma\tau\eta)g(\sigma\tau, \eta)g(\sigma, \tau)^\eta.$$

Similarly, $\lambda(\sigma)(\lambda(\tau)\lambda(\eta)) = \lambda(\sigma\tau\eta)g(\sigma, \tau\eta)g(\tau, \eta)$. By the associative law we have, using the additive notation for A,

$$g(\sigma, \tau\eta) + g(\tau, \eta) = g(\sigma\tau, \eta) + g(\sigma, \tau)^\eta. \tag{7.12}$$

A function $g : \pi \times \pi \to A$ satisfying (7.12) is called a *factor system* or a *2-cocycle.* If λ chooses the identity as representative of A, then $g(\sigma, 1) = 0$ for all $\sigma \in \pi$. From this, setting $\tau = 1$ in (7.12), it follows $g(1, \sigma) = 0$. If $g(\sigma, 1) = 0$ for all $\sigma \in \pi$, the factor system g is said to be *normalized.*

If μ is another choice function of representatives, then $\mu(\sigma) = \lambda(\sigma)h(\sigma)$, for $h(\sigma) \in A$, and being as above $\mu(\sigma)\mu(\tau) = \mu(\sigma\tau)g'(\sigma, \tau)$ we have:

$$\begin{aligned}\mu(\sigma)\mu(\tau) &= \lambda(\sigma)h(\sigma)\lambda(\tau)h(\tau) = \lambda(\sigma)\lambda(\tau)h(\sigma)^\tau h(\tau)\\ &= \lambda(\sigma\tau)g(\sigma, \tau)h(\sigma)^\tau h(\tau)\end{aligned}$$

and $\mu(\sigma\tau)g'(\sigma, \tau) = \lambda(\sigma\tau)h(\sigma\tau)g'(\sigma, \tau)$. Hence we have the following relation between g' and g (additive notation):

$$g'(\sigma, \tau) = g(\sigma, \tau) - h(\sigma\tau) + h(\sigma)^\tau + h(\tau). \tag{7.13}$$

We say that two 2-cocycles g' and g are *equivalent* when there exists a function $h : \pi \to A$ such that (7.13) is satisfied. Thus, an extension realizing π determines a class of equivalent 2-cocycles.

If g is not normalized (i.e. $\lambda(1)$ is not the identity of A), we have $\lambda(1)\lambda(\tau) = \lambda(\tau)g(1,\tau)$, that is $g(1,\tau) = \lambda(\tau)^{-1}\lambda(1)\lambda(\tau) = \lambda(1)^{\tau}$. If μ chooses the identity of A as representative of A, by the above we have $1 = \mu(1) = \lambda(1)h(1)$, i.e. $\lambda(1) = h(1)^{-1}$, or $\lambda(1) = -h(1)$ in additive notation. From (7.13) it follows: $g'(1,\tau) = g(1,\tau) - h(\tau) + h(1)^{\tau} + h(\tau) = g(1,\tau) + h(1)^{\tau} = \lambda(1)^{\tau} - \lambda(1)^{\tau} = 0$. Since g' as defined by (7.13) is by definition equivalent to g, we have that a non-normalized factor system is equivalent to a normalized one.

If, for a certain $h : \pi \to A$, one has

$$g(\sigma,\tau) = -h(\sigma\tau) + h(\sigma)^{\tau} + h(\tau) \tag{7.14}$$

then g is a 2-*cobord*. It is clear that if g is a 2-cobord then $-g$ is also a cobord, where $-g$ is determined by $-h$ defined by $(-h)(\sigma) = -h(\sigma)$. Then two 2-cocycles are equivalent when their difference is a 2-cobord.

Let us now show that given a 2-cocycle $g : \pi \times \pi \to A$ it is possible to construct an extension E of A that realizes π. As set E let us take the cartesian product $\pi \times A$ and let us introduce the operation

$$(\sigma, a_1)(\tau, a_2) = (\sigma\tau, g(\sigma,\tau) + a_1^{\tau} + a_2). \tag{7.15}$$

The pair (1,0) is the identity and $(\sigma,a)^{-1} = (\sigma^{-1}, -a^{\sigma^{-1}} - g(\sigma^{-1},\sigma)^{\sigma^{-1}})$. The mapping $(\sigma, a) \to \pi$ is a surjective homomorphism with kernel $A^* = \{(1,a), a \in A\} \simeq A$ e $E/A^* \simeq \pi$. If we choose as representatives of A^* the pairs $(\sigma,0)$ we have $(\sigma,0)^{-1}(1,a)(\sigma,0) = (\sigma^{-1}, -g(\sigma^{-1},\sigma)^{\sigma^{-1}})(1,a)(\sigma,0) = (\sigma^{-1}, -g(\sigma^{-1},\sigma)^{\sigma^{-1}} + a)(\sigma,0) = (1, g(\sigma^{-1},\sigma) - g(\sigma^{-1},\sigma) + a^{\sigma}) = (1,a^{\sigma})$, so that, by identifying $(1,a)$ with a, the group E realizes π.

With our choice of representatives of A^*, i.e $\lambda(\sigma) = (\sigma,0)$, the group E reproduces the 2-cocycle g used to construct it. Indeed, we have $(\sigma,0)(\tau,0) = (\sigma\tau,0)(1,g_1(\sigma,\tau))$, for a certain $g_1(\sigma,\tau) \in A^*$. From the product $(\sigma,0)(\tau,0)$ $=(\sigma\tau,0)(1,g(\sigma,\tau))$, we have $g_1(\sigma,\tau) = g(\sigma,\tau)$, for all $\sigma,\tau \in \pi$, and $g_1 = g$.

If $g' \sim g$, the groups with the operation (7.15) obtained by means of g and g' are isomorphic, an isomorphism being given by $(\sigma,a) \to (\sigma, -h(\sigma) + a)$, where h is the function of (7.14) giving the equivalence. Clearly, the given mapping is bijective. Moreover, $(\sigma, -h(\sigma) + a_1)(\tau, -h(\tau) + a_2) = (\sigma\tau, g'(\sigma,\tau) - h(\sigma)^{\tau} + a_1^{\tau} - h(\tau) + a_2) = (\sigma\tau, g(\sigma,\tau) - h(\sigma\tau) + h(\sigma)^{\tau} + h(\tau) - h(\sigma)^{\tau} + a_1^{\tau} - h(\tau) + a_2) = (\sigma\tau, -h(\sigma\tau) + g(\sigma,\tau) + a_1^{\tau} + a_2)$, which is the image of $(\sigma\tau, g(\sigma,\tau) + a_1^{\tau} + a_2) = (\sigma, a_1)(\tau, a_2)$.

In this way we have determined all the extensions of A that realize π. As already observed, we do not have a one-to-one correspondence between the classes of equivalent 2-cocycles and the classes of extensions that as groups are isomorphic, because it may happen that two non equivalent 2-cocycles give rise to groups E and E' that are isomorphic as groups but not as extensions, as the following example shows.

Example 7.4. Let $\pi = \{1, \sigma, \tau\}$ and $A = \{0, x_1, x_2\}$, both isomorphic to the cyclic group of order 3. Since $\mathbf{Aut}(A) \simeq C_2$, the action of π on A can only be trivial. An extension E of A that realizes the trivial action is the cyclic group C_9, obtained by means of the 2-cocycle:

$$g_1(\sigma, \tau) = \begin{cases} 0, & \text{if } \sigma = 1 \text{ or if } \tau = 1, \\ x_1, & \text{otherwise.} \end{cases}$$

The pair (σ, x_1) is a generator for E (one has $(\sigma, x_1)^2 = (\tau, g_1(\sigma, \sigma) + 2x_1) = (\tau, 0), \ldots, (\sigma, x_1)^8 = (\tau, x_1), (\sigma, x_1)^9 = (1, 0)$). The group E' obtained in the same way by using the 2-cocycle

$$g_2(\sigma, \tau) = \begin{cases} 0, & \text{if } \sigma = 1 \text{ or if } \tau = 1, \\ x_2, & \text{otherwise,} \end{cases}$$

is also cyclic of order 9 (the pair (σ, x_2) is a generator here too). However, g_1 and g_2 are not equivalent. Indeed, suppose there exists h such that:

$$g_2(\sigma, \tau) = g_1(\sigma, \tau) + h(\sigma) + h(\tau) - h(\sigma\tau), \tag{7.16}$$

for all pairs $\sigma, \tau \in \pi$. Then for $\sigma = 1$ one has $0 = 0 + h(1) + h(\tau) - h(\tau)$, and therefore $h(1) = 0$. Now consider (7.16) for $\tau = \sigma$ (recall that $\sigma^2 = \tau$):

$$g_2(\sigma, \sigma) = g_1(\sigma, \sigma) + h(\sigma) + h(\sigma) - h(\tau). \tag{7.17}$$

If $h(\sigma) = 0$, from (7.17) we have $x_2 = x_1 - h(\tau)$, i.e. $h(\tau) = x_2$, so that (7.16) becomes $x_2 = x_1 + x_2$ and $x_1 = 0$, which is absurd. If $h(\sigma) = x_1$, (7.17) yields $x_2 = 3x_1 - h(\tau)$, and since $3x_1 = 0$ we have $h(\tau) = x_1$; but then from (7.16) it follows $x_2 = 3x_1 = 0$, which also is absurd. Finally, if $h(\sigma) = x_2$, from (7.17) we have $x_2 = x_1 + 2x_2 - h(\tau)$, and $h(\tau) = 0$, and (7.16) gives $x_2 = x_1 + x_2$ and $x_1 = 0$, which is again absurd. Therefore, the equivalence relation established between the 2-cocycles is in general finer then that of isomorphism.

The function g such that $g(\sigma, \tau) = 0$ for all $\sigma, \tau \in \pi$ is a 2-cocycle, the *null* 2-cocycle. With this g, (7.15) becomes $(\sigma, a_1)(\tau, a_2) = (\sigma\tau, a_1^\tau + a_2)$, and the extension E is the semidirect product (the extension splits). Moreover, if g' is equivalent to the null 2-cocycle g then it is of the form (7.14), i.e it is a 2-cobord, and the groups obtained by means of g and g' are isomorphic. Hence:

Theorem 7.8. *An extension E of A realizing π splits if and only if the 2-cocycle relative to E is a 2-cobord.*

Remark 7.2. Let $\{1\} \rightarrow A \xrightarrow{\varphi} G \xrightarrow{\psi} B \rightarrow \{1\}$, be an exact sequence of groups, and let $\lambda : B \rightarrow G$ be a homomorphism such that $\psi\lambda = id_B$. This is the case for a semidirect product (split extension): there is a choice function $\lambda : B \rightarrow G$ of representatives of A which is an injective homomorphism (the chosen representatives form a subgroup isomorphic to B). Conversely, if there is a homomorphism such that $\psi\lambda = id_B$, then the extension splits. Indeed, let $g \in G$; then $\psi(g) = b \in B$, and $\psi(g(\lambda(b^{-1})) = \psi(g)\psi(\lambda(b^{-1})) = \psi(g)b^{-1} = 1$. Hence $g\lambda(b^{-1}) = a \in A$, so $g = a\lambda(b^{-1})^{-1}$, i.e $G = A\lambda(B)$. Let $g \in A \cap \lambda(B)$; then $\psi(g) \in \psi(\varphi(A)) = \{1\}$ (the sequence is exact at G). But $g = \lambda(b)$, some $b \in B$, so from $\psi(g) = 1$ it follows $1 = \psi\lambda(b) = b$, and $g = \lambda(1) = 1$. The presence of a 2-cocycle relative to an extension which is not a 2-cobord witnesses the fact that λ is not a homomorphism.

To the set of 2-cocycles it can be given the structure of an abelian group with the obvious operation of addition, and also to the set of 2-cobords. The former is denoted by $Z^2(\pi, A)$, and the latter by $B^2(\pi, A)$.

Definition 7.3. The quotient: $H^2(\pi, A) = Z^2(\pi, A)/B^2(\pi, A)$ is the *second cohomology group of π with coefficients in A.*

The above discussion may now be summarized as follows:

Theorem 7.9. *There exists a one-to-one correspondence between the elements of $H^2(\pi, A)$ and the equivalence classes of extensions of A realizing π. This correspondence sends the zero of $H^2(\pi, A)$ to the class of the semidirect product.*

If π and A are finite, there is only a finite number of functions $\pi \times \pi \to A$, and so a finite number of 2-cocycles. In particular, $H^2(\pi, A)$ is finite. (That the number of extensions E is finite also follows from the fact that if $E/A \simeq \pi$, then $|E| = |\pi||A|$, so that there is only a finite number of groups E.)

As in the case of H^1 we have, for all A,

Theorem 7.10. *If $|\pi| = n$, then $nH^2(\pi, A) = 0$.*

Proof. The proof is similar to that for H^1. Fix τ and η in (7.12) and sum over σ: $\sum_\sigma g(\sigma, \tau\eta) + \sum_\sigma g(\tau, \eta) = \sum_\sigma g(\sigma\tau, \eta) + \sum_\sigma g(\sigma, \tau)^\eta$. Setting $\sum_\sigma g(\sigma, \tau) = \gamma(\tau)$, and taking into account that $\sigma\tau$ runs over all elements of π as σ varies in π, we have

$$\gamma(\tau\eta) + ng(\tau, \eta) = \gamma(\eta) + \gamma(\tau)^\eta, \tag{7.18}$$

and $ng(\tau, \eta) = \gamma(\eta) + \gamma(\tau)^\eta - \gamma(\tau\eta) \in B^2(\pi, A)$ (set $h = \gamma$ in (7.14)). ♢

As we have done for H^1, we consider here too the case A finite, $|A| = m$, and $(n, m) = 1$. There exists r such that $rn \equiv 1 \bmod m$. Multiplying (7.18) by r we obtain $g(\tau, \eta) = r\gamma(\eta) + (r\gamma(\tau))^\eta - r\gamma(\tau\eta)$, and setting $r\gamma = h$ we have $g(\tau, \eta) \in B^2(\pi, A)$: in this case, we already have $H^2(\pi, A) = 0$. It follows:

Theorem 7.11 (Schur-Zassenhaus). *Let A be a finite abelian group, E an extension of A such that $(|A|, |E/A|) = 1$. Then E is a semidirect product.*

Proof. Apply the previous argument with $\pi = E/A$. ♢

Remark 7.3. We know that, under the hypotheses of the theorem, we also have $H^1(\pi, A) = 0$, so that two complements of A in the semidirect product of π by A are conjugate. Another case in which every extension of A by π splits for any A is the one in which π is a free group (cf. *Ex.* 7.5, 3 below).

Theorem 7.11 also holds if A is not abelian:

Corollary 7.5. *Let E be an extension of N with $(|N|, |E/N|) = 1$. Then E is a semidirect product.*

Proof. We reduce to the abelian case, and then apply the preceding theorem. Note that it is sufficient to prove that E contains a subgroup of order $|E/N|$. Induct on $|N|$. If $|N| = \{1\}$ there is nothing to prove. Let $n = |N| > 1$ and S a p-Sylow subgroup of N. By the Frattini argument, if $H = \mathbf{N}_G(N)$ then $E = NH$ and $m = |E/N| = |NH/N| = |H/(N \cap H)|$. The group H/S contains $(H \cap N)/S$ as a normal subgroup, of order a proper divisor of n (since $|S| > 1$) and of index m in H/S. By induction, H/S contains a subgroup L/S of order m. $Z = \mathbf{Z}(S)$ is normal in L, S/Z is normal in L/Z and has index m, and since $p \nmid m$, by induction L/Z contains a subgroup K/Z of order m. Therefore, K is an extension of the abelian group Z, of order a power of p, by K/S, of order m prime to p. By the preceding theorem K, and therefore E, contains a subgroup of order m. $\diamondsuit$

If N or E/N is solvable, the hypothesis $(|N|, |E/N|) = 1$ implies that two complements for N are conjugate (Theorems 7.4 and 7.5)[2].

Examples 7.5. 1. As in the case of H^1, we have $H^2(\pi, A) = 0$ if π is finite and A is divisible and torsion free. The proof is the same as that for H^1.

2. With reference to *Ex.* 7.4, recall that for any prime p there are two groups of order p^2 i.e. Z_{p^2} (which is not a semidirect product) and $Z_p \times Z_p$ (which is a semidirect product, and in fact direct). It follows $|H^2(Z_p, Z_p)| > 1$, and since $pH^2(Z_p, Z_p) = 0$, the group $H^2(Z_p, Z_p)$ has at least p elements. Therefore, if $p > 2$ there certainly exist extensions of Z_p by Z_p that are isomorphic but not equivalent. (Also in this more general case, since $\mathbf{Aut}(Z_p) \simeq Z_{p-1}$, an action Z_p on Z_p can only be trivial, so that an extension realizing it can only be central).

3. If $\pi = \mathbf{Z}$, then $H^2(\pi, A) = 0$ all A. Indeed, if E realizes π and $E/A \simeq \pi$, then $E/A = \langle Ae \rangle$ for a certain $e \in E$, and Ae being of infinite order, it cannot be $e^n \in A$ for $n \neq 0$. Hence $\langle e \rangle \cap A = \{1\}$, $E = \langle e \rangle A$, and E is the semidirect product of $\mathbf{Z} \simeq \langle e \rangle$ by A. The proof is similar in case π is nonabelian free. Indeed, $E/A \simeq \langle e_\lambda,\ \lambda \in \Lambda \rangle \simeq \pi$, with the Ae_λ free generators. π being

[2] Since $(|N|, |E/N|) = 1$ one of the two groups has odd order and therefore, by the Feit–Thompson theorem, already quoted in Chapter 5, is solvable. Hence, any two complements for N are always conjugate.

free, no product $Ae_{\lambda_1}Ae_{\lambda_2}\cdots Ae_{\lambda_k} = Ae_{\lambda_1}e_{\lambda_2}\cdots e_{\lambda_k}$ can equal A, so that no product $e_{\lambda_1}e_{\lambda_2}\cdots e_{\lambda_k}$ belongs to A. Then $A\cap\langle e_\lambda\ \lambda\in\Lambda\rangle = \{1\}$ and therefore $E = A\langle e_\lambda\ \lambda\in\Lambda\rangle \simeq A\pi$, a semidirect product.

4. If $A = C_2$, an action of a group π on A is trivial. Let $\pi = V$, the Klein group; then an extension of C_2 by V has order 8, and is central. There are five groups of order 8; one is cyclic, and this cannot have V as a quotient. The other four, $C_2\times C_2\times C_2, C_4\times C_2, D_4$ e Q, are all central extensions of C_2 by V. This is obvious for the first group. The second one is $U(15)$, with $C_4 = \{1,2,4,8\}$ and $C_2 = \{1,11\}$. The latter group C_2 is not the one we need because the quotient is C_4; we must take $C_2 = \{1,4\}$. For the dihedral group and the quaternion group take the respective centers. The groups of order 8 can also be obtained as extensions (either central or not) of V and C_4 by C_2 (see *Ex.* 2.10, 5).

7.3.1 H^1 and Extensions

If E is an extension of the abelian group A we have seen how A becomes an E/A-modulo. We want to determine the structure of $H^1(E/A, A)$ starting from E. Consider the subgroup D of the automorphism group of E consisting of the automorphisms fixing A and E/A elementwise: $a^\alpha = a$, $(Ax)^\alpha = Ax$ for all $a\in A$ and $x\in E$. The second equality is equivalent to $x^{-1}x^\alpha\in A$. We shall see that $H^1(E/A, A)$ is isomorphic to a quotient of D. The group D is abelian: given $\alpha,\beta\in D$ we have, for all $x\in E$, $x^{-1}x^\alpha = a$ and $x^{-1}x^\beta = a'$, some $a, a'\in A$. Hence $x^{\alpha\beta} = (xa)^\beta = x^\beta a = xa'a$ and $x^{\beta\alpha} = (xa')^\alpha = x^\alpha a' = xaa'$.

The function $f_\alpha : Ax \to x^{-1}x^\alpha$, $\alpha\in D$ is a 1-cocycle of the E/A-module A. Firstly, it is well defined: if $Ax = Ay$, then $y = ax$ and $y^{-1}y^\alpha = x^{-1}a^{-1}(ax)^\alpha = x^{-1}a^{-1}ax^\alpha = x^{-1}x^\alpha$. Moreover, $f_\alpha(Ax\cdot Ay) = (xy)^{-1}(xy)^\alpha = y^{-1}x^{-1}x^\alpha y^\alpha = y^{-1}x^{-1}x^\alpha yy^{-1}y^\alpha = (x^{-1}x^\alpha)^y(y^{-1}y^\alpha) = f_\alpha(Ax)^{Ay}f_\alpha(Ay)$.

The correspondence $D\to Z^1(E/A, A)$ given by $\alpha\to f_\alpha$, is in fact an isomorphism: if $\beta\to f_\beta$, then $f_{\alpha\beta}(Ax) = x^{-1}x^{\alpha\beta} = x^{-1}x^\alpha(x^{-1})^\alpha x^{\alpha\beta} = x^{-1}x^\alpha(x^{-1})^\alpha x^{\beta\alpha} = x^{-1}x^\alpha(x^{-1}x^\beta)^\alpha = x^{-1}x^\alpha(x^{-1}x^\beta) = f_\alpha(Ax)f_\beta(Ax)$, recalling that α is trivial on A. If $f_\alpha = f_\beta$, then $x^{-1}x^\alpha = x^{-1}x^\beta$, $x^\alpha = x^\beta$ for all $x\in E$, and therefore $\alpha = \beta$. This shows that the correspondence is one-to-one. It is also onto: given a 1-cocycle f, we define $\alpha : x\to xf(Ax)$ and show that α is an automorphism of E. It is bijective: on each coset xA of A, multiplication by $f(xA)$ is a permutation of the elements of that coset. Indeed, if x_i is a representative of xA, and $y\in xA$, then $y = x_ia$, some a, and since $a = a'f(xA)$ for some a', y is the image of the (unique) element x_ia'. Hence α is a permutation of E. Moreover, $(xy)^\alpha = xyf(AxAy) = xyf(Ax)^yf(Ay) = xf(Ax)\cdot yf(Ay) = x^\alpha y^\alpha$. Thus α is an automorphism of E. Finally, if $x\in A$, then $f(Ax) = 1$ and $x^\alpha = x$. By definition $x^{-1}x^\alpha = f(Ax)\in A$, so α is trivial on A and E/A.

If α corresponds to a cobord, $f(Ax) = a^xa^{-1}$ for a fixed $a\in A$, and therefore $x^\alpha = xa^xa^{-1} = axa^{-1}$. Thus α is conjugation by a^{-1}, $\alpha = \gamma_{a^{-1}}$.

Conversely, conjugation by an element a of A belongs to D since it fixes all the elements of A (A being abelian) and $x^{-1}x^a = x^{-1}a^{-1}xa = (x^{-1}a^{-1}x)a \in A$ (A being normal). Therefore, the group D contains a subgroup D_0 isomorphic to A (through the isomorphism $a \to \gamma_{a^{-1}}$). We can conclude as follows:

Theorem 7.12. *If E is an extension of the abelian group A, the first cohomology group $H^1(E/A, A)$ is isomorphic to the quotient group D/D_0, where D is the group of automorphisms of E fixing A and E/A elementwise, and $D_0 \simeq A$ is the subgroup of the inner automorphisms induced by the elements of the group A.*

7.3.2 $H^2(\pi, A)$ for π Finite Cyclic

Let $\pi = \{1, \sigma, \sigma^2, \ldots, \sigma^{n-1}\}$, A a π–module and E an extension of A realizing π. In addition, let $E/A = \{A, Ax, Ax^2, \ldots, Ax^{n-1}\} \simeq \pi$, in the isomorphism $\sigma^i \to Ax^i$, and λ the function that chooses the representative x^i of the coset associated with σ^i, $\lambda(\sigma^i) = x^i$ (note that $\lambda(\sigma^i) = \lambda(\sigma)^i$).

If $i+j < n$, the product $x^i x^j$ equals x^{i+j}, which is the representative of the coset associated with the product $\sigma^i\sigma^j$: $\lambda(\sigma^i)\lambda(\sigma^j) = \lambda(\sigma^i\sigma^j)$. If $i+j \geq n$, the element x^{i+j} is no longer a representative; the coset to which it belongs is represented by x^k,where $i+j = n+k$: $x^i x^j = x^{i+j} = x^{n+k} = x^n x^k$, and therefore $\lambda(\sigma^i)\lambda(\sigma^j) = a\lambda(\sigma^k) = a\lambda(\sigma^i\sigma^j)$, where $a = x^n$. Hence the choice function λ determines the 2-cocycle:

$$g(\sigma^i, \sigma^j) = \begin{cases} 1, \text{ if } i+j < n \\ a, \text{ if } i+j \geq n. \end{cases} \tag{7.19}$$

Now $a = x^n$ commutes with x, and since E realizes π, we have $a^\sigma = \lambda(\sigma)^{-1}a\lambda(\sigma) = x^{-1}ax = a$, and therefore a is fixed by all the elements of π, i.e. $a \in A^\pi$. Conversely, for each $a \in A^\pi$ the function g of (7.19) is a 2-cocycle. Indeed, let us show that $g(\sigma^i, \sigma^j\sigma^k)g(\sigma^j, \sigma^k) = g(\sigma^{i+j}, \sigma^k)g(\sigma^i, \sigma^j)$, for all $i, j, k < n$ (which is (7.12) written multiplicatively, and recalling that π is trivial on a and therefore on the function $g(\sigma^i, \sigma^j)$ whose value is 1 or a). Let us write the previous equality as $uv = zt$, and consider various cases:

1. $i+j < n$; then $t = 1$.

1*a*. $j+k < n$. Then $v = 1$; if $i+j+k \geq n$, then $u = z = a$; if $i+j+k < n$, then $u = z = 1$. In any case $uv = zt$.

1*b*. $j+k \geq n$, and therefore $v = a$; moreover, $(i+j)+k = i+(j+k) \geq n$, so that $z = a$. Now $\sigma^j\sigma^k = \sigma^{j+k} = \sigma^{j+k-n}$, and we have necessarily $i+j+k-n < n$ because $i+j < n$ and $k < n$. Hence $u = 1$, and in this case also $uv = zt$.

2. $i+j \geq n$; $t = a$.

2*a*. $j+k < n$; $v = 1$. Moreover $z = 1$ because, i and $j+k$ being less than n, we have $i+j-n+k < n$, and $u = a$ because $i+j+k = (i+j)+k \geq n+k \geq n$.

2*b*. $j+k \geq n$; $v = a$. Moreover $i + (j+k-n) = (i+j-n)+k$, and therefore they are both less or both greater than n; hence $u = z$.

In this way we have a one-to-one mapping between A^π and $Z^2(\pi, A)$, and it is immediate to see that it is an isomorphism: $A^\pi \simeq Z^2(\pi, A)$.

If g is a 2-cobord, relative to some function $h : \pi \longrightarrow A$, we have $g(\sigma^i, \sigma^j) = h(\sigma^i)^{\sigma^j} h(\sigma^j) h(\sigma^{i+j})^{-1}$, and since (7.19) should hold,

$$h(\sigma^i)^{\sigma^j} h(\sigma^j) h(\sigma^{i+j})^{-1} = \begin{cases} 1, \text{ if } i+j < n \\ a, \text{ if } i+j \geq n, \end{cases} \tag{7.20}$$

with $a \in A^\pi$. Let $h(\sigma) = b \in A$. From the first alternative of (7.19) it follows, for $i = 0$ and any j, that $h(1) = 1$. For $i = 1$ $j = 1$, $b^\sigma b = h(\sigma^2)$; for $i = 1$ and $j = 2$, $b^{\sigma^2} h(\sigma^2) = h(\sigma^3)$, which together with the previous one implies $h(\sigma^3) = b^{\sigma^2} b^\sigma b$. In general, for $i = 1, 2, \ldots, n-1$, and recalling that A is abelian, $h(\sigma^i) = b b^\sigma b^{\sigma^2} \cdots b^{\sigma^{i-1}}$. Hence, the value of h on π is determined by the first alternative of (7.19). Now given h as in (7.20) and with $h(1) = 1$, we have, if $i + j < n$,

$$\begin{aligned} h(\sigma^i)^{\sigma^j} h(\sigma^j) &= b^{\sigma^j} b^{\sigma^{j+1}} b^{\sigma^{j+2}} \cdots b^{\sigma^{j+i-1}} \cdot b b^\sigma b^{\sigma^2} \cdots b^{\sigma^{j-1}} \\ &= b b^\sigma b^{\sigma^2} \cdots b^{\sigma^{j+i-1}} \\ &= h(\sigma^{i+j}). \end{aligned}$$

If instead $i + j \geq n$, let us evaluate $h(\sigma^i)^{\sigma^j}$:

$$\begin{aligned} h(\sigma^i)^{\sigma^j} &= (b b^\sigma b^{\sigma^2} \cdots b^{\sigma^{i-1}})^{\sigma^j} \\ &= b^{\sigma^j} b^{\sigma^{j+1}} \cdots b^{\sigma^{j+i-1}} \\ &= b^{\sigma^j} b^{\sigma^{j+1}} \cdots b^{\sigma^{j+(n-j)}} b^{\sigma^{j+(n-j+1)}} b^{\sigma^{j+(n-j+k-1)}} \\ &= b^{\sigma^j} b^{\sigma^{j+1}} \cdots b^{\sigma^{n-1}} \cdot b^{\sigma^n} \\ &= b \cdot b^\sigma \cdots b^{\sigma^{k-1}}, \end{aligned}$$

where $i+j = n+k$ (and therefore $i = n+k-j$). With $h(\sigma^j) = b b^\sigma b^{\sigma^2} \cdots b^{\sigma^{j-1}}$ and $h(\sigma^{i+j})^{-1} = h(\sigma^{n+k})^{-1} = (b b^\sigma b^{\sigma^2} \cdots b^{\sigma^{k-1}})^{-1}$ we have:

$$\begin{aligned} &h(\sigma^i)^{\sigma^j} h(\sigma^j) h(\sigma^{i+j})^{-1} = \\ &b^{\sigma^j} b^{\sigma^{j+1}} \cdots b^{\sigma^{n-1}} b b^\sigma \cdots b^{\sigma^{j-1}} \cdots b^{\sigma^{k-1}} \cdot b b^\sigma \cdots b^{\sigma^{j-1}} \cdot (b b^\sigma \cdots b^{\sigma^{k-1}})^{-1} = \\ &b b^\sigma \cdots b^{\sigma^{n-1}}, \end{aligned}$$

and with $a = b b^\sigma b^{\sigma^2} \cdots b^{\sigma^{n-1}} = N(b)$, the norm of b, (7.19) holds. If a is of the form $N(b)$ for some b, then the 2-cobord (7.19) is determined, with h defined by (7.20), and therefore:

$$B^2(\pi, A) \simeq \{N(b),\ b \in A\}.$$

By composing the isomorphism $A^\pi \simeq Z^2(\pi, A)$ with the projection

$$Z^2(\pi, A) \to Z^2(\pi, A)/B^2(\pi, A)$$

we obtain a surjective homomorphism whose kernel is precisely the set $\{N(b),\ b \in A\}$.

Hence we may conclude with the following theorem.

Theorem 7.13. *If π is a finite cyclic group of order n and A a π–module, then:*

$$H^2(\pi, A) \simeq \frac{A^\pi}{\{N(a),\ a \in A\}}.$$

In particular, if A is a trivial π–module, then $H^2(\pi, A) \simeq A/A^n$. Moreover, if A is n-divisible, i.e. $A^n = A$, then $H^2(\pi, A) = \{1\}$.

Remark 7.4. The elements of the form (7.20) rise when the representatives of the cosets of A are changed, and precisely when one takes as representatives the elements $z^i = (bx)^i$. Indeed, with such a choice we have:

$$\begin{aligned} z^n &= (bx)^n = bxbx \cdots bx = b \cdot xbx^{-1} \cdot x^2bx \cdots bx \\ &= b \cdot xbx^{-1} \cdot x^2bx^{-2} \cdot x^3bx^{-3} \cdots x^{(n-1)}bx^{-(n-1)} \cdot x^n \\ &= bb^{x^{-1}}b^{x^{-2}} \cdots b^{x^{-(n-1)}}x^n \\ &= bb^xb^{x^2} \cdots b^{x^{n-1}}x^n \end{aligned}$$

since x^{-i} runs over the whole π as i varies. Moreover, recalling that E realizes π, we have $b^{x^i} = b^{\sigma^i}$. If g and g' are the 2-cocycles relative to the choices of representatives x^i and z^i, respectively, we have:

$$g'g^{-1} = \begin{cases} 1, & \text{if } i + j < n \\ z^nx^{-n} = bb^\sigma b^{\sigma^2} \cdots b^{\sigma^{n-1}}, & \text{if } i + j \geq n. \end{cases}$$

Exercises

With reference to *ex.* 8, let f be a 1-cocycle $\pi \to C$, $f(\sigma) = c_\sigma = \beta(b_\sigma)$. Then $\beta(b_{\sigma\tau}) = c_{\sigma\tau} = c_\sigma^\tau + c_\tau = \beta(b_\sigma)^\tau + \beta(b_\tau) = \beta(b_\sigma^\tau) + \beta(b_\tau)$ and therefore $b_{\sigma\tau} - b_\sigma^\tau - b_\tau \in ker(\beta) = A$. Hence there exists $a_{\sigma,\tau}$ such that $a_{\sigma,\tau} = b_{\sigma\tau} - b_\sigma^\tau - b_\tau$.

14. Show that $g : \pi \times \pi \to A$ defined by $(\sigma, \tau) \to a_{\sigma,\tau}$, where $a_{\sigma,\tau}$ is determined as above by a 1-cocycle $f : \pi \to C$, is a 2-cocycle of π in A. Moreover if two cocycles are determined by f and f' with $f' \sim f$ then they are equal.

A different choice of inverse images of c_σ gives rise to the equivalence relation between the 2-cocycles we know. Indeed, let $c_\sigma = \beta(b_\sigma) = \beta(b'_\sigma)$. Then $b_\sigma - b'_\sigma \in ker(\beta) = A$ and we have $b_\sigma - b'_\sigma = a_\sigma$. If $a'_{\sigma,\tau}$ and $a_{\sigma\tau}$ are determined by b' and b then:

$$\begin{aligned} a'_{\sigma,\tau} &= b'_{\sigma\tau} - b'^\tau_\sigma - b'^\tau = b_{\sigma\tau} - b_\sigma^\tau - a_{\sigma\tau} + a_\sigma^\tau + a_\tau \\ &= a_{\sigma,\tau} - a_{\sigma\tau} + a_\sigma^\tau + a_\tau. \end{aligned}$$

Let us associate with a class $[f]$ of 1-cocycles of π in C the class $[g]$ of 2-cocycles of π in A, where g is obtained from f as seen above, $\delta : [f] \rightarrow [g]$ (this mapping δ is called *transgression.*)

15. Prove that the sequence:

$$\cdots \rightarrow H^1(\pi, B) \xrightarrow{\overline{\beta}} H^1(\pi, C) \xrightarrow{\delta} H^2(\pi, A)$$

is exact at $H^1(\pi, C)$.

If g is a 2-cocycle of π in A, by composing g with α we obtain a 2-cocycle of π in B; in this way α induces $\overline{\alpha} : [g] \rightarrow [\alpha g]$.

16. Prove that the sequence:

$$\cdots \rightarrow H^1(\pi, C) \xrightarrow{\delta} H^2(\pi, A) \xrightarrow{\overline{\alpha}} H^2(\pi, B)$$

is exact at $H^2(\pi, A)$.

Finally, consider the sequence:

$$\cdots \rightarrow H^2(\pi, A) \xrightarrow{\overline{\alpha}} H^2(\pi, B) \xrightarrow{\overline{\beta}} H^2(\pi, C)$$

where $\overline{\beta}$ is defined by composition: $\overline{\beta}[g] = [\beta g]$.

17. Prove that the sequence above is exact at $H^2(\pi, B)$.

The exercises above may be resumed as follows: *the sequence*

$0 \rightarrow H^0(\pi, A) \xrightarrow{\alpha^{\pi}} H^0(\pi, B) \xrightarrow{\beta^{\pi}} H^0(\pi, C) \xrightarrow{\delta} H^1(\pi, A) \xrightarrow{\overline{\alpha}} H^1(\pi, B) \xrightarrow{\overline{\beta}} H^1(\pi, C) \xrightarrow{\delta}$
$H^2(\pi, A) \xrightarrow{\overline{\alpha}} H^2(\pi, B) \xrightarrow{\overline{\beta}} H^2(\pi, C)$

is exact (this is the *exact cohomology sequence*).

18. *i)* Prove that if $H \trianglelefteq \pi$, and A is a π-module, then A^H is a π/H module via the obvious action $a^{\sigma H} = a^{\sigma}$;
ii) prove that A^H is also a π-module with the π-module structure induced by the canonical homomorphism $\lambda : \pi \rightarrow G/H$;
iii) by the above we have a map $H^1(\pi/H, A^H) \rightarrow H^1(\pi, A^H)$ given by $[\phi] \rightarrow [\phi\lambda]$;
iv) the inclusion of π-modules $\iota : A^H \rightarrow A$ induces a map $H^1(\pi, A^H) \rightarrow H^1(\pi, A)$ given by $[\phi] \rightarrow [\iota\phi]$;
v) the composition $[\phi] \rightarrow [\iota\phi] \rightarrow [\iota\phi\lambda]$ yields the *inflation map*

$$\inf : H^1(\pi/H, A^H) \rightarrow H^1(\pi, A).$$

19. Prove that the sequence

$$\{0\} \rightarrow H^1(\pi/H, A^H) \xrightarrow{\inf} H^1(\pi, A)$$

is exact. [*Hint*: if $[\phi] \in ker(\inf)$, then $\iota\phi\lambda$ is principal, i.e. there exists $a \in A$ such that $\iota\phi\lambda(\sigma) = a^{\sigma} - a$. Prove that $a \in A^H$.]

The injection $j : H \rightarrow \pi$ induces the *restriction map*

$$\text{res} : H^1(\pi, A) \rightarrow H^1(H, A)$$

given by $[\psi] \rightarrow [\psi j]$.

20. (*The inflation–restriction sequence*) The sequence

$$\{0\} \to H^1(\pi/H, A^H) \stackrel{\text{inf}}{\to} H^1(\pi, A) \stackrel{\text{res}}{\to} H^1(H, A)$$

is exact. [*Proof.* We have $[\iota\phi\lambda\mu(h)] = [\iota\phi\lambda(h)] = [\iota\phi(H) = 0]$ (being $\phi(H) = \{0\}$). This proves that $Im(\text{inf}) \subseteq ker(\text{res})$. For the other inclusion, let $[\phi] \in ker(\text{res})$, i.e. $[\phi\mu] = 0$ (so $\phi\mu$ is principal). Hence there exists $a \in A$ such that $\phi\mu(h) = \phi(h) = a^h - a$, for all $h \in H$. With this a, let $\eta : \pi \to A$ be the principal crossed homomorphism $\eta(\sigma) = a^\sigma - a$; then $(\phi - \eta)\mu(h) = \phi\mu(h) - \eta\mu(h) = \phi(h) - \mu(h) = (a^h - a) - (a^h - a) = 0$. But $[\phi - \eta] = [\phi]$, so we may assume that ϕ restricted to H is zero, $\phi\mu = 0$. It follows $\phi(\sigma h) = \phi(\sigma) + \phi(h)^\sigma = \phi(\sigma)$ and $\phi(h\sigma) = \phi(h) + \phi(\sigma)^h = \phi(\sigma)^h$. Therefore, $\phi(\sigma)^h = \phi(h\sigma) = \phi(\sigma h') = \phi(\sigma)$, i.e. $\phi(\sigma) \in A^H$. Hence define $\phi_1(\sigma) : \pi \to A^H$ by $\phi_1(\sigma) = \phi(\sigma)$, and for $\iota : A^H \to A$ we have $\iota\phi_1 = \phi$. Then we can define $\psi : \pi/H \to A^H$ by $\psi\lambda(\sigma) = \phi_1(\sigma)$, i.e. $\psi\lambda = \phi_1$. It follows $\iota\psi\lambda = \iota\phi_1 = \phi$, and therefore $[\phi] = \text{inf}([\psi])$, as required. In dimension zero, $H^0(\pi/H, A^H) = (A^H)^{\pi/H} = A^\pi$, $H^0(\pi, A) = A^\pi, H^0(H, A) = A^H$, the inflation is the identity $A^\pi \to A^\pi$, and the restriction is the inclusion $A^\pi \to A^H$; hence the sequence $\{0\} \to A^\pi \stackrel{\text{id}}{\to} A^\pi \stackrel{\iota}{\to} A^H$ cannot be exact ($ker(\iota) = 0 \neq Im(\text{id}) = A^\pi$).

21. Prove that if π is finite, and $\mathbf{Z}$ and $\mathbf{R}(+)/\mathbf{Z}$ are trivial π-modules then $H^2(\pi, \mathbf{Z}) \simeq \mathbf{Hom}(\pi, \mathbf{R}/\mathbf{Z})$. Hence, if π is cyclic of order n, then $H^2(\pi, \mathbf{Z})$ is also cyclic of order n.

22. Prove that $B^2(\pi, \mathbf{C}^*)$ (trivial action) has a complement M in $Z^2(\pi, \mathbf{C}^*)$. [*Hint:* it suffices to show that $B^2(\pi, \mathbf{C}^*)$ is divisible (Corollary 4.5).] (The group $M \simeq Z^2(\pi, \mathbf{C}^*)/B^2(\pi, \mathbf{C}^*)$ is isomorphic to the Schur multiplier $M(\pi)$; see next section.)

7.4 The Schur Multiplier

Definition 7.4. Let π be a group, $\mathbf{C}^*$ the multiplicative group of non zero complex numbers considered as a trivial π-module. The *Schur multiplier* $M(\pi)$ of the group π is the second cohomology group of π with coefficients in the group $\mathbf{C}^*$:

$$M(\pi) = H^2(\pi, \mathbf{C}^*).$$

Lemma 7.6. *Let π, be finite, $|\pi| = n$. Then, for each 2-cocycle g of π in $\mathbf{C}^*$ there exists a 2-cocycle $g' \sim g$ whose values are n-th roots of unity.*

Proof. Set $\prod_{\tau\in\pi} g(\sigma, \tau) = \gamma(\sigma)$, and let $h(\sigma)$ be an n-th root of $\gamma(\sigma)^{-1}$, with $h(1) = 1$. Define $g'(\sigma, \tau) = g(\sigma, \tau)h(\sigma)h(\tau)h(\sigma\tau)^{-1}$. We have $g' \sim g$ and $g'(\sigma, \tau)^n = g(\sigma, \tau)^n h(\sigma)^n h(\tau)^n h(\sigma\tau)^{-n}$. From (7.18) (written multiplicatively), and taking into account that the action of π is trivial, we have: $g(\sigma, \tau)^n = \gamma(\sigma)\gamma(\tau)\gamma(\sigma\tau)^{-1} = h(\sigma)^{-n}h(\tau)^{-n}h(\sigma\tau)^n$, $g'(\sigma, \tau)^n = 1$, for all $\sigma, \tau \in \pi$. ◇

Theorem 7.14. *If π is finite, then $M(\pi)$ is also finite.*

Proof. If $|\pi| = n$, and since there are n n-th roots of unity, and n^{n^2} functions from $\pi \times \pi$ to an n-element set, there are at most n^{n^2} 2-cocycles like the g' of the lemma. ◇

From Theorem 7.10 it follows:

Corollary 7.6. *If π is a finite p-group, so also is $M(\pi)$.*

From Theorem 7.13 we have:

Corollary 7.7. *If π is finite cyclic, then $M(\pi) = \{1\}$.*

If π is infinite cyclic, we know that the second cohomology group of π is the identity group, no matter what the π-module of the coefficients is.

7.4.1 Projective Representations

The Schur multiplier arises in connection with the study of central extensions of groups, which in turn are related to projective representations. A *projective representation* of degree n of a finite group π is a homomorphism

$$\theta : \pi \to PGL(V)$$

into the group $PGL(V) = GL(V)/K^*$ of projective transformations of a vector space of dimension n over a field K. Clearly, a projective representation θ may be obtained by composing an ordinary (linear) representation $\pi \to GL(V)$ with the natural projection $p : GL(V) \to PGL(V)$. The question arises whether it is always the case that a projective representation is obtained in this way from an ordinary representation. In other words, is it always possible to lift θ to an ordinary (linear) representation which composed with p reproduces θ? In general the answer is no; if we try to lift θ we face an obstruction due to the presence of 2-cocycles as we now see.

Let $\lambda : \pi \to GL(V)$ be a lifting; then for $\sigma \in \pi$, $\theta(\sigma) = \lambda(\sigma)K^*$. Let $\sigma, \tau \in \pi$; from $\theta(\sigma)\theta(\tau) = \theta(\sigma\tau)$ it follows

$$\lambda(\sigma)\lambda(\tau) = \alpha(\sigma,\tau)\lambda(\sigma\tau) \tag{7.21}$$

$\alpha(\sigma,\tau) \in K^*$. By the associative property α is a 2-cocycle; if $\lambda(1) = I$, α is normalized (see Section 7.3). Thus a lifting determines a 2-cocycle α[3]; unless $\alpha = 1$, it is this cocycle that prevents the lifting from being a homomorphism. If λ' is another choice of representatives, then $\lambda'(\sigma) = h(\sigma)\lambda(\sigma)$, and if α' is the cocycle relative to λ' then α and α' differ by the coboundary determined by $h : \pi \to K^*$.

[3] Called a *multiplier* of the projective representation.

In this way, the representation θ determines a class $[\alpha]$ of cocycles, i.e. an element of $H^2(\pi, K^*)$, with the trivial action of π. Hence, a projective representation has a lifting to an ordinary representation if, and only if, $H^2(\pi, K^*)$ is the identity group. On the other hand, a mapping

$$\lambda : \pi \to GL(V) \tag{7.22}$$

satisfying (7.21) reproduces θ when composed with the projection $GL(V) \to GL(V)/K^*$. Thus we may take (7.22) such that (7.21) holds as an alternative definition of a projective representation of π. We say that this representation is *irreducible* if there are no non trivial subspaces of V invariant under $\lambda(\sigma)$ for all $\sigma \in \pi$. Two projective representations θ_1 and θ_2 on two spaces V_1 and V_2 are *projectively equivalent* if there exists an invertible linear transformation ϕ of V_1 in V_2 such that

$$\theta_1(\sigma) = h(\sigma)\phi^{-1}\theta_2(\sigma)\phi, \quad \sigma \in \pi,$$

for a certain $h : \pi \to K^*$. It easily seen that this equivalence induces that of the cocycles α_1 and α_2 relative to θ_1 and θ_2 by means of the coboundary determined by h. If $h(\sigma) = 1$, then θ_1 and θ_2 are *linearly equivalent.*

If α is equivalent to the null cocycle, that is if α is a coboundary, then the corresponding projective representation θ is equivalent to an ordinary one. Indeed, let $\alpha(\sigma, \tau) = h(\sigma)h(\tau)h(\sigma\tau)^{-1}$. Then with $\theta_2(\sigma) = h(\sigma)\theta_1(\sigma)$ we have a representation θ_2 equivalent to θ_1 (with ϕ the identity), and such that $\theta_2(\sigma)\theta_2(\tau) = \theta_2(\sigma\tau)$, and therefore θ_2 is ordinary.

Every cocycle α of π in K^* determines a projective representation of π. Indeed, let V be a vector space over K of dimension equal to the cardinality of π, and let $\{v_\eta\}, \eta \in \pi$ be a basis of V. Define, for $\sigma \in \pi$, $v_\eta\rho(\sigma) = \alpha(\eta, \sigma)v_{\eta\sigma}$; then

$$\begin{aligned} v_\eta\rho(\sigma)\rho(\tau) &= \alpha(\eta, \sigma)v_{\eta\sigma}\rho(\tau) = \alpha(\eta, \sigma)\alpha(\eta\sigma, \tau)v_{\sigma\eta\tau} = \alpha(\eta, \sigma\tau)\alpha(\sigma, \tau)v_{\eta\sigma\tau} \\ &= \alpha(\sigma, \tau)v_\eta\rho(\sigma\tau), \end{aligned}$$

and therefore $\rho(\sigma)\rho(\tau) = \alpha(\sigma, \tau)\rho(\sigma\tau)$. Hence ρ is a projective representation.

Example 7.6. Let $\pi = \{1, a, b, ab\}$, the Klein group, and let α be the normalized cocycle:

$$\begin{aligned} &\alpha(a, b) = \alpha(b, ab) = \alpha(ab, a) = 1, \\ &\alpha(x, y) = -1, \text{ otherwise.} \end{aligned}$$

Then,

$$\begin{aligned} v_1\rho(a) &= \alpha(1, a)v_{1\cdot a} &&= 1 \cdot v_a, \\ v_a\rho(a) &= \alpha(a, a)v_{aa} &&= -1 \cdot v_1, \\ v_b\rho(a) &= \alpha(b, a)v_{ba} &&= -1 \cdot v_{ab}, \\ v_{ab}\rho(a) &= \alpha(ab, a)v_{aba} &&= 1 \cdot b, \end{aligned}$$

and similarly for the other elements of π.

Projective representations naturally arise in the problem of the extension of an ordinary representation of a normal subgroup N of a group π to a representation of π. Indeed, let $\rho : N \to GL(V)$; then ρ^x defined by $\rho^x(h) = \rho(h^x) = \rho(x^{-1}hx)$, $h \in N$ is also a representation of N. Suppose that, for all $x \in \pi$, ρ^x is equivalent to ρ; then there exists $\phi_x \in GL(V)$ such that $\rho(h)\phi_x = \phi_x\rho^x(h)$, for all $h \in N$. The mapping $\phi : \pi \to GL(V)$ given by $x \to \phi_x$ is not in general a representation of π; however, it is a projective representation of π. In fact, for $x, y \in \pi$ we have:

$$\phi_{xy}^{-1}\rho(h)\phi_{xy} = \rho^{xy}(h) = \phi_y^{-1}\phi_x^{-1}\rho(h)\phi_x\phi_y,$$

from which it follows, for $h \in N$, $\phi_x\phi_y\phi_{xy}^{-1}\rho(h) = \rho(h)\phi_x\phi_y\phi_{xy}^{-1}$, i.e. the transformation $\alpha(x,y) = \phi_{xy}^{-1}\phi_x\phi_y$ commutes with ρ. Now if ρ is irreducible, and $K = \mathbf{C}$, then by the Schur lemma $\alpha(x,y)$ is a scalar transformation. Since $1 \in N$, we may take $\phi_1 = I$. Hence $\phi_x\phi_y = \alpha(x,y)\phi_{xy}$, and $x \to \phi_x$ is a projective representation of π. (Note that the above holds for any pair of groups N and π with π acting on N by automorphisms.)

Let G be a group containing a central subgroup A and such that $G/A \simeq \pi$. If ρ is an ordinary irreducible representation of G over $\mathbf{C}$, by the Schur lemma $\rho(a)$ is scalar, $a \in A$, so $\rho(a)$ belongs to $\mathbf{C}^*$. Hence ρ induces a homomorphism $G/A \to PGL(V)$, that is a projective representation of π.

We have seen that it is not always possible to lift a projective representation of a group to an ordinary one. The question arises as to whether there exists a group such that all projective representations of π lift to it. By a result of Schur, over the complex numbers and π a finite group, at least one such group exists, it is finite, and is a central extension of an abelian group A by π. In other words, there exist a group G and an epimorphism $\eta : G \to \pi$ such that for any a projective representation θ of π, there exists an ordinary representation ρ of G such that the following diagram:

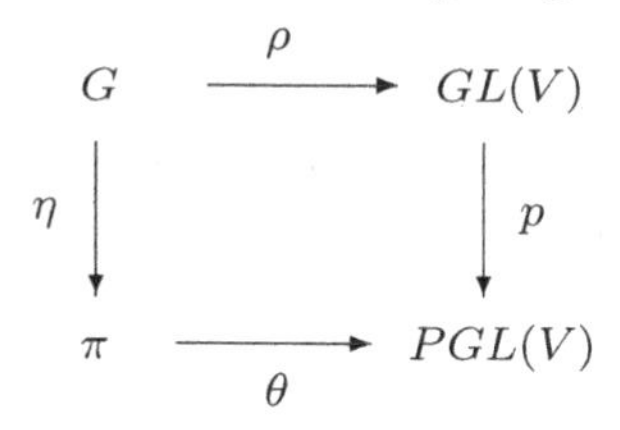

commutes (p is the natural projection). This will be proved in the next section. A group G with the above properties will be called a *covering group* of the group π.

7.4.2 Covering Groups

Before proving the Schur theorem we consider the general case of not necessarily finite groups. With the notation of the preceding subsection, and with K =$\mathbf{C}$, let us now look for the conditions that a groups G must satisfy in order that θ lifts to G. Since G depends on a class of cocycles $[g]$ of π in A, and θ on a class $[\alpha]$ cocycles of π in $\mathbf{C}^*$, it is clear that these conditions will involve relations between $[g]$ and $[\alpha]$.

Let us begin with a necessary condition. If θ lifts to G we have:

$$\rho(\lambda(\sigma)a) = \mu(\sigma, a)\theta(\sigma) \tag{7.23}$$

where λ is a function that chooses the representatives of the cosets of A, $\lambda(\sigma)A \to \sigma$ is a homomorphism, and $\mu(\sigma, a) \in \mathbf{C}^*$. Let g be a cocycle relative to the extension G, α one relative to θ. From (7.23) it follows:

$$\begin{aligned}\rho(\lambda(\sigma)a \cdot \lambda(\tau)b) &= \rho(\lambda(\sigma)a)\rho(\lambda(\tau)b) = \mu(\sigma, a)\mu(\tau, b)\theta(\sigma)\theta(\tau)\\ &= \mu(\sigma, a)\mu(\tau, b)\alpha(\sigma, \tau)\theta(\sigma\tau)\end{aligned}$$

and $\rho(\lambda(\sigma)a \cdot \lambda(\tau)b) = \rho(\lambda(\sigma)\lambda(\tau)g(\sigma,\tau)ab) = \mu(\sigma\tau, g(\sigma,\tau)ab)\theta(\sigma\tau)$; hence $\mu(\sigma, a)\mu(\tau, b)\alpha(\sigma, \tau) = \mu(\sigma\tau, g(\sigma, \tau)ab)$. For $\sigma = \tau = 1$ from this equality one obtains $\mu(1, a)\mu(1, b) = \mu(1, ab)$, and therefore a homomorphism (a character) $\chi : A \to \mathbf{C}^*$ given by $\chi(a) = \mu(1, a)$ is determined. For $a = b = 1$,

$$\mu(\sigma, 1)\mu(\tau, 1)\alpha(\sigma, \tau) = \mu(\sigma\tau, g(\sigma, \tau)), \tag{7.24}$$

and, for $\tau = 1$ and $a = 1$, $\mu(\sigma, 1)\mu(1, b) = \mu(\sigma, b)$, from which:

$$\mu(\sigma\tau, g(\sigma, \tau)) = \mu(\sigma\tau, 1)\mu(1, g(\sigma, \tau)).$$

Hence from (7.24) it follows:

$$\chi(g(\sigma, \tau)) = \alpha(\sigma, \tau)\mu(\sigma, 1)\mu(\tau, 1)\mu(\sigma\tau, 1)^{-1},$$

that is $\chi(g(\sigma, \tau)) \sim \alpha(\sigma, \tau)$. It is clear that the composition χg defined by $(\chi g)(\sigma, \tau) = \chi(g(\sigma, \tau))$ is a cocycle of π in $\mathbf{C}^*$. Moreover, if $g' \sim g$ by means of h, then $\chi g' \sim \chi g$ by means of χh. Therefore, the relation we seek between $[g]$ and $[\alpha]$ in order that θ lifts to G is the existence of a character χ of A such that $[\chi g] = [\alpha]$. We have seen that $[\alpha] \in M(\pi)$ determines a projective representation of π. We summarize the above discussion in the following lemma.

Lemma 7.7. *Let G be a central extension of A with $G/A \simeq \pi$, and let $[g]$ be the class of cocycles relative to this extension. Then a necessary condition that every projective representation of π lifts to G is that the mapping $\delta : \widehat{A} \to M(\pi)$ given by $\chi \to [\chi g]$ be surjective.*

Note that since $\widehat{A} = H^1(A, \mathbf{C}^*)$ (trivial action) the above mapping $\delta : \widehat{A} \to M(\pi)$ is the transgression $\delta : H^1(A, \mathbf{C}^*) \to H^2(\pi, \mathbf{C}^*)$ we met in *ex.* 15.

The condition of this lemma is also sufficient.

Lemma 7.8. *Let G and $[g]$ be as in the previous lemma. Then, if δ is surjective, every projective representation θ of π lifts to G.*

Proof. Let $[\alpha]$ be relative to θ, and let $\chi \in \widehat{A}$ such that $[\chi g] = [\alpha]$. We have $\chi g(\sigma, \tau) = \alpha(\sigma, \tau)h(\sigma)h(\tau)h(\sigma\tau)^{-1}$. Thus, let us define:

$$\rho(\lambda(\sigma)a) = \chi(a)h(\sigma)\theta(\sigma),$$

and let us check that it is a homomorphism. We have:

$$\begin{aligned}\rho(\lambda(\sigma)a \cdot \lambda(\tau)b) &= \rho(\lambda(\sigma\tau)g(\sigma, \tau)ab) = \chi(g(\sigma\tau)ab)h(\sigma\tau)\theta(\sigma\tau)\\ &= \chi(g(\sigma\tau))\chi(ab)h(\sigma\tau)\theta(\sigma\tau),\end{aligned}$$

and

$$\begin{aligned}\rho(\lambda(\sigma)a)\rho(\lambda(\tau)b) &= \chi(a)h(\sigma)\theta(\sigma)\chi(b)h(\tau)\theta(\tau)\\ &= \chi(a)\chi(b)h(\sigma)h(\tau)\theta(\sigma)\theta(\tau).\end{aligned}$$

Moreover, $\chi(a)h(\sigma) \in \mathbf{C}^*$ and therefore $\chi(a)h(\sigma)I$ belongs to the center of $GL(n, \mathbf{C})$. It follows that $\rho(\lambda(\sigma)a)\mathbf{C}^* = \theta(\sigma)$, i.e. θ lifts to G. ◇

We summarize the two previous lemmas in the following theorem.

Theorem 7.15. *Let G be a group, $A \subseteq Z(G)$, $G/A \simeq \pi$, $[g]$ the class of cocycles relative to the extension G. Then every projective representation of π lifts to G if, and only if, the transgression $\delta : \widehat{A} \to M(\pi)$ is surjective.*

If G is finite, then $|A| = |\widehat{A}| \geq |M(\pi)|$, and therefore $|G| = |A| \cdot |\pi| \geq |M(\pi)| \cdot |\pi|$. If equality holds, and this happens when $|\widehat{A}| = |M(\pi)|$, then G is a *covering group*[4] of π (or a *representation group* of π).

If the transgression δ is surjective, the condition $|\widehat{A}| = |M(\pi)|$ implies that it is also injective. In the lemma to follow we see that δ is always a homomorphism; a necessary and sufficient condition for δ to be injective is also given.

Lemma 7.9. *Let g be a 2–cocycle of a finite group π in the abelian group A (not necessarily finite), G the corresponding central extension of A. Then:*

i) the transgression $\delta : \chi \to [\chi g]$ is a homomorphism of $\widehat{A}$ in $M(\pi)$ whose kernel consists of the characters of A whose value on the elements of $A \cap G'$ is 1: $ker(\delta) = \{\chi \in \widehat{A} \mid \chi(a) = 1,\ a \in A \cap G'\}$;

ii) $\delta(\widehat{A}) \simeq A \cap G'$, a finite group;

iii) δ is injective if, and only if, $A \subseteq G'$.

Proof. *i)* Let $\chi \in ker(\delta)$, i.e. $\chi g \in B^2(\pi, \mathbf{C}^*)$. Then, for a certain $h : \pi \to \mathbf{C}^*$, $\chi g(\sigma, \tau) = h(\sigma)h(\tau)h(\sigma\tau)^{-1}$. Consider the mapping $\psi : G \to \mathbf{C}^*$ given by $a\lambda(\sigma) \to \chi(a)h(\sigma)$. It is a homomorphism; indeed:

$$\begin{aligned}\psi(a\lambda(\sigma) \cdot b\lambda(\tau)) &= \chi(ab)\chi(g(\sigma, \tau))h(\sigma\tau) = \chi(a)\chi(b) \cdot h(\sigma)h(\tau)h(\sigma\tau)^{-1}\\ &= \chi(a)h(\sigma) \cdot \chi(b)h(\tau) = \psi(a\lambda(\sigma)) \cdot \psi(b\lambda(\tau)).\end{aligned}$$

---

[4] Schur's *Darstellungsgruppe*.

Moreover, ψ coincides with χ on A. Since $\mathbf{C}^*$ is abelian, the kernel of ψ contains G'. Now, $\lambda(1)$ represents A; hence $\lambda(1) \in A$, and we have:

$$1 = \psi(\lambda(1)^{-1}\lambda(1)) = \chi(\lambda(1)^{-1})h(1).$$

It follows, for each $a \in A$,

$$\psi(a) = \psi(a\lambda(1)^{-1}\lambda(1)) = \chi(a)\chi(\lambda(1)^{-1})h(1) = \chi(a)$$

and therefore, if $a \in A \cap G'$, we have $1 = \psi(a) = \chi(a)$.

Conversely, let χ be such that $\chi(a) = 1$ for $a \in A \cap G'$; then we can extend χ to G as follows. First extend χ to AG' by defining $\beta : AG' \to \mathbf{C}^*$ according to $\beta : ax \to \chi(a)$, $a \in A$, $x \in G'$. The mapping β is well defined because if $ax = a_1y$, then $a_1^{-1}a = yx^{-1} \in G'$, $a_1^{-1}a \in A \cap G'$ and therefore $\chi(a_1^{-1}a) = 1$ e $\chi(a_1) = \chi(a)$. Moreover, β is a homomorphism:

$$\beta(ax \cdot a_1y) = \beta(aa_1 \cdot x'y) = \chi(aa_1) = \chi(a)\chi(a_1) = \beta(ax) \cdot \beta(a_1y).$$

$\mathbf{C}^*$ being abelian, $ker(\beta) \supseteq G'$, and therefore β factorizes through a homomorphism β_1 of AG'/G' in $\mathbf{C}^*$. The latter group being divisible, β_1 extends to a homomorphism η of G/G' in $\mathbf{C}^*$, and this one, composed with the canonical homomorphism $G \to G/G'$, to one of γ of G in $\mathbf{C}^*$. Let us summarize the homomorphisms that appear in this discussion in the following picture:

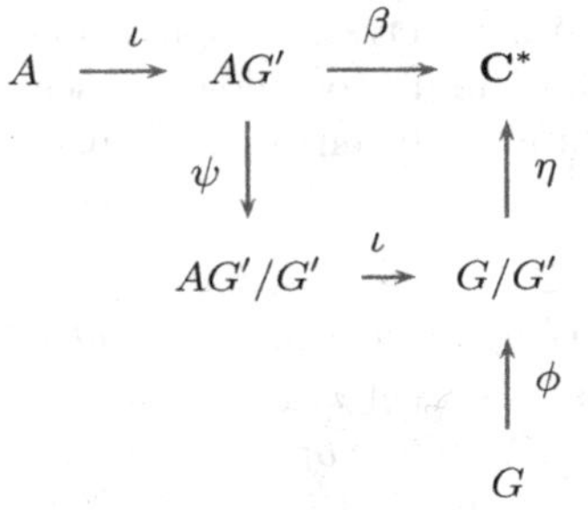

$$\begin{array}{ccccc}
A & \xrightarrow{\iota} & AG' & \xrightarrow{\beta} & \mathbf{C}^* \\
 & & \psi \downarrow & & \uparrow \eta \\
 & & AG'/G' & \xrightarrow{\iota} & G/G' \\
 & & & & \uparrow \phi \\
 & & & & G
\end{array}$$

where ϕ and ψ denote the canonical homomorphisms and ι the inclusion; moreover, $\beta\iota = \chi$ and $\eta\psi = \gamma$. Thus, for $a \in A$, we have:

$$\gamma(a) = \eta\psi(a) = \eta(aG') = b_1(ag') = \beta_1\psi(a) = \beta(a) = \chi(a),$$

and therefore γ extends χ. In particular, this holds for $a = g(\sigma, \tau)$. It follows:

$$\gamma(\lambda(\sigma))\gamma(\lambda(\tau)) = \gamma(g(\sigma,\tau)\lambda(\sigma\tau)) = \gamma(g(\sigma,\tau)\gamma(\lambda(\sigma\tau)) = \chi(g(\sigma,\tau))\gamma(\lambda(\sigma\tau))$$

and therefore $\chi g(\sigma, \tau) = \gamma(\lambda(\sigma))\gamma(\lambda(\tau))\gamma(\lambda(\sigma\tau))^{-1}$, i.e. $\chi g \in B^2(\pi, \mathbf{C}^*)$.

ii) The restriction of a character χ of A to the subgroup $A \cap G'$ yields a character χ_1 of $A \cap G'$, and the mapping $f : \widehat{A} \to \widehat{A \cap G'}$, given by $\chi \to \chi_1$, is a homomorphism. On the other hand, $\mathbf{C}^*$ being divisible and $A \cap G' \subseteq A$, every character of $A \cap G'$ extends to one of A, so f is surjective. The kernel is given by the χ's such that χ_1 is the identity $\widehat{1}$ of $A \cap G'$, i.e. by the χ's whose value on $A \cap G'$ is 1. It follows $ker(f) = ker(\delta)$, and therefore $\widehat{A \cap G'} \simeq \widehat{A}/ker(\delta) = Im(\delta)$. Since $\widehat{A \cap G'} \simeq A \cap G'$ the result follows. (G/A being finite and $A \subseteq \mathbf{Z}(G)$, by Schur's theorem (Theorem 2.38) G' is finite, and therefore $A \cap G'$ is also finite.

iii) Since the identity character $\widehat{1}$ of A has value 1 on the whole of A, $ker(\delta) = \{\widehat{1}\}$ if, and only if, $A \cap G' = A$, i.e. $A \subseteq G'$. ◊

Example 7.7. Lemma 7.9 can be used to show that the multiplier of certain groups is not trivial. Let $\pi = V$ be the Klein group and let G be either the dihedral group D_4 or the quaternion group. In both cases G is an extension of $A = \mathbf{Z}_2$ by V, and since $A = \mathbf{Z}(G) = G'$, G is a central extension of A with $A \subseteq G'$. By *iii)* of the lemma, δ is injective and therefore $M(V)$ contains a subgroup isomorphic to $\mathbf{Z}_2$, and so is nontrivial. (We shall see later (*Ex.* 7.8) that $M(V) \simeq \mathbf{Z}_2$.)

Let us now come to the existence of a covering. As we have seen, we must have $|A| = |M(\pi)|$, or what is the same because of Theorem 7.15 and Lemma 7.9, $A \simeq M(\pi)$. The group $M(\pi)$ is a direct product of cyclic groups:

$$M(\pi) = C_1 \times C_2 \times \cdots \times C_d,$$

with $|C_i| = e_i$, $i = 1, 2, \ldots, d$. Now take a group A isomorphic to $M(\pi)$:

$$A = A_1 \times A_2 \times \cdots \times A_d,$$

where $A_i = \langle a_i \rangle$, $o(a_i) = e_i$. In order to have a covering we need a cocycle g, that we construct as follows. A generator of C_i is a class of cocycles; as representative of this class we can take a cocycle whose values are e_i-th roots of unity (Lemma 7.6). If w_i is a primitive e_i-th root of unity we have:

$$g_i(\sigma, \tau) = w_i^{n_i(\sigma,\tau)}, \tag{7.25}$$

where $n_i(\sigma, \tau)$ is an integer depending on σ and τ. The property of the cocycles translates into the following property of the integers n_i:

$$n_i(\sigma, 1) \equiv 0 \equiv n_i(1, \sigma) \bmod e_i,$$
$$n_i(\sigma, \tau\eta) + n_i(\tau, \eta) \equiv n_i(\sigma\tau, \eta) + n_i(\sigma, \tau) \bmod n_i.$$

Then set $g(\sigma, \tau) = \prod a_i^{n_i(\sigma,\tau)}$. The g thus defined is a 2-cocycle of π in A under the trivial action of π on A:

$$g(\sigma, \tau\eta)g(\tau, \eta) = \prod a_i^{n_i(\sigma,\tau\eta)+n_i(\tau,\eta)}, \; g(\sigma\tau, \eta)g(\sigma, \tau) = \prod a_i^{n_i(\sigma\tau,\eta)+n_i(\tau,\eta)}$$

and the two products are equal due to the properties of the n_i's.

In order that the central extension G of A relative to g be a covering it is necessary, by Theorem 7.15, that the transgression δ be surjective. Let us then show that if $[g'] \in M(\pi)$ there exists $\chi \in \widehat{A}$ such that $\delta(\chi) = [\chi g] = [g']$. We have $[g'] = \prod [g_i]^{k_i}$, with the g_i given by (7.26) and for certain k_i. The character χ defined by $\chi(a_i) = w_i^{k_i}$ is such that, composed with g, gives:

$$\chi g(\sigma, \tau) = \chi(\prod a_i^{n_i(\sigma,\tau)}) = \prod \chi(a_i)^{n_i(\sigma,\tau)} = \prod w_i^{n_i(\sigma,\tau)k_i} = \prod g_i(\sigma, \tau)^{k_i},$$

and since $[\prod g_i^{k_i}] = \prod [g_i]^{k_i} = [g']$, we have $\chi g \sim g'$.

Theorem 7.16 (Schur). *Let π be a finite group. Then there exists a covering group G of π, i.e. an extension G of an abelian group A such that:*

i) $A \subseteq \mathbf{Z}(G) \cap G'$;

ii) $A \simeq M(\pi)$;

iii) $G/A \simeq \pi$.

Every projective representation of π lifts to an ordinary representation of G.

Covering groups need not be unique. With $\pi = V$, the Klein group, both the dihedral group of order 8 and the quaternion group are central extensions of a subgroup $A = Z_2$ satisfying *i)* and *iii)* of the theorem. In *Ex.* 7.8 it will be proved that $M(V) \simeq Z_2$, hence *ii)* will also be satisfied. Thus we have two non isomorphic groups that are covering groups of V. We will see in Theorem 7.18 a sufficient condition for a covering group to be unique.

7.4.3 $M(\pi)$ and Presentations of π

Let $F/R \simeq \pi$ be presentation of π with F free of finite rank. If $\pi \simeq G/A$ we have a surjective homomorphism $F \to G/A$ which, by the property of free groups, lifts to a homomorphism $F \to G$. In general, the latter is not surjective any more; it is surjective if $A \subseteq \mathbf{\Phi}(G)$, as we now show.

Lemma 7.10. *Let $F/R \simeq \pi \simeq G/A$, with F free of finite rank, and $A \subseteq \mathbf{\Phi}(G)$. Then the homomorphism $\mu : F \to G$ extending the canonical homomorphism $\nu : F \to \pi$ is surjective.*

Proof. Consider the diagram:

$$\begin{array}{ccccccccc} & & & & & & G & & \\ & & & & & \overset{\mu}{\nearrow} & \downarrow \gamma & & \\ 1 & \to & R & \to & F & \xrightarrow{\nu} & \pi & \to & 1 \end{array}$$

Let $F = \langle y_1, y_2, \ldots, y_n \rangle$, and let $g_1, g_2, \ldots, g_n \in G$ be such that $\gamma(g_i) = \nu(y_i)$. Then G/A is generated by the images g_iA, so that G is generated by the g_i and A, and since $A \subseteq \mathbf{\Phi}(G)$, by the g_i alone. Moreover,

$$\gamma\mu(y_i) = \nu(y_i) = \gamma(g_i),$$

from which $\mu(y_i)g_i^{-1} \in ker(\gamma) = A$, and $\mu(y_i) = x_i g_i, x_i \in A$. Thus:

$$\mu(F) = \langle x_i g_i,\ i = 1, 2, \ldots, n\rangle = G,$$

where the second equality follows from Lemma 5.5. (Note that it is not necessary that A be abelian.) ♢

If $A \subseteq \mathbf{Z}(G) \cap G'$ as in Theorem 7.16, then the condition $A \subseteq \mathbf{\Phi}(G)$ is satisfied because of the following result.

Lemma 7.11 (Gaschütz). *In any group G,*

$$\mathbf{Z}(G) \cap G' \subseteq \mathbf{\Phi}(G).$$

Proof. If $\mathbf{\Phi}(G) = G$ there is nothing to prove. Otherwise, let M be maximal in G not containing $\mathbf{Z}(G) \cap G'$. Then $G = M(\mathbf{Z}(G) \cap G')$ by the maximality of M and $M \trianglelefteq G$ because $\mathbf{Z}(G) \cap G' \subseteq \mathbf{Z}(G)$. Then $G/M \simeq \mathbf{Z}_p$, p a prime, so that $G' \subseteq M$, and a fortiori $\mathbf{Z}(G) \cap G' \subseteq M$, a contradiction. ♢

Lemma 7.12. *Let $F/R \simeq \pi \simeq G/A$, with F free of finite rank and $A \subseteq \mathbf{Z}(G) \cap G'$. Then we have a surjective homomorphism $F/[R, F] \to G$ and an induced homomorphism $R \cap F'/[R, F] \to A$.*

Proof. Let $w \in F$; then $\mu(w) \in A$ if, and only if, $\gamma\mu(w) = 1 = \nu(w)$, and therefore if, and only if, $w \in R$. It follows $\mu(R) = A$ and

$$\mu([R, F]) = [\mu(R), \mu(F)] = [A, G] = 1.$$

Then we have a surjective homomorphism $\overline{\mu} : F/[R, F] \to G$, and in this homomorphism

$$\overline{\mu}(\frac{R \cap F'}{[R, F]}) = \mu(R) \cap \mu(F') = A \cap G' = A.$$

♢

Corollary 7.8. *Let π be finite. With $A = M(\pi)$ in the preceding lemma we have a surjective homomorphism $R \cap F'/[R, F] \to M(\pi)$.*

Proof. By Theorem 7.16 there exists a central extension G of $M(\pi)$ by π, with $M(\pi) \subseteq G'$. By the preceding lemma we have the result. ♢

Theorem 7.17. *Let π be a finite group, $\pi = F/R$ with F free of rank n. Let*

$$H = \frac{F}{[R, F]},\quad I = \frac{R}{[R, F]}.$$

Then:

i) *I is a f.g. abelian group, a subgroup of $\mathbf{Z}(H)$, and $(R \cap F')/[R, F]$ is the torsion subgroup $T(I)$ of I, and therefore finite;*

ii) *I is of (torsion free) rank n;*

iii) $|(R \cap F')/[R, F]| \leq M(\pi)$.

Proof. *i*) It is clear that $I \subseteq \mathbf{Z}(H)$; moreover:

$$|H/\mathbf{Z}(H)| = |(H/I)/(\mathbf{Z}(H)/I| \leq |H/I| = |F/R| = |\pi| < \infty,$$

and therefore $\mathbf{Z}(H)$ has finite index in H. By Schur's theorem (Theorem 2.38) the derived group H' is finite. Now $H' = F'/[R,F]$, and $I \cap H' = R \cap F'/[R,F]$ is finite, and therefore is contained in the torsion subgroup of I. But

$$(R/[R,F])/((R \cap F')/[R,F]) \simeq (R/R \cap F') \simeq F'R/F', \tag{7.26}$$

and the latter group, being a subgroup of F/F' is free abelian. It follows that the torsion subgroup of I is contained in $R \cap F'/[R,F]$ and therefore:

$$T(I) = \frac{R \cap F'}{[R,F]}.$$

ii) RF/F' is free abelian and has finite index in F/F':

$$[F/F' : RF'/F'] = |F/RF'| \leq |F/R| = |\pi| < \infty,$$

and therefore has the same rank n as F/F'. But $I/T(I)$ is isomorphic to RF'/F', so that I also has rank n.

iii) The torsion subgroup of a f.g. abelian group is a direct factor:

$$\frac{R}{[R,F]} = \frac{R \cap F'}{[R,F]} \times \frac{S}{[R,F]}$$

for a certain S. Now $R/[R,F]$ is contained in the center of $F/[R,F]$ and therefore its subgroups are normal in $F/[R,F]$; in particular, $S/[R,F] \trianglelefteq F/[R,F]$ and $S \trianglelefteq F$. Hence F/S is a central extension of $R/S \simeq (R \cap F')/[R,F]$ by π, and since $R/S \subseteq \mathbf{Z}(F/S) \cap (F/S)'$, the transgression $\delta : \widehat{R/S} \to M(\pi)$ is injective (Lemma 7.9, *iii*)). In particular:

$$\left|\frac{R \cap F'}{[R,F]}\right| = \left|\frac{R}{S}\right| = \left|\widehat{\frac{R}{S}}\right| \leq |M(\pi)|,$$

as required. ◇

From this theorem and from the previous corollary it follows:

Corollary 7.9. *For a finite group* π,

i)

$$M(\pi) \simeq \frac{R \cap F'}{[R,F]}$$

(Hopf's formula);

ii) *with the notation of* Theorem 7.17 (proof of *iii*)), F/S *is a covering group of* π.

Example 7.8. Let us show that the Schur multiplier $M(V)$ of the Klein group V has order 2. We know that $|M(V)| \geq 2$ (*Ex.* 7.7). Consider the presentation $F/R \simeq V$, with $F = \langle x, y \rangle$ free of rank 2. V being abelian, $F' \subseteq R$, and therefore $M(V) \simeq F'/[R, F]$. Now,

$$M(V) \simeq \frac{F'}{[F', F]} / \frac{[R, F]}{[F', F]},$$

and $F'/[F', F] = \Gamma_2(F)/\Gamma_3(F)$ is cyclic, generated by the class $[x, y]$ modulo $\Gamma_3(F)$ (Theorem 5.19). Hence $M(V)$, as a quotient of a cyclic group, is also cyclic. Let us show that $[x, y]^2 \in [R, F]$. We have $[\overline{x}, \overline{y}] \in \mathbf{Z}(\overline{F})$ (the bar means modulo $[R, F]$), since by Theorem 7.17, *i*), $\overline{R} \subseteq \mathbf{Z}(\overline{F})$, and moreover $\overline{x}^2 \in \mathbf{Z}(\overline{F})$ because $x^2 \in R$ and therefore $[x^2, g] \in [R, F]$, for all $g \in F$. It follows $[\overline{x}, \overline{y}]^2 \overline{x}^2 = ([\overline{x}, \overline{y}]\overline{x})^2 = (\overline{yx})^{-1}\overline{x}^2(\overline{yx}) = \overline{x}^2$, and cancelling $\overline{x}^2$ we have $[\overline{x}, \overline{y}]^2 = \overline{1}$, i.e. $[x, y]^2 \in [R, F]$. In the cyclic group $M(V)$ the square of every element is 1, so that either $M(V) = \{1\}$ or $M(V) = \mathbf{Z}_2$. The former is excluded because $|M(V)| \geq 2$, hence the latter holds.

Note that in this example it is proved that the Schur multiplier of a 2-generated finite abelian group is cyclic (this is also true for 2-generated infinite abelian groups (cf. Remark 7.5, 2, below).

Corollary 7.10. *If F/R e F_1/R_1 are finite presentations of a finite group, then:*

$$\frac{R \cap F'}{[R, F]} \simeq \frac{R_1 \cap F_1'}{[R_1, F_1]}.$$

Remarks 7.5. 1. The previous corollary also holds without the assumption that the group F/R and the rank of the free groups be finite.

2. If π is infinite, one defines $M(\pi)$ directly as

$$M(\pi) = \frac{R \cap F'}{[R, F]}.$$

But in the infinite case $M(\pi)$ and $H^2(\pi, \mathbf{C}^*)$ are in general different groups. As an example, consider the infinite group $\pi = \mathbf{Z} \times \mathbf{Z}$. We know that $\pi = F/F'$, with F free of rank 2 (see p. 188), and therefore we have a presentation F/R of π with $R = F'$; it follows

$$\frac{R \cap F'}{[R, F]} = \frac{F'}{[F', F]},$$

and as observed in the previous example this group is infinite cyclic (the quotient groups of the lower central series of a free group are torsion free):

$$M(\mathbf{Z} \times \mathbf{Z}) \simeq \mathbf{Z}.$$

Now $\mathbf{Z}$ is not isomorphic to its dual. Indeed, $\mathbf{Hom_Z}(\mathbf{Z}, \mathbf{C}^*) \simeq \mathbf{C}^*$, in the isomorphism which associates an element h of $\mathbf{Hom_Z}(\mathbf{Z}, \mathbf{C}^*)$ with the value it has at 1:

$h \to h(1)$. However, it can be shown[5] that $H^2(\pi, \mathbf{C}^*)$ is isomorphic to the dual of $M(\pi)$: $H^2(\pi, \mathbf{C}^*) \simeq \mathbf{Hom_Z}(M(\pi), \mathbf{C}^*)$, and therefore

$$H^2(\mathbf{Z} \times \mathbf{Z}, \mathbf{C}^*) \simeq \mathbf{C}^*.$$

The Heisenberg group is an example of an extension of $M(\mathbf{Z} \times \mathbf{Z}) \simeq \mathbf{Z}$ by $\mathbf{Z} \times \mathbf{Z}$ (*Ex.* 4.3).

Corollary 7.11. *If π is a finite group, generated by n elements with r defining relations, and if d is the minimum number of generators of $M(\pi)$, then*

$$d \leq r - n \tag{7.27}$$

(if $r-n < 0$ set $M(\pi) = \{1\}$). In particular, if $r = n$ then $d = 0$, and therefore $M(\pi) = \{1\}$, and if $r = n+1$ then $M(\pi)$ is cyclic.

Proof. Let $\pi = F/R$, with F free. R is generated by the conjugates of the the r relators, and therefore $R/[R, F]$, being central in $F/[R, F]$, is generated by the images of the relators. If π is finite, and generated by n elements, then $I/M(\pi)$ is of rank n (Theorem 7.17, *ii*)), and therefore the minimum number of generators for I is $n + d$. ◇

More precisely:

Theorem 7.18. *Let π be presented by n generators and r defining relations, and let ρ be the (torsion free) rank of π/π'. Then $M(\pi)$ is generated by $d \leq r$ elements and*

$$n - r \leq \rho - d. \tag{7.28}$$

Proof. $R/R\cap F' \simeq F'R/F'$ is free abelian (a subgroup of F/F'), and (7.26) shows that it is a quotient of $R/[R, F]$ (by $M(\pi)$), hence a direct summand of it (Corollary 4.4):

$$R/[R, F] \simeq M(\pi) \oplus F'R/F'.$$

It follows $d(M(\pi)) + \rho(F'R/F') \leq r$, and $\rho(F'R/F') \leq r - d(M(\pi))$. Consider now the derived group π' of π. From $\pi = F/R$ we have $\pi' = F'R/R$, so $\pi/\pi' = (F/R)/(F'R/R) \simeq F/F'R$, and therefore their (torsion free) rank is the same: $\rho(\pi/\pi') = \rho(F/F'R)$. From $(F/F')/(F'R/F') \simeq F/F'R$ we have $\rho(F/F') - \rho(F'R/F') = \rho(F/F'R)$, that is, $n - \rho(\pi/\pi') = \rho(F'R/F') \leq r - d(M(\pi))$, from which the result follows. ◇

In the case of a finite group the torsion free rank of π/π' is zero, and we have (7.27). Inequality (7.28) gives an upper bound for the deficiency of a finitely presented group.

[5] By means of a result known as the "universal coefficient theorem".

We end this chapter, and the book, by showing that if a group π is *perfect*, i.e. it equals its commutator subgroup, then up to isomorphisms a covering group G of π is unique. This will be a consequence of the fact that the derived group G' of G is determined by the presentation $F/R \simeq \pi$ of π, as we now prove.

Theorem 7.19. *Let G be a covering group of a finite group π, $G/A \simeq \pi$, and let H be as in* Theorem 7.17. *Then $G' \simeq H'$. In particular, if $\pi = \pi'$, then all covering groups of π are isomorphic.*

Proof. From $A \subseteq \mathbf{Z}(G) \cap G'$ it follows $G'/A \simeq G'A/A \simeq \pi'$. Hence $|G'| = |A||\pi'| = |M(\pi)||\pi'|$. By Lemma 7.12 we have a surjective homomorphism $I = F/[R,F] \to G$. In particular, H' maps onto G', so if we show that $|G'| = |H'|$ we are done. Indeed, $\pi' = H'I/I \simeq H'/I \cap H'$. But $H' = F'[R,F]/[F,R] = F'/[R,F]$, so $I \cap H' = R/[R,F] \cap F'/[R,F] = (R \cap F')/[R,F] \simeq M(\pi)$. Hence $|H'| = |\pi'||I \cap H'| = |\pi'||M(\pi)| = |G'|$, as required. ◊

Corollary 7.12. *If π is a finite nonabelian simple group, then a covering group of π is unique up to isomorphisms.*

8

Solution to the exercises

8.1 Chapter 1

1. $a^{-1}b$ and ba^{-1} solve the two equations. Conversely, if $ax = a$ has a solution, then there exists e_r such that $ae_r = a$; if $b \in G$, then $b = ya$, some y, and therefore $be_r = (ya)e_r = y(ae_r) = ya = b$; hence e_r is neutral on the right for all $b \in G$, and similarly there exists e_l, neutral on the left. Multiplication of the two gives $e_le_r = e_l$, $e_le_r = e_r$, and *ii)* follows. As for *iii)*, solving $ax = e$ and $ya = e$, if $ax = az$ we have $y(ax) = y(az)$ from which $(ya)x = (ya)z$, i.e. $ex = ez$ and $x = z$; hence the right inverse is unique, and so is the left inverse. If $ax = e$, then $xa = exa = yaxa = y(ax)a = yea = ya = e$ and $ax = e = xa$.

2. If $o(x)$ is infinite, and $o(\varphi(x)) = n$, then $\varphi(x)^n = \varphi(x^n) = \varphi(1) = 1$, so x^n and 1 have the same image. This implies $x^n = 1$, contrary to assumption, and $o(\varphi(x))$ is infinite. If $o(x) = n$, then $1 = \varphi(1) = \varphi(x^n) = \varphi(x)^n$, so $o(\varphi(x))|n$. The same argument with φ^{-1} implies $n|o(\varphi(x))$.

3. *i)* $(ab)^2 = abab$, $a^2b^2 = aabb$: cancelling yields the result. *ii)* $(ab)^{n+1} = (ab)^n ab = a^n b^n ab = a^{n+1}b^{n+1}$, from which $b^n a = ab^n$; substituting $n+1$ for n we have $b^{n+1}a = ab^{n+1}$, i.e. $b \cdot b^n a = ab^{n+1}$. But $b^n a = ab^n$, so $b \cdot ab^n = ab^{n+1}$, and by cancelling $ba = ab$.

4. $o(a^d) = o(a)/(o(a), d) = o(a)/d, o(b^d) = o(b)/d$ and $(o(a)/d, o(b)/d) = 1$. It follows $o((ab)^d) = o(a^d b^d) = o(a)/d \cdot o(b)/d = o(a)o(b)/d^2 = m/d$.

5. $a_i a_{i+1} \cdots a_n$ is the inverse of $a_1 a_2 \cdots a_{i-1}$.

6. If a is unique of its order, since $o(a^{-1}) = o(a)$ we have $a = a^{-1}$, $a^2 = 1$ and the result.

7. If no nonidentity element coincides with its inverse, then by pairing one element and its inverse we see that the member of nonidentity elements is even. Adding the identity we have an odd number of elements.

9. We only consider the order 3. Let $1, a, b$ be the three distinct elements of G. The group must contain the product ab: if $ab = a$, then $b = 1$, which is excluded, and if $ab = b$ then $a = 1$, which is also excluded. Then $ab = 1$, i.e. $b = a^{-1}$. Next, a^2

Machì A.: Groups. An Introduction to Ideas and Methods of the Theory of Groups.
DOI 10.1007/978-88-470-2421-2_8, Springer-Verlag Italia 2012

must belong to the group: if $a^2 = a$, then $a = 1$,which is excluded; if $a^2 = 1$, then $a = a^{-1} = b$, which is also excluded. Then $a^2 = a^{-1}$, and the group is $\{1, a, a^2\}$, which is therefore cyclic, generated by a (and by a^{-1}).

10. The operation is not the same.

11. In C_6, $o(a) = 6$, $b = a^2$, $o(b) = 3$, $o(ab) = o(a^3) = 2$, whereas $\text{lcm}(o(a), o(b))$=6.

12. If the elements all have finite period, then the subgroups they generate are infinite in number, otherwise the group would be finite. If there is an element of infinite period, then it generates a group isomorphic to the the integers, and this group has an infinite number of subgroups.

13. If $a \in H$, then with $b = a$ we have $aa = a^2 \in H$, and therefore with $b = a^2$ we have $a^2 a = a^3 \in H$, etc. Hence, if $o(a) = n$, then $a^n = 1 \in H$, and $a^{-1} = a^{n-1} \in H$.

14. By theorems 1.33 and 1.35 a cyclic group of order n has $\varphi(d)$ elements of order d for all $d|n$, and these are all the elements of the group.

15. *i)* Let $n = rs$, $(r, s) = 1$, $ur + vs = 1$, $a = a^1 = a^{ur+vs} = a^{ur}a^{vs}$. Then $o(a^{ur}) = \frac{n}{(n,ur)} = \frac{rs}{(rs,ur)} = \frac{rs}{r} = s$, and similarly $o(a^{vs}) = r$. With $x = a^{vs}$ and $y = a^{ur}$ we have $a = xy$, which is of the required type. If $a = tz$, with t and z of order r and s, respectively, and commuting, we have $a^r = x^r y^r = t^r z^r = y^r = z^r$, and being $ur \equiv 1 \bmod s$ and $o(y) = o(z) = s$ we have $y = y^{ur} = z$ and $x = x^{vs} = t^{vs} = t$. The general case follows immediately.

ii) $\varphi(n)$ is the number of generators of a cyclic group of order n, and $\varphi(p_i^{h_i})$ that of the generators of the unique subgroup of order $p_i^{h_i}$. A product $x_1 x_2 \cdots x_r$, $o(x_i) = p_i^{h_i}$, $i = 1, 2, \ldots, r$, is an element of order n, and by *i)* an element of order n admits a unique such expression. As to the second equality, observe that the elements that generate a cyclic group of order p^n are those that do not belong to the subgroup of order p^{n-1}.

16. If $n \geq 4$, a permutation interchanging two consecutive vertices of the polygon and fixing the other ones is not a symmetry.

19. $j = \begin{pmatrix} -i & 0 \\ 0 & i \end{pmatrix}$, $k = \begin{pmatrix} 0 & 1 \\ -1 & 0 \end{pmatrix}$.

22. A rational number may be expressed as a product of primes with integer exponents, and a polynomial with integer coefficients may be expressed as a linear combination of the monomials $1, x, x^2, \ldots$ Order the prime numbers, e.g. in increasing order: $p_0 = 2, p_1 = 3, p_2 = 5, \ldots$ An ordered n-tuple of integers $m_0, m_1, \ldots, m_{n-1}$ determines a unique element of Q: $r/s = p_0^{m_0} p_1^{m_1} \cdots p_{n-1}^{m_{n-1}}$, and a unique polynomial $p(x) = m_0 + m_1 x + \cdots + m_{n-1}x^{n-1}$. The mapping $\frac{r}{s} \to p(x)$ is the required isomorphism. For example, to $\frac{7}{18} = 2^{-1}3^{-2}7$ there corresponds the polynomial $-1 - 2x + 7x^3$.

26. A permutation matrix is obtained from the identity matrix by exchanging rows and columns. Since the identity matrix has determinant 1, and since an exchange of two rows or columns changes the sign of the determinant, we have the result. Moreover, the product of two matrices with determinant 1 has again determinant 1, and the inverse of a matrix with determinant 1 has again determinant 1. Hence the matrices with determinant 1 form a subgroup, A say, of the group of permutation matrices (*Ex.* 1.3, 3); the latter is therefore a union $A \cup A'$, where A' is the set of matrices with determinant -1. Multiplication of the elements of A by a matrix with determinant -1 yields $|A|$ matrices with determinant -1, so that $|A'| > |A|$. Similarly, multiplication of the elements of A' by a matrix with determinant -1 yields $|A'|$ matrices with determinant 1, so that $|A| > |A'|$. It follows $|A| = |A'|$.

27. The generators are all contained in a subgroup of the chain.

28. If $H < C_{p^\infty}$ and is infinite, then it contains primitive p^k-th roots of unity for any k (each C_{p^n} may be written as a union of p^m-th roots of unity with $m \leq n$). Then H contains all the p^n-th roots for all $n = 1, 2, \ldots$, and therefore it coincides with C_{p^∞}. If H is finite, $H \subseteq C_{p^k}$, some k, and therefore $H = C_{p^h}$, some $h \leq k$.

29. If not, $G = H \cup \langle S \rangle$, contrary to Theorem 1.6.

30. $a\rho b \Rightarrow a^{-1}a\rho a^{-1}b$, and therefore $a^{-1}b\rho 1$. Let us show that the the elements equivalent to 1 form a subgroup. If $x\rho 1$, then $x^{-1}x\rho x^{-1}$, and so also $x^{-1}\rho 1$; if $a \sim 1, b\rho 1$ then $ab\rho a\rho 1$.

31. It is the only one containing the identity.

32. $Hab^{-1} = K$, a subgroup (see the previous *ex.*).

33. $a = nq + r$, $0 < r < n$, $a^{\varphi(n)} = (nq + r)^{\varphi(n)} = \sum_{k=0}^{\varphi(n)} \binom{\varphi(n)}{k} (nq)^k r^{\varphi(n)-k}$. If $k > 0$, then all the summands are multiples of n, and therefore are zero mod n. There remains the term for $k = 0$, i.e. $r^{\varphi(n)}$. But $r < n$ and $(r, n) = 1$, and therefore r belongs to the group $U(n)$ whose order is $\varphi(n)$, so that $r^{\varphi(n)} = 1$.

34. $\varphi(p) = p - 1$ (see the previous *ex.*).

37. $1 = x^{2n-1} = x^{2n}x^{-1}$; with $y = x^n$ we have $x = y^2$.

38. If $H = \{h_1, h_2, \ldots, h_n\}$ and $x \notin H$, $Hx = \{h_1x, h_2x, \ldots, h_nx\}$, then $Hx \cap H = \emptyset$ because if $h_i = h_jx$ then $x = h_j^{-1}h_i \in H$, which is excluded. If $y \notin H \cup Hx$, $Hy = \{h_1y, h_2y, \ldots, h_ny\}$ has empty intersection with Hx (because if $h_ix = h_jy$, then $yx^{-1} = h_j^{-1}h_i \in H$, $y = h_kx$, which is excluded) and with H (as above), etc. Thus, the group is partitioned into a disjoint union of subsets all having the same cardinality, namely $|H|$.

40. Let $[G{:}H] < \infty$ and K infinite. If $H \cap K = \{1\}$, the distinct elements of K belong to distinct cosets of H ($Hk = Hk_1 \Rightarrow kk_1^{-1} \in H \cap K = \{1\}$ and $k = k_1$) and H has at least as many cosets as there are elements of K.

8.2 Chapter 2

4. Let H be another subgroup of order coprime to its index. Then $HN \leq G$ ($N \trianglelefteq G$), $|HN|$ divides $|G|$, $|HN/N|$ divides $|G/N|$, so that $(|HN/N|, |N|) = (|HN/N|, |H|) = 1$. If $|HN/N| \neq 1$, let p be a prime dividing $|HN/N|$; then $p||H/(H \cap N)|(= |HN/N|)$, and therefore $p||H|$, which is absurd.

8. H, K are maximal (index p) and normal, $HK \leq G$, $HK = G$ by the maximality of H (or K); $p = |G/K| = |H/(H \cap K)| = |H|$, and similarly $|K| = p$. G is not cyclic because it has two subgroups of index p.

9. If $L/H = (A/H)(B/H)$, $AB = L$ is the sought-for subgroup. Indeed, if $g \in L$, then $gH = aH \cdot bH = abH$, $g = ab'$ ($H \subseteq B$) and $L \subseteq AB$; conversely, $A/H \subseteq L/H$ and $A \subseteq L$. Similarly, $B \subseteq L$ and $AB = L$.

12. Since $n \geq 3$, there exist i, j, k distinct. If $\sigma \neq 1$, let $\sigma(i) = j$; then σ does not commute with τ such that $\tau(j) = k$ and $\tau(i) = i$.

13. If $x = ar^i$ is central, then $ar^i \cdot r = r \cdot ar^i$, $r^{i+1} = r^{i-1}$, $r = r^{-1}$, excluded. Hence $x = r^i$, $a \cdot r^i = r^i \cdot a$, $r^i = r^{-i}$, $o(r^i) \leq 2$. If n is odd, then $x = r^i = 1$; if n is even, then $r^i = \frac{n}{2}$ and the center is $\{1, r^{\frac{n}{2}}\}$.

16. $H = \langle x \rangle$ is normal, and as in the previous *ex.*, $xy^{p-1} = y^{p-1}x$, and $y^{p-1} \in \mathbf{C}_G(x)$. But $(p-1, o(y)) = 1$ since there are no divisors of the order of the group less than p, and therefore y itself belongs to $\mathbf{C}_G(x)$.

17. $x^n = (xy \cdot y^{-1})^n = (y^{-1}xy)^n = y^{-1}x^n y$, and $x^n \in \mathbf{Z}(G)$. Conversely, $ab \sim ba$ and therefore also $(ab)^n \sim (ba)^n$; the latter being central, they are equal.

18. If $x \sim y$, then $x = ab$ and $y = ba$; hence $xy = abba = ab^2a = a^2b^2$ by the previous *ex.*; similarly, $yx = a^2b^2$.

23. If $H \leq \mathbf{Z}(G)$, and $G/H = \langle Ha \rangle$, then $G = \langle H, a \rangle$ so that G is generated by commuting elements.

24. $G/\mathbf{Z}(G) \simeq \mathbf{I}(G)$ cyclic implies G abelian (*ex.* 23), and therefore G admits the automorphism $\sigma : a \to a^{-1}$ of order 2. In both cases ($\mathbf{Aut}(G) = \mathbf{Z}$ or $\mathbf{Aut}(G)$ of odd order) $\sigma = 1$ and $a = a^{-1}$ for all $a \in G$, so that G is a vector space over $\mathbf{F}_2$. If it has more than two elements it has elements of order 2 (for example, those interchanging two basis vectors and fixing the remaining ones).

25. If $\gamma_y = (\gamma_x)^k = \gamma_{x^k}$ for all x, then $x^{\gamma_y} = x^{-k}xx^k = x$ for all x.

26. $\alpha \in \mathbf{Aut}(S^3)$ permutes the three elements of order 2 of S^3, and we have a homomorphism $\mathbf{Aut}(S^3) \to S^3$, which is injective (if α and β induce the same permutation, then $\alpha\beta^{-1}$ induces the identity on the three elements, and therefore on the whole S^3 which is generated by these). But $\mathbf{Z}(S^3) = \{1\}$, and therefore $\mathbf{I}(S^3) \simeq S^3$. It follows, $\mathbf{Aut}(S^3) = \mathbf{I}(S^3) \simeq S^3$.

27. $Z = \mathbf{Z}(G) \neq \{1\}$; if $Z \not\subseteq M$, then $MZ = G$. Let $m' \in M$ and $x \in G$; hence $x = mz$ and we have:

$$x^{-1}m'x = (mz)^{-1}m'(mz) = z^{-1}m^{-1}m'mz = z^{-1}m''z = m'' \in M,$$

so that $M \trianglelefteq G$. Moreover $|G/M| = p$ because G/M has no proper subgroups (Theorem 1.17). If $Z \subseteq M$, then M/Z is maximal in G/Z, and therefore, by induction, normal and of index p. Now apply Theorem 2.5.

29. $|\mathbf{cl}_H(x)| = [H : \mathbf{C}_H(x)] = [G : \mathbf{C}_H(x)]/[G : H] = [G : \mathbf{C}_G(x)][\mathbf{C}_G(x) : \mathbf{C}_H(x)]/2$ $= |\mathbf{cl}_G(x)|[\mathbf{C}_G(x) : \mathbf{C}_G(x) \cap H]/2 = |\mathbf{cl}_G(x)||\mathbf{C}_G(x)H/H|$, and since $\mathbf{C}_G(x)H/H \leq G/H|$, we have $|\mathbf{C}_G(x)H/H| = 1, 2$, and the result.

30. If $x \notin H$, then $\mathbf{C}_G(x) \cap H = \{1\}$ (if $xh = hx$, then $x \in \mathbf{C}_G(h) \subseteq H$) and therefore $|\mathbf{C}_G(x)| = 2$, from which $o(x) = 2$, $|\mathbf{cl}(x)| = [G : \mathbf{C}_G(x)] = \frac{|G|}{2}$, and therefore $\mathbf{cl}(x) = G \setminus H$ (the elements $\mathbf{cl}(x)$ are the $h^{-1}xh$, $h \in H$).

31. Following the hint, assume now that $(xy)^\alpha = (xy)^{-1}$; then $(xy)^\alpha = y^{-1}x^{-1} = x^\alpha y^\alpha = x^{-1}y^{-1}$ so $y \in \mathbf{C}_G(x)$, against the choice of y. Therefore, for any $y \in G \setminus \mathbf{C}_G(x)$ that maps to its inverse, there is an element (i.e. xy) that does not. Hence the latter fact happens for at least half of the elements of $G \setminus \mathbf{C}_G(x)$, i.e. for at least $\frac{1}{2}|G \setminus \mathbf{C}_G(x)| \geq \frac{1}{4}|G|$ elements, while more than $\frac{1}{4}$ of the elements go to their inverses. It follows $G = \mathbf{Z}(G)$. As to the example, consider the identity automorphism of D_4.

32. If all the elements of G have order 2, then G is a vector space over $\mathbf{F}_2$, which if it has more than two elements has a non trivial automorphism group. If there exists $x \neq x^{-1}$ and G is abelian, the automorphism $g \to g^{-1}$ is not the identity because it does not fix x. If G is not abelian, and $xy \neq yx$, the inner automorphism induced by y is not the identity.

33. We prove more: if α commutes with all inner automorphism then it is the identity. Indeed, if $\alpha^{-1}\gamma_x\alpha = \gamma_x$, then $\gamma_{x^\alpha} = \gamma_x$, $x^{-1}x^\alpha \in \mathbf{Z}(G) = \{1\}$ and $x^\alpha = x$.

34. *i)* Setting $\delta_{a,b^2} = 1$, if $a = b^2$ and zero otherwise, we have $\rho(a) = \sum_b \delta_{a,b^2}$, from which $\rho(a)^2 = \sum_b \rho(a)\delta_{a,b^2}$ and therefore $\sum_a \rho(a)^2 = \sum_b \sum_a \rho(a)\delta_{a,b^2}$. If $a \neq b^2$ the inner summands are zero; only those for which $a = b^2$ remain, and therefore $\sum_a \rho(a)^2 = \sum_b \rho(b^2)$; letting b vary as a in G yields the result.

ii) By *i)*, the sum on the right equals

$$\sum_{a \in G} \rho(a^2) = \sum_c \sum_a \delta_{b^2,c^2} = \sum_c \sum_b \delta_{(bc)^2,c^2} = \sum_b \sum_c \delta_{(bc)^2,c^2},$$

and $(bc)^2 = c^2$ if and only if $c^{-1}bc = b^{-1}$. Hence the inner sum equals $|I(b)|$, and the total sum $\sum_b |I(b)|$ (note that this sum is never zero because $b = 1$ is also counted, and being $1^{-1} = 1$, it is inverted by all the elements of G: $|I(b)| = |G|$).

iii) If $a \sim a^{-1}$, then $|I(a)| = |\mathbf{C}_G(a)|$. If $x \in \mathbf{cl}(a)$, then $|\mathbf{C}_G(x)| = |\mathbf{C}_G(a)|$, and the contribution of a conjugacy class to the sum is $|\mathbf{C}_G(a)| \cdot |\mathbf{cl}(a)| = |G|$; if there are $c'(G)$ classes, the contribution is $|G| \cdot c'(G)$.

iv) In a group of odd order every element is a square.

36. *i)* If k divides $2n$, then $k = 2, k$ or $2k$, where k divides n. If k divides n, then in the cyclic group C_n there exists a subgroup of order k. This being unique, it is characteristic in C_n and therefore normal in D_n. Multiplication with a subgroup of order 2 not in C_n yields a subgroup of order $2k$.

ii) Let $\{1\} \neq K \subseteq H$ and $y \in \mathbf{N}_G(K)$. Then $\{1\} \neq K = K^y \subseteq H \cap H^y$, so that $H = H^y$ and $y \in \mathbf{N}_G(K)$. Hence $\mathbf{N}_G(K) \subseteq \mathbf{N}_G(H)$. Conversely, if $y \in \mathbf{N}_G(H)$, then $y \in \mathbf{N}_G(K)$ since K is characteristic in H.

37. Let $xy^{-1} = hk$; then $x = hky$ and $H^x = H^{hky} = H^{ky}$, and with $g = ky$ we have $H^xK^y = H^gK^g = (HK)^g = G^g = G$.

39. Let $G = \bigcup H_i$, $H_i \cap H_j = \{1\}$. If $|H_i| = h_i$, each conjugacy class of the H_i contains $\frac{|G|}{h_i}$ subgroups, since $\mathbf{N}_G(H_i) = H_i$. Therefore, being $H_i \cap H_j = \{1\}$, the contribution of a class to the number of elements of G is $\frac{|G|}{h_i}(h_i - 1)$; adding the identity, $|G| = 1 + \sum_{i=1}^s \frac{|G|}{h_i}(h_i - 1)$, where s is the number of classes. But $\frac{h_i}{(h_i-1)} \geq \frac{1}{2}$ (since $h_i \geq 2$), and therefore $|G| \geq 1 + s|G| \cdot \frac{1}{2}$. If $s \geq 2$, $|G| \geq 1 + |G|$, which is absurd. Thus $s=1$, $|G| = 1 + \frac{|G|}{h_1}(h_1 - 1)$, $h_1(|G|-1) = |G|(h_1-1)$, $|G| = h_1$ and $G = H_1$.

41. If $H \cap H^x = \{1\}$, then $|HH^x| = |H||H^x| = n^2$ and $HH^x = G$, which is absurd (*ex.* 35).

42. *i)* $H^xK = (K^{x^{-1}})^x = (K^{x^{-1}}H)^x = KH^x$;

ii) $HH^x = H^xH$, so that $HH^x \leq G$ and therefore it commutes with every subgroup of G. An inclusion $H \subset HH^x$ is against the maximality of H, and $HH^x = G$ is impossible; hence $HH^x = H$, i.e. $H^x = H$ and $H \lhd G$.

43. Let $g \in \mathbf{N}_G(K)$; then $g \in \langle H, g^{-1}Hg\rangle \subseteq \langle K, g^{-1}Kg\rangle = K$, and therefore $\mathbf{N}_G(K) = K$, and *i)* follows. Let $H \subseteq K \cap gKg^{-1}$; then $g \in \langle H, g^{-1}Hg\rangle \subseteq \langle K, K \cap gKg^{-1}\rangle = K$, so that $gKg^{-1} = K$, and *ii)* follows. Conversely, if $K = \langle H, g^{-1}Hg\rangle$, then $H, g^{-1}Hg \subseteq K$, that is $H \subseteq gKg^{-1}$, i.e. $H \subseteq K \cap gKg^{-1}$. By *ii)*, $K = gKg^{-1}$, and since $K \supseteq H$ and $g \in \mathbf{N}_G(K)$, by *i)* we have $g \in K = \langle H, g^{-1}Hg\rangle$.

45. *i)* If $\alpha \neq 1$ commutes with γ_x, then $x^{-1}x^\alpha \in \mathbf{Z}(G) = \{1,-1\}$ and $x^\alpha = x, -x, \forall x \in G$. If $i^\alpha = i$, then $j^\alpha \neq j$ otherwise $\alpha = 1$; hence $j^\alpha = -j$, and similarly $k^\alpha = -k$. Therefore α coincides with the conjugation induced by i on the generators j and k, so that $\alpha = \gamma_i$.

ii) Since H is a Klein group, the suggested automorphisms belong to a coset of H different from H; it follows that G/H has order divisible by 2 and 3, and so by 6, and from $G/H \leq S^3$ it follows $G/H \simeq S^3$, and $|G| = |V||S^3| = 24$.

46. If $A \subset \mathbf{C}_G(A)$, and $x \in \mathbf{C}_G(A) \setminus A$, $\langle A, x\rangle$ is abelian and properly contains A. A maximal abelian then implies $A = \mathbf{C}_G(A)$. Conversely, let $A \subset H$, with H abelian; then H centralizes A, so $\mathbf{C}_G(A) = A \subset H \subseteq \mathbf{C}_G(A)$, which is absurd.

47. If for all $x \in G$ we have $\alpha^{-1}\gamma_x\alpha = \gamma_{x^\alpha} = \gamma_x$, then $x^\alpha x^{-1} \in \mathbf{Z}(G) = \{1\}$, $x^\alpha = x$ and $\alpha = 1$.

48. Following the hint, consider a coset of H_i, $i \neq 1$ not appearing in the union, and repeat the operation. After a finite number of steps we reach an expression of G as a union of the cosets of one of the H_i, which therefore has finite index. (This result may be used to prove that a vector space V over an infinite field K cannot be a finite set–theoretic union of subspaces, otherwise one of these subspaces W would be of finite index, being a subgroup of $V(+)$, and the space V/W would contain a finite number of vectors. But if K is infinite and $v \notin W$, there exist infinite multiples of $v + W$.)

49. The mapping $(H \cap K)h \to (H^x \cap K^x)x^{-1}hx$ is bijective.

50. If $x \in H_{i-1}$, then $H_i^x \subseteq G_i^x = G_i$ because G_i is normal in G_{i-1}; it follows $H_i^x \subseteq G_i \cap H = H_i$, and similarly for x^{-1}. Moreover, $H_{i-1}/H_i = H_{i-1}/(G_i \cap H_{i-1}) \simeq H_{i-1}G_i/G_i \leq G_{i-1}/G_i$.

51. *i)* If $\{G_i\}$ is a normal series of G containing H, then $\{G_i \cap K\}$ is a normal series of K containing $H \cap K$. *ii)* If $\{H_i\}$ is a normal series of G containing K, then $G = H_0 \supseteq H_1 \supseteq \ldots \supseteq K \supseteq G_1 \cap K \supseteq \ldots \supseteq \{1\}$ is a normal series of G containing $H \cap K$. *iv)* Consider the subgroups $\{1, a\}$ and $\{1, ar\}$ of D_4; their product is not a subgroup. *v)* If $G = G_0 \supseteq G_1 \supseteq \ldots \supseteq G_i = H$, H is normal and of order coprime to the index in G_{i-1}, it is unique of its order in G_{i-1} and therefore it is characteristic in G_{i-1} and thus normal in G_{i-2}, etc.

54. It is sufficient to consider the case $n = 2$. *i)* Let $x = h_1h_2 \in K$, $o(h_1) = m_1, o(h_2) = m_2$. Then $x^{m_2} = h_1^{m_2} \in K \cap H_1$, $x^{m_1} = h_2^{m-1} \in (K \cap H_2)$, and if $m_1r + m_2s = 1$ then $x = x^{m_1r}x^{m_2s} \in (K \cap H_1)(K \cap H_2)$ and the product is direct. *ii)* the two factors are characteristic (normal and of order coprime to the index) and therefore $\alpha \in \mathbf{Aut}(G)$ induces an automorphism α_i on H_i, and the mapping $\alpha \to (\alpha_1, \alpha_2)$ is an isomorphism $\mathbf{Aut}(G) \to \mathbf{Aut}(H_1) \times \mathbf{Aut}(H_2)$. The example of the group $V = C_2 \times C_2$ shows that the hypothesis $(|H_1|, |H_2|) = 1$ is necessary.

59. We show something more, that is, that if H is complete, and $H \leq G$, then $G = H \times \mathbf{C}_G(H)$. If $g \in G$, the mapping $h \to g^{-1}hg$ is an automorphism of H, and therefore is inner: $g^{-1}hg = h_1^{-1}hh_1, \forall h \in H$. It follows $gh_1^{-1} \in \mathbf{C}_G(H), g \in \mathbf{C}_G(H)H$ and $G = \mathbf{C}_G(H)H$. But $\mathbf{C}_G(H) \cap H = \mathbf{Z}(H) = \{1\}$ and $\mathbf{C}_G(H) \trianglelefteq G$ because $H \trianglelefteq G$.

61. If H is maximal, then $\{1\} \neq A_1 \lhd A$, $A_1 \subset A$. Hence $H \subset HA_1$, from which $G = HA_1$. But $|A| = |G/H| = |HA_1/H| = |A_1/A_1 \cap H| = |A_1|$, and $A = A_1$. Conversely, let A be minimal normal, and let $H \supset M$. Then $G = MA$, $G/A = MA/A \simeq M/(M \cap A)$. But $M \cap A$ is normal in A, because A is abelian, and also in M (because $A \lhd G$), and therefore $M \cap A \lhd G$. By minimality, $M \cap A = \{1\}$, so $G/A \simeq M$; but we also have $G/A \simeq H$, so $|M| = |H|$ and $M = H$.

62. $(1,2)^{(2,3)} = (1,3)$, $(1,3)^{(3,4)} = (1,4)$, etc. In this way one obtains the transpositions $(1,i), i = 2,3,\ldots,n$, that generate S^n.

63. *i)* Conjugation of (1,2) with the powers of $(1,2,\ldots,n)$ yields the transpositions $(i, i+1)$ that generate S^n (cf. the previous *ex.*).

67. *i)* 3 divides 12, and an element of order 3 of S^4 is a 3-cycle; hence a subgroup of order 12 contains a 3-cycle, and being normal (index 2) it contains all the 3-cycles (they are all conjugate). It follows that this subgroup is A^4. By *Ex.* 1.7, 6, there are three D_4 that meet in the Klein group $V = \{I, (1,2)(3,4), (1,3)(2,4), (1,4)(2,3)\}$, which is also the Klein group of A^4. The subgroups D_4 contain altogether three C_4 and four Klein groups. In S^4 an element of order 4 is a 4-cycle, and there are $(4-1)! = 6$ of them, divided into three pairs each containing an element and its inverse. Therefore there are at most three groups C_4, and since each D_4 contains one of these, exactly three. There are nine elements of order 2 (the six transpositions and the three products of two transpositions) and they are contained in the four Klein groups of the D_4. If H is a subgroup of order 4, and $H \not\subset A^4$, then $2 = |S^4/A^4| = |HA^4|/|H \cap A^4| = 4/|H \cap A^4|$, from which $|H \cap A^4| = 2$. An element of order 2 of A^4 belongs to the Klein group of A^4, and therefore to all the groups D_4. An element of order 2 of H belongs to some D_4, and therefore H is contained in this D_4, and hence it has already been counted. This proves *iv)* and *vi)*. If $|H| = 8$, then $HA^4 = S^4$ and $|H \cap A^4| = 4$, so that H contains the Klein group of A^4. The elements of H cannot all be of order 2, otherwise it will have seven subgroups of order 4, whereas in the whole group S^4 there are only four such subgroups. Hence H contains an element of order 4, and so a subgroup of order 4, and this is contained in a D_4. Since $H \cap D_4$ is also contained in this D_4, we have $|H \cap D_4| > 4$, and $H = D_4$, and we have *ii)*. There are eight elements of order 3 divided into four pairs contained into four subgroups of order 3, and we have *v)*. There are four S^3

obtained by fixing one of the digits. If $|H| = 6$, a 3-element of it is a 3-cycle; an element of order 2 is not an even permutation because $|H \cap A^4| = 3$, and therefore is a transposition; this must only involve digits already present in a 3-cycle of H, otherwise the product with such a 3-cycle would yield an element of order 4 (example: (1,2,3)(1,4)=(1,2,3,4)); but 4 does not divide 6. On the other hand, H has only one subgroup of order 3 (as we already know, but this also follows from $|H \cap A^4| = 3$), and therefore the elements of H are a 3-cycle, its inverse, and transpositions in which the digit missing in the 3-cycle does not appear. Hence this digit is fixed by all the elements of H, and H is one of the S^3 already considered. *iii)* follows.

68. There are seven partitions of 5, and therefore there are seven conjugacy classes in S^5; the elements are shown in the following table:

classes	*cycle structure*	*# elements*	*order*	*parity*
C_1	(1)(2)(3)(4)(5)	1	1	even
C_2	(1,2)(3)(4)(5)	10	2	odd
C_3	(1,2,3)(4)(5)	20	3	even
C_4	(1,2,3,4)(5)	30	4	odd
C_5	(1,2,3,4,5)	24	5	even
C_6	(1,2)(3,4)(5)	15	2	even
C_7	(1,2,3)(4,5)	20	6	odd

The number of k-cycles can be determined using *ex.* 4. For the other elements, consider for instance those of type $(1,2)(3,4)$. There are five ways of choosing the first digit, four ways for the second, three for the third and two for the fourth; a transposition can be written in two ways, and so can the product of two transpositions: $2 \cdot 2 \cdot 2 = 8$ ways, for a total of $5 \cdot 4 \cdot 3 \cdot 2/8 = 15$. The 5-cycles cannot be all conjugate in A^5 because $24 \nmid 60 = |A^5|$, and therefore (*ex.* 29) A^5 has two conjugacy classes of 5-cycles, each containing 12 elements. A 5-cycle is not conjugate to its square in A^5: an element that conjugates them is necessarily odd. Two 3-cycles are conjugate in A^5: if they have only one digit in common, then $(i,j,k)^{(j,h)(k,l)} = (i,h,l)$; if they have two, then $(i,j,k)^{(i,j)(k,l)} = (i,j,l)$. The 15 elements of C_6 cannot be divided into two equal parts, so they are all conjugate in A^5 (again by *ex.* 29). Summing up, A^5 has five conjugacy classes with 1,20,12,12 and 15 elements. That A^5 is simple can be seen by observing that a non trivial normal subgroup ia a union of conjugacy classes, and so its order is obtained as a sum of integers among the ones listed above. However no sum of these equals a proper divisor of 60.

69. *i)* We have, for the subgroup of homotheties, the two affinities $\varphi_{1,0} = I$ and $\varphi_{5,0} = (1,5)(2,4)(0)(3)$, the cyclic group of order 2. As for the translations, we have the subgroup of order 6 generated by $\varphi_{1,1}$, i.e. by the cycle $(0,1,2,3,4,5)$. $\varphi_{5,0}$ inverts $\varphi_{1,1}$, and we have D_6.

ii) with $a = 1$, $\varphi_{1,0} = I$, $\varphi_{1,1} = (0,1)(x, x+1)$, $\varphi_{1,x} = (0,x)(1,x+1)$ and $\varphi_{1,x+1} = (0,x+1)(1,x)$. These four affinities form a group isomorphic to the additive group of the field (the Klein group V); these are the translations. With $b = 0$ we have the three homotheties: $\varphi_{1,0} = I$, $\varphi_{x,0} = (1,x,x+1)(0)$, $\varphi_{x+1,0} = (1,x+1,x)(0)$, that make up a cyclic group of order 3, isomorphic to the multiplicative group of the field C_3. The semidirect product of C_3 by V by is the group A^4.

70. If $D_n = \langle a, r\rangle$, then $\langle r^2\rangle$ is characteristic in $\langle r\rangle$ and therefore normal in D_n. But $o(r^2) = n$ or $\frac{n}{2}$, so $D_n/\langle r^2\rangle$ has order 2 or 4 and is abelian. Moreover, $r^2 = [a, r]$, etc.

72. $a^n = y^{-1}ay$ implies $a^{n-1} = a^{-1}y^{-1}ay$.

74. If $x_1, \ldots, x_n$ are the elements of G, then $x_1x_2\cdots x_nG' = G'x_1G'x_2G'\cdots x_nG'$. The x_iG' commute, so we can move $x_i^{-1}G'$ close to x_iG' so that the latter product equals 1; the result follows.

76. From $|G|r + ns = 1$ we have $abG' = (abG')^{|G|r+ns} = (abG')^{ns} = (a^nb^nG')^s = (cG')^{ns} = (cG')^{-|G|r+1} = cG'$, because $o(cG')$ divides $|G|$.

77. $[x, y] = ((y^{-1})^x)^2(x^{-1})^2(xy)^2$.

78. If the center has index p, the center and an element not belonging to it generate the whole group, so the group is abelian. The quotient w.r.t. a normal subgroup of order p^2 (which exists, cf. *Ex.* 2.3, 2) has order p^2, is abelian (Theorem 2.15), and therefore contains G'.

79. *i)* A conjugate of x^α is the image under α of a conjugate of x:

$$y^{-1}x^\alpha y = (y^{-\alpha^{-1}}xy^{\alpha^{-1}})^\alpha,$$

and therefore $|\mathbf{cl}(x^\alpha)| \leq |\mathbf{cl}(x)|$. *ii)* If $x_1, x_2, \ldots, x_n$ generate H, $[H : \mathbf{C}_G(x_i)]$ is finite for all i, and therefore $\mathbf{Z}(G) = \bigcap \mathbf{C}_G(x_i)$ also has finite index. By Schur's theorem (Theorem 2.38) H' is finite. *iii)* If such a subgroup H exists, an element $h \in H$ of prime order p has all its conjugates in H, and therefore $h \in \Delta(G)$. Conversely, let $h \in \Delta(G)$, and let H be generated by the conjugates of h, also of order p. Since the conjugate of a product is the product of the conjugates of the factors, we have $H \subseteq \Delta(G)$ and $H \trianglelefteq G$. By *ii)*, H' is finite, and so is H/H' (it is abelian and finitely generated by elements of finite order). Therefore, H is finite, and since it contains an element of order p, its order is divisible by p.

80. By *ex.* 18, $b^{-1}ab$ commutes with a and therefore with a^{-1}. It follows $[a, b]^2 = [a, b][b, a^{-1}] = a^{-1}b^{-1}a^2ba^{-1} = 1$ because $a^2 \in \mathbf{Z}(G)$ (*ex.* 17). But a commutator is a product of squares (*ex.* 77), and therefore $G' \subseteq \mathbf{Z}(G)$. The commutators being in the center and having order 2, G' is elementary abelian.

82. $G' \subseteq \mathbf{Z}(G)$ because it is normal and of order 2. If $x_1x_2 = x_2x_1$ or $x_2x_3 = x_3x_2$ there is nothing to prove. Then $x_1x_2x_3 = x_1x_3x_2c$. If $x_1x_3 = x_3x_1$, then $x_1x_3x_2c = x_3x_1x_2c = x_3x_2x_1c^2 = x_3x_2x_1$, and if $x_1x_3 \neq x_3x_1$, then $x_1x_3x_2c = x_3x_1cx_2c = x_3x_1x_2$. If the given property holds, let us consider $y^{-1}xy$, whose value under the six permutations equals either x or yxy^{-1}, i.e. $y^{-2}xy^2 = x$. Hence, either $y \in \mathbf{Z}(G)$, and a fortiori $y^2 \in \mathbf{Z}(G)$, or $y^2 \in \mathbf{Z}(G)$.

8.3 Chapter 3

3. *i)* Under the action of K, the stabilizer of H is $K_H = H \cap K$, and the orbit is $\{Hk, k \in K\}$, that is, the set of the cosets of H contained in HK has cardinality $[K : H \cap K]$. *ii)* $[\langle H \cup K\rangle : H] \geq [HK : H] = [K : H \cap K]$.

5. Under the action of G on the cosets of a subgroup H of index 3 we have a homomorphism $G \to S^3$, whose kernel is contained in H. If the kernel is $\{1\}$, $G \simeq S^3$; if it equals H, which has order 2, it commutes elementwise with the subgroup of order 3, which is normal because it has index 2, and then G is cyclic.

6. Let $[G:H]=n$; then the action of G on the cosets of H yields a homomorphism $G \to S^n$ with kernel K. Then G/K is isomorphic to a subgroup of S^n, and K is the normal subgroup we seek.

7. *i)* They are the orbits of the action of K on the right cosets of H. *ii)* The sought-for number is the index of the stabilizer of Ha in K; but $Hak = Ha$ if and only if $k \in H^a$, and therefore the index is $[K : H^a \cap K]$. Each coset contains $|H|$ elements. *iv)* $|H^{xk} \cap K| = |(H^{xk} \cap K)^{k^{-1}}| = |H^x \cap K|$. *v)* Following the hint, assume that Hy is not one of the latter cosets, and $[K : H^y \cap K] = t$, then $Hy, Hyk_2, \ldots, Hyk_t$ are t more distinct cosets. Thus the representatives x_i divide up into blocks containing $t_1, t_2, \ldots, t_s$ elements, where $[K : H^{x_i} \cap K] = t_i$. (This generalizes the case $K = G$, $[K : H^{x_i} \cap K] = [G : H^{x_i}] = t$ for all i, in which there is only one block). *vi)* In S^3, with $H = \{I, (1,2)\}, K = \{I, (1,3)\}$, we have two double cosets, of cardinality 2 and 4: $\{(2,3),(1,2,3)\}$ $\{I,(1,2),(1,3),(1,3,2)\}$. *vii)* The number we seek is the index of the stabilizer of aK in H. But $haK = aK \Leftrightarrow a^{-1}ha \in K \Leftrightarrow h \in aKa^{-1}$, and therefore the number is $[H : aKa^{-1}] = [a^{-1}Ha : a^{-1}Ha \cap K]$.

9. $\{I, (1,2)(3,4), (1,3)(2,4), (1,4)(2,3)\}$ and $\{I, (1,2)(3)(4), (1)(2)(34), (1,2)(3,4)\}$: the first one is transitive, the second is not.

10. With $\varphi : Ha \to a^{-1}H$ and θ the identity.

11. $\varphi : (H^x)^g \to \mathbf{N}_G(H)xg$ and $\theta = 1$.

12. *i)* If $\alpha \in \Gamma$ and $a \in \alpha$, then $1 = aa^{-1} \in \alpha a^{-1} = \beta \in \Gamma$, and if $x \in G$ then $x = 1 \cdot x \in \beta x = \gamma \in \Gamma$. *ii)* The subset α_i containing 1 is the desired subgroup. Let $1 \in \alpha$; if $x, y \in \alpha$, then $1 = yy^{-1} \in \alpha y^{-1}$, and therefore $\alpha y^{-1} = \alpha$, and $1 \in \alpha$. Moreover, $xy^{-1} \in \alpha y^{-1} = \alpha$.

13. Let $\alpha \in \Delta$, $g \in \mathbf{N}_G(H)$. Let $h \in H$ fix α; then $(\alpha^g)^{g^{-1}hg} = (\alpha^h)^g = \alpha^g$ and $\alpha^g \in \Delta$ since $g^{-1}hg \in H$.

19. *i)* Let $K = H \cap \mathbf{Z}(G)$. We have $K \neq \{1\}$ (Corollary 3.5), and $H/K \trianglelefteq G/K$; by induction, G/K contains a subgroup H_1/K of index p in H/K and normal in G/K. Hence $[H : H_1] = p$ and $H_1 \trianglelefteq G$.

ii) If A is the unique abelian subgroup of index p of H, then it is characteristic in H and therefore normal in G. Let A_1 be another such subgroup; the elements of $A \cap A_1$ commute with the elements of A and A_1 and therefore with $H = AA_1$; it follows, setting $Z = \mathbf{Z}(H)$, $A \cap A_1 \subseteq Z$. There are two cases: *a)* $A \cap A_1 \subset Z$. Since $[H : A \cap A_1] = p^2$, we have $[H : Z] \leq p$ and H abelian. Let $K = H \cap \mathbf{Z}(G)$; K is contained in a maximal subgroup A' of H, so A'/K is maximal in H/K. By induction, there exists $B/K \subseteq H/K$ of index p in H/K and normal in G/K, from which $B \trianglelefteq G$, and is abelian because is contained in H. *b)* $A \cap A_1 = Z$ which is normal in G. Then A/Z has index p in H/Z, and by induction there exists B/Z of index p in H/Z and normal in G/Z. Hence $B \trianglelefteq G$ and $[H : B] = p$. $A \cap A_1 = Z$ has order p^{n-2}, and as $Z \subseteq B$ we have $Z \subseteq \mathbf{Z}(B)$, so that $[B : \mathbf{Z}(B)] \leq p$ and B is abelian.

20. If $S \cap G' < S$, then $\{1\} \neq S/(S \cap G') \simeq SG'/G' \leq G/G'$, and p divides the order of the abelian group G/G', which therefore has a subgroup K/G' of index p. Then K is normal in G and has index p.

23. $\Omega \setminus \Gamma$ either is empty, or it is a union of orbits of cardinality divisible by p.

24. Following the hint, $\langle x \rangle$ acts on Δ_1 and Δ_2, that are orbits, and by the previous *ex.* $|\Delta_1| \equiv 1 \bmod p$, $|\Delta_2| \equiv 0 \bmod p$. Similarly, $|\Delta_1| \equiv 0 \bmod p$ and $|\Delta_2| \equiv 1 \bmod p$, a contradiction. Hence there is only one orbit.

25. As in the proof of Sylow's theorem (part *iii*), *a*), S_0 fixes no point in the set Ω of the other Sylow p-subgroups. Therefore for the sets Γ and Ω of *ex.* 23, we have $|\Gamma| = 0$ and $|\Omega| = kp$. It follows $n_p = 1 + kp$. G acts by conjugation on the set Ω of the p-Sylows; in the previous *ex.*, taking as H a p-Sylow S, S is the unique point of Ω fixed by S. Hence the action of G is transitive, that is, the Sylow p-subgroups are all conjugate.

27. If $x^g = y$, x and y have the same order, and $x = x_1x_2\cdots x_m$, $y = y_1y_2\cdots y_m$, $o(x_i) = o(y_i) = p_i^{h_i}$, $x_ix_j = x_jx_i$, $y_iy_j = y_jy_i$. If $s_i = o(\prod_{j\neq i} x_j) = o(\prod_{j\neq i} y_j)$ there exists r_i such that $r_is_i \equiv 1 \bmod o(x_i)$, and therefore also $r_is_i \equiv 1 \bmod o(y_i)$; hence $y_i = y_i^{r_is_i} = y^{r_is_i} = ((x_1x_2\cdots x_m)^g)^{r_is_i} = ((x_1x_2\cdots x_m)^{r_is_i})^g = (x_i^{r_is_i})^g = x_i^g$. If $x_i \in P$, $y_i \in Q$, let $t \in G$ be such that $P = Q^t$. Then $y_i^t = x_i^{gt} \in P$, and the conjugation of x and y reduces to that of x_i and x_i^{gt}, both belonging to P.

30. There are $(p-1)!$ p-cycles, and they divide up $p-1$ by $p-1$ in subgroups of order p. The number we seek is therefore $(p-1)!/(p-1) = (p-2)!$ The number of Sylow p-subgroups of S^p is congruent to 1 mod p, so $(p-2)! \equiv 1 \bmod p$; multiplying by $p-1$ we have $(p-1)! \equiv p-1 \equiv -1 \bmod p$.

31. *ii*) The action of G on the cosets of H yields a homomorphism $G \to S^p$ with kernel K. $|G/K|$ divides $p!$ and therefore $p^2 \nmid |G/K|$; from $[G:H][H:K] = [G:K]$ we have that $p \nmid [H:K]$. Hence every p-subgroup of H is contained in K, and in particular so does $O_p(H)$, which therefore is contained in $O_p(K)$. But $K \trianglelefteq H$, $O_p(K)$ is characteristic in K and hence normal in H, from which it follows $O_p(K) \subseteq O_p(H)$, $O_p(K) = O_p(H) \trianglelefteq G$.

32. Let $H = O^p(G)$; if p divides $|H/H'|$, let $o(H'x) = p$, $x = x_1x_2\cdots x_r$, with $p \nmid o(x_i)$, $H'x = \prod H'x_i$, and $o(H'x)|$ lcm $o(H'x_i)$ (the $H'x_i$ commute because H/H' is abelian). But $o(Hx_i)|o(x_i)$, hence $p \nmid o(H'x_i)$ for any i.

33. S is either cyclic or elementary abelian, and therefore $\mathbf{Aut}(S)$ has order either $p(p-1)$ or $(p^2-1)(p^2-p) = (p-1)^2p(p+1)$; if $C = \mathbf{C}_G(S)$, $|G/C|$ divides $|\mathbf{Aut}(S)|$ but $p \nmid |G/C|$. If q, a prime, divides $|G/C|$, $q > p$ and $p \nmid (p-1)$; if $q \nmid (p+1)$, then $q = p+1$, excluded. It follows $G/C = \{1\}$ i.e. $S \subseteq \mathbf{Z}(G)$.

34. $|\alpha^S| = [S : S_\alpha] = [S : S \cap G_\alpha] = p^h$, and since $[G : S \cap G_\alpha] = [G:S][S : S \cap G_\alpha]$, p^h is the largest power of p that divides $[G : S \cap G_\alpha] = [G : G_\alpha][G_\alpha : S \cap G_\alpha]$. If p^k divides $|\alpha^G| = [G : G_\alpha]$, then p^k divides $p^h = |\alpha^S|$.

35. $o(x) = o(y) = 2, y \in Sx \Rightarrow y = sx$, $sxsx = 1$, $xsx = s^{-1} \in S \cap S^x = \{1\}$, $y = x$.

36. If $x, y \in S_1$, $xy^{-1} \in S_1$ so also a p-element; but $xy^{-1} \in \mathbf{N}_G(S)$ and therefore $xy^{-1} \in S$, $xy^{-1} \in S \cap S_1 = \{1\}$ and $x = y$. Thus, two p-elements of a coset of $\mathbf{N}_G(S)$ belong to two distinct Sylow p-subgroups, and so the number of p-elements of $\mathbf{N}_G(S)$ is at most equal to that of the p-Sylow subgroups, and since none of these p-elements belongs to S, their number is at most $n_p - 1$.

37. *i)* $|\Omega| = [G : M]$, so $|\Omega| = 2[G : S] = 2m$, m odd; *ii)* If $Mgx = Mg$ then $gxg^{-1} \in M$; *iii)* the permutation induced by x on Ω consists of m transpositions, so is odd; *iv)* the image of G in S^Ω has a subgroup of index 2 (i.e. its intersection with A^Ω); so G has a subgroup od index 2, H say, not containing x. But $H \supseteq G'$, so $x \notin G'$.

38. $n_q = 1, p, p^2$. If $n_q = p \equiv 1 \bmod q$, then $q|(p-1)$, whereas $q > p$. If $n_q = p^2 \equiv 1 \bmod q$, then $q|(p^2-1) = (p-1)(p+1)$ and $q|(p+1)$; but $q \geq p+1$ and therefore $q = p+1$, from which $p = 2$ and $q = 3$.

39. Let $x \in S$ and $y \in S' \neq S$ be involutions. If $x \not\sim y$, x and y centralize the involution $z = (xy)^k$. But either $(xz)^2 = xzxz = 1 \Rightarrow zxz = x \Rightarrow x \in S \cap S^z \Rightarrow x = 1$, which is excluded, or $S = S^z$ and $z \in S$; similarly, $z \in S'$, and therefore $z = 1$, excluded. If $x, y \in S$, and $x \in S^g, \forall g \in G$, then $x \in S \cap S^g = \{1\}$, and either $x = 1$, excluded, or $S^g = S$, $\forall g$, and $S \trianglelefteq G$, excluded. Then there exists $g \in G$ such that $g^{-1}xg \notin S$; by the above, $g^{-1}xg \sim y$ and therefore $x \sim y$.

40. If $y^{-1}xy = x^{-1}$, then $y^{-2}xy^2 = x$, i.e. $y^2 \in \mathbf{C}_G(x)$; if $o(y)$ is odd, then $y \in \mathbf{C}_G(x)$, $x = x^{-1}$, excluded. If $o(y)$ is even, $o(y) = 2^k r$, $2 \nmid r$, $y = tu$ with $t = y^r$, $o(t) = 2^k$, $o(u) = r$. Now, $y \in \mathbf{N}_G(\langle x \rangle)$, so $t = y^r \in \mathbf{N}_G(\langle x \rangle)$, and therefore t belongs to a Sylow 2-subgroup of $\mathbf{N}_G(\langle x \rangle)$, which like $\langle x \rangle$ is normal there. The intersection of the two subgroups is $\{1\}$ ($o(x)$ is odd), and therefore t centralizes x. It follows $x^{-1} = y^{-1}xy = (tu)^{-1}x(tu) = u^{-1}t^{-1}xtu = u^{-1}xu$, with $o(u)$ odd, already excluded.

43. If $p > 11$, $n_p = 1$, and if $p = 11$ then $n_p = 1, 12$. If $n_p = 12$, there are $10 \cdot 12 = 120$ elements of order 11, and in this case it cannot be $n_3 = 22$ (too many elements), but if $n_3 = 1, 4$, G is not simple. If $p < 11$, then $p = 7$, and a 7-Sylow is normal.

44. $180 = 2^2 \cdot 3^2 \cdot 5$. If $n_5 = 6$ and G is simple, then G embeds in A^6 with index 2, which is absurd. Let $n_5 = 36$; then we have $(5-1)36 = 144$ 5-elements. If $n_2 = 5$, and G is simple, then G embeds in A^5, which is also absurd. $n_3 = 10$. If the maximal intersection is $\{1\}$, then there are $(9-1)10 = 80$ 3-elements which, added to the previous ones, give 220 elements: too many. Then the maximal intersection has order 3, its normalizer has order at least $9 \cdot 4 = 36$ and therefore index at most 5, and if G is simple it embeds in A^5, which is absurd.

$288 = 2^5 \cdot 3^2$. $n_2 = 9, n_3 = 16$. Since $16 \not\equiv 1 \bmod 3^2$, we have two 3-Sylows whose intersection is not $\{1\}$, and so is a C_3. $N = \mathbf{N}_G(C_3)$ contains at least four 3-Sylows, so $|N| \geq 4 \cdot 9 = 36$. If $|N| = 36$, the 3-Sylow being not normal, the 2-Sylow P is normal (*Ex.* 3.4, 9), and has order 4. But a group of order 4 is normalized by a 2-subgroup of order at least 8, and therefore $|\mathbf{N}_G(P)|$ has order at least $9 \cdot 8 = 72$ and therefore index at most 4.

$315 = 3^2 \cdot 5 \cdot 7$. $n_3 = 7 \not\equiv 1 \bmod 9$, $P \cap Q = C_3$, $N = \mathbf{N}_G(C_3)$, $9||N|$, and since N has more than one 3-Sylow, it has seven of them. Hence $7||N|$, so $7 \cdot 9 = 63||N|$ and $[G : N] \leq 5$. If it equals 5 and G is simple, G embeds in A^5, which is absurd.

$400 = 2^4 \cdot 5^2$, $n_5 = 16 \not\equiv 1 \bmod 5^2$, $S_5 \cap S_5' = C_5$, and the normalizer N of C_5 has 16 Sylow 5-subgroups, it has order divisible by 16 and 25, and therefore by 400. It follows that C_5 is normal.

$900 = 2^2 \cdot 3^2 \cdot 5^2$, $n_5 = 1, 6, 36$. If $n_5 = 6$, and G is simple, G embeds in A^6, which is absurd. Hence $n_5 = 36 \not\equiv 1 \bmod 5^2$, $S_5 \cap S_5^{'} = C_5$, the normalizer C_5 has at least six 5-Sylows and therefore order at least $6 \cdot 5^2 = 150$ and so index 2,3 or 6.

46. An element of order 7 has a cycle of length 7 and a fixed point, and there are $8!/7$ of them, giving rise to $(8!/7)/6 = 8 \cdot 5!$ subgroups of order 7, all conjugate. It follows $|\mathbf{N}(S_7)| = |A^8/8 \cdot 5! = 21$. Now, in a group of order $336 = 2^4 \cdot 3 \cdot 7$, $n_7 = 1$ or 8, and if $n_7 = 8$ then $|\mathbf{N}(S_7)| = 42$ and G cannot be embedded in A^8.

Similarly, the elements of order 11 in A^{12} are given by an 11-cycle and a fixed point, so they are $12!/11 = 12 \cdot 10!$ in number, and divide up 10 by 10 in subgroups of order 11, all conjugate. Hence, the index of $\mathbf{N}(S_{11})$ is $(12!/11)/10 = 12 \cdot 9!$ so its order is $|A^{12}|/12 \cdot 9! = (12!/2)/12 \cdot 9! = 55$. Now $264 = 2^3 \cdot 3 \cdot 11$, and if $|\mathbf{N}(S_{11})| = 22$ the group cannot be embedded in A^{12}.

51. If α fixes the four 3-Sylows, let $x^{-1}P_i x = P_j$; then $(x^{-1})^\alpha P_i x^\alpha = P_j^\alpha = P_j = x^{-1}P_i x$, and $x^\alpha x^{-1} \in C_G(P_i) = \{1\}$, i.e., $x^\alpha = x$, and this holds for all $x \in G$. Hence, if α fixes the four 3-Sylows, it also fixes all the elements of S^4. Therefore, if two automorphisms induce the same permutation of the 3-Sylows, they are equal; it follows that $\mathbf{Aut}(S^4)$ has at most 24 automorphisms, and the center being the identity S^4 has at least 24 inner automorphisms. It follows $\mathbf{Aut}(S^4) = \mathbf{I}(S^4)$.

53. a^2 commutes with s^t, $\forall t$, and $a^t s^k = s^{-k} a^t$, $t = \pm 1$. Moreover, $s^k a^{-1} = s^{k-n} s^n a^{-1} = s^{k-n} a^2 a^{-1} = s^{k-n} a$. A product of powers of s and of a may then be reduced to the form $s^k a^t$, $0 \leq k < 2n$, $t = 0, 1$. It follows $|G| = 4n$. The factor group w.r.t. $\langle a^2 \rangle$ is generated by $\bar{s}$ and $\bar{a}$ such that $\bar{s}^n = \bar{a}^2 = \bar{1}$ and $\bar{a}^{-1}\bar{s}\bar{a} = \bar{s}^{-1}$, and this is the group D_n.

54. If $H \leq S^n$ has index k, we have $G \longrightarrow S^k$ whose kernel is different from $\{1\}$ (because $k < n$) and is contained in S^k, and therefore is A^n, the unique normal subgroup of S^n. It follows $k = 2$. The S^{n-1} obtained by fixing a digit has index n.

55. A conjugate G_i^x is the G_j to which belongs $\varphi_i^{x^{-1}yx}$, so conjugating by an element x amounts to a permutation of the G_i; hence $K^x = K$, all x, and K is normal.

57. S is the kernel of the homomorphism sending a matrix A of B to the diagonal matrix given by the diagonal of A.

58. There are ten Sylow 3-subgroups in the group A^5, and four in $A^4 < A^5$.

59. $p^r m = \sum_{i=1}^m p^r p^r/d_i$, hence $m = \sum_{i=1}^m p^r/d_i$. But $K \cap a_i^{-1}Ha_i \subseteq K$, so $d_i | p^r$, d_i is a power of p, and the previous sum is a sum of powers of p that cannot all be different from 1 because $p \nmid m$. It follows that $p^r/d_i = 1$ for at least one i, $d_i = p^r$ i.e. $|K \cap a_i^{-1}Ha_i| = p^r = |K|$, and $K = a_i^{-1}Ha_i$.

60. $p^r m = \sum_{i=1}^m \frac{p^r p^r}{d_i}$ from which $m = \sum_{i=1}^m \frac{p^r}{d_i}$. But $K \cap a_i^{-1}Ha_i \subseteq K$, hence $d_i | p^r$, d_i is a power of p, and the previous sum is a sum of powers of p that cannot all be different from 1 because $p \nmid m$. Then, for at least one i we have $\frac{p^r}{d_i} = 1$, $d_i = p^r$, i.e. $|K \cap a_i^{-1}Ha_i| = p^r = |K|$, and $K = a_i^{-1}Ha_i$.

62. *i)* There are $\varphi(\frac{n}{d})$ elements of order n/d and these are the generators of $\langle g^d \rangle$; by Theorem 3.19, *ii)*, these elements fix the same number of points. Any element of G has order $\frac{n}{d}$ for a suitable divisor d of n.

ii) The relevant group is $C_n = \langle g \rangle$. If $d|n$, the element x^d partitions the pearls into d groups of $\frac{n}{d}$ pearls each (those that go onto one another under the rotation g^d). In order that a necklace be fixed by g^d, the pearls of each group must have the same color, and there being a choices for each group, g^d fixes a^d necklaces. By *ex.* 61, the desired number is $\frac{1}{n}\sum_{d|n} a^d\varphi(n/d)$.

64. Following the hint, if $g \notin H$ then $\mathbf{C}_H(g) < \mathbf{C}_G(g)$, from which

$$c(H) < (1/|H|)\sum_{g\in G}|\mathbf{C}_G(g)| = [G:H](1/|G|)\sum_{g\in G}|\mathbf{C}_G(g)| = [G:H]c(G).$$

By *ex.* 4, $|\mathbf{C}_G(x)| \leq [G:H]|\mathbf{C}_H(x)|$; consequently

$$|G|c(G)\sum_{x\in G}|\mathbf{C}_G(x)| \leq [G:H]\sum_{x\in G}|\mathbf{C}_H(x)|.$$

But the last expression equals

$$[G:H]\sum_{y\in H}|\mathbf{C}_G(y)| \leq [G:H]^2\sum_{x\in G}|\mathbf{C}_H(x)| = |H|c(H),$$

and therefore $[G:H][G:H]|H|c(H) = |G|[G:H]c(H)$, from which $c(G) \leq [G:H]c(H)$. If equality holds, two conjugate elements of G are conjugate by an element of H, and this implies $H \trianglelefteq G$ (if $h \in H$, then $g^{-1}hg = h_1^{-1}hh_1 \in H$). As to the counterexample, in D_4 we have $c(G) = 5, c(V) = 4, [G:V] = 2$ and $5 < 2\cdot 4$.

65. If $\chi(g) \geq N$, then $\forall g$, $\sum_{g\in G}\chi(g) \geq |G|N$. But $\chi(1) = |\Omega|$ implies $\sum_{1\neq g\in G}\chi(g) \geq (|G|-1)N + |\Omega| = |G|N + (|\Omega| - N)$. $|\Omega| > N$ is against (3.6), and if $|\Omega| = N$ the action is trivial.

66. The elements of G fixing some point of Ω belong to $\bigcup_{\alpha\in\Omega} G_\alpha$, and therefore they are at most $\sum_{\alpha\in\Omega}(|G_\alpha| - 1) + 1$. Now $|G_\alpha| = |G|/|\Omega|$, so that the above sum equals $|\Omega|(|G|/|\Omega| - 1) + 1 = |G| - |\Omega| + 1$, and the difference between $|G|$ and this number is $|\Omega| - 1$.

67. *i*) The sum on the left hand side of (3.8) equals $a_1 + n$, since $a_n = 1$ and $a_2 = a_3 = \ldots = a_{n-1} = 0$. On the right hand side, with $N = 1$, the sum then equals $a_0 + a_1 + 1$, from which $a_0 = n - 1$. Conversely, if $a_0 = n - 1$, then $n - 1 + a_1 + \cdots + a_{n-1} + 1 = n + a_1 + \cdots a_{n-1} = a_1 + 2a_2 + \cdots + (n-1)a_{n-1} + n$, from which $a_2 + 2a_3 + \cdots + (n-2)a_{n-1} = 0$; but $a_i \geq 0$ implies $a_2 = a_3 = \ldots = a_{n-1} = 0$.

ii) By transitivity, $p||G|$, and therefore there exists a subgroup of order p that in S^p is generated by a p-cycle whose elements fix no point, they are $p-1$ in number, and therefore they are the only ones.

69. α fixes $\mathbf{cl}(g)$ but none of its points; hence $\mathbf{cl}(g)$ splits into orbits of cardinality dividing $o(\alpha)$ and hence of cardinality p. It follows $|\mathbf{cl}(g)| = kp$, and so $p||G|$.

76. *i*) If G contains $(1,2)$, being 2-transitive it contains σ such that $\sigma(1) = 1, \sigma(2) = i, i = 3, \ldots, n$; then $(1,2)^\sigma = (1,i) \in G$, and these transpositions generate S^n (Corollary 2.7).

ii) If $(1,2,3) \in G$, there exists $\sigma \in G$ such that $\sigma(3) = 3, \sigma(1) = i$, $i = 4, \ldots, n$. Then, if $\sigma(2) \neq 2$, $\tau = (1,2,3)^\sigma = (i, 2^\sigma, 3)$ and $(1,2,3)^\tau = (1,2,i)$, and if $\sigma(2) = 2$, $(1,2,3)^\sigma = (1,3,4)$, and the product $(1,2,3)(1,3,4) = (1,2,4) \in G$. These 3-cycles generate A^n (Theorem 2.33).

78. In S^n, the nonidentity elements of the two subgroups consist of p^2 cycles of length p. If $\sigma \in S^n$ has this form, then among its conjugates we find the elements of the two subgroups, so that the intersections of the two subgroups with $\mathbf{cl}(\sigma)$ has $p^3 - 1$ elements. If $\sigma = 1$, then these intersections equal 1, and in all other cases they equal zero.

79. If $G = \{1, a, b, c\}$ let, in the first action,

$$a = (1,2)(3,4)(5,6)(7,8),\ b = (1,3)(2,4)(5,7)(6,8),\ c = (1,4)(2,3)(5,8)(6,7),$$

and, in the second,

$$a = (5,6)(7,8)(9,10)(11,12),\ b = (1,2)(3,4)(9,10)(11,12),\ c = (1,2)(3,4)(5,6)(7,8).$$

In both cases $\chi(1) = 12, \chi(a) = \chi(b) = \chi(c) = 4$, but the two actions are not equivalent because in the first one there are four points fixed by every element of G, whereas in the second this is not the case.

81. *i)* If τ and σ contain c, then $\tau\sigma^{-1} \in H$;
ii) if σ fixes c, σ must contain c or one of its powers, and conversely. Then G_c has the stated form because $\sigma^k \in H$;
iii) follows from *ii)*;
iv) the contribution of each orbit to the sum is $|G|/k|H| \cdot k|H| = |G|$;
vi) the required number is the sum

$$\sum_{k=1}^{n} P_k = \sum_{k=1}^{n}((1/|G|)\sum_{\sigma\in G} kz_k(\sigma)) = (1/|G|)\sum_{\sigma\in G}(\sum_{k=1}^{n} kz_k(\sigma)) = (1/|G|)|G|n = n.$$

(**Remark.** the sum $\sum kz_k(\sigma)$ over k is n, whereas over $\sigma \in G$ is $P_k|G|$.)

82. $y \in K \Rightarrow \chi(xyx^{-1}) = |\Omega|$. It follows $\chi^G(y) = (1/|H|)|G| \cdot |\Omega|$, which is the degree of G. In the action induced to G, y then fixes all the elements of $\Omega \times T$, so it belongs to the kernel K_1 of this action.

83. Let Γ be the set on which H acts, Ω that on which G does. Then $g \in G$ acts on $\Omega \times (\Gamma \times T)$ as $(\alpha, (\gamma, x_i))^g = (\alpha^g, (\gamma^h, x_j))$, where $x_i g = hx_j$, and on $(\Omega \times \Gamma) \times T$ as $((\alpha, \gamma), x_i)^g = ((\alpha^h, \gamma^h), x_j)$. If in the first action g fixes $(\alpha, (\gamma, x_i))$, then $\alpha^h = \alpha$, $\gamma^h = \gamma$ and $x_j = x_i$, and therefore $g = x_i^{-1}hx_i$. It follows that g fixes $((\alpha^{x_i}, \gamma), x_i)$ in the second one, and conversely.

84. The degrees of the two actions are the same: $[G : H] = [K : H \cap K]$, and as representatives of the cosets of H in G one may take the representatives k_i of the cosets of $H \cap K$ in K. It follows $k_i k = h_j k_j$ and $(\alpha, k_i)^k = (\alpha^{h_j}, k_j)$, with $h_j = k_i k k_j^{-1} \in H \cap K$.

85. The cosets of H contained in HaK have the form ak_i. Then the k_i are representatives of the cosets of $a^{-1}Ha \cap K$ in K, since $ak_i(ak_j)^{-1} = ak_ik_j^{-1}a^{-1} \in H$ if and only if $k_ik_j^{-1} \in a^{-1}Ha \cap K$. With $ak_ik = h_{i,j}ak_j$, $(\alpha, ak_i)^k = (\alpha^{h_{i,j}}, ak_j)$ one has *i)*, and *ii)* follows from $k_ik = a^{-1}h_{i,j}ak_j$, $(\alpha, k_i)^k = (\alpha^{a^{-1}h_{i,j}a}, ak_j) = (\alpha^{h_{i,j}}, ak_j)$.

86. (Cf. (3.11)) From the hint, $\chi(x_i)$ is repeated $[H : \mathbf{C}_H(x_i)]$ times, so $\chi(xgx^{-1})$ equals $\chi(x_i)$ a number of times equal to $|\mathbf{C}_G(x_i)|[H : \mathbf{C}_H(x_i)]$.

88. *i*) ⇒ *ii*) Let $|H| = p$. If the kernel of the action of G on the cosets of H is not H itself, then it equals 1, and the action, which is faithful and transitive, is not regular. ($|G| > |\Omega| = [G : H]$). *ii*) ⇒ *i*) The action is similar to that on the cosets of a subgroup H, and the kernel contains a subgroup of order p (Cauchy) which is therefore normal, and the action is not faithful.

91. Let $\alpha^h = \alpha$, $\beta \in \Omega$, $\alpha^x = \beta$ where $x \in \mathbf{C}_G(H)$ (transitivity). Then $\beta = \alpha^x = \alpha^{hx} = \alpha^{xh} = (\alpha^x)^h = \beta^h$, and $h = 1$.

93. $\mathbf{C}_G(\mathbf{C}_G(H)) \supseteq H$, which is transitive, so $\mathbf{C}_G(H))$ is semiregular (*ex.* 91), and therefore regular. But if $\mathbf{C}_G(H)$ is transitive, H is semiregular, and so regular. Consequently, $|H| = |\Omega| = |\mathbf{C}_G(H)|$; but $H \supseteq \mathbf{C}_G(H)$ implies $\mathbf{C}_G(\mathbf{C}_G(H)) = H$.

95. If x is regular, $x = (\alpha_1, \alpha_2, \ldots, \alpha_k)(\beta_1, \beta_2, \ldots, \beta_k) \cdots (\gamma_1, \gamma_2, \ldots, \gamma_k)$, with d cycles, then $x = (\alpha_1, \beta_1, \ldots, \gamma_1, \alpha_2, \beta_2, \ldots, \gamma_2, \ldots, \alpha_k, \beta_k, \ldots, \gamma_k)$. Conversely, $x = (1, 2, \ldots, n)^h$ is a product of (n, h) cycles of length $\frac{n}{(n,h)}$.

96. $\emptyset \neq (\Delta \cap \Delta_1) \cap (\Delta \cap \Delta_1)^x \Rightarrow \emptyset \neq (\Delta \cap \Delta^x) \cap (\Delta_1 \cap \Delta_1^x) \Rightarrow \Delta = \Delta^x$ and $\Delta_1 = \Delta_1^x$.

98. If Δ is a block, $\alpha, \beta \in \Delta$ and $\gamma \notin \Delta$. By 2-transitivity, there exists $g \in G$ such that $\alpha^g = \alpha$, $\beta^g = \gamma$. It follows $\alpha \in \Delta \cap \Delta^g \neq \emptyset$, so it must be $\Delta = \Delta^g$. But $\gamma \notin \Delta$ and $\gamma \in \Delta^g$, and $\Delta \neq \Delta^g$. Hence $\Delta \neq \Delta^g$.

99. *i*) $\emptyset \neq \alpha^H \cap (\alpha^H)^x \Rightarrow \alpha^h = \alpha^{h_1 x} \Rightarrow h_1 x h^{-1} \in G_\alpha \subseteq H \Rightarrow x \in H$, from which $\alpha^H = (\alpha^H)^x$.

ii) If $\alpha^H = \alpha^K$, for all $k \in K$ there exists $h \in H$ such that $\alpha^h = \alpha^k$, $hk^{-1} \in G_\alpha \subseteq H$ and $k \in H$.

iii) $x, y \in \theta(\Delta) \Rightarrow \alpha^x \in \Delta \Rightarrow \alpha^{xy^{-1}} \in \Delta^{y^{-1}}$; $\alpha^{-1} \in \Delta \Rightarrow \alpha \in \Delta^{y^{-1}}$. It follows $\Delta \cap \Delta' \neq \emptyset \Rightarrow \Delta = \Delta^{y^{-1}} \Rightarrow \alpha^{xy^{-1}} \in \Delta^{y^{-1}} = \Delta$.

iv) $\theta'\theta(h) = \{x \in G \mid \alpha^x \in \alpha^H\}$ and $\alpha^x \in \alpha^H \Leftrightarrow$ there exists $h \in H$ such that $\alpha^x = \alpha^h$, i.e. $\alpha^{hx^{-1}} = \alpha, hx^{-1} \in G_\alpha \subseteq H \Rightarrow x \in H$, and $\theta'\theta = I$. $\theta\theta'(\Delta) \subseteq \Delta$. Let $\beta \in \Delta$; there exists $x \in G$ such that $\beta = \alpha^x \Rightarrow x \in \theta'(\Delta)$; then $\alpha^x \in \theta\theta'(\Delta)$ and $\theta\theta' = I$.

103. If $Z = \mathbf{Z}(G) \subseteq G_\alpha$, then $Z \subseteq \bigcap_{\alpha \in \Omega} G_\alpha = \{1\}$; if $ZG_\alpha = G$ (G_α is maximal), $G_\alpha \trianglelefteq G$ and $G_\alpha = \{1\}$. Then $\{1\}$ is maximal, and $|G| = p$.

105. N normal and nontrivial is transitive (*ex.* 102), and if it is not regular then $\{1\} \neq N_\alpha = N \cap G_\alpha \trianglelefteq G_\alpha$, and $N \cap G_\alpha = G_\alpha$ because of the simplicity of G_α. If $N = G_\alpha$, then $G_\alpha = \{1\}$ and $N = \{1\}$, which is excluded because of the transitivity of N. The maximality of G_α then implies $N = G$, and G is simple.

106. *i*) $G_\alpha \cap H_1 = \{1\}$ (H_1 is regular) and $G = G_\alpha H_1$ (H_1 is transitive); then $G_\alpha \simeq G_\alpha/(G_\alpha \cap H_1) = G_\alpha H_1/H_1 = G/H_1 \simeq H_2$ (and similarly $G_\alpha \simeq H_1$).

ii) Let us show that G_α is maximal. Let $K \supseteq G_\alpha$; we have $G/H_1 = KH_1/H_1 \simeq K/(K \cap H_1)$, and therefore if $K \cap H_1 = \{1\}$, $G_\alpha \simeq K$ and $K = G_\alpha$. Let $L = K \cap H_1 \neq \{1\}$; we have $L \trianglelefteq G$ because from $G = G_\alpha H_2 = KH_2$ and $g \in G$ it follows $g = kh, k \in K, h \in H_2$ and $L^g = L^{kh} = L^h = L$ ($L \triangleleft K$ and $L \subseteq H_1$ commutes with the elements of H_2). Then $L = K \cap H_1 \trianglelefteq H_1$, $K \cap H_1 = H_1$, and $H_1 \subseteq K$. Similarly, $H_2 \subseteq K$, $K = G$, and G_α is maximal.

107. Let $n > 5$; the stabilizer of a digit is A^{n-1}, which is simple by induction, and A^n is 2-transitive and therefore primitive. By *ex.* 105, if A^n is not simple it contains a normal regular subgroup N. Let $x \in N$, $i^x = j$, $h^x = k$, all distinct. If $y = (i,j)(h,k,l,m)$ (plus possible 1-cycles), $1 \neq xx^y \in N$ fixes i, contrary to the regularity of N.

108. *i)* The kernel of the given action is $\{1\}$ (it cannot be A^5). By identifying the six Sylow 5-subgroups with the digits from 1 to 6, S^6 contains a transitive group isomorphic to S^5, and indeed it contains six such subgroups since the normalizer of this S^5 in S^6 has index at most 6, and hence exactly 6.

ii) Let $(1,2,3,4,5)$ be a 5-cycle; if (i,j) is a transposition, by transitivity there exists a permutation σ taking j to 6, $j^\sigma = 6$, so that $(i,j)^\sigma = (i^\sigma, 6)$, with $i^\sigma = k \neq 6$. Conjugation of $(k,6)$ by the powers of the 5-cycle yields the transpositions $(1,6),(2,6),(3,6),(4,6),(5,6)$ that generate S^6.

iii) The kernel of φ is $\{1\}$, so φ is injective, hence surjective, and therefore an automorphism.

iv) If $(H^\sigma)^\tau = H^\sigma$, then $\sigma\tau\sigma^{-1} \in \mathbf{N}_G(H) = H$, and H contains the transposition $\sigma\tau\sigma^{-1}$, against *ii)*.

v) $\varphi^{-1}\psi \in \mathbf{I}(S^5)$, so the factor group only contains two cosets. If φ and ψ are not inner, $\varphi^{-1}\psi$ fixes the class of transpositions. Apply Lemma 3.6.

vi) $\mathbf{I}(S^5) \simeq S^5$, so $\mathbf{Aut}(S^6)$ has order 1440.

109. If $i < j$, since in the sequence $12\ldots j \ldots i \ldots n$ the digits k, $i < k < j$, are each inverted once and there are $(j-i)-1$ of them, we have at least this number of inversions. Since digit i is less than the above digits k, it is inverted at least $(j-i)-1$ times, but since it is also inverted w.r.t.j the number of inversions of k is $j-i$. The other digits are not inverted. It follows that (i,j) has $(j-i)-1+(j-i) = 2(j-i)-1$ inversions.

110. *i)* If σ is a product of k transpositions of type $(i, i+1)$, the same happens for σ^{-1}. Thus the number is the same.

ii) If a digit is not inverted in one of the two permutations, then it is inverted in the other. If i is inverted t_i times in the first permutation and t'_i in the second, then the sum $t_i + t'_i$ is the total number of possible inversions of digit i, i.e. $n - i$. It follows $\sum_i (t_i + t'_i) = \sum_i (n-i) = \frac{n(n-1)}{2}$, so that if the first permutation has $\sum_i t_i = t$ inversions, the second has $\sum_i t'_i = \frac{n(n-1)}{2} - t$.

111. By the previous exercise, a permutation and its transpose have $\frac{n(n-1)}{2}$ inversions between them, so the average number of inversions in the permutation and in its transpose is half this number, i.e. $\frac{n(n-1)}{4}$. If all permutations are equally probable, this is also the average number of inversions in a random permutation.

112. *i)* The $k-1$ digits may be ordered in $(k-1)!$ ways, and the remaining $n-k$ in $(n-k)!$ ways, for a total of $\binom{n-1}{k-1}(k-1)!(n-k)! = (n-1)!$ ways. Hence the required probability is $\frac{(n-1)!}{n!} = \frac{1}{n}$, which is independent of k. Other solution[1]: write a permutation σ as a product of cycles (including the fixed points) in such a way that each cycle ends with the smallest digit, and these last elements are in increasing order. Hence the first cycle ends with 1. By removing the parenthesis of the cycles we have another permutation σ', from which one can recover σ uniquely: the first

---

[1] Lovász L.: Combinatorial Problems and Exercises. North-Holland (1979), p. 197.

cycle ends with 1, the second with the smallest digit not belonging to the first cycle, and so on. The length of the cycle of σ containing 1 is determined by the position of 1 in σ', and it is clear that this can be any of the n positions with the same probability. Hence the probability is $1/n$.

ii) Besides counting as above one may consider here too the two permutations σ and σ'; then 1 and 2 belong to the same cycle of σ if and only if 2 appears before 1 in σ'. Since 2 can appear before or after 1 with the same probability, the required probability is $\frac{1}{2}$. A different proof: let A be the set of permutations in which 1 and 2 belong to the same cycle, and let $\tau = (1,2)$. By multiplying the elements of A by τ we obtain the set of permutations in which 1 and 2 belong to different cycles (Serret's lemma). Obviously, $S^n = A \cup A\tau$, $A \cap A\tau = \emptyset$ and $|A| = |A\tau| = \frac{n!}{2}$, so that a permutation belongs to A with probability 1/2.

113. Let $\sigma \in S^n$. If σ is even, then $\sigma(n+1)(n+2)$ is again even; if it is odd, then $\sigma(n+1, n+2)$ is even. In both cases we have an even permutation.

114. If $\alpha = \gamma^{-1}\sigma\gamma$, $\gamma \notin A^n$, let $\tau = (i,j)$, where i and j are fixed by σ. Then $\tau\gamma \in A^n$ and $(\tau\gamma)^{-1}\sigma\tau\gamma = \gamma^{-1}\tau\sigma\tau\gamma = \gamma^{-1}\sigma\gamma = \alpha$.

117. Write down 6. Since 5 is not inverted, write 5 to the left of 6. Since 4 has one inversion, write 4 in between 5 and 6, etc.

119. *i*) If all the transpositions disconnecting α also disconnect σ, then the cycles of α are contained in those of σ, contrary to transitivity.

ii) Let $\tau = (i,j)$ as in *i*). Then α has the form $(h,k,\ldots,i,\ldots,j,\ldots)\cdots$. If $k \neq i,j$, then take $\gamma = (\alpha\tau)^{-1}$. If $k = i$, then $\alpha = (h,i,\ldots,j,\ldots)\cdots$, so that $\alpha\tau = (h,j,\ldots)(i,\ldots)\cdots$. Therefore h and j belong to the same cycle of $\alpha\tau$, and since i and j belong to the same cycle of $\sigma\tau$ (τ connects σ) we may take $\gamma = (\sigma\tau)^a(\alpha\tau)^b$, some a and b.

iii) If $z(\sigma) = 1$, the result is obvious. Let $z(\sigma) > 1$ and let τ be as in *i*). Then by *ii*) the group $\langle\sigma\tau, \alpha\tau\rangle$ is transitive, and since $z(\sigma\tau) = z(\sigma) - 1$ we have, by induction, $z(\sigma\tau) + z(\alpha\tau) \leq n+1$; with $z(\alpha\tau) = z(\alpha) + 1$ the result follows.

iv) The group $\langle\sigma, \alpha\sigma\rangle$ is the same as $\langle\sigma, \alpha\rangle$, so it is transitive. Let τ connect σ and disconnect $\alpha\sigma$. The group $\langle\sigma\tau, \alpha\rangle$ is transitive, and $z(\sigma\tau) = z(\sigma) - 1$. By induction, $z(\sigma\tau) + z(\alpha) + z(\alpha\sigma\tau) \leq n+2$, and since $z(\alpha\sigma\tau) = z(\alpha\sigma) + 1$ we have the result.

v) Write the nonnegative difference between the right and left sides of *iv*) as $(n - z(\sigma)) + (n - z(\alpha)) + (n - z(\alpha\sigma)) - 2n + 2$. Of the three permutations either one or all three are even, and therefore so is the sum of the first three summands.

vi) Following the hint, if all pairs r,s belonging to the same cycle of σ_1 also belong to the same cycle of the product $\sigma_1\sigma_2\cdots\sigma_k\sigma_{k+1}$, then the cycles of this product are contained in those of σ_1. The group $\langle\sigma_1, \sigma_2, \cdots, \sigma_k, \sigma_{k+1}\rangle = \langle\sigma_1\sigma_2\cdots\sigma_k, \sigma_1\sigma_2\cdots\sigma_k\sigma_{k+1}\rangle$ being transitive, the group $\langle\sigma_1, \sigma_2, \cdots, \sigma_k\rangle$ would also be transitive, contrary to the choice of k. Hence there exist r and s belonging to the same cycle of $\sigma_1\sigma_2\cdots\sigma_k\sigma_{k+1}$ and to two different cycles of σ_1. Let $\tau = (r,s)$; then $z(\tau\sigma_1) = z(\sigma_1) - 1$, so by induction $z(\tau\sigma_1) + z(\sigma_2) + \cdots + z(\sigma_{k+1}) = kn + 2$. Hence from $z(\tau\sigma_1\sigma_2\cdots\sigma_{k+1}) = z(\sigma_1\sigma_2\cdots\sigma_{k+1}) + 1$ it follows $\sum_{i=1}^{k+1} z(\sigma_i) + z(\sigma_1\sigma_2\cdots\sigma_{k+1}) \leq kn + 2$. Now $z(\sigma_1\sigma_2\cdots\sigma_m) = z(\sigma_1\sigma_2\cdots\sigma_{k+1}\sigma_{k+2}\cdots\sigma_m) \leq z(\sigma_1\sigma_2\cdots\sigma_{k+1}) + (m -$

$k-1)n - \sum_{i=k+1}^{m} z(\sigma_i)$, and finally

$$\sum_{i=1}^{m} z(\sigma_i) + z(\sigma_1\sigma_2\cdots\sigma_m) \le \sum_{i=1}^{k+1} z(\sigma_i) + z(\sigma_1\sigma_2\cdots\sigma_{k+1}) + \sum_{i=k+2}^{m} z(\sigma_i) +$$
$$+(m-k-1)n - \sum_{i=1}^{k+1} z(\sigma_i) \le kn + 2 + (m-k-1)n = (m-1)n + 2,$$

as required. Here too the difference of the two sides of the inequality is even. Since the parity of a permutation γ is the same as that of $n - z(\gamma)$, if an even number of the σ_i are odd permutations, the product is even, and so the difference is even; if an odd number are odd, the first two summands are odd, and again the difference is even.

120. If Γ_1 is an orbit of G, and $y \in G_1$, then Γ_1^y is an orbit of G because $(\Gamma_1^y)^x = (\Gamma_1^x)^y = \Gamma_1^y$, for all $x \in G$. Thus G_1 acts on the orbits of G, and similarly G acts on those of G_1. If Γ is an orbit of G_1 on those of G, $\Gamma = \{\Gamma_1, \Gamma_2, \ldots, \Gamma_t\}$, then G_1 acts on $\Omega = \bigcup \Gamma_i$, and the set $\Delta = \{\Delta_1, \Delta_2, \ldots, \Delta_s\}$ of orbits of this action is an orbit of G. G acts on Δ: let $x \in G$, and let Δ_i^x be defined as follows. Pick $k \in \Delta_i$; then $k \in \Omega$, $k \in \Gamma_j$, some j, and so $k^x \in \Gamma_j$ because Γ_j is an orbit of G. If $k^x \in \Delta_l$, define $\Delta_i^x = \Delta_l$. This does not depend on the choice of k, because if $h \in \Delta_i$, then $h^y = k$, some $y \in G_1$, and h^x also belongs to Δ_l since $(h^x)^y = (k^y)^x = k^x \in \Delta_l$. Δ_l being an orbit of G_1, h^x also belongs to Δ_l. This action is transitive: let $u \in \Delta_i, v \in \Delta_l$; then $u, v \in \Omega$. If they belong to different Γ_i, then $u^y = v$, some $y \in G_1$, since Γ is an orbit of G_1; thus $\Delta_i = \Delta_l$. If $\Delta_i \ne \Delta_l$, then u and v belong to the same orbit of G, so that $u^x = v$, some $x \in G$, and as seen above this implies $\Delta_i^x = \Delta_l$. The mapping $\Gamma \to \Delta$ provides the required bijection.

121. If $i, j \in \Gamma_1 \in \Gamma$, then $\sigma(i)$ and $\sigma(j)$ belong to the same orbit because if $y \in G$, such that $y(i) = j$, then $y\sigma(i) = \sigma y(i) = \sigma(j)$. Hence σ acts on Γ, and the same holds for α.

123. From (3.14) we have, if $\gamma > 0$, that $2g - 2 \ge |G|(2\gamma - 2)$, hence $g \ge \gamma$. If $\gamma = 0$, then G being non negative, $g \ge \gamma$.

125. $g = 0$ implies $\gamma = 0$. For $G = \langle\phi\rangle$, (3.15) reads $2(o(\phi) - 1) = \sum_{i=1}^{o(\phi)-1} \chi(\phi^i)$, and since $\chi(\phi^i) \ge \chi(\phi)$, all i, we have $2(o(\phi)-1) \ge (o(\phi)-1)\chi(\phi)$, so that $\chi(\phi) \le 2$. Similarly, $\chi(\phi^i) \le 2$. If $\chi(\phi^i) < 2$ for some i, then the above sum cannot be equal to $2(o(\phi) - 1)$, so that $\chi(\phi^i) = 2$ all i.

126. $\bar\phi$ is a power of c, since c is a cycle and $\bar\phi$ commutes with c. Thus $\bar S = \{\bar\phi, \phi \in S\} \subseteq \langle c\rangle$, and the mapping $S \to \bar S$ is a homomorphism. If $\bar\phi = 1$, then ϕ being regular we have $\phi = 1$, and the mapping is injective. Thus S is cyclic.

127. From (3.14) we have $\chi(\phi) \le 2 + \frac{2g}{o(\phi)-1} - \frac{2\gamma o(\phi)}{o(\phi)-1}$.

128. Let $\psi \ne \phi$ be an involution fixing $2g + 2$ points, and let $o(\phi\psi) = n$. If n is even, there are two conjugacy classes of involutions, each one containing $n/2$ elements, and one class with a central involution; if n is odd, there is only one class of n elements. Since conjugate elements fix the same number of points, in both cases there are at least $(2g + 2)n$ points fixed by involutions. From (3.14) we have

$2g-2 \geq 2n(2\gamma-2)+(2g-2)n$. It follows $\gamma = 0$, $2g-2 \geq (2g-2)n$, and since $g \geq 2$ this implies $n = 1$ and $\psi = \phi$.

130. For G acting on the set of cycles of σ we have $\sum_{I \neq \phi \in G} \chi(\phi) = \sum_c (|G_c| - 1)$, where G_c is the stabilizer of the cycle c. If c and c' belong to the same orbit of G, then $|G_c| = |G_{c'}|$, and the sum becomes $\sum_{c_i} \frac{|G|}{|G_{c_i}|}(|G_{c_i}|-1)$, where the c_i constitute a set of representatives of the orbits of G. Do the same for α and $\alpha\sigma$. The result follows.

131. If $r = 4$, then $\frac{2g-2}{|G|} = 2 - \sum_{i=1}^{4} 1/n_i$, and we have the following table:

n_1	n_2	n_3	n_4	$2 - \sum_{i=1}^{4}(1/n_i) \geq$	$\lvert G\rvert \leq$
>2	>2	>2	>2	$2/3$	$3(g-1)$
$=2$	>2	>2	>2	$1/2$	$4(g-1)$
$=2$	$=2$	>2	>2	$1/3$	$6(g-1)$
$=2$	$=2$	$=2$	>2	$1/6$	$12(g-1)$

If $r = 3$ (we can assume $n_1 \leq n_2 \leq n_3$) the corresponding table is:

n_1	n_2	n_3	$1 - \sum_{i=1}^{3}(1/n_i) \geq$	$\lvert G\rvert \leq$
>3	>3	>3	$1/4$	$8(g-1)$
$=3$	>3	>3	$1/6$	$12(g-1)$
$=3$	$=3$	>3	$1/12$	$24(g-1)$
$=2$	>4	>4	$1/10$	$20(g-1)$
$=2$	$=4$	>4	$1/20$	$40(g-1)$
$=2$	$=3$	>6	$1/42$	$84(g-1)$

The result follows.

132. A conjugate xgx^{-1} of g fixes u if and only if g fixes u^x, so $|\mathbf{cl}(g) \cap H|$ equals the number of fixed points of g. But this number is the same for g and its transpose g^t, so $|\mathbf{cl}(g) \cap K| = |\mathbf{cl}(g^t)^t \cap H^t| = |\mathbf{cl}(g^t) \cap H| = |\mathbf{cl}(g) \cap H|$. This holds more generally for the group $SL(n,q)$, with H the subgroup stabilizing the line $(t,0,0,\ldots,0)$, and K that stabilizing the hyperplane $(0,x_2,\ldots,x_n)$.

137. Since $n \geq 3$, there exist distinct i,j,k. Then

$$E_{i,j}(\alpha) = E_{i,k}(\alpha)E_{k,j}(1)E_{i,k}(-\alpha)E_{k,j}(-1).$$

8.4 Chapter 4

1. The mapping $G \to nG$ given by $x \to nx$ is a surjective homomorphism with kernel $G[n]$.

2. $o(a)b + nc = 1 \Rightarrow o(a)ba + nca = a \Rightarrow a = n(ca)$.

3. *i)* If G is finite, $|G|y = 0$ for all y, and therefore no $x \neq 0$ is divisible by $|G|$. Moreover, given n and the coset $x+H$, $x \notin H$, we have $x = ny$ and $x+H = n(y+H)$.

ii) If $o(x) = m$ is finite, and $x = ny$, then $0 = mx = mny$, and so the order of y is also finite.

v) One direction is obvious. As to the other direction, if $pG = G$ then $p^2G = p(pG) = pG = G$, and so for all powers of p. If $n = qp$, $nG = qpG = q(pG) = qG = G$, etc.

vi) If G has a maximal subgroup M, the quotient G/M is finite (of prime order), and therefore not divisible, and so G is not divisible either. Conversely, if G is not divisible, then, for some p, $pG < G$. In G/pG all the elements have order p, so that G/pG is a direct sum $\sum_\lambda G_\lambda/pG$, where the summands are all isomorphic to $\mathbf{Z}_p$ (a vector space over the field $\mathbf{Z}_p$). Let $H/pG = \sum_{\lambda\neq\mu} G_\lambda/pG$; then $G/H \simeq (G/pG)/(H/pG) \simeq G_\mu/pG \simeq \mathbf{Z}_p$, and H is maximal.

viii) If n is an integer, a and b are divisible by n, $a = na'$ and $b = nb'$, and so is $a+b$: $a + b = n(a' + b')$.

4. C_{p^∞} has no maximal subgroups. A direct proof is the following (additive notation; see *Ex.* 1.6, 3). Let $g = s\omega_k$, $n = p^hm$, $(p, m) = 1$. We have $p^kd + mf = 1$, $p^kdg + mfg = g$, and $g = mfg$ because $p^kg = p^ksa_k = 0$, and if h is such that $p^h\omega_{k+h} = \omega_k$, then $g = mfg = mfs\omega_k = mfsp^h\omega_{k+h} = p^hm(fs\omega_{k+h}) = n(fs\omega_{k+h})$, and $fs\omega_{k+h}$ is the sought-for element.

5. By *ex.* 1, with $A = C_{p^\infty}$, we have $p^nA \simeq A/A[p^n]$, and A being divisible, $p^nA = A$.

6. Since $\mathbf{Q} = \bigcup_k \langle \frac{1}{k!} \rangle$, every finite set of elements is contained in some $\langle \frac{1}{k!} \rangle$.

9. An element of G determines a finite subset of the integers (the exponents of the generators needed to write it). It follows that the cardinality of G equals that of the set of finite subsets of a countable set, and therefore is countable.

10. We have $x_ix_j = n_{i,j}x_k, x_ix_jx_l = n_{i,j}x_kx_l = n_{i,j}n_{k,l}$, etc.; the $n_{i,j} \in N$ generate a finite group and therefore so do the x_i.

13. Let M be the maximal subgroup, and let $x \notin M$. If $\langle x \rangle \neq G$, then $\langle x \rangle$ is contained in a maximal subgroup (Theorem 4.3), and this must be M, excluded. Then $\langle x \rangle = G$, and G is cyclic. If G is infinite, then it has an infinite number of maximal subgroups. Hence G is finite, and if its order is divisible by more than one prime then it has more than one subgroup of prime index, and therefore maximal.

14. C_{p^∞} has no maximal subgroups.

15. An integer belonging to the Frattini subgroup is divisible by all primes, so is zero.

16. M maximal has index p, a prime, and therefore order n/p. Hence it is generated by x^p. Moreover, $\langle x^p \rangle \cap \langle x^q \rangle = \langle x^{pq} \rangle$, etc.

22. $\mathbf{Z}$ and $\mathbf{Z}_{13}$, respectively.

23. If $C_2 = \langle a \rangle$ and $C_8 = \langle b \rangle$, then $H = \langle ab^2 \rangle$ has order 4 and only admits the trivial decomposition $H \times \{1\}$ as a direct product. However, it is not contained in any cyclic subgroup of order 8 (the only possible decomposition of G is that with a cyclic group of order 8 and one of order 2).

24. If G is a p-group, let $p^{h_1} \geq p^{h_2} \geq \ldots \geq p^{h_t}$ be the elementary divisors of G. Hence, if $p^{k_1} \geq p^{k_2} \geq \ldots \geq p^{k_t}$ are the elementary divisors of G/H then $h_i \geq k_i$, so that G

contains a subgroup with elementary divisors p^{k_i}, $i = 1, 2, \ldots, t$. Proceeding in this way for all Sylow p-subgroups of G/H we have the result.

26. Since $x \neq 0$, there exists a system of generators x_i such that $x = h_1x_1 + h_2x_2 + \cdots + h_nx_n$, where the h_i are not all zero. But the $-x_i$ are also a system of generators, hence in some system of generators one of the h_i is positive; the smallest of these is the $h(x)$ we seek. As for the second part, if $\{x_i\}$ is a system of generators for which $x = h(x)x_1 + \cdots h_2x_2 + \cdots + h_nx_n$, then setting $h_i = q_ih(x) + r_i$, $0 \leq r_i \leq h(x)$, we have $x = h(x)y + r_ix_i$, where $y = x_1 + q_2x_2 + \cdots + q_nx_n$, and $y, x_2, \ldots, x_n$ are a system of generators of the required type. It follows $r_i = 0$ and $y = h(x)y$. If $S = \{x_i\}$ and $y = h_1x_1 + h_2x_2 + \cdots h_nx_n$, then $x = h(x)y = h_1h(x)x_1 + h_2h(x)x_2 + \cdots + h_nh(x)x_n$, so that $h(x)$ divides all the coefficients and therefore their gcd. Conversely, if $x = k_1x_1 + k_2x_2 + \cdots + k_nx_n$ and h divides all the k_i, then $k_i = hs_i$ and $x = h(s_1x_1 + s_2x_2 + \cdots + s_nx_n) = hz$, and if $z = m_1y + m_2y_2 + \cdots + m_ny_n$, then $x = hz = m_1hy + m_2hy_2 + \cdots + m_nhy_n$, so $m_1h = h(x)$, and $h(x)$ divides h.

28. Let $x_1, x_2 \in X$, $\sigma(x_1) = \sigma(x_2)$, G any group, and $g_1 \neq g_2$ two elements of G. With $f(x_1) = g_1$, and $f(x_2) = g_2$ we have $\phi\sigma(x_1) = \phi\sigma(x_2)$, so $f(x_1) = f(x_2)$ and $g_1 = g_2$.

30. Let $G = A/R$ with A free, $A = \sum^{\oplus} \mathbf{Z}_\lambda$, and embed each copy $\mathbf{Z}_\lambda$ in a copy $\mathbf{Q}_\lambda$ of $\mathbf{Q}$. Then $G = A/R = (\sum^{\oplus} \mathbf{Z}_\lambda)/R \subseteq (\sum^{\oplus} \mathbf{Q}_\lambda)/R$, and the latter group is divisible.

33. If G is cyclic, apply *Ex.* 1.9, 1, *i*), and *ex.* 15 of the same chapter. If G is elementary abelian G is a vector space over $\mathbf{F}_p$, and *Ex.* 1.3, 2 applies; otherwise, G is a direct product of cyclic groups $S_1 \times S_2 \times \cdots \times S_t$, and an automorphism α of order p of S_1 extends to the whole group through $\alpha'(x_1x_2\cdots x_t) = \alpha(x_1)x_2\cdots x_t$, and $o(\alpha') = p$.

35. Following the hint, by Schur's theorem G' is finite, and having finite index and G being infinite it equals $\{1\}$; hence G is abelian. G is torsion free (the torsion subgroup would be $f.g$ and therefore finite, and having finite index G would be finite). By Corollary 4.1, G is a direct sum of copies of $\mathbf{Z}$; but if there is more than one summand one of them has infinite index. Hence there is only one summand, and $G \simeq \mathbf{Z}$.

36. *ii*) any mapping f from 1 to a group G extends to a homomorphism $\phi : \mathbf{Z} \to G$ given by $\phi(n) = f(1)^n$. *iii*) Let X be a free basis of F, $|X| \geq 2$, let $x, y \in X$, and consider the mapping $f : F \to \mathbf{Q}$ of F in the quaternion group (or any nonabelian group) given by $f(x) = i, f(y) = j$. Then f extends to a homomorphism ϕ, and $\phi(xy) = \phi(x)\phi(y) = ij = k$ and $\phi(yx) = \phi(y)\phi(x) = ji = -k$. However, $xy = yx$ implies $\phi(xy) = \phi(yx)$, which is absurd. Hence $|X| = 1$, and F is cyclic. *iv*) If X is a basis of the group G, and $f : X \to \mathbf{Z}$ a function not sending all the elements of X to 0, then f cannot be extended to a homomorphism $G \to \mathbf{Z}$ because the unique homomorphism of a finite group to $\mathbf{Z}$ (or to any torsion free group) is the trivial one.

38. If $F/N \simeq \mathbf{Z}_2$, then a homomorphism $F \to \mathbf{Z}_2$ is determined once the images of the two generators u and v are known. There are three possibilities: $f(u) = 1, f(v) = 0$. In this case $u^{h_1}v^{k_1}u^{h_2}v^{k_2}\ldots u^{h_t}v^{k_t} \to (\sum_i h_i)f(u) + (\sum_i k_i)f(v) = \sum_i h_i$, and the kernel N consists of the elements of F in which the sum of the exponents of u is zero. If $f(u)=0, f(v)=1$ then the kernel N consists of the elements of F in which the sum of the exponents of v is zero, and if $f(u)=1, f(v)=1$ then the kernel N consists of the elements of F in which the sum of the exponents of u and v is zero. (If $f(u)=0, f(v)=0$ then the kernel is F.)

39. Let F be free on X, $\iota : X \to F$ and $f : X \to G$ the inclusions. Let ϕ extend f. Then ϕ is surjective because X generates G, and is injective because if the images of two words of F were equal to the same element $g \in G$, then these images would be two different spellings of g.

40. Let $w = x_1^{\epsilon_1}x_2^{\epsilon_2}\ldots x_m^{\epsilon_m}$ be reduced, and let $w^n=1$, $n>0$. If w is cyclically reduced, the reduced form of w^n is

$$w^n = x_1^{\epsilon_1}x_2^{\epsilon_2}\ldots x_m^{\epsilon_m}x_1^{\epsilon_1}x_2^{\epsilon_2}\ldots x_m^{\epsilon_m}\ldots x_1^{\epsilon_1}x_2^{\epsilon_2}\ldots x_m^{\epsilon_m},$$

whose length is mn. But this length is zero because $w^n=1$, so either $n=0$, which is excluded, or $m=0$, i.e. $w=1$. If w is not cyclically reduced, let

$$w = y_r^{\eta_r}y_{r-1}^{\eta_{r-1}}\ldots y_1^{\eta_1}x_1^{\epsilon_1}x_2^{\epsilon_2}\ldots x_m^{\epsilon_m}y_1^{-\eta_1}y_2^{-\eta_2}\ldots y_r^{-\eta_r}$$

where $\eta=\pm 1$. Then

$$w^n = y_r^{\eta_r}y_{r-1}^{\eta_{r-1}}\ldots y_1^{\eta_1}(x_1^{\epsilon_1}x_2^{\epsilon_2}\ldots x_m^{\epsilon_m})^n y_1^{-\eta_1}y_2^{-\eta_2}\ldots y_r^{-\eta_r}$$

so the reduced form of w^n has length $2r+mn$, and as above $r=0$ and $m=0$.

43. The cases $n=1,2$ are obvious. Denote the generators by a,b,c,d. For $n=3$, $ba=c$ implies $c=ba^{-1}$, and from $ab=c$ it follows $ab=b^{-1}a$; from $b=a^{-1}c$, multiplication by a yields $ba=a^{-1}b$ (a,b,c behave like the units i,j,k, and we have the quaternion group). As to $n=4$, multiplying by $ab=c$ and $bc=d$ we have $ab^2c=cd=a$ and $c=b^{-2}$, $ab=b^{-2}$, $a=b^{-3}$, $d=bc=bb^{-2}=b^{-1}$; $ad=b$ implies $b^{-3}b^{-1}=b$ and $b^5=1$. Hence, the group has at most five elements. The mapping $G\to C_5=\langle x\rangle$ given by $a\to x^2, b\to x, c\to x^3, d\to x^4$ is onto, so $G\simeq C_5$.

45. From $x^{-1}yx=y^2$ it follows $y=(y^{-1}x^{-1}y)x=x^{-2}x=x^{-1}$; substituting x^{-1} for y in the previous equality we have $x=1$, and so also $y=1$ (more generally, the identity group is obtained with the relations $x^{-1}y^nx=y^{n+1}$ and $y^{-1}x^ny=x^{n+1}$, for all n).

46. If N is generated by y_i and G/N by Nx_j, then $x_j^{-1}y_ix_j\in N$, so the latter is a word in the y_i: $x_j^{-1}y_ix_j=u_{i,j}$. A relation r_k of G/N is a product of cosets Nx_j equal to N, so the inverse image r'_k of r_k is a word in the y_i: $r'_k=w_k$. Hence, G is generated by y_i and x_j; the relations are those of N to which $x_j^{-1}y_ix_ju_{i,j}^{-1}=1$ and $r'_kw_k^{-1}=1$ must be added.

47. *i)* The subgroup generated by u and v is free, and being abelian it is cyclic $H=\langle w\rangle$ (*ex.* 36, *iii*)). Hence, $u=w^h$ and $v=w^k$. If two powers of u and v commute, then the subgroup generated by u and v, which is of rank at most 2, cannot be of rank 2. As above, it is cyclic.

48. By *ex.* 47, $x\rho y$ implies $x = u^r, y = u^s$, and $y\rho z$ that $y = v^p, z = v^q$. Then $u^s = v^p$ (they are both equal to y) implies, by the previous exercise, that u and v are powers of the same element w: $u = w^h, v = w^k$. It follows $x = w^{hr}$ and $z = w^{kq}$, so $x\rho z$.

50. If $y^{-1}uy = u^{-1}$, then y^2 centralizes u (this holds in any group, cf. Chapter 2, *ex.* 14), and since y^2 and u commute, y and u also commute (*ex.* 47).

51. Subgroups: let $H \leq G$ the restriction to H of the mapping ϕ defined on G with values in a finite group and such that $\phi(h) = 1$ does the job. Cartesian products: consider $x \in G = \prod G_\lambda$, and its projection $1 \neq x_\lambda \in G_\lambda$; then x_λ has a nonidentity image in a finite group under a mapping ϕ, and so does x by composing ϕ with the projection onto G_λ. Direct products: a direct product of groups is a subgroup of a cartesian product (of course this can be proved directly as for the cartesian product).

52. *i*) Such a group is a direct sum of copies of the integers and of finite groups.
ii) Theorem 4.40.

53. Every homomorphism of the group to a finite group is trivial.

55. Map $GL(n, \mathbf{Z})$ to $GL(n, \mathbf{Z}_m)$ by reducing the entries $a_{i,j}$ mod m; the image of A is not the identity matrix.

56. Consider the free group $\langle A, B\rangle$ of *Ex.* 4.5. As a subgroup of $GL(2, \mathbf{Z})$ it is residually finite, and so are all its subgroups (see Corollary 4.7). (This result also holds for free groups of uncountable rank.)

57. Let $\{x_1, x_2, \ldots, x_n\}$ be a free basis of F; the mapping $x_i \longrightarrow a_i$ extends to a homomorphism $\phi : F \longrightarrow F$, which is onto because the a_i generate F. Hence $F \simeq F/ker(\phi)$, and F being hopfian it follows $ker(\phi) = \{1\}$, and ϕ is an isomorphism.

58. *ii*) The homomorphisms $\phi : \mathbf{Q} \longrightarrow \mathbf{Q}$ are linear transformations over $\mathbf{Q}$ (cf. *Ex.* 1.9, 3), therefore if ϕ is surjective, dim($\mathbf{Q}$)=dim($\mathbf{Q}$)–dim($ker(\phi)$), i.e. $1 = 1 -$ dim($ker(\phi)$), $ker(\phi) = \{0\}$ and ϕ is injective. With the same proof $\mathbf{Q}$ is cohopfian. Similarly, $\mathbf{Q} \oplus \mathbf{Q}$ is hopfian.
vi) The proper subgroups of $\mathbf{C}_{p^\infty}$ are finite, so the group is cohopfian. By *ex.* 5, it is not hopfian.

59. Let G be the group. Since the root of the tree must remain fixed, G permutes the vertices at each level n. Let P be the finite group obtained by restricting G to the vertices at level n. If g is not the identity on this level, then under the homomorphism $G \longrightarrow P$, obtained by restriction, the image of g is not the identity.

60. With $C_2 = \langle a\rangle$ and $C_4 = \langle b\rangle$, the subgroup $\{(1, 1), (a, b), (a, b^2), (a, b^3)\}$ is a subdirect product; the subgroup $\{(1, 1), (a, 1), (1, b^2), (a, b^2)\}$ is not a subdirect product.

8.5 Chapter 5

4. *i)* The congruence is true for $m = 1$. Assume it is true for $m > 1$; then:

$$(1+p)^{p^m} = ((1+p)^{p^{m-1}})^p = (1+kp^m)^p = \sum_{i=0}^{p} \binom{p}{i} (kp^m)^i$$
$$= 1 + \sum_{i=1}^{p-1} \binom{p}{i} (kp^m)^i + k^p p^{mp}.$$

The terms of the sum are multiple of p^{mi+1} and therefore also of p^{m+1}. Moreover, $mp \geq m+1$, and therefore the last term is a multiple of p^{m+1}. The result follows.

5. *i)* Were G nilpotent, it would have class c, say, and every subgroup would have class at most c. But in G there are subgroups of arbitrarily high class.

7. If $x \in Z_2$, then $[x,g] = x^{-1}g^{-1}xg \in \mathbf{Z}(G)$, for all $g \in G$. But if the product of two elements belongs to the center, the two elements commute (Chapter 2, *ex.* 19 *i)*).

9. Let N be the subgroup. If $m = 1$, then $N \subseteq Z = Z_1$ (Theorem 5.5). Induct on m; since $N \cap Z \neq \{1\}$, $NZ/Z| = |N/N \cap Z| < |N|$, so $|NZ/Z| = p^k$, $k < m$, and by induction $NZ/Z \subseteq Z_k(G/Z) \subseteq Z_{m-1}(G/Z) = Z_m/Z$. Hence $NZ \subseteq Z_m$, so $N \subseteq Z_m$.

12. $A \subseteq \mathbf{C}_G(A) = C$. If $A \neq C$, then $C/A \neq \bar{1}$ is normal in G/A (it is $C \trianglelefteq G$) and therefore (Theorem 5.5) $C/A \cap \mathbf{Z}(G/A) \neq \bar{1}$. Let $x \notin A$; $K/A = \langle Ax \rangle \subseteq C/A \cap \mathbf{Z}(G/A)$. Since $K \subseteq C$, we have $A \subseteq \mathbf{Z}(K)$, and K/A being cyclic, K is abelian. But $K/A \trianglelefteq G/A$, as a subgroup of the center, implies $K \trianglelefteq G$, and K properly contains A, contrary to the maximality of A.

14. *i)* Otherwise in G/G', which is abelian, there exists at most one subgroup for each possible order, and G/G' would be cyclic (Theorem 2.4), and therefore $G/\mathbf{\Phi}(G)$ would also be cyclic. But then $G = \langle \mathbf{\Phi}(G)x \rangle = \langle x \rangle$.

ii) If $\langle x \rangle \neq G$, then $\langle x \rangle \subseteq M$, M maximal and therefore normal. Then all the conjugates of x belong to M.

15. If p divides $|M|$ and $[G:M]$, let $P \in Syl_p(M)$, $P \subseteq S \in Syl_p(G)$. But $\mathbf{N}_S(P) \supset P$, and $M \subseteq \mathbf{N}_G(P)$ (it is $P \trianglelefteq M$). It follows that both M and $\mathbf{N}_G(P)$ are contained in $\mathbf{N}_G(P)$, and by the maximality of M, $\mathbf{N}_G(P) = G$ and $P \trianglelefteq G$.

17. H exists because there is at least $H = G$. If $N \cap H \not\subseteq \mathbf{\Phi}(H)$, let $M \leq H$ be maximal and $N \cap H \not\subseteq M$. Since $N \trianglelefteq G$, we have $N \cap H \trianglelefteq H$, and therefore $(N \cap H)M$ is a subgroup of H, it properly contains M and therefore is the whole H. Then $G = NH = N(N \cap H)M = NM$, with $N < H$, against the minimality of H.

18. Let H be minimal such that $G = NH$. Let $G/N \in \mathcal{C}$. Since $G = NH$ we have $G/N = HN/N \simeq H/(H \cap N) \in \mathcal{C}$. Now $H/\mathbf{\Phi}(H) \simeq (H/(N \cap H))/(\mathbf{\Phi}(H)/(N \cap H))$, a quotient of a group of $\mathcal{C}$, and so $H \in \mathcal{C}$ ($N \cap H \subseteq \mathbf{\Phi}(H)$ by the previous *ex.*

19. $c \in C \subseteq AB \Rightarrow c = ab \Rightarrow b = a^{-1}c \in B \cap C$ and $C \subseteq A(B \cap C)$, and since both A and $B \cap C$ are contained in C, we have the other inclusion.

20. *i)* If $\mathbf{\Phi}(N) \not\subseteq \mathbf{\Phi}(G)$, there exists M maximal in G and $\mathbf{\Phi}(N) \not\subseteq M$, and being $\mathbf{\Phi}(N) \trianglelefteq G$ because it is characteristic in $N \trianglelefteq G$, we have $\mathbf{\Phi}(N)M = G$. But with $A = \mathbf{\Phi}(N), B = M$ and $C = N$ we have, from Dedekind's identity, $N = N \cap G = N \cap \mathbf{\Phi}(N)M = \mathbf{\Phi}(N)(M \cap N)$, i.e. $N = \langle \mathbf{\Phi}(N), M \cap N\rangle$, $N = M \cap N$, $N \subseteq M$, $\mathbf{\Phi}(N) \subseteq M$, contrary to the choice of M.

ii) C_4 is not normal in G, $\mathbf{\Phi}(C_4) = C_2$ and $\mathbf{\Phi}(G) = \{1\}$.

iii) N nilpotent implies $N' \subseteq \mathbf{\Phi}(N)$ and N normal in G implies $\mathbf{\Phi}(N) \subseteq \mathbf{\Phi}(G)$ (this follows from *i*)). Hence $G/\mathbf{\Phi}(G) \simeq (G/N')/(\mathbf{\Phi}(G)/N')$, a factor group of a nilpotent group, and therefore nilpotent; now apply Theorem 5.9, *ii*).

22. G is nilpotent because $SZ/Z \trianglelefteq G/Z$, $SZ \trianglelefteq G$, $S \trianglelefteq G$. If p and q are primes that divide $|G|$, P and Q the corresponding Sylow subgroups, then $PQ/Q \leq G/Q$ is abelian; but $PQ/Q \simeq P/(P \cap Q) \simeq P$, so all Sylow subgroups are abelian, and being normal G is abelian. It follows that there is only one prime dividing $|G|$, and G is a p-group.

23. If $\alpha, \beta \in A$, then $(x^{-1}x^{\alpha})^{\beta} = (x^{\beta})^{-1}x^{\alpha\beta}$. But $x^{-1}x^{\alpha} \in H$ implies $(x^{-1}x^{\alpha})^{\beta} = x^{-1}x^{\alpha}$. It follows $xx^{-\beta} = x^{\alpha}x^{-\alpha\beta}$. But $xx^{\beta} \in H$ (with $x = y^{-1}$), and therefore $(xx^{-\beta})^{\alpha} = xx^{-\beta}$, i.e. $x^{\alpha}x^{-\beta\alpha} = x^{\alpha}x^{-\alpha\beta}$, $x^{-\beta\alpha} = x^{-\alpha\beta}$, for all $x \in G$, so $\alpha\beta = \beta\alpha$.

25. For $g \in G$ we have $N_1^g \trianglelefteq H_n$ because if $h \in H_n$ then $(N_1^g)^h = (N_1^{h'})^g = N_1^g$. Moreover, N_1^g is minimal in H_n (if $N_2 \subseteq N_1^g$ and normal in H_n, then $N_2^{g^{-1}} \trianglelefteq H_n$ and $N_2^{g^{-1}} < N_1$). By induction, $N_1^g \subseteq N_G(H)$, and this holds for all $g \in G$. But $K = \langle N_1^g, g \in G\rangle \neq \{1\}$ is normal in G and contained in N, and therefore $K = N$ and $N \subseteq N_G(H)$.

27. If G is nilpotent, take $K = G$. Otherwise there exists a maximal subgroup M which is not normal. By induction, $M = \langle K^g,\ g \in M\rangle$, and therefore $\langle K^g,\ g \in G\rangle \supseteq M$, but $\langle K^g,\ g \in G\rangle$ being normal it cannot equal M; hence it properly contains M, and by the maximality of M it equals G.

28. *i)* G is residually nilpotent if and only if $\bigcap K_{\lambda} = \{1\}$ for all K_{λ} such that G/K_{λ} is nilpotent, and K_{λ} contains some Γ_i. It follows $\bigcap \Gamma_i = \bigcap K_{\lambda} =$ and the result.

ii) D_{∞} contains a sequence of normal subgroups $\langle 2^k\rangle$, $k = 1, 2, \ldots$ such that the quotient groups are dihedral of order 2^{k+1}, hence nilpotent. Since $\bigcap_k \langle 2^k\rangle = \{1\}$, D_{∞} is residually nilpotent.

29. *ii)* Let $G = SK$, $H \leq G$. Then $H \cap K$ is normal in H and of order is not divisible by p. Now $H/H \cap K \simeq HK/K \leq G/K$, whose order is a power of p. Hence $H \cap K$ is the required normal p-complement of H. As to a factor group G/N, $KN/N \simeq K/K \cap N$; therefore $|KN/N|$ is not divisible by p and its index $[G/N : KN/N]$, being equal to $|G/KN|$, is a power of p. KN/N is the required normal p-complement of G/N.

30. If $\{y_i\}$ is another transversal, let $y_i = a_i x_i$, $i = 1, 2, \ldots, n$. Then $y_i g = a_i x_i g = a_i h_i x_{\sigma(i)} = a_i h_i a_{\sigma(i)}^{-1} y_i$, so that, in the first case, $V(g) = \prod_i h_i$, and in the second $V(g) = \prod_i a_i h_i a_{\sigma(i)}^{-1} = \prod_i a_i \prod_i h_i \prod_i a_{\sigma(i)}^{-1} = \prod_i a_i \prod_i a_{\sigma(i)}^{-1} \prod_i h_i = \prod_i h_i$.

31. $G = SK$, $p||N| \Rightarrow N \not\subseteq K \Rightarrow N \cap K = \{1\}$ (minimality of N), $NK/K \simeq N/(N \cap K) \simeq N$, so that N is isomorphic to a subgroup of $G/K = SK/K \simeq S$, a Sylow p-subgroup, and therefore is a p-group. Being normal, it is contained in all the

Sylow p-subgroup, and if $N \subseteq S$ we have $N \cap \mathbf{Z}(S) \neq \{1\}$ and normal in $SK = G$. By the minimality of N, we have $N \subseteq \mathbf{Z}(S)$ and therefore $N \subseteq \mathbf{Z}(G)$.

32. $\mathbf{Z}(S)K/K$ is the center of $SK/K = G/K$.

34. *iii*) By *ex.* 32, $\mathbf{Z}(S)K \trianglelefteq G$, and if $x \in G$ then $(\mathbf{Z}(S)K)^x = \mathbf{Z}(S)^x K$. Let $\mathbf{Z}(S)^x \subseteq S$; then $\mathbf{Z}(S)^x = K\mathbf{Z}(S)^x \cap S = K\mathbf{Z}(S) \cap S = \mathbf{Z}(S)$.

iv) Each of the three involutions of the Klein group that the three 2-Sylows have in common constitutes, together with the identity, the center of one of the 2-Sylows.

35. Let A be maximal abelian in the p-Sylow S. If $A < S$, then $A < \mathbf{N}_S(A) = \mathbf{N}_G(A) \cap S = \mathbf{C}_G(A) \cap S$, so there exists $x \in \mathbf{C}_G(A) \cap S$ and $x \notin A$. Then $\langle A, x\rangle$ is abelian, it contains A and therefore $\langle A, x\rangle = S$ and S is abelian. Since $\mathbf{N}_G(S) = \mathbf{C}_G(S)$, G is p-nilpotent for every p, and therefore is nilpotent. The Sylow subgroups being abelian, G is abelian.

36. *i*) The only fixed point is S (if $S_1^x = S_1$, then $(S_1)^S = S_1$, $S \subseteq \mathbf{N}_G(S_1)$ and $S = S_1$); the other orbits have cardinality p.

ii) Let $N = \mathbf{N}_G(S), C = \mathbf{C}_G(S)$, $y \in N \setminus C$, $P^y = P, P_1^y = P_1$, with P and P_1 in the same orbit as S. There exists $z \in S$ such that $P^z = P_1$; it follows $P^{zy} = P_1^y = P_1 = P^z = (P^y)^z = P^{yz}$, $[y, z]$ fixes P and belongs to S ($[y, x] = (y^{-1}x^{-1}y)x \in S$ because $y \in N$). It follows $[y, z] = 1$ and $y \in C$, contrary to assumption; it cannot be that P be the fixed point of S because P_1 is also in its orbit.

37. $|G| = 264 = 2^3 \cdot 3 \cdot 11$, $|\mathbf{N}_G(S_{11})| = 22$. $x \in S_{11}$ acts on the 11-Sylows with two orbits, of length 1 and 11, respectively. If $1 \neq y \in N \setminus C$, $o(y) = 2$, y fixes S_{11} and a p-Sylow in the other orbit. On the latter orbit, y acts as transpositions on the remaining 10 elements, so its representation in S^{12} is an odd permutation. Hence $N = C$ (but one may proceed as in the next proof; cf. also *ex.* 46 of Chapter 3).

$|G| = 420 = 2^2 \cdot 3 \cdot 5 \cdot 7$, $n_7 = 15$, $|\mathbf{N}_G(S_7)| = 28$. An element x of order 7 acts on the 15 7-Sylows with orbits of length 1,7 and 7. If an element y of order 2 centralizes $\langle x\rangle$ then $z = xy$ has order 14, so z^2 has the same representation as x. Hence z is represented as one fixed point and a 14-cycle, so is odd. Now $N(\langle x\rangle)/C(\langle x\rangle)$ divides 6 ($= |\mathbf{Aut}(\langle x\rangle)|$), so $|N/C| = 1, 2$; hence $|C| = 28$ or $|C| = 14$, and in any case there is an element of order 2 centralizing $\langle x\rangle$. It follows that G does not embed in A^{15}.

As to $|G| = 720$, let an element of order 19 act on the Sylow 19-subgroups.

38. $|G| = 1008 = 2^4 \cdot 3^2 \cdot 7$, $n_7 = 36$, $|N_G(S)| = 28, |C_G(S)| = 14$, $x \in C_G(S)$, $o(x) = 14$. x^2 acts with one fixed point and five orbits of length 7, and therefore x acts with one orbit of length 1, one of length 7 and two of length 14. It follows $\chi(x^7) = 8$. If $y \in N \setminus C$, $\chi(y) \leq 6$, and therefore x^7 is not conjugate to y. But there is only one element of order 2 in $C_G(S)$ (which is cyclic), so $|\mathbf{cl}(x^7) \cap N_G(S)| = 1$. Then (3.9) implies

$$|\mathbf{cl}(x^7)| = \frac{|\mathbf{cl}(x^7) \cap N_G(S)| \cdot |G : N_G(S)|}{\chi(x^7)} = \frac{1 \cdot 36}{8},$$

which is not an integer.

$2016 = 2^5 3^2 7$, and if $n_7 = 8$ and G is simple then G embeds in A^8; but $|N/C| = 1, 2, 3, 6$ (N is the normalizer and C the centralizer of C_7), 2 divides $|N|$, and therefore there exists an element of order 14 that cannot belong to A^8. With $n_7 = 36$, $1 \neq x \in S \in Syl_7(G)$ is represented on the set of the 7-Sylows with one fixed point and five orbits of length 7. $y \in N \setminus C$ of order 3 has seven conjugates in N, it acts on the orbits of $\langle x \rangle$ and therefore fixes two orbits plus the fixed point, and hence a point in each orbit (it cannot fix more than one point), and we have $\chi(y) = 3$, or else all the orbits, and $\chi(y) = 6$. From

$$\mathbf{cl}(x) = \frac{|\mathbf{cl}(x) \cap N| \cdot [G : N]}{\chi(x)}$$

it then follows $\mathbf{cl}(x) = 7 \cdot 36/3 = 7 \cdot 12$, and the centralizer of x, which has index $7 \cdot 12$ would be of order 24. But the order of this centralizer must be divisible by 9 since it contains the 3-Sylow. Or else $\mathbf{cl}(x) = 7 \cdot 36/6 = 7 \cdot 6 = 42$, and the centralizer of x would be of order 48, again not divisible by 9.

39. Let G be a counterexample. $p \nmid |\mathbf{Aut}(G)|$, so $p \nmid |\mathbf{I}(G)| = |G/\mathbf{Z}(G)|$. Then the center of G contains a p-Sylow S; by Burnside (Theorem 5.30) we have $G = NS$, $N \trianglelefteq G$, $N \cap S = \{1\}$ and $G = S \times N$. $p^2 || S|$, so S admits an automorphism α of order p (*ex.* 33 of Chapter 4). If $x = st, s \in S, t \in N$, then $\alpha'(x) = \alpha(s)t$ is an automorphism of order p of $|G|$, and $p||\mathbf{Aut}(G)|$.

42. $y = x^g \in \mathbf{C}_G(H^g) \Rightarrow H^g, H \subseteq \mathbf{C}_G(y)$; if P is a p-Sylow of $\mathbf{C}_G(y)$ containing H, $z \in \mathbf{C}_G(y)$ is such that $(H^g)^z \subseteq P$, and $u \in G$ is such that $P^u \subseteq S$, then $H^{gzu} \subseteq S$. By assumption, $H^{gzu} = H = H^u$, from which $gzu, u \in \mathbf{N}_G(H)$, and so $gz \in \mathbf{N}_G(H)$; moreover, $x^{gz} = y^z = y$.

43. $1 \neq y = x^g \in S^g \Rightarrow y \in S \cap S^g$. If $S^g \neq S$, $y \in S \cap S^g = \{1\}$, impossible. Then $S^g = S$ and $g \in \mathbf{N}_G(S)$.

44. *i)* Let $G = SK$, $S^g = S_1$, $g = sx$, $s \in S, x \in K$ and $S^x = S_1$. Let $y \in S \cap S_1$; then $y^x \in S_1$, $y^{-1}y^x = [y,x] \in S_1 \cap K = \{1\}$ ($K \triangleleft G \Rightarrow [y,x] \in K$), $y^x = y$ and $x \in \mathbf{C}_G(S \cap S_1)$.

ii) If $x^{uv} = y, u \in S, v \in K$, then $x^u = y$. Indeed, $(x^u)^{-1}y \in S$; but $(x^u)^{-1}y = (x^u)^{-1}x^{uv} = ((x^u)^{-1}v^{-1}x^u)v \in K$, so $(x^u)^{-1}y = 1$.

iii) $N = \mathbf{N}_G(P)$ is p-nilpotent, with $N \cap K$ as a normal p-complement. Let $x \in P$, $y \in N \cap K$; then $x^{-1}(y^{-1}xy) \in P$ and $(x^{-1}y^{-1}x)y \in K$, from which $x^{-1}x^y = 1$, $y \in C = \mathbf{C}_G(P)$ and $N \cap K \subseteq C$. But $N/C \simeq (N/(N \cap K)/(C/(N \cap K)$, a quotient of a p-group since $N/(N \cap K) \simeq NK/K \leq G/K$, a p-group.

45. n is squarefree (if $n = p^k m$, $k > 1$, then $\varphi(n) = (p-1)p^{k-1}\varphi(m)$ and p divides both n and $\varphi(n)$), so the p-Sylows are cyclic of prime order. G is p-nilpotent, where p is the smallest prime dividing $|G|$ (Theorem 5.31, *i)*), $G = C_pK$. If $|K| = m$, since $(m, \varphi(m)) = 1$, by induction K is cyclic. C_p acts by conjugation on K; but $p \nmid \varphi(m) = |\mathbf{Aut}(K)|$, so C_p centralizes K, and G is cyclic. Conversely, if $(n, \varphi(n)) \neq 1$, then $p^2|n$ for some p, $n = p^2m$, and we have the non cyclic group $C_p \times C_p \times C_m$.

46. The subgroups realizing the $\mathcal{F}$-conjugation between A and $A^g = B$ also realize that between C and C^g.

47. Obvious for $m = 1$. By induction, A_1 is $\mathcal{F}$-conjugate to A_m by the element $h_1h_2 \cdots h_{m-1} = h$. If $y_1, y_2, \ldots, y_r$ and $U_1, U_2, \ldots, U_r$ realize the $\mathcal{F}$-conjugation between A_1 and A_m by h, and $z_1, z_2, \ldots, z_s$ and $V_1, V_2, \ldots, V_s$ the one between A_m and A_{m+1} by h_m, than $x_1, x_2, \ldots, x_n$ and $T_1, T_2, \ldots, T_n$ realize the $\mathcal{F}$-conjugation between A_1 and A_m by $hh_m = h_1h_2 \cdots h_{m-1}h_m$, where $n = r + S$ and

$$x_i = \begin{cases} y_i, & \text{se } 1 \leq i \leq r \\ z_{i-r}, & \text{se } r+1 \leq i \leq n \end{cases}, \quad T_i = \begin{cases} U_i, & \text{se } 1 \leq i \leq r \\ V_{i-r}, & \text{se } r+1 \leq i \leq n. \end{cases}$$

48. The normalizer $N_G(V)$ of the Klein group $V = \langle x, z \rangle$ in G is isomorphic to S^4, and in this group the three involutions are conjugate. Similarly, $V_1 = \langle y, z \rangle$, and in $N_G(V_1)$ the three involutions of V_1 are conjugate. Let $x^h = z$, $h \in N_G(V)$, and $z^k = y$, $k \in N_G(V_1)$; then $x^{hk} = y$.

49. Let $x^\sigma = g^{-1}xg$, and let $g = y^{-1}y^\sigma$. It follows $x^\sigma = y^{-\sigma}yxy^{-1}y^\sigma$, $(yxy^{-1})^\sigma = yxy^{-1}$, so that $yxy^{-1} = 1$ and $x = 1$.

50. We have $y^\sigma = x^\sigma x^{\sigma^2} \cdots x^{\sigma^{n-1}} x^{\sigma^n} = x^{-1}yx$, and by the previous *ex.* $y = 1$.

51. By the previous *ex.*, $xx^\sigma x^{\sigma^2} = 1 = x^{\sigma^2}x^\sigma x$, and xx^σ and $x^\sigma x$ are both equal to $(x^{\sigma^2})^{-1}$.

52. If γ_a is inner, and $x^{\sigma\gamma_a} = x$, then $a^{-1}x^\sigma a = x$, and $x = 1$ (*es.* 49). This may also be seen from the fact that the $\sigma\gamma_a$ are all conjugate to σ: given a, there exists b such that $a = b^\sigma b^{-1}$; then $\sigma\gamma_a = \gamma_b\sigma\gamma_b^{-1}$.

53. Let $N = \mathbf{N}_G(H)$; σ induces a f.p.f. automorphism of N, and let P_1 the invariant p-Sylow of N. $H \subseteq P_1$ because $H \trianglelefteq N$, and by the maximality $P_1 = H$. If S is a p-Sylow of G containing H, we have $H = N \cap S = N_S(H)$. S being a p-group, this implies $H = S$.

54. Let $S^\sigma = S$; then $S^\tau = S^{\sigma\tau} = S^{\tau\sigma} = (S^\tau)^\sigma$, so $S^\tau = S$.

55. If $x \in \mathbf{C}_G(\alpha)$, then $(S^x)^\alpha = (S^\alpha)^x = S^x$, and $S^x = S$. It follows $\mathbf{C}_G(\alpha)S \leq G$ and S is normal there and so unique, and $\mathbf{C}_G(\alpha) \cap S$ is Sylow in $\mathbf{C}_G(\alpha)$. If $x \in \mathbf{C}_G(\alpha)$, then $(\mathbf{C}_G(\alpha) \cap S)^x = \mathbf{C}_G(\alpha)^x \cap S^x = \mathbf{C}_G(\alpha) \cap S$, and $\mathbf{C}_G(\alpha) \cap S \trianglelefteq G$.

56. If $x \in D_\infty$, then $x = bcbc \cdots bc$. The mapping interchanging b and c is a f.p.f automorphism of order 2, but D_∞ is nonabelian.

58. If $N \cap H = \{1\}$, then $N \cap H^g = N^g \cap H^g = (N \cap H)^g = \{1\}$ for all $g \in G$. Hence an element of N fixes no point an therefore belongs to K.

61. *i)* $HN/N \simeq H/(H \cap N)$ and N are solvable, and therefore so is HN.

62. $H \neq G$ because $\sigma \neq 1$. If $a \notin H$, then $(ah)^\sigma = a^{-1}h = (ah)^{-1} = h^{-1}a^{-1}$, so $a^{-1}ha = h^{-1}$, and $H \trianglelefteq G$. The automorphism induced by conjugation by $a \in G$ inverts the elements of H, and the one induced by σ on G/H inverts the elements Ha, $a \notin H$: $(Ha)^\sigma = Ha^\sigma = Ha^{-1} = (Ha)^{-1}$. Hence H and G/H are abelian, and G is solvable.

63. Since $\sigma \neq 1$, there exists $g \in G$ such that $g^\sigma = gx$ with $x \neq 1$. For all $g \in G$, the mapping $h \to g^{-1}hg$ is an automorphism of F. Let $h \in F$, $h = g^{-1}h_1g$; then $h = h^\sigma = (g^{-1}h_1g)^\sigma = (g^\sigma)^{-1}h_1^\sigma g^\sigma) = (gx)^{-1}h_1(gx) = x^{-1}g^{-1}h_1gx = x^{-1}hx$, and $x \in \mathbf{C}_G(F) \subseteq F$, so that $x^\alpha = x$. It follows, $g^{\sigma^2} = (g^\sigma)^\sigma = (gx)\sigma = g^\sigma x^\sigma = g^\sigma x = gx \cdot x = gx^2$, and $g^{\sigma^k} = gx^k$ for all k. If $k = o(\sigma)$, $g = g^{\sigma^n} = gx^n$ and $x^n = 1$, so $o(x)|k$. Hence $o(x) \neq 1$ and it divides both $o(\sigma)$ and $|G|$.

64. *i*) $\Rightarrow$ *ii*) A solvable simple group is of prime order, and therefore abelian. *ii*) $\Rightarrow$ *i*) If G has odd order, by assumption it is not simple. If it is abelian, it is solvable; otherwise, it admits a proper normal subgroup N. But N and G/N have odd order and so are solvable by induction, and it follows that G is solvable.

66. H exists (the second term of a composition series), and by the Frattini argument we have $G = H\mathbf{N}_G(Q) = HQ = H$, which is absurd. It follows that q does not exist, and G is a p-group.

67. N minimal normal is transitive and is an abelian p-group, and therefore regular and of order equal to the degree.

68. The action of the group on the cosets of a maximal subgroup is primitive. Apply the previous *ex.* Another proof: with M maximal, and N minimal normal and $N \not\subseteq M$ we have $[G : M] = [MN : M] = [N : M \cap N]$, a divisor of $|N|$. If $N \subseteq M$, by induction, $[G/N : M/N]$ is a power of a prime, and so is $[G : M]$.

69. If the indices are not coprime they are powers of the same prime p. It follows that if $q \neq p$, then q divides the orders of the two subgroups.

70. Let N be minimal normal. *a*) If $N \subseteq \Phi$, then a maximal subgroup M/Φ of G/Φ has order coprime to the index. By induction, there exists a p-Sylow $S\Phi/\Phi$ normal in G/Φ and therefore $S \trianglelefteq G$. *b*) If $N \not\subseteq \Phi$, let M be maximal not containing N. Then $G = MN$, so $[G : M] = [MN : M] = [N : M \cap N]$ a power of p, so $p \nmid |M|$ and therefore $|M \cap N| = 1$, $|G| = |M||N|$, and N is Sylow.

71. In the factor group w.r.t. G' the cosets of the x_i commute, the order of their product is the product of their orders, and therefore they are all equal to the identity, i.e. G'. In other words, $x_i \in G'$ for all i. The same argument applied to G' takes the x_i to G'', and so on.

72. Let $S_1S_2 \cdots S_t$ be the product in some order, and let N be a minimal normal subgroup, $|N| = p^k$. By induction, the product of the Sylow subgroups S_iN/N is the whole G/N, hence the product of the S_iN is G. Moving the various copies of N close to the p-Sylow S_j containing it, we have $S_jN = S_j$ and G is the product of the S_i.

73. $H' = G^{i-1}/G^{i+1}$, $H'' = G^i/G^{i+1}$, cyclic, and $(H'/H'' = G^{i-1}/G^{i+1})/(G^i/G^{i+1}) \simeq G^{i-1}/G^i$, cyclic, so $H''' = \{1\}$. Now $H/\mathbf{C}_H(H'')$ is abelian (isomorphic to a subgroup of $\mathbf{Aut}(H'')$), and therefore $H' \subseteq \mathbf{C}_H(H'')$ and $H'' \subseteq \mathbf{Z}(H')$. It follows $H'/\mathbf{Z}(H')$ cyclic (isomorphic to the factor group $(H'/H'')/(\mathbf{Z}(H')/H'')$ of the cyclic group H'/H''); hence H' is abelian so $H'' = \{1\}$. But $H'' = G^i/G^{i+1}$.

74. *i*) If p is the smallest divisor of $|G|$, G is p-nilpotent (Theorem 5.31, *i*)), $G = SK$, K is solvable by induction, G/K is solvable (cyclic), and so is G.

ii) The derived series has abelian quotients with cyclic Sylow subgroups and so these quotients are cyclic. In particular, G'/G'' and G''/G''' are cyclic. By *ex.* 73, $G'' = G'''$, and solvability implies $G'' = \{1\}$, so G' is also cyclic.

75. Let H be maximal solvable, $N = \mathbf{N}_G(H)$. If $N \neq H$, let $p||N/H|$, p prime, and let $K/H \leq G/H$ of order p. K/H is solvable (it is cyclic), H is solvable, and therefore K is solvable, against the maximality of H.

76. In S^p a subgroup of order p is a C_p generated by a p-cycle, and is self centralizing. G/C_p is cyclic (automorphisms of C_p), hence G is metacyclic and therefore solvable (*ex.* 74). Moreover, since G normalizes C_p, G is a subgroup of the normalizer of C_p whose order is $p(p-1)$. Conversely, if G is solvable, let N be minimal normal; N is transitive (Theorem 3.8), so $p||N|$. But N is a p-group, so $|N| = p$.

77. If $G'' \subset H$, we have the result. If $G'' \not\subseteq H$, then $G''H = G$ and $G''H/G'' \simeq H/(H \cap G'')$, abelian, from which $G' \subseteq G''$, $G'' = G'$ and $G' = \{1\}$.

78. If a p-Sylow has order p^n, there exists a composition series in which the last n composition factors have order p: $\ldots \supseteq H_1 \supseteq H_2 \supseteq \ldots \supseteq H_{n+1} = \{1\}$. Then H_1 is Sylow, it is subnormal because it appears in a normal series, and therefore is normal, and G is nilpotent. Conversely, if G is nilpotent, G has a normal subgroup for each divisor of its order (*ex.* 26), and if $p_{i_1}, p_{i_2}, \ldots, p_{i_n}$ is a sequence of the primes dividing $|G|$ then G has a normal subgroup H of order $p_{i_2}, \ldots, p_{i_n}$, and so of index p_{i_1}; H has a normal subgroup of index p_{i_2}, etc.

79. Induct on the length of n of the derived series of the group G. If $n = 1$, G is abelian and therefore finite, and so is G/G' which is abelian and torsion. Hence G' is f.g., and by induction it is finite, and so in G.

80. We prove that A is maximal. Let $A \subset M$; then there exists $x = ab^k a' \in M$, $a, a' \in A$, $b^k \neq 1$, $\langle x \rangle = B$, and since $A \subset M$, we have $b^k \in M$, and so also $b \in M$, from which $ABA \subseteq M$ and $M = G$.

81. *i*) Obviously $A \subseteq \mathbf{C}_G(A)$. If $A \subset \mathbf{C}_G(A)$, let $x \in \mathbf{C}_G(A) \setminus A$; then $\langle A, x \rangle \supset A$ and is abelian. Hence, if A maximal abelian, then $A = \mathbf{C}_G(A)$. Conversely, if $A = \mathbf{C}_G(A)$ and $A \subset M$, with M abelian, then $M \subseteq \mathbf{C}_G(A)$, so $\mathbf{C}_G(A) = A \subset M \subseteq \mathbf{C}_G(A)$, which is absurd;

ii) the Fitting subgroup $F = \mathbf{F}(G)$ contains all the normal abelian subgroups;

iii) if F is abelian and G is solvable, then $F = \mathbf{C}_G(F)$; apply *i*).

82. $[G : \mathbf{N}_G(Q)] = [G : M][M : \mathbf{N}_G(Q)]$, $[G : \mathbf{N}_G(Q)] \equiv 1 \bmod q$; similarly, $[M : \mathbf{N}_G(Q)] = [M : \mathbf{N}_M(Q)] \equiv 1 \bmod q$, and so also $[G : M] \equiv 1 \bmod q$. But $[G : M] = p$, a prime, and $p \equiv 1 \bmod q$ implies $q|(p-1)$, which is absurd because $p \leq q$. As to solvability, let $[M : Q]$ be maximal in G/Q; then $[G : M] = p$, every maximal subgroup of G/Q has prime index, and by induction G/Q is solvable, and Q being solvable so is G.

83. *i*) Let $L = H \cap K \neq \{1\}$; then, for $g \in G$, we have $g = hk$ and $L^g = L^{hk} = L^k \subseteq K$. Hence $\{1\} \neq \langle L^g \rangle \subseteq K$ and $\langle L^g \rangle$ is a proper normal subgroup of G.

ii) Assume by contradiction that G is simple. Let Q be a Sylow q-subgroup, $N = \mathbf{N}_G(Q)$. If $Q < N$, then $p||N|$, and for a certain Sylow p-subgroup P, $\{1\} \neq P \cap N$ is Sylow in N. It follows $G = PQ = PN$, but by *i*) this is not possible. Then $N = Q$, and so $\mathbf{C}_G(Q) = N$; by Burnside, G has a normal q-complement, $G = QK$, $K = P \lhd G$ contrary to G simple.

84. Every group is a quotient of a free group, and quotients of solvable groups are solvable.

8.6 Chapter 6

1. If $A = (a_{i,j})$ and $A^* = (b_{i,j})$, with bases e_i and the dual ϵ_i, then $[\rho_s^*(\epsilon_i)](\rho_s(e_j)) = \epsilon_i(e_j) = \delta_{i,j}$, and since $\rho_s^*(\epsilon_i) = \sum_h b_{h,i}e_h$ e $\rho_s(e_j) = \sum_k a_{k,j}e_j$, we have $[\rho_s^*(\epsilon_i)]$ $(\rho_s(e_j)) = \sum_h b_{h,i}a_{h,j} = \delta_{i,j}$. The result follows.

2. W^0 is invariant: if $f \in W$ and $v \in W^0$, $f(\rho_s(v)) = (\rho_s^{*-1}f)(v) = 0$, because $\rho_s^{*-1}f \in W$. It follows that either $W^0 = \{0\}$ or $W^0 = V$, and from the dimension formula we have the result.

3. The kernel is the set of integral multiples of π.

6. Let N be minimal normal in the p−group P. Then N is the unique minimal normal subgroup ($N \subseteq \mathbf{Z}(P)$ and has order p, and is unique because $\mathbf{Z}(P)$ is cyclic). If $1 \neq x \in N$, and ρ is faithful, then $x \notin ker(\rho)$ and therefore, for some i, $x \notin ker(\rho_i)$; hence $N \notin ker(\rho_i)$. If ρ_i is not faithful, its kernel must contain N, excluded.

7. From $\sum_{i=1}^k n_i^2 = p^2$, with $n_1 = 1$ and $n_i | p^2$, it follows $n_i = 1$ all i, and the group is abelian.

8. *i)* $\varphi = \sum_{i=1}^r \alpha_i\chi_i$, where the χ_i are irreducible characters (Theorem ??). If the α_i are integer, then φ is the character of the representation obtained by summing α_i times χ_i, $i = 1, 2, \ldots, r$. If φ is a character, $(\varphi, \chi_i) = \alpha_i$ is the multiplicity of χ_i in φ, and so is an integer.
ii) $[\varphi^G, \chi] = [\varphi, \chi_H]$; apply *i)*.

9. If χ is the character of the representation as a permutation group of the given group G, then it splits in a representation of degree 1, on which χ is worth 1, plus one of degree $|G| - 1$. If ν is the character of the latter then $\chi = 1 + \nu$.

10. We have:

$$(\chi\lambda, \chi\lambda) = \frac{1}{|G|}\sum_{s\in G}\chi(s)\lambda(s)\overline{\chi(s)\lambda(s)} = \frac{1}{|G|}\sum_{s\in G}\chi(s)\overline{\chi(s)} = (\chi, \chi) = 1,$$

recalling that $\lambda(s)$ is a root of unity. Apply Theorem 6.6, *ii)*.

11. S^4 has five conjugacy classes, i.e. the classes 1, (1,2), (1,2)(3,4), (1,2,3) and (1,2,3,4). Besides the two representations of degree 1 (unit and alternate), we have the representation of degree 3 on the hyperplane W (*Ex.* ??, 2), so that the sum of the squares of the degrees gives $1^2 + 1^2 + 3^2 + a^2 + b^2 = 24$, $a^2 + b^2 = 13$, and $a = 2$ and $b = 3$:

	1	(1,2)	(1,2)(3,4)	(1,2,3)	(1,2,3,4)
χ_1	1	1	1	1	1
χ_2	1	−1	1	1	−1
χ_3	2				
χ_4	3	1	−1	0	−1
χ_5	3				

The product of χ_4 and χ_2 (previous *ex.*) yields the values of χ_5: 3, −1, −1, 0, 1, and column orthogonality allows us to complete the table:

	1	(1,2)	(1,2)(3,4)	(1,2,3)	(1,2,3,4)
χ_1	1	1	1	1	1
χ_2	1	−1	1	1	−1
χ_3	2	0	2	−1	0
χ_4	3	1	−1	0	−1
χ_5	3	−1	−1	0	1

12. Let $\chi_i \neq 1_G$ be an irreducible character of G, $i = 2, \ldots, r$. Then $(1_H^G - 1_G, \chi_i) = (1_H^G, \chi_i) - (1_G, \chi_i) = m_i - 0$, and $m_i \neq 0$ for some i because, being $1_H^G \neq 1_G$, some χ_i, $i \neq 1$, must appear in 1_H^G. Moreover, $(1_H^G - 1_G, 1_G) = (1_H^G, 1_G) - (1_G, 1_G) = 1 - 1 = 0$. This proves that $1_H^G - 1_G$ is a character (*ex.* 8).

14. *i*) We know that $g \sim g^{-1}$, but $g \not\sim g^2$ and therefore we have two representatives of the conjugacy classes of H (that in this case contain only one element) contained in a class of G. It follows that, on the class of g, the value of μ is $|\mathbf{C}_G(g)|(\frac{\epsilon}{|\mathbf{C}_H(g)|} + \frac{\epsilon^4}{|\mathbf{C}_H(g)|}) = 5 \cdot (\frac{\epsilon}{5} + \frac{\epsilon^4}{5}) = \epsilon + \epsilon^4$. Similarly, on the class of ϵ^2 the value of μ^G is $\epsilon^2 + \epsilon^3$. Since there are no elements of order 2 or 3 in H, the value of μ^G on these elements is zero. Since $(\mu^G, \mu^G) = \frac{1}{60}(1 \cdot 12 + 12\overline{(\epsilon + \epsilon^4)}(\epsilon + \epsilon^4) + 12\overline{(\epsilon^2 + \epsilon^3)}(\epsilon^2 + \epsilon^3)) = \frac{1}{60}(144 + 12(\epsilon + \epsilon^2 + \epsilon^3 + \epsilon^4 + 4)) = \frac{1}{60}(144 + 12 \cdot 3) = 3$, μ^G is not irreducible.

ii) $(\mu^G, \chi_3) = \frac{1}{60}(1 \cdot 5 \cdot 12) = 1$, so $(\mu^G - \chi_3, \chi_3) = (\mu^G, \chi_3) - (\chi_3, \chi_3) = 1 - 1 = 0$; if $\chi \neq \chi_3$, $(\mu^G - \chi, \chi) = (\mu^G, \chi) - (\chi, \chi) = m - 1$, where m is the number of times χ appears in μ^G. It follows (*ex.* 8) that $\mu^G - \chi_3$ is a character; its values on the five classes are 7, –1, 1, $\epsilon + \epsilon^4$, $\epsilon^2 + \epsilon^3$, in this order. Since $(\mu^G - \chi_3, \mu^G - \chi_3) = \frac{1}{60}(7 \cdot 7 + 15 + 20 + 36) = 2$, $\mu^G - \chi_3$ is not irreducible.

iii) $(\mu^G, \chi_2) = \frac{1}{60}(48 - 12(\epsilon + \epsilon^4) - 12(\epsilon^2 + \epsilon^3) = \frac{1}{60}(48 - 12 \cdot -1) = 1$, from which $(\mu^G - \chi_3 - \chi_2, \chi_2) = (\mu^G, \chi_2) - (\chi_3, \chi_2) - (\chi_2, \chi_2) = 1 + 0 - 1 = 0$, and similarly for the product with χ_3. If χ is irreducible and different from χ_2 and from χ_3, the value of the product with χ is (μ^G, χ), which is an integer because χ is a character. The values of $\mu^G - \chi_3 - \chi_2$ on the five classes are 3, –1, 0, $\epsilon + \epsilon^4 + 1$, $\epsilon^2 + \epsilon^3 + 1$, in this order, and the product of this character by itself is $\frac{1}{60}(9 + 15 + 36) = 1$, so that it is irreducible; let it be χ_4.

iv) Interchanging ϵ and ϵ^2 in the values of χ_4 we obtain the fifth row of the table:

	1	(1,2)(3,4)	(1,2,3)	(1,2,3,4,5)	(1,3,5,2,4)
χ_1	1	1	1	1	1
χ_2	4	0	1	−1	−1
χ_3	5	1	−1	0	0
χ_4	3	−1	0	α	β
χ_5	3	−1	0	β	α

where $\alpha = \epsilon + \epsilon^4 + 1 = \frac{1+\sqrt{5}}{2}$, $\beta = \epsilon^2 + \epsilon^3 + 1 = \frac{1-\sqrt{5}}{2}$.

14. The matrix of ρ_t is a permutation matrix whose trace $\chi(t)$ is the number of points fixed by t under conjugation; hence $\chi(t) = |\mathbf{C}_G(t)|$. The number m_i of times in which the representation ρ_i, with character χ_i, appears in ρ is:

$$m_i = \frac{1}{|G|}\sum_t \chi(t)\chi_i(t) = \frac{1}{|G|}\sum_j k_j\chi(s_j)\chi_i(s_j) = \frac{1}{|G|}\sum_j k_j\frac{|G|}{k_j}\chi_i(s_j),$$

since $\chi(s_j) = |\mathbf{C}_G(s_j)| = \frac{|G|}{k_j}$, and the last sum equals $\sum_j \chi_i(s_j)$, which, being equal to m_i, is an integer.

8.7 Chapter 7

2. *i)* $x^{-1}(y^\sigma y^{-1})x = (x^{-\sigma}x)^{-1} \cdot [(x^{-1}y)^\sigma (x^{-1}y)^{-1}]$, a product of two elements of $[G, \pi]$. *ii)* $(x^\sigma x^{-1})^\tau = (x^{\sigma\tau}x^{-1})(x^\tau x^{-1})^{-1}$. *iii)* If π is trivial on G/H, then $(Hg)^\sigma = Hg^\sigma = Hg$, for all $g \in G$, so $g^\sigma g^{-1} \in H$ and $[G, \pi] \subseteq H$.

4. Under the first action there is only one nontrivial crossed homomorphism, i.e. $f(\sigma) = c$: $1 = f(1) = f(\sigma\sigma) = f(\sigma)^\sigma f(\sigma) = c \cdot c = 1$, and this is principal because $f(\sigma) = a^\sigma a = ba = c$. If the action is trivial, $f(\sigma) = a$ is a crossed homomorphism: $1 = f(\sigma\sigma) = f(\sigma)^\sigma f(\sigma) = a \cdot a = 1$. Similarly for b and c.

5. *i)* Set $H = [G, \pi]$; if $g \in G$, then $g \in Hg$ and π fixes Hg; by Corollary 7.2 there exists $hg \in Hg$ with $hg \in G^\pi$. It follows $g \in HG^\pi$, $G = HG^\pi = G^\pi H$. *ii)* The inclusion $[G, \pi, \pi] \subseteq [G, \pi]$ follows from the fact that $[G, \pi]$ is π-invariant. As to the other inclusion, observe that for $g \in G$, it follows from *i)* that $g = hx$ with $h \in [G, \pi]$ and $x \in G^\pi$. Then $g^\sigma g^{-1} = (hx)^\sigma (hx)^{-1} = h^\sigma h^{-1} \in [G, \pi, \pi]$.

7. If $c_1 = \beta(b_1)$, $c_2 = \beta(b_2)$ and $c_1 \xrightarrow{\delta} [f_1]$, $c_2 \xrightarrow{\delta} [f_2]$, since $c_1 + c_2 = \beta(b_1 + b_2)$ we have the 1-cocycle of π in A

$$f(\sigma) = (b_1 + b_2)^\sigma - (b_1 + b_2) = (b_1^\sigma - b_1) + (b_2^\sigma - b_2) = f_1(\sigma) + f_2(\sigma)$$

where f_1 and f_2 are determined by b_1 and b_2. Then the class determined by $c_1 + c_2$ is $[f_1 + f_2]$, which is the sum $[f_1] + [f_2]$; it follows that δ is a homomorphism.

Let us now show that the sequence is exact at C^π, the exactness at the other points being already known. If $c \in Im(\beta^\pi)$, then $c = \beta(b)$ with $b \in B^\pi$, so that $f(\sigma) = b^\sigma - b = b - b = 0$, and therefore c determines the class $[0]$. Hence,

$$Im(\beta^\pi) \subseteq ker(\delta).$$

Conversely, let $c \in ker(\delta)$; then $f : \sigma \to b^\sigma - b$ is a 1-cobord of π in A, and therefore for some $a \in A$, we have $b^\sigma - b = a^\sigma - a$, from which $(b-a)^\sigma = b - a$ or $b - a \in B^\pi$. But $\beta^\pi(b - a) = \beta(b - a) = \beta(b) - \beta(a) = \beta(b) = c$, since $a \in A = ker(\beta)$. Hence $c \in ker(\delta)$ comes under β from the element $b - a$, and therefore:

$$ker(\delta) \subseteq Im(\beta^\pi).$$

It is clear that $H^1(\pi, A)$ is not the only group that makes the sequence exact: if G is any group, then the product $H^1(\pi, A) \times G$ also makes the sequence exact. Moreover, from the previous proof we see that if β^π is surjective, the image of δ is $\{0\}$.

8. The sequence is exact at $H^1(\pi, A)$. Indeed, if $[f]$ comes from $c \in C^\pi$, then $f(\sigma) = b^\sigma - b$, and therefore it is a 1-cobord of π in B. It follows $Im(\delta) \subseteq ker(\overline{\alpha})$. If αf is a cobord, its image under β is such that $\beta\alpha(f(\sigma)) = b^\sigma - b = 0$, since $\beta\alpha = 0$. Hence $\beta(b)^\sigma = \beta(b)$ and $\beta(b) \in C^\pi$. Setting $c = \beta(b)$, c determines an f such that $\alpha(f(\sigma)) = b^\sigma - b$. Now $\delta(c) = [f]$, by definition of δ, and therefore $[f]$ comes from $c \in C^\pi$. Hence $ker(\overline{\alpha}) \subseteq Im(\delta)$.

The sequence is exact at $H^1(\pi, B)$. From $\overline{\alpha}([f]) = [\alpha f] \xrightarrow{\overline{\beta}} [\beta\alpha f] = [0]$ it follows $Im(\overline{\alpha}) \subseteq ker(\overline{\beta})$. An element of the kernel of $\overline{\beta}$ is of type $[\beta f]$, where βf is a 1-cobord: $\beta f : \sigma \to c^\sigma - c$, $c \in C$. By the surjectivity of β, $c = \beta(b)$ and therefore $\beta f(\sigma) = \beta(b)^\sigma - \beta(b)$ and $\beta(f(\sigma) - b^\sigma + b) = 0$, i.e. $f(\sigma) - b^\sigma + b \in ker(\beta) = A$. Then there exists $a_\sigma \in A$ such that $\alpha(a_\sigma) = f(\sigma) - b^\sigma + b$, the correspondence $f' : \pi \to B$ that associates with σ the element $f(\sigma) - (b^\sigma - b)$ is a 1-cocycle equivalent to f. Now $g : \sigma \to a_\sigma$ is a 1-cocycle $\pi \to A$, and $\alpha g : \sigma \to \alpha(g(\sigma)) = \alpha(a_\sigma) = f'(\sigma)$. Hence $\alpha g = f'$ e $\overline{\alpha}[g] = [\alpha g] = [f'] = [f]$, so that $ker(\overline{\beta}) \subseteq Im(\overline{\alpha})$. The exactness at the other points has already been observed.

14. A simple calculation shows that $a_{\sigma,\tau\eta} + a_{\tau,\eta} = a_{\sigma\tau,\eta} + a^\eta_{\sigma,\tau}$. As for the second part, we have $f'(\sigma) = x^\sigma - x + f(\sigma)$, $\forall \sigma \in \pi$, for some $x \in C$. Let $f'(\sigma) = \beta(b')$ and $x = \beta(b)$; then:

$$\beta(b') = x^\sigma - x + f(\sigma) = \beta(b^\sigma) - \beta(b) + \beta(b_\sigma) = \beta(b^\sigma - b + b_\sigma),$$

and so $b'_\sigma - b^\sigma + b - b_\sigma \in ker(\beta) = A$. Then there exists $a_\sigma \in A$ such that $b'_\sigma = a_\sigma + b^\sigma - b + b_\sigma$, from which

$$b'_{\sigma\tau} = b^{\sigma\tau} - b + b_{\sigma\tau} + a_{\sigma\tau} = b'^{\tau}_{\sigma} + b'^{\tau} = b^{\sigma\tau} - b^\tau + b^\tau_\sigma + a^\tau_\sigma + b^\tau - b + b_\tau - a_\tau,$$

and therefore $a_{\sigma\tau} = a^\tau_\sigma + a_\tau$. Setting $a'_{\sigma,\tau} = b'^{\sigma\tau} - b'^{\tau}_{\sigma} - b'_{\tau}$, we have

$$\begin{aligned} a'_{\sigma,\tau} &= b'_{\sigma\tau} - b'^{\tau}_{\sigma} - b'^{\tau} \\ &= a_{\sigma,\tau} + b^{\sigma\tau} - b + b_{\sigma\tau} - a^\tau_\sigma - b^{\sigma\tau} + b^\tau - b^\tau_\sigma - a_\tau - b^\tau + b - b_\tau \\ &= a_{\sigma,\tau} - a^\tau_\sigma - a_\tau + b_{\sigma\tau} - b^\tau_\sigma - b_\tau \\ &= b_{\sigma\tau} - b^\tau_\sigma - b_\tau \\ &= a_{\sigma\tau}, \end{aligned}$$

as required.

15. If $[g] = [0]$, g is of the form (7.15); if $f(\sigma) = c_\sigma = \beta(b_\sigma)$, setting $f_1(\sigma) = b_\sigma$ yields $\beta(f_1(\sigma)) = f(\sigma)$ and $f_1(\sigma\tau) - f_1(\sigma)^\tau - f_1(\tau) = g(\sigma, \tau) = h(\sigma\tau) - h(\sigma)^\tau - h(\tau)$. Thus $f_1 - h$ is a 1-cocycle of π in B. We have:

$$\overline{\beta}[f_1 - h] = [\beta(f_1 - h)] = [\beta f_1 - \beta h] = [\beta f_1] = [f],$$

and therefore, if $[f] \in ker(\delta)$, then $[f] \in Im(\overline{\beta})$, and we have the first inclusion $ker(\delta) \subseteq Im(\overline{\beta})$. Conversely, if $[f'] \in H^1(\pi, B)$, for the g determined by $\beta f'$ we have $f'(\sigma\tau) - f'(\sigma)^\tau - f'(\tau) = g(\sigma, \tau)$. But the left hand side is zero (f is a 1-cocycle), and so also the right hand side, and therefore $g(\sigma, \tau) = 0$. Thus $\delta[\beta f'] = 0$, and we have the other inclusion $Im(\overline{\beta}) \subseteq ker(\delta)$.

16. If $[g] \in Im(\delta)$, then $[g] = \delta[f]$, $\alpha g(\sigma, \tau) = f(\sigma\tau) - f(\sigma)^\tau - f(\tau)$, and therefore $[\alpha g] = [0]$. By definition, $[\alpha g] = \overline{\alpha}[g]$, and thus $[g] \in ker(\overline{\alpha})$.

Conversely, let $\overline{\alpha}[g] = [0]$, i.e. $[\alpha g] = [0]$. Then gh is of the form (7.15), and βh is a cocycle of π in C since

$$\beta h(\sigma\tau) - \beta h(\sigma)^\tau - \beta h(\tau) = \beta\alpha g(\sigma\tau) = 0,$$

because $\beta\alpha = 0$. Moreover, the image of $[\beta h] \in H^1(\pi, C)$ is precisely $[g]$, so that $[g] \in Im(\delta)$. Hence the sequence is exact at $H^2(\pi, A)$.

17. If $g \in Z^2(\pi, A)$, we have $\overline{\beta}[\alpha g] = [\beta\alpha g] = [0]$, because $\beta\alpha = 0$, and therefore $Im(\overline{\alpha}) \subseteq ker(\overline{\beta})$.

As to the other inclusion, let $g \in Z^2(\pi, B)$ be such that $\overline{\beta}[g] = [0]$; then $[\beta g] = [0]$ and $\beta g(\sigma, \tau) = h(\sigma, \tau) - h(\sigma)^\tau - h(\tau)$, where $h : \pi \to C$. β is surjective; let us choose for each σ an element of B whose image under β is $h(\sigma)$ (i.e. let us choose a representative for the class $h(\sigma)$). In this way we define a function f of π in B such that $\beta f(\sigma) = h(\sigma)$. We have:

$$\beta g(\sigma, \tau) = \beta f(\sigma\tau) - \beta f(\sigma)^\tau - \beta f(\tau).$$

By definition, the function g' defined as $g'(\sigma, \tau) = g(\sigma, \tau) - f(\sigma\tau) + f(\sigma)^\tau + f(\tau)$ is a 2-cocycle equivalent to g. If f_1 yields another choice of representatives, a g'' equivalent to g is obtained; hence the class $[g]$ is the same. Now, $\beta g'(\sigma, \tau) = 0$, and so $g'(\sigma, \tau) \in ker(\beta) = Im(\alpha)$, $g'(\sigma, \tau) = \alpha(a_{\sigma,\tau})$. The function $g_1 : \pi \times \pi \to A$ given by $(\sigma, \tau) \to a_{\sigma,\tau}$ is well defined (α is injective), and it is a 2-cocycle because $g' = \alpha g$, and the latter is a 2-cocycle and α is injective. Hence, $[\alpha g_1] = [g'] = [g]$, and by definition $[\alpha g_1] = \overline{\alpha}[g_1]$. It follows $[g] = \overline{\alpha}[g_1] \in Im(\overline{\alpha})$, so the sequence is exact at $H^2(\pi, B)$.

19. For $\sigma \in H$, $\iota\phi\lambda(\sigma) = \iota\phi(H) = 0$ (ϕ is a crossed homomorphism and $\phi(1) = 0$), so $\sigma \in H$ implies $\iota\phi\lambda(\sigma) = a^\sigma - a = 0$, i.e. $a^\sigma = a$. Hence $a \in H$. Now $a^\sigma = \iota(a^{\sigma H})$, and $a = \iota(a^H)$; it follows $\iota\phi\lambda(\sigma) = \iota(a^{\sigma H} - a^H)$, and ι being injective, $\phi\lambda(\sigma) = a^{\sigma H} - a^H$, i.e. $\phi(\sigma H) = a^{\sigma H} - a$, with $a \in$, and ϕ is principal.

21. From the exact sequence $0 \to \mathbf{Z} \to \mathbf{R} \to \mathbf{R}/\mathbf{Z} \to 0$ we have the exact sequence:

$$H^1(\pi, \mathbf{R}) \to H^1(\pi, \mathbf{R}/\mathbf{Z}) \to H^2(\pi, \mathbf{Z}) \to H^2(\pi, \mathbf{R}).$$

But the first and the last terms are zero because $\mathbf{R}$ is divisible and torsion free; it follows $H^2(\pi, \mathbf{Z}) \simeq H^1(\pi, \mathbf{R}/\mathbf{Z})$. The action of π being trivial, the crossed homomorphisms are homomorphisms, and we have the result. Moreover, $\mathbf{R}/\mathbf{Z}$ is isomorphic to the multiplicative group of the complex numbers of modulus 1. It follows that if π is cyclic of order n, then $H^2(\pi, \mathbf{Z})$ is also cyclic of order n.

22. If $g(\sigma, \tau) = h(\sigma\tau)^{-1}h(\sigma)h(\tau)$ is a coboundary, let $k(\sigma)^n = h(\sigma)$. Then $g_1 = k(\sigma\tau)^{-1}k(\sigma)k(\tau)$ is also a coboundary, and $g_1^n = g$.

References

General

Carmichael R.D.: Introduction to the theory of groups of finite order. Dover, New York (1956).

Gorenstein D.: Finite groups. Harper & Row, New York (1968).

Hall M.: The theory of groups. McMillan, New York (1959).

Huppert B.: Endliche Gruppen I. Springer-Verlag, Berlin Heidelberg New York (1967).

Kaplansky I.: Infinite abelian groups. The University of Michigan Press, Ann Arbor (1954).

Kargapolov M., Merzliakov I.: Éléments de la théorie des groupes. Éditions Mir, Moscou (1985).

Macdonald I.D.: The theory of groups. Clarendon Press, Oxford (1968).

Rotman J.J.: An introduction to the theory of groups. Fourth Edition. Springer-Verlag, New York (1995).

Zassenhaus H.: The theory of groups. Chelsea, New York (1958).

Geometric and combinatorial theory

de la Harpe P.: Topics in geometric group theory. The University of Chicago Press (2000).

Johnson D.L.: Presentations of groups. LMS Lecture Notes Series, New York and London (1976).

Lyndon R.C., Schupp P.E.: Combinatorial Group Theory. Springer-Verlag, Berlin Heidelberg New York (1977).

Magnus W., Karrass A., Solitar D.: Combinatorial Group Theory. Dover, New York (1976).

Permutation groups

Cameron P.: Permutation groups. LMS Student texts 45, Cambridge University Press (1999).

Wielandt H.: Finite permutation groups. Academic Press, New York (1964).

Representations and characters

Curtis C.W., Reiner I.: Representation theory of finite groups and associative algebras. Interscience Publ., London (1962).

Isaacs I.M.: Character theory of finite groups. Dover, New York (1976).

Kirillov A.A.: Elements of the theory of representations. Springer-Verlag, New York (1976)

Ledermann W.: Introduction to group characters. Second Edition. Cambridge University Press (1987).

Najmark M.A., Stern A.I.: Theory of group representations. Springer-Verlag, Berlin Heidelberg New York (1982).

Cohomology of groups

Brown K.S.: Cohomology of groups. Springer-Verlag, Berlin Heidelberg New York (1982).

Karpilovsky G.: The Schur multiplier. Claredon Press, Oxford (1987).

Serre J.-P.: Corps locaux. Hermann, Paris (1968).

Index

Collana Unitext La Matematica per il 3+2

As of 2004, the books published in the series have been given a volume number. Titles in grey indicate editions out of print.
As of 2011, the series also publishes books in English.

A. Bernasconi, B. Codenotti
Introduzione alla complessit computazionale
1998, X+260 pp, ISBN 88-470-0020-3

A. Bernasconi, B. Codenotti, G. Resta
Metodi matematici in complessit computazionale
1999, X+364 pp, ISBN 88-470-0060-2

E. Salinelli, F. Tomarelli
Modelli dinamici discreti
2002, XII+354 pp, ISBN 88-470-0187-0

S. Bosch
Algebra
2003, VIII+380 pp, ISBN 88-470-0221-4

S. Graffi, M. Degli Esposti
Fisica matematica discreta
2003, X+248 pp, ISBN 88-470-0212-5

S. Margarita, E. Salinelli
MultiMath - Matematica Multimediale per l Universit
2004, XX+270 pp, ISBN 88-470-0228-1

A. Quarteroni, R. Sacco, F.Saleri
Matematica numerica (2a Ed.)
2000, XIV+448 pp, ISBN 88-470-0077-7
2002, 2004 ristampa riveduta e corretta
(1a edizione 1998, ISBN 88-470-0010-6)

13. A. Quarteroni, F. Saleri
    Introduzione al Calcolo Scientifico (2a Ed.)
    2004, X+262 pp, ISBN 88-470-0256-7
    (1a edizione 2002, ISBN 88-470-0149-8)

14. S. Salsa
    Equazioni a derivate parziali - Metodi, modelli e applicazioni
    2004, XII+426 pp, ISBN 88-470-0259-1

15. G. Riccardi
    Calcolo differenziale ed integrale
    2004, XII+314 pp, ISBN 88-470-0285-0

16. M. Impedovo
    Matematica generale con il calcolatore
    2005, X+526 pp, ISBN 88-470-0258-3

17. L. Formaggia, F. Saleri, A. Veneziani
    Applicazioni ed esercizi di modellistica numerica
    per problemi differenziali
    2005, VIII+396 pp, ISBN 88-470-0257-5

18. S. Salsa, G. Verzini
    Equazioni a derivate parziali   Complementi ed esercizi
    2005, VIII+406 pp, ISBN 88-470-0260-5
    2007, ristampa con modifiche

19. C. Canuto, A. Tabacco
    Analisi Matematica I (2a Ed.)
    2005, XII+448 pp, ISBN 88-470-0337-7
    (1a edizione, 2003, XII+376 pp, ISBN 88-470-0220-6)

20. F. Biagini, M. Campanino
    Elementi di Probabilit   e Statistica
    2006, XII+236 pp, ISBN 88-470-0330-X

21. S. Leonesi, C. Toffalori
    Numeri e Crittografia
    2006, VIII+178 pp, ISBN 88-470-0331-8

22. A. Quarteroni, F. Saleri
    Introduzione al Calcolo Scientifico (3a Ed.)
    2006, X+306 pp, ISBN 88-470-0480-2

23. S. Leonesi, C. Toffalori
    Un invito all Algebra
    2006, XVII+432 pp, ISBN 88-470-0313-X

24. W.M. Baldoni, C. Ciliberto, G.M. Piacentini Cattaneo
    Aritmetica, Crittografia e Codici
    2006, XVI+518 pp, ISBN 88-470-0455-1

25. A. Quarteroni
    Modellistica numerica per problemi differenziali (3a Ed.)
    2006, XIV+452 pp, ISBN 88-470-0493-4
    (1a edizione 2000, ISBN 88-470-0108-0)
    (2a edizione 2003, ISBN 88-470-0203-6)

26. M. Abate, F. Tovena
    Curve e superfici
    2006, XIV+394 pp, ISBN 88-470-0535-3

27. L. Giuzzi
    Codici correttori
    2006, XVI+402 pp, ISBN 88-470-0539-6

28. L. Robbiano
    Algebra lineare
    2007, XVI+210 pp, ISBN 88-470-0446-2

29. E. Rosazza Gianin, C. Sgarra
    Esercizi di finanza matematica
    2007, X+184 pp,ISBN 978-88-470-0610-2

30. A. Mach
Gruppi   Una introduzione a idee e metodi della Teoria dei Gruppi
2007, XII+350 pp, ISBN 978-88-470-0622-5
2010, ristampa con modifiche

31 Y. Biollay, A. Chaabouni, J. Stubbe
Matematica si parte
A cura di A. Quarteroni
2007, XII+196 pp, ISBN 978-88-470-0675-1

32. M. Manetti
Topologia
2008, XII+298 pp, ISBN 978-88-470-0756-7

33. A. Pascucci
Calcolo stocastico per la finanza
2008, XVI+518 pp, ISBN 978-88-470-0600-3

34. A. Quarteroni, R. Sacco, F. Saleri
Matematica numerica (3a Ed.)
2008, XVI+510 pp, ISBN 978-88-470-0782-6

35. P. Cannarsa, T. D Aprile
Introduzione alla teoria della misura e all analisi funzionale
2008, XII+268 pp, ISBN 978-88-470-0701-7

36. A. Quarteroni, F. Saleri
Calcolo scientifico (4a Ed.)
2008, XIV+358 pp, ISBN 978-88-470-0837-3

37. C. Canuto, A. Tabacco
Analisi Matematica I (3a Ed.)
2008, XIV+452 pp, ISBN 978-88-470-0871-3

38. S. Gabelli
Teoria delle Equazioni e Teoria di Galois
2008, XVI+410 pp, ISBN 978-88-470-0618-8

39. A. Quarteroni
Modellistica numerica per problemi differenziali (4a Ed.)
2008, XVI+560 pp, ISBN 978-88-470-0841-0

40. C. Canuto, A. Tabacco
Analisi Matematica II
2008, XVI+536 pp, ISBN 978-88-470-0873-1
2010, ristampa con modifiche

41. E. Salinelli, F. Tomarelli
Modelli Dinamici Discreti (2a Ed.)
2009, XIV+382 pp, ISBN 978-88-470-1075-8

42. S. Salsa, F.M.G. Vegni, A. Zaretti, P. Zunino
Invito alle equazioni a derivate parziali
2009, XIV+440 pp, ISBN 978-88-470-1179-3

43. S. Dulli, S. Furini, E. Peron
Data mining
2009, XIV+178 pp, ISBN 978-88-470-1162-5

44. A. Pascucci, W.J. Runggaldier
Finanza Matematica
2009, X+264 pp, ISBN 978-88-470-1441-1

45. S. Salsa
Equazioni a derivate parziali Metodi, modelli e applicazioni (2a Ed.)
2010, XVI+614 pp, ISBN 978-88-470-1645-3

46. C. D Angelo, A. Quarteroni
Matematica Numerica Esercizi, Laboratori e Progetti
2010, VIII+374 pp, ISBN 978-88-470-1639-2

47. V. Moretti
Teoria Spettrale e Meccanica Quantistica Operatori in spazi di Hilbert
2010, XVI+704 pp, ISBN 978-88-470-1610-1

48. C. Parenti, A. Parmeggiani
Algebra lineare ed equazioni differenziali ordinarie
2010, VIII+208 pp, ISBN 978-88-470-1787-0

49. B. Korte, J. Vygen
Ottimizzazione Combinatoria. Teoria e Algoritmi
2010, XVI+662 pp, ISBN 978-88-470-1522-7

50. D. Mundici
Logica: Metodo Breve
2011, XII+126 pp, ISBN 978-88-470-1883-9

51. E. Fortuna, R. Frigerio, R. Pardini
Geometria proiettiva. Problemi risolti e richiami di teoria
2011, VIII+274 pp, ISBN 978-88-470-1746-7

52. C. Presilla
Elementi di Analisi Complessa. Funzioni di una variabile
2011, XII+324 pp, ISBN 978-88-470-1829-7

53. L. Grippo, M. Sciandrone
Metodi di ottimizzazione non vincolata
2011, XIV+614 pp, ISBN 978-88-470-1793-1

54. M. Abate, F. Tovena
Geometria Differenziale
2011, XIV+466 pp, ISBN 978-88-470-1919-5

55. M. Abate, F. Tovena
Curves and Surfaces
2011, XIV+390 pp, ISBN 978-88-470-1940-9

56. A. Ambrosetti
Appunti sulle equazioni differenziali ordinarie
2011, X+114 pp, ISBN 978-88-470-2393-2

57. L. Formaggia, F. Saleri, A. Veneziani
Solving Numerical PDEs: Problems, Applications, Exercises
2011, X+434 pp, ISBN 978-88-470-2411-3

58. A. Machì
Groups. An Introduction to Ideas and Methods of the Theory of Groups
2011, XIV+372 pp, ISBN 978-88-470-2420-5

The online version of the books published in this series is available at SpringerLink.
For further information, please visit the following link:
http://www.springer.com/series/5418